CONTROL SYSTEM PRINCIPLES AND DESIGN

Ernest O. Doebelin

Department of Mechanical Engineering
Ohio State University

John Wiley & Sons

New York   Chichester   Brisbane   Toronto   Singapore

Library of Congress Cataloging in Publication Data:

Doebelin, Ernest O.
  Control system principles and design.

  Includes index.
  1. Automatic control.   I. Title.
TJ213.D585   1985   629.8   85-3204
ISBN 0-471-08815-3

Printed in the United States of America

10 9 8 7 6 5 4 3 2 1

Preface

A teaching career that started in 1954 has allowed me to observe and participate in many developments in the field of control engineering. I feel this historical perspective provides a background particularly useful in evaluating alternative approaches and in creating a new control text at this time. In 1954, outside of electrical engineering, elective control courses were extremely rare and required undergraduate courses unheard of. At that time, the Ohio State Department of Mechanical Engineering instituted a required undergraduate course in control systems, which led in due course to the writing of my first textbook.* Over the years, I came to view the areas of system modeling, measurement, and control as a coherent body of knowledge of great utility in engineering practice and developed a total of eight courses (and seven associated laboratories) spanning the range from introductory (sophomore level) system dynamics courses (we require two), through separate required courses in measurement and control principles, to senior elective/beginning graduate courses in measurement system design and control system design, and ending finally with two graduate courses in system modeling and response. Texts† for all these courses were generated along the way, the present book being designed for the two control courses. (The first, 1962, control text was written for a required introductory course.)

Although enthusiasts like myself have for many years encouraged the introduction of required control courses into the mechanical engineering curriculum, and many schools have done this, the practice is not yet universal. It is hoped that the current national interest in industrial revitalization and productivity enhancement through application of flexible manufacturing systems, robotics, microprocessor-based controls, and computer-aided design and manufacturing will serve as a final push to make a required course in control as certain as a required course in thermodynamics. I have tried to design this new text to meet these needs in such a way that faculties will be encouraged to provide the needed curricular space for this important material. This is best done, I believe, by presenting control, not as a narrowly specialized mathematical exercise, but as a widely applied technology with close connections to, and important impacts on, many other areas of engineering design.

---

*E. O. Doebelin, *Dynamic Analysis and Feedback Control*, McGraw–Hill, 1962.

†E. O. Doebelin, *Measurement Systems,* McGraw–Hill, 1966, 1975, 1983; E. O. Doebelin, *System Dynamics,* Merrill, 1972; E. O. Doebelin, *System Modeling and Response,* Wiley, 1983.

iii

Those familiar with my earlier works will find the same spirit has guided the present effort:

a. A balanced presentation of the mathematical and physical aspects of the subject.

b. Emphasis on time-tested approaches useful now and in the future in engineering practice.

c. A conscious and persistent effort to relate specialized topics to their proper role in the larger scene of engineering design.

d. Use of innumerable realistic application examples to develop hardware familiarity and an appreciation of the distinction between math models and real equipment, and to demonstrate the practical relevance of the methods being presented.

Although this philosophical attitude sets the tone of the work, technical and pedagogical details are also important. The book is designed, and has been used here at Ohio State, for both a required introductory course and a second, design-oriented elective course. It is thus suitable also for an elective course that is the student's first control course. The practical flavor of the text and its emphasis on careful and complete explanations should also make it appealing for self-study by practicing engineers needing information in the general area or on some specific topic. My earlier control text (1962) included considerable background material on system dynamics and this seems today to be still desirable since many schools do not yet require a separate system dynamics course. For those, such as my own students, who *do* have a good system dynamics preparation, this text material can be left for a quick outside-of-class review, giving more time to spend on strictly control topics. As for mathematical tools, I still find no real advantage to using the Laplace transform in an introductory required course, and leave this for the second course. I realize many prefer to use the Laplace transform from the outset, so have included this coverage early in the text for them. Since the Laplace transform is basically a means of evaluating the *response* of linear, constant-coefficient systems, the ready availability of digital simulation languages, which handle *all* kinds of systems (linear, nonlinear, time-varying, sampled-data, random inputs, etc.), makes using the Laplace transform less necessary for detailed performance calculations.

As in my 1962 book, nonlinear aspects of control systems, both intentional and parasitic, are considered an integral part of even an introductory course and are not to be left for a "second" course (which many students in a required first course would never take). This position may have been a little radical in 1962, but today, with digital simulation languages being widely available and inexpensive to use, there is no excuse for leaving students ignorant of nonlinear behavior. With regard to analytical techniques for nonlinear systems I present only the describing-function approach; other approaches have not demonstrated any great practical utility in design. Although there is a separate chapter on nonlinear systems, many nonlinear effects pertinent to control performance are also sprinkled throughout the text, where appropriate, and are treated using simulation.

The relative roles of "classical" and "modern" control theory in a text at this level is the subject of some controversy. Authors often allot each approach a roughly equal portion of space but then offer the student no real guidance as to the relative practical utility of each method. Since the provision of such guidance is, in my view, a major function of teachers and authors, I feel the need to provide in this text my best effort in that direction. In most current texts actual real-world applications of modern control, complete with hardware/ software details, are almost never given, whereas it is easy to give hundreds such applications for classically designed systems. The reason for this is that the significant applications (and there are some very significant ones) of modern control are not only rare but are also complex and not easily explained. As long as this is the situation, I feel there is little justification for presenting modern control concepts in our first few control courses since one is unable to show students convincing evidence of their application, and most students at this level will never need these tools. Rather, we should reserve the presentation of modern control for advanced courses where significant applications can be shown and understood.

At the risk of alienating some potential users of this book, I have therefore opted for a concentration on classical methods, augmented with computer simulation. As one who was present during the creation and development of modern control theory, I find it necessary to emphasize that the theory is now 30 years old and there has been ample time and more than ample effort expended on its application for some judgements to be reached on its utility. Lack of on-line computing power, sometimes claimed as a main reason for a dearth of applications, becomes a less convincing argument with the appearance of each new generation of microcomputers. My own conclusion is that modern control theory certainly is a suitable subject for advanced courses and research in the control field, and that a few very significant applications have been made, but that the vast majority of practical control system designs are satisfactorily worked out using classical analytical methods augmented by computer simulation and experimental development. It is also important to note that this situation is not about to change. Some proponents of modern control approaches have over the years fostered the impression that classical methods would "shortly" be superseded, thus it was necessary to prepare students for this revolution by replacing classical methods with "modern." I believe enough time has passed to make it safe to reject this view and restore in academic circles the view that classical methods (augmented with simulation) are the analytical methods of choice in most system design work.

A "modern" development that I feel is really more significant to practical design work is the wide availability and low cost of general-purpose digital simulation languages (such as CSMP and ACSL) for computer-aided design. These are of course used for many noncontrol purposes but their application in the control field is changing design procedures in important ways. These high-level languages are very easy to learn and use and they apply to every kind of control system, including new concepts or combinations that a designer might daily conceive for specific applications. Since digital computers are being used more and more as system controllers, a digital simulation of a proposed new control system design also begins to look more and more like the actual system itself, making the transition from design calculations to operating hardware/ software much more direct and rapid. Simulation studies without an adequate theoretical background can of course become aimless gropings and extremely wasteful of computer resources; however it is equally short-sighted to overlook the potential for replacing tedious and approximate analytical procedures with straightforward and accurate numerical evaluations. A good example here is the classical gain-setting procedures (Nichols charts, root locus, etc.) for linear systems. This book still explains these methods since they give a general understanding of the effects of design changes, in addition to providing a specific numerical result. I also, however, immediately show how easy it is to run a CSMP simulation with gain as a multivalued parameter to explore its effect and make a choice. Every other aspect of design can be similarly and efficiently explored with these simulation tools. Nonlinear and/or time-varying effects that defy analytical treatment are readily evaluated. These simulation tools are felt to be such an important accessory to the basic analytical methods that they are used heavily throughout the text. Although a practical designer is handicapped without access to such tools, and students definitely benefit from their hands-on use, the text examples are presented in such a way that readers get a large share of the benefits even though they may not have personal access to a simulation facility.

In addition to computer-aided design, the other main impact of computer technology on control is of course the use of digital computers as "components" (generally controllers) in operating control systems. Although many control specialists will want to take one or two courses devoted specifically to digital control systems, a quite useful capability in this area can be developed with a more modest effort. I have tried to provide such a treatment in Chapter 14. This chapter contains a condensed treatment of basic analytical methods, but (perhaps more importantly) explains in considerable detail how many digital systems can be understood, analyzed, designed, and applied using the basic methods for continuous systems developed earlier in the text. Simulation methods are again helpful in reaching these goals.

At this point, it is appropriate to go systematically through the chapters in sequence, giving advice (to those desiring it) on the selection of topics for certain types of courses. Our own curriculum has a required first course that concentrates on basic principles and develops some modest design competence, using a format of three hours of lecture and two hours of lab per week, for one quarter (about ten weeks), in the senior year. Students have had two system dynamics courses (three hours of lecture and two hours of lab each) at sophomore/junior level and a measurement systems course (one hour of lecture and four hours of lab) at senior level. The second control course extends in breadth and depth from the first, is a combined senior-elective and beginning-graduate course, and uses a three hour lecture and two hours of lab format. Its specific purpose is to prepare students for practical control system design in industry. The two courses together would cover essentially everything in the book.

Chapter 1, the introductory chapter, is much longer than most introductions since instead of just introducing the basic feedback principle it also gives an overview of all the important applications areas and all the basic categories of control systems. This is necessary if we want students to see control not just as another narrow topic but as a pervasive influence throughout engineering design, with a highly developed hierarchy of concepts and methods. Most of this chapter can be left for outside reading, with class discussion concentrating on Sections 1.3 and 1.4. We would assign the entire chapter for detailed reading in our first course and have students quickly review it at the beginning of the second course.

Chapter 2 combines some mathematical background with physical system modeling. Our first course would skip the Laplace transform material; our second course would cover it *quickly*. Due to our students' extensive system dynamics background, Chapter 2 would be left mainly for a quick review. Some class time would be spent on the hydraulic servo components, their digital simulation, and on dead-time effects. Chapter 3 is very design oriented and would not be given thorough treatment in our first course and covered in detail in the second. Chapter 4 would be covered in our first course, with particular emphasis on the programmable logic controller, the most widely used device in factory automation. Chapters 5, 6, and 7 would all be covered in detail in our first course.

In Chapter 8 the first course would cover the basic describing function method, apply it to on–off controllers, and check it with digital simulation. The rest of the chapter would be covered in the second course. Sections 9.1 to 9.4 plus the unity-feedback part of Section 9.6 and all of Section 9.7 would be covered in the first course, the rest of Chapter 9 in the second. In Chapter 10, Sections 10.1 to 10.4 are included in the first course, 10.5 to 10.7 in the second. Sections 11.1 to 11.4 would be covered in the first course, 11.5 to 11.7 in the second. In Chapter 12, ideal PI, PD, and PID modes are discussed in the first course, leaving most of the chapter for the second. Chapters 13 and 15 are reserved entirely for the second course; whereas Sections 14.1 to 14.3 are covered in the first course, and 14.4 to 14.6 in the second.

The intelligent design and use of control systems requires that one have:

a. Knowledge of the basic configurations (modes of control) that have been devised and the characteristic performance features of each. This allows one to initially select one or more alternative design concepts that have potential for success.

b. Familiarity with available hardware so that commercially available components to implement the design concepts can be selected.

c. Competence in modeling of physical systems with suitable equations, using judicious assumptions.

d. Facility in the use of analytical, simulation, and experimental techniques for determination of system response and suggesting design changes.

This text provides an integrated treatment of all these aspects of control engineering and thus prepares the student for early productivity upon entering industrial practice. I have attempted to make the coverage of item "a" particularly complete since this storehouse of basic ideas is what one draws on when conceiving a design for a new application. The vast majority of practical systems are designed in this way rather than by some mathematical synthesis procedure. The list of topics includes not only fundamental control modes such as on–off, proportional, integral, derivative, phase lead, phase lag, lead/lag, and cancellation compensation, but also more specialized schemes such as command and disturbance feedforward, cascade control, state-variable feedback, reset-windup compensation, Smith predictors and sampling controllers for dead-time process, phase-lock servos, intentional nonlinearities such as mode switching for conditional stability/saturation problems, adaptive-gain proportional controllers, square-root velocity servos, secondary feedback in on–off thermostats, open-loop digital control using digital actuators (step motors, etc.), and noninteracting control for multivariable processes. Many of these specialized (but important) topics are difficult to present convincingly since their nonlinearity or other complications frustrate analytical treatment; however digital simulation easily reveals their features and also provides a practical design capability.

In closing, I would like to acknowledge the support of the Department of Mechanical Engineering at Ohio State in providing assistance in manuscript preparation. Typing was capably performed by Elizabeth Fisher and Barbara Dolatabady.

Ernest O. Doebelin

Table of Contents

CHAPTER 1
Introduction

1.1 ROLE OF CONTROL SYSTEMS IN ENGINEERING DESIGN

Engineering students might find themselves using this book in various contexts. At my school it is used for two courses. The first is a senior-level course that is required of all mechanical engineering students. My philosophy for such a course emphasizes breadth, practicality, and concentration on fundamental principles. The major design techniques can be introduced and illustrated, however restrictions of time will ordinarily limit the development of design competence to a modest level. I believe this is a proper compromise for such a group of students since most of them will not become control specialists and this will be their "one and only" control course. Since I feel such a course should be required in *every* mechanical engineering curriculum (and most other engineering curricula also) and since this goal has not yet been attained, this section will present arguments for this viewpoint. These arguments serve to illuminate the important role played by control systems in almost every area of engineering design and are thus of particular significance to students who do not consider themselves control specialists and may "need convincing" that this is an area of study worthy of their time.

The book also includes material at a level (and in a quantity) suitable for a control specialist course emphasizing design. Such a specialist course would normally be elective rather than required, but could be either a "first" or "second" control course. Our course is used as a senior elective by undergraduates and as a graduate course by beginning graduate students. For graduate students coming from our undergraduate program, this would be a second course; however, graduate students from "outside" might take it as their first control course. Thus introductory material such as that in this section is still appropriate, though it might be left for background reading rather than class discussion. Finally, the text is intended to be useful for those practicing engineers whose academic training did not include control studies but who encounter assignments that require self-study of specific topics in the field. This group of readers is, of course, brought to the text by the greatest possible incentive, a current and pressing practical need, and thus the motivational aspect of this chapter is of secondary significance to them.

I now wish to justify both required and elective courses in control by demonstrating the widespread and significant impact of control concepts on many aspects of engineering, at the same time laying the foundation for the specific technical developments that come later in the text. I will show that all types of engineered products and services are tending to depend more and more on

1

associated control systems for their optimum functioning. In contrast with earlier years, these control systems are increasingly considered to be an integral part of the overall system rather than afterthought "add-ons." Although the "invention" of such overall systems does not necessarily require the talents of a control specialist, it is unlikely to be created by someone totally ignorant of control principles. That is, there needs to be a melding of control basics with the engineering lore of the specific product area. Recognition of this need is the basis of my conviction that required courses in control are easily justified in today's curricula.

Although the above argument is sufficient, even further benefits accrue when a required control course is instituted. Students completing such a course will be in a good position to decide whether their talents and interests lie in this area, making the choice whether to continue with an elective course more rational, and allowing the elective course to be of sufficient breadth and depth to develop a practical level of design competence. When control courses are taught with a heavy hardware emphasis (as this text encourages), they become particularly good capstone design courses in a curriculum since they require the application and integration of many previously developed subject areas such as applied mathematics, computer-aided design, electronics, system dynamics, hydraulics, pneumatics, measurement systems, and the like. Finally, some of the techniques taught in control courses turn out to have important application in non-control-contexts.[1] For example, the design of hydrostatic fluid bearings is not conventionally considered to be a part of control, yet the dynamic instabilities of such bearings are studied with the same techniques developed for stability considerations in control system courses. A similar situation exists for machine tool chatter, combustion oscillations, and, in fact, all self-excited oscillatory phenomena.

We have up to this point been using the words *control* and *control system* as if the reader perfectly understood their meaning. Although individual readers may indeed have rather specific meanings in mind, I cannot leave such fundamental issues to chance and now proceed to develop some definitions systematically. It may be helpful to categorize some of the many applications of control systems according to the type of device being controlled, one possible listing being as follows.

Energy Sources Nuclear and fossil-fueled steam power plants, internal combustion engines and turbines, hydroelectric power plants, wind turbines, solar power plants.

Materials Production Facilities Open hearth furnaces, rolling mills, petrochemical plants, cement plants, lumber mills.

---

[1]K. N. Chen et al., A System Approach to the Dynamic Characteristics of Hydrostatic Bearings Used on Machine Tools, *Int. J. Mach Tool Des. Res.* Vol. 20, pp. 287–297; S. A. Jaliwala et al., An Application of Control Theory to Optimize Automotive Cam Design, ASME Paper 80-DGP-13, 1980.

Vehicles and Transport Systems   Railways, aircraft, spacecraft, ships, automobiles, pipelines, conveyors, elevators.

Construction Equipment   Excavators, graders, tile-laying machines, tunneling machines, cranes.

Manufacturing Equipment   Machine tools, industrial robots, foundry equipment, forging and stamping presses, assembly machines, automatic test equipment, plastic molding presses, textile machinery, packaging machines, heating, ventilating, and air conditioning system.

Agricultural Equipment   Tillage, planting, and cultivation machines, spraying equipment, harvesters, environmental control equipment.

Consumer Goods, Appliances   Refrigerators, washing machines, ranges, water softeners, furnaces, mixers and blenders, cameras, sewing machines.

Computers and Peripherals   Analog, digital, and hybrid computers, tape drives, disk drives, printers, card and tape punches, plotters.

Weapon Systems   Missiles, launchers, tracking systems, submarines, torpedoes, hydrofoil craft, tanks.

Communications   Videotape recorders/reproducers, copying machines, printing presses.

Measurement Systems   Servomanometers, hot-wire anemometers, servo accelerometers, pen, optical, thermal, and electrostatic recorders.

Medical Equipment   Heart and lung machines, dialysis machines, pacemakers, artificial hearts, prosthetic devices.

This list is not exhaustive, yet it seems to show that "everything" needs a control system. Our definition of control is at this point intentionally broad enough to encourage exactly such a viewpoint. Section 1.2 will try to bring some focus to this somewhat diffuse picture by separating control systems into functional types.

In each of the above-listed applications there is some "device to be controlled" which often has a use aside from any associated control system. In control parlance this entity is given various names: *process, plant,* and *controlled system* being perhaps the most common, my preference being *process.* As seen in Fig. 1.1, process *inputs* are flows of energy and/or material that cause the process to react or respond. Mathematically, inputs are considered to be known or assumed and are classified into *manipulated inputs* (subject to our control) and *disturbance inputs* (undesirable and unavoidable effects beyond our control). If a process needs a control system it may be because of the presence of significant disturbance inputs. If we decide to implement a control system we must be sufficiently clever in our management of the manipulated inputs so as to counteract the effects of the disturbances. This implies that there are some *response variables* associated with the process, which we require to behave in some specified fashion. At this point we note that to cause a desired process response,

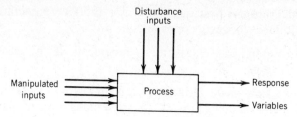

FIGURE 1.1 *Process input/output configuration.*

control of the manipulated inputs might be required even if disturbances were not present, thus the need for controls can arise from a requirement for command following, disturbance rejection, or both. To make these definitions and concepts more concrete, consider the familiar example of a home heating system. Indoor temperature is the response variable of major interest and it is clearly affected by the main disturbance input, outdoor temperature; however the manipulated input (gas and air flow to the furnace) might be changed in order to achieve a new desired room temperature, as well as to counteract a varying outdoor temperature.

This attempt to define the role of control systems in engineering design has been somewhat circuitous and to have a full appreciation of it you may need the more detailed familiarity you will gain as we progress through the text. However, let us try to draw the threads together as best we can before leaving this section. The design of every engineered process initially proceeds from specifications, a list of the functions to be performed. Following the cardinal rule of good design, to strive for simplicity, one attempts to configure the process so that specifications may be met with minimum equipment and no (or only the most rudimentary) controls. The original or early forms of most inventions (particularly the more "ancient" ones) will usually fit this pattern. If specifications cannot be met or, as seems to be inevitable, customers soon demand improved specifications, the process design must be refined. This refinement always runs into practical or theoretical limitations and then the addition of suitable controls often allows significant further improvements to be made that would have been unavailable by other means. Since this evolutionary process of refinement of specifications and designs applies to all products and services we should not be surprised that, as shown in the earlier lists of applications, "everything needs controls". Whereas in earlier days the need for controls may have taken years to make itself apparent, and may in fact have often been unrecognized because of process designers' unfamiliarity with control possibilities, today the "original invention" often includes control aspects that are vital to performance. Since conservation of energy and materials, and continual increase in labor productivity are agreed upon as vital to the maintenance and improvement of world living standards, and since these goals invariably require improved performance of all technical processes, the increasingly important role of control in the overall picture of engineering design begins to appear. Some critics of modern society do not take such a benign view of technical progress. Engineers might be well advised to sometimes temper their professional enthusiasm with consideration of such alternative viewpoints.[2]

As one of many possible examples of this evolutionary process, consider the metal-cutting lathe of Fig. 1.2. In one of its early forms, Fig. 1.2*a*, both longitudinal and transverse motions of the tool were controlled by the human operator using handwheels to actuate rack-and-pinion or screw positioning mechanisms, which we here prefer to consider as "controls" although others might include them as intrinsic parts of the lathe. In such a machine, accuracy depends critically on the skill and care of the human operator and production speed is clearly limited. Provision of motor driven slide motions was a logical next improvement. Since conical bodies of revolution (tapers) are often-needed machine parts, an additional "control" in the form of a taper attachment (Fig. 1.2*b*) was soon invented. The hydraulic tracer concept of Fig. 1.2*c* was a real breakthrough in that it allowed rapid and accurate production of parts of complex shapes by relatively unskilled operators. The stylus which traces over the template has only a few ounces of spring loading, maintaining accurate contact without causing significant errors due to elastic deflection or wear, yet the cutting tool is powerfully positioned by the hydraulic cylinder fed by the precision servovalve. An even more recent development is the numerically controlled lathe of Fig. 1.2*d* where desired x and y motions are input digitally, actual slide positions are measured with displacement sensors, and errors are driven toward zero by hydraulic or electric servodrives on the slides. The latest refinements being proposed involve concepts such as laser measurement of actual workpiece dimension (most present methods measure tool-slide position) and measurement of cutting forces and temperatures to allow "adaptive" changes in speeds and feeds to keep them near the allowed maxima at all times.

1.2   CLASSIFICATION OF CONTROL SYSTEM TYPES

The previous section has tried to convince the reader that a basic knowledge of control concepts will turn out to be practically useful for a large percentage of engineers, irrespective of whether they consider themselves control specialists or not. Except for some details related to the process being controlled, however, no real attempt has been made to give definitions and descriptions of the various types of control systems found in actual practice. Since such a broad overview is a desirable prerequisite to the technical details forthcoming in later chapters, we now undertake to develop it.

Perhaps the broadest overall classification separates control systems into two fundamental types, open-loop and closed-loop (feedback). Figure 1.3 shows

---

[2]D. F. Noble, Is Progress What it Seems to Be?, Datamation, Nov. 15, 1984, pp. 140–154.

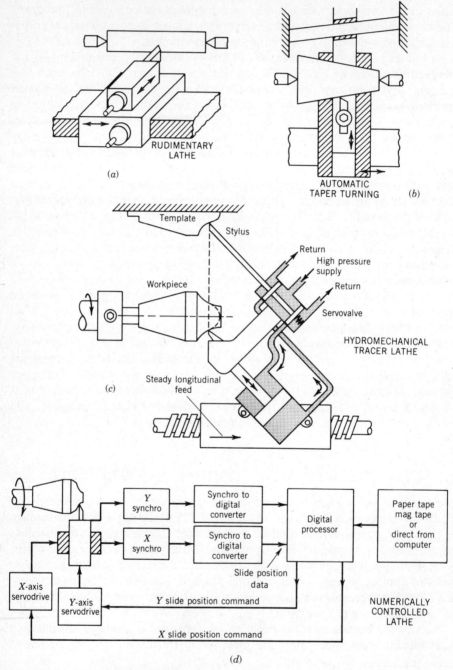

FIGURE 1.2   *Evolution of lathe controls.*

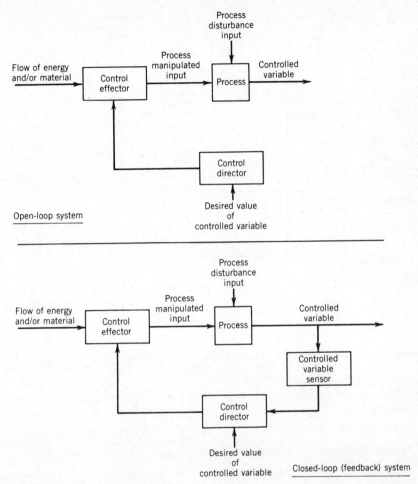

FIGURE 1.3   *Functional operation of open-loop and closed-loop systems.*

functional block diagrams of each. In *open-loop* systems the process response variable of interest (now called *controlled variable*) is determined by the combined effects of the disturbance inputs and manipulated input. The manipulated input (flow of energy and/or material) is varied by the *control effector* (also called final control element or actuator) in response to signals from the *control director* (also called the controller). The control director receives information input as to the *desired value* of the controlled variable and translates this into a control signal for the control effector by implementing the *control law* (also called control algorithm) that is "built into" the control director. Open-loop systems of the basic type shown in Fig. 1.3 are often satisfactory if disturbances are not too great, changes in desired value not too severe, and/or performance specifications not too stringent. Figure 1.4 shows a *disturbance-compensated*

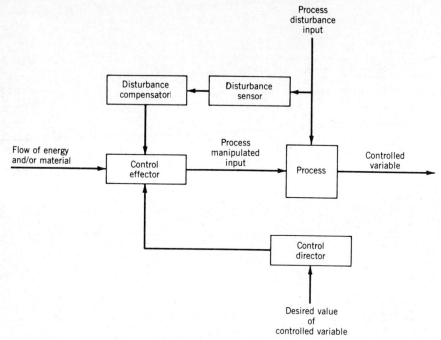

FIGURE 1.4 *Disturbance-compensated open-loop system.*

open-loop system, a refinement in which the process manipulated input is derived not only from the desired value command but partially from a measurement of a disturbance. Implementation of such a scheme requires that:

1. We must be able to measure the disturbance.
2. We must be able to estimate the effect of the disturbance on the controlled variable, so we can compensate for it.

Disturbance compensation can be used by itself or in combination with feedback control. A practical example of the latter scheme is found in some heating and air conditioning systems for buildings, where temperature sensors inside the building are used in a feedback configuration while temperature sensors outside are used to compensate for ambient temperature disturbances. Although not as common as disturbance compensation, *command-compensated open-loop systems* (and combinations of these with feedback schemes) also exist. Here, based on knowledge of process characteristics, the desired value input (command) is augmented by the command compensator to produce improved performance. Thus, in the example of Fig. 1.5, even though the desired process output is simply a new constant value, knowledge of process lags makes the command compensator apply an initial "overcorrection," giving faster response. Disturbance/command compensated systems are also called *feedforward* systems.

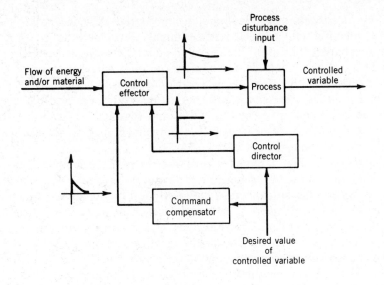

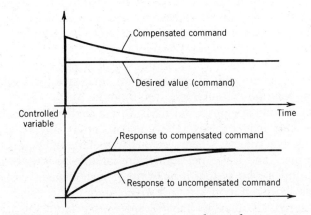

FIGURE 1.5 *Command-compensated open-loop system.*

Open-loop systems without disturbance or command compensation are generally the simplest, cheapest, and most reliable control schemes and should be considered first for any control task. If specifications cannot be met, disturbance and/or command compensation should be considered next. As we will see in Chapter 7, design of these various types of open-loop systems does not generally require any specialized control theory, basic system dynamics[3] being sufficient in most cases. When conscientious implementation of open-loop techniques by a knowledgeable designer fails to yield a workable system, the more powerful

[3]E. O. Doebelin, *System Dynamics,* Merrill, Columbus, Ohio, 1972; *System Modeling and Response,* Wiley, New York, 1980., E. O. Doebelin, *Measurement Systems,* McGraw-Hill, New York, 1983.

closed-loop (feedback) methods should be considered. A wide array of these is available, together with a specialized and highly developed theory for each of the major subareas. Since this theory is not generally presented elsewhere, the vast majority of the hundreds of control texts found in a large library will be devoted almost entirely to a study of feedback systems. Although the analytical design of feedback systems is currently based on a well-developed mathematical theory, the basic concept is quite obvious and reasonable. Thus the original invention and much of the early practical implementation were accomplished almost entirely without the aid of mathematics.

Consideration of Fig. 1.3 makes clear the operating principle of feedback systems and their basic advantages over open-loop systems. We see that an open-loop system can be converted to closed-loop by adding the functions of *measurement of the controlled variable* and *comparison of measured and desired values of the controlled variable.* Thus if we wish to control the temperature of a room, it is entirely reasonable that we first measure the temperature, compare this with the desired temperature, and make changes in the room heating rate only when desired and measured values differ. Note that errors between commanded and actual values of the controlled variable will tend to be corrected *no matter what their source.* This includes errors due to changing commands, process disturbances, disturbances to equipment other than the process, and changes in hardware parameter values. *The only exception (which is often critical, however) is the controlled variable sensor.* If this sensor gives wrong information, the feedback system has no way of correcting for this, thus feedback control depends vitally on accurate measurement (see footnote 3).

The fact that open-loop systems never measure the controlled variable is the basis of their possible inadequacies. They fundamentally rely on conditions staying close to design values. When system parameters and/or disturbances depart from "normal" and cause the controlled variable to wander from the desired value, the open-loop system is unaware of such changes. (Even a disturbance-compensated open-loop system corrects *only* for the disturbance or disturbances measured; all others go uncorrected. Furthermore, changes due to wear, aging, environmental effects, and the like in the disturbance sensor/compensator, control effector, and/or process cause the compensation to be imperfect.) The reader at this point should begin to appreciate the potential power of the simple feedback concept, a potential which has been realized in countless practical applications. An associated problem, possible instability, accompanies the many benefits of feedback, however this phenomenon is reasonably well understood and is controllable by proper design.

Another classification involves broad areas of application: *servomechanisms* and *process control.* The accepted definition of a servomechanism is a feedback control system in which the controlled variables are mechanical motions; however I would like to extend this to include forces and torques. In the so-called process industries (chemicals, petroleum, steam power, food, etc.) one repeatedly encounters the need to control temperature, flow rate, liquid level in vessels, pressure, humidity, chemical composition, and the like; such applications are generally considered process control. When the desired value is more or less fixed and the main problem is to reject disturbance effects, the control system is sometimes called a *regulator*. Historically, the wide practical application of control first took place in the process area although isolated applications of feedback can be traced into antiquity.[4] Most of the basic concepts were developed and brought to successful practical realization in the process area by the intuitive and experimentally oriented engineering methods typical of the 1915 to 1940 era. The technical needs of military systems around the time of the second world war led to more "scientific" approaches in all engineering fields, particularly control, where a comprehensive mathematical theory (couched in the language of servomechanisms and aimed mainly at such applications) was developed around 1940 to 1950. Since this time, terminology and design methods of the two areas have gradually converged so that a control engineer has little difficulty in moving among companies that specialize in one or the other area.

Figure 1.6 shows a *servomechanism*[5] application; the positioning of the read/write head on a moving-head disk drive for a computer memory. Data is magnetically stored in 200 concentric circular tracks separated by 0.01 in. on a 12-in. diameter disk rotating at 2400 RPM. Each track is organized circumferentially into 24 sectors, each sector holding 256 8-bit bytes of data. The head-positioning servomechanism gets from the drive controller its desired value command as a digital signal ("set cylinder") specifying which one of the 200 tracks is to be read or written. Head position is measured digitally by an optical encoder sending pulses to an up–down counter whose contents thus represent actual head position at any instant. Comparison of this signal with the desired value stored in the cylinder address register gives the error, which is processed through the velocity curve generator to produce an analog voltage proportional to the square root of the distance to be moved. This voltage serves as a command to a velocity feedback loop and the square root relation gives a motion with constant acceleration/deceleration, a desirable form of mechanical motion. A "voice coil" translational dc motor (similar to a loudspeaker, in principle, but with a longer stroke) is driven by a servo amplifier whose input is the velocity error signal developed by subtracting measured motor velocity from the velocity command. Allowable total error in head position is ±0.0015 in., and since there are several sources of error in the error budget, the servo system was allowed only 20% of the total. Since these particular drives are intended for mobile applications, disturbances in the form of frame vibration and tilt (pitch/roll) must be designed against. The final servo design provides a "stiffness" against disturbing forces of 10 lb per 0.001 in., allowing the system to maintain the 0.0003-in. position error in the face of 3-g "bumps" and ±30° pitch/roll. Speed of response is 7 ms for a one-track move and 55 ms for the maximum possible move of 200

---

[4] O. Mayr, *The Origins of Feedback Control*, MIT Press, Cambridge, Mass., 1970.

[5] H. Chestnut and R. W. Mayer, *Servomechanisms and Regulating System Design*, Vol. 1, 1951, Vol. 2, 1955, John Wiley, New York.

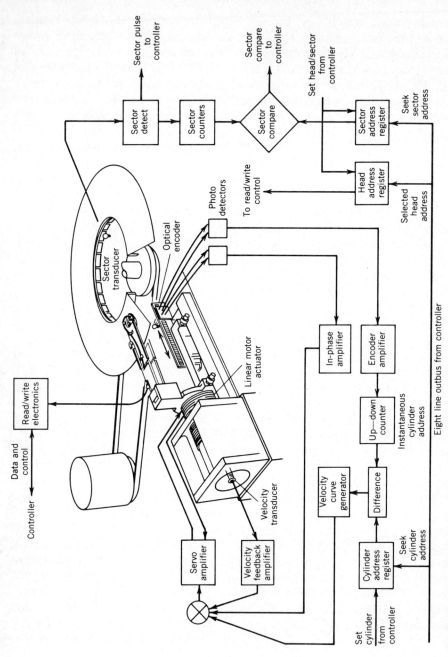

FIGURE 1.6 *Disk memory head-positioning servomechanism.*
Source: J. E. Herlinger and W. J. Lloyd, Inside the 7900 Disc Drive, Hewlett-Packard Journal, Vol. 23, No. 9, May 1972, Palo Alto, California.

tracks. The "loop within a loop" (velocity feedback system embedded within a position feedback system) configuration utilized in this application is a classical scheme called *cascade control* in the process field and *minor-loop feedback* (or *state-variable feedback*) in servomechanisms.

Although the power level of the disk drive just discussed is less than 1 kW, our next servomechanism example, the grinding machine of Fig. 1.7, is rated at 300 hp (224 kW). Between stages of rolling alloy steel billets, it is necessary

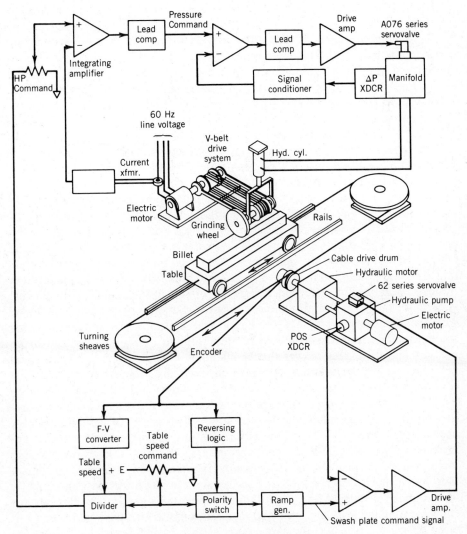

FIGURE 1.7 *Electrohydraulic billet-conditioning servosystem.*
Source: T. Peter Neal, A High-Performance Electrohydraulic Control System for Large Grinding Machines, Technical Bulletin 146, Moog Inc., East Aurora, New York, 1980.

to remove surface impurities and defects, the traditional method being to use a scarfing torch and acid bath. Since this approach uses large amounts of energy (natural gas), produces noxious fumes and wastes considerable steel, the development in the late 1960s of zirconia/alumina abrasive wheels suitable for mechanical cleanup of billets was welcomed as an alternative. The control system of our example was designed to implement the new grinding approach in optimum fashion. A billet is fastened to a rail-guided cart that reciprocates beneath the axially fixed grinding wheel, an arrangement similar to a conventional surface grinder. This grinder's goal is, however, quite different since we do *not* wish to produce a plane billet surface but rather want to remove the minimum material necessary to clean up the surface in preparation for the next rolling operation, which can tolerate considerable surface unevenness. We thus need a scheme that can "skin off" surface defects while tending to follow the gross surface contours of the rough billets. Study of typical billet shapes led to mathematical formulation of three types of idealized surface irregularities: a straight 8% slope, a 1-in. sinusoidal dip having a 24-in. wavelength, and a bent-end billet (1 in. deflection over the last 4 in. of length). These contours must be successfully traversed at a cart speed of 5 ft/s.

Two separate (but interrelated) control systems were devised to solve this design problem. Each uses hydraulics as the actuating means, not unusual in high-power, fast response servo systems. One feedback system controls the cart's reciprocating motion whereas the other manipulates the grinding wheel depth-of-cut motion. It was decided that grinding horsepower, rather than depth of cut, should be the controlled variable in the feedback loop that positions the grinding wheel. Grinding horsepower is nominally proportional to the rate of material removal and if we can keep it at its maximum allowable level (determined by the power rating of the grinding machine) we can remove material at the maximum feasible rate. Since frictional effects in the grinding-wheel drive system are fairly small and relatively constant, grinding power can be inferred with adequate accuracy by measuring electrical input power to the motor, using a convenient current transformer. This measurement is compared with a voltage proportional to desired power that is manually dialed in on an hp-command potentiometer by the operator. A minor-loop feedback (cascade control) scheme is employed within the main horsepower loop by using the integral of the hp-loop error signal as a differential-pressure command for the hydraulic cylinder that positions the grinding wheel. A differential-pressure (ΔP) transducer connected across the two sides of the power piston gives a measurement that approximates the grinding force, thus this minor loop tries to keep the grinding force at a value specified by the pressure command. This minor force-control loop has a much faster response than the main horsepower loop and is able, for instance, to quickly feed the grinding wheel into a gap in the billet surface (thus maintaining grinding force and material removal) if such a gap is encountered during a traverse of the cart.

The cart-motion system uses a fixed-displacement rotary hydraulic motor supplied by a variable-displacement pump ("hydrostatic transmission") as the driving means. The pump displacement is controlled by the angle of a swash-plate mechanism and can be smoothly varied from maximum flow in one direction, through zero flow, to maximum flow in the reverse direction, thus giving continuous control of motor speed and direction. A swash-plate position transducer is used to effect closed-loop control of swash-plate position in response to a table speed command voltage from a manually set potentiometer. Thus a given command voltage produces a proportional pump displacement and nominal motor (cart) speed; however cart speed is *not* measured and compared with the speed command, thus the cart-speed control is open-loop. At the endpoints of cart travel, where velocity must go to zero and then reverse, an encoder position signal triggers reversing logic and a polarity switch to reverse the swash-plate position command, while a ramp generator smoothly blends the transition from positive velocity, through zero, to negative.

Although table speed is not fed back for comparison with the speed command, it *is* measured (using the encoder pulse rate and a frequency-to-voltage converter) to implement an "antipenciling" feature. As table speed falls off during table reversal, enforcement of the constant-grinding-horsepower command would lead to increased depth of cut, resulting in a "penciled" billet end. To avoid this, the horsepower command is automatically reduced at the reversal points by exciting the command potentiometer with a voltage that reduces with table speed. This voltage is obtained by dividing in an analog divider the measured table speed by the speed command.

The automatic controls on this grinding machine do not eliminate the need for a human operator, they simply allow the operator to perform the grinding task with greater speed, efficiency, and accuracy with less demand on operator skill. The basic strategy is to set the desired grinding horsepower at the machine's capacity and then adjust manually the width of cut until the desired depth of cut (judged visually) is produced. If the resulting combination of cutting width and depth is judged to be too coarse, the operator resets the desired horsepower at a lower value. Once these controls are set to the operator's satisfaction, the servo system automatically compensates for wheel wear, billet hardness, and surface irregularities by adjusting depth of cut to hold grinding horsepower constant. A comprehensive failure-detection logic control system (not shown in Fig. 1.7). is also provided to protect the operator and machine.

Although *industrial robots* constitute a somewhat specialized class of control applications, popular interest and technical potential warrant a brief discussion here. These position control devices range from simple open-loop mechanical "pick-and-place" mechanisms of limited versatility to fully programmable, computer controlled, multiaxis servomechanisms capable of duplicating intricate human motions. They can replace people in tasks such as spray painting, spot and arc welding, machine feeding, and assembly, where necessary precision, unhealthful environments, muscle fatigue and/or monotonous repetition make human operation impossible, undesirable, or inefficient. Some robots have a "learning mode" in which, say, a skilled human paint sprayer leads the robot arm through the desired spraying motions on an actual part while the robot memorizes the entire sequence of motions; it can then reproduce them over and over without human intervention. Arc welding robot systems (where gravity effects on the weld puddle are significant) may use a coordinated "two-hand" configuration in which one robot hand moves the welding torch while another manipulates the welded parts so as to prevent gravity-induced "dripping" from the weld puddle (Fig. 1.8a). Figure 1.8b shows a minicomputer-controlled six-axis robot coordinated with two numerically controlled turning centers. The robot selects randomly sized, randomly delivered parts from a stop-station conveyor, loads them into the machine, removes finished parts, and presents them to a laser gaging station which relays any part-size variations to the machine

FIGURE 1.8 *Industrial robots.* (a) *Coordinated two-hand configuration.* (b) *Minicomputer-controlled six-axis robots.*
Sources: Advanced Robotics Corp., Newark, Ohio, Cincinnati Milacron Corp., Cincinnati, Ohio.

FIGURE 1.8 (Cont.)

control that adjusts the cutting tool to compensate for wear, thermal expansion, and the like.

Turning now to *process controls,*[6] Fig. 1.9 shows a *self-operated controller* for regulating the water level in a boiler steam drum. Most of the feedback controls developed in the early days of process control were of the self-operated type, and although more sophisticated controls are now necessary for many applications, self-operated controls are still the optimum solution for simpler problems. Self-operated controllers are available for many different controlled variables and share a number of characteristic features. The "self-operated" terminology derives from the fact that the control system needs no separate power supply; all necessary operating energy is obtained from the controlled process. This one aspect is responsible for much of the simplicity, compactness, reliability, and low cost typical of such controllers. Furthermore, although all the functions of a feedback system (as in Fig. 1.3) are present, the hardware is designed and marketed as an integrated unit, rather than as a separate sensor, a control director, and a control effector. In non-self-operated systems, these three components are distinct entities and are, in fact, often purchased from *different* manufacturers. In terms of control system design, both "component"

[6]F. G. Shinskey, *Process-Control Systems*, McGraw–Hill, New York, 1979.

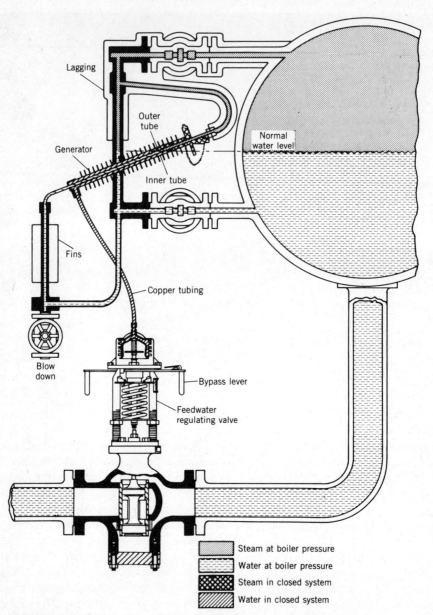

Lagging

Outer
tube

Normal
water level

Generator

Inner tube

Fins

Copper tubing

Blow
down

Bypass lever

Feedwater
regulating valve

Steam at boiler pressure

Water at boiler pressure

Steam in closed system

Water in closed system

FIGURE 1.9   *Self-operated process controller for boiler liquid level.*
Source: Feedwater Control Systems, Bulletin 530, Bailey Meter Co., Wickliffe, Ohio, 1968.

design and system integration have been largely worked out by the manufacturer of self-operated controllers, leaving to the user only the optimum "tuning" (after installation) of certain adjustable parameters. These adjustments are often rather limited and lead to a lack of versatility, which of course is a disadvantage of this approach to system control.

The feedwater regulator of Fig. 1.9 is suitable for boilers with relatively steady loads less than about 75,000 lb. of steam per hour. Unsteady loads make control more difficult owing to a steam drum dynamic response characteristic called "shrink and swell." When demand for steam suddenly increases, drum pressure drops and the combustion control system increases the boiler heat input. Both these actions cause a volume increase ("swell") in the steam bubbles below the liquid surface, causing a rise in liquid level even though the amount of liquid in the drum has not changed. Since the *ultimate* effect of an increased steam flow is a *reduction* in liquid level, the initial response in the opposite direction is misleading. Furthermore, a corrective increase in (cool) feedwater flow rate also has an initial reverse effect since steam bubbles condense, causing a momentary *lowering* of liquid level. These effects (and their opposites, called shrink, which occur on load decrease) are a confusing aspect of boiler dynamics which may require a more sophisticated control system.

The actual operation of this thermohydraulic regulator is, briefly, as follows. When the regulator is installed, the sealed system consisting of the generator outer tube, valve-actuating metal bellows, and connecting copper tubing is partially filled with water to establish a liquid level partway along the slanting outer tube. When the boiler is operating, the inner tube of the generator will exhibit a steam/water interface that accurately follows the drum water level. Also, the originally "empty" space in the outer tube will now be filled with steam due to the heating effect of steam in the inner tube. (This may require an initial bleed off of air trapped in the outer tube.) We have now established steam/water interfaces in both the inner and outer generator tubes. When these two positions coincide (as in Fig. 1.9) an equilibrium situation exists in which the pressure in the outer tube stays fixed, causing a fixed opening of the feedwater valve. This valve can be manually biased so that it supplies exactly the amount of water needed to provide the boiler's design steam rate at this equilibrium condition, thus also giving drum level equilibrium. If drum level should now decrease for any reason, the inner tube's steam/water interface moves down (left), exposing outer tube water to inner tube steam (separated, of course, by the inner tube metal wall) and causing some of this water to flash into steam, raising outer tube pressure, which expands the bellows and opens the feedwater valve. This increases water inflow rate and tends to restore water level toward its desired value. The sensitivity of this correction is influenced by (among other things) the slope of the generator. (A shallow slope gives a large change in generator heat transfer area for a small drum level change, giving a large valve correction for a small error.) Regulators of this type cannot maintain drum level exactly constant at different steady loads, but come close enough to be adequate in some applications.

For boilers with larger capacities and/or severe load fluctuations, control schemes of several different levels of complexity and performance have been devised, the so-called three-element control of Fig. 1.10 being perhaps the most sophisticated in common use. Since variations in steam demand are the major disturbance, a steam flowmeter is used to implement a *disturbance compensation* (disturbance feedforward) scheme. Feedwater flow rate is also measured, and compared with steam flow rate, since equality of these two flows encourages a material balance in the steam drum. *Cascade control* concepts are also used since any difference between steam and water flow rates causes a rapid correction of the feedwater control valve (secondary controller or "inner loop") while a slower primary loop using a liquid level signal maintains the long-term average drum level constant. Such an approach handles the shrink–swell phenomenon nicely since short-term level changes are essentially ignored while true drum water inventory is adequately controlled by matching the two flow rates. The hardware used to implement this control scheme (in contrast with the self-operated controller approach) employs distinct and separate measuring, control, and actuating devices, however these are readily interconnected into a complete system since all produce and/or accept standard pneumatic signals (3 to 15 psig)

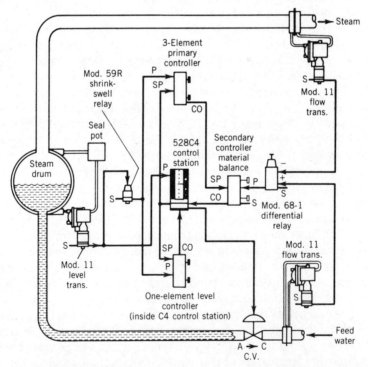

FIGURE 1.10   *Boiler level control system using disturbance feedforward and cascade control.*
Source: Moore Combustion Control Systems, Bulletin 5000, Moore Products Co., Spring House, Pennsylvania, 1974.

as inputs and outputs. This pneumatic control technology originally dominated the process control field and has been highly developed to provide a wide range of devices suitable for almost any application. Today, complete lines of hardware performing similar functions but based on electronics technology and communicating with standard electrical signals (4 to 20 mA, etc.) are also available and share the process control market with the more traditional pneumatic approach. Early electronic controllers used analog methods based on operational amplifiers, but now digital methods based on microprocessors are common. Use of built-in A/D and D/A converters at the input and output of the microprocessor may make these controllers externally indistinguishable from their analog counterparts.

Electric power utilities use boiler steam to drive turbine/generators that feed the electric power grid. Traditionally, the control systems for the boiler and the turbine were largely independent; however, comprehensive *digital computer control systems*[7] are now capable of integrating the actions of the two subsystems, making possible significant performance improvements. Figure 1.11 shows the

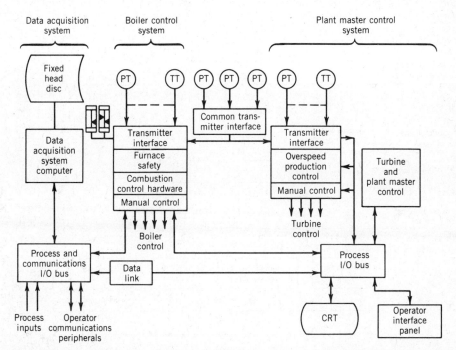

FIGURE 1.11   *Digital computer system for integrated control of steam/electric power generation.*
Source: M. H. Binstock, Using Advanced Control to Improve Power Plant Performance, *In Tech,* Jan. 1980, pp. 45–48.

---

[7]G. F. Franklin and J. D. Powell, *Digital Control of Dynamic Systems*, Addison–Wesley, Reading, Massachusetts 1980.

configuration of a proposed system of this type. The use of digital computers in control systems extends from the single-loop microprocessor controllers (functionally equivalent to earlier analog types) previously discussed, to large mainframe computers that oversee the operation of entire plants. All these systems are so-called *sampled-data systems* since the computers accept only intermittent discrete data, whereas analog systems work with continuous signals. If the sampling rate is sufficiently fast (relative to system speed of response), the sampled-data system's behavior can be closely predicted from a continuous model; otherwise special sampled-data, discrete system, and digit control theory must be used. The important application area called *numerical control*[8] in the manufacturing field is devoted to digital control of machine tools and is thus part of the overall digital control scene.

The most significant control applications of computers are those in which total operation of some complex process is supervised by the computer so as to continuously achieve optimum economic performance. Design and analysis of such systems is facilitated by adopting a *multilevel* or *hierarchical viewpoint of control strategy*. Such a strategy can be configured and implemented in various ways; Fig. 1.12 gives one possible interpretation and an example of a practical

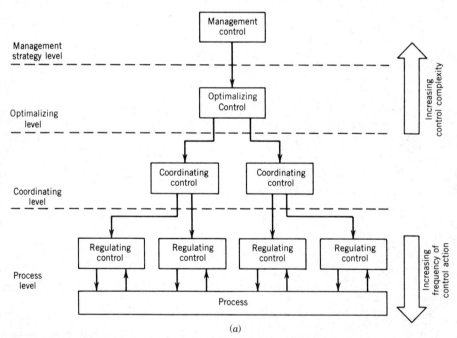

FIGURE 1.12   *Multi-level ("hierarchical") control strategy and example:* **(a)** *Multilevel hierarchical control structure.* **(b)** *Paper machine as an example of multilevel control structure.*

---

[8]R. S. Pressman and J. E. Williams, *Numerical Control and Computer-Aided Manufacturing*, Wiley, New York, 1977.

application.[9] In large and complex processes the multilevel approach subdivides the system into a hierarchy of simpler control design problems, rather than attempting direct synthesis of a single comprehensive controller for the entire process. Thus the controller on a given "level of control" can be less complex due to the existence of lower-level controllers that remove higher-frequency disturbances. At each higher level in the hierarchy, the complexity of the control algorithms increases, but the required frequency of execution decreases.

On the lowest level of control (process level), selected variables such as flows and speeds are regulated to specific setpoints using conventional feedback concepts. In the paper machine application (Fig. 1.12*b*), six such variables were

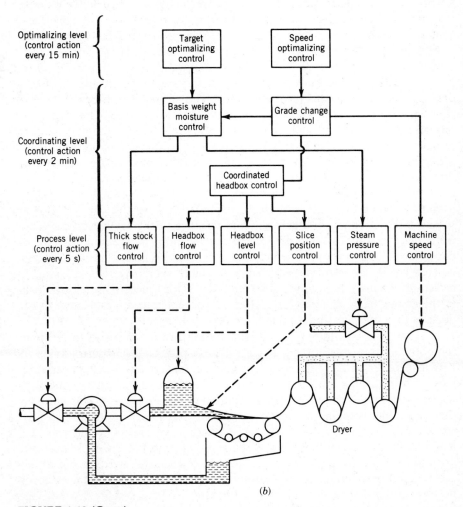

(*b*)

FIGURE 1.12 (Cont.)

[9]J. E. Eickelberg and W. L. Adams, *A Multi-Level Paper Machine Control System,* Industrial Nucleonics Corp., Columbus, Ohio, 1971.

identified. The computer samples these variables and updates control action every five seconds, an interval appropriate to the dynamic response of the physical processes involved. Either *centralized* or *distributed computer architectures* might be employed. That is, all levels of control could be directly managed by one large computer, or smaller (microprocessor-based) computers could be remotely located near specific process equipment and interfaced with a central computer that manages the overall operation. To accommodate memory requirements in the most economical way, process control computers for large systems generally require both rapid access (but expensive) memory (such as semiconductor) and slower, cheaper mass memory, such as disk. Algorithms for process-level control are executed most frequently but are relatively simple, so expensive rapid-access memory is appropriate, whereas complex high-level control algorithms may reside on disk since they are needed less frequently.

Although the process-level controls may be successful in the sense that the controlled variables adhere well to their desired values ("setpoints"), overall process operation may not be optimum unless these setpoints are adjusted to obtain desired *product* quality and quantity. The second level of control, (called coordinating) could also be called the "product" level since it involves interpreting product setpoints and manipulating the process-level setpoints to satisfy product specifications. In the paper machine, the coordination level includes a coordinated headbox control, a grade-change control, and a basis weight/moisture control. The grade-change control allows a smooth transition when the machine is required to change from one grade of paper to another (this is done *without* stopping the machine). Basis weight is the weight (lb_f/ft^2) of paper as it leaves the machine; moisture is the percentage of this weight that is water. Both these quantities must be held to specifications of the purchaser of the paper, thus their control, though necessarily indirect, is important. The coordinated headbox control provides smooth, stable operation by suppressing interactions among the three variables it directs. This is accomplished by storing in the computer an analytical model of the process behavior and using its predictions to counteract undesirable interactions.

The function of the third (process optimalizing) level of control is to determine the optimum setpoints of the product-level and process-level control loops and may include both steady-state and dynamic optimalization. The paper machine includes two distinct functions at this level. The target optimalizing control is a scheme that measures the statistical variation of basis weight and moisture and then shifts the setpoints so that a predetermined (small) percentage of off-specification product is always produced, irrespective of process changes. This is a desirable strategy since many manufacturing processes are most profitably operated when a certain small percentage of rejects is allowed, because there is a tradeoff between the money lost by throwing away rejects and the expense necessary to install and/or operate a "tighter" control system. In the paper machine it takes about 15 min. to gather a statistically significant sample (about 1000 readings) and thus this algorithm need be executed on the computer only at these intervals. The other optimalizing-level control involves machine speed.

A paper machine represents an investment of several million dollars and the more paper it produces per hour, the more return there is on investment. Speed optimalization is achieved by an algorithm that gradually increases machine speed until the limit of performance of some part of the machine is reached. It then runs at this speed until process changes allow a greater speed or require a reduction. Thus operation is at all times at or near the performance limits without danger of machine damage.

At the top of the hierarchy we find the management strategy level of control. Combined in this level are the control activities involving human decision making, often based on complex or unquantifiable variables. It is on this level that management implements the production strategy, for example, selecting operating points based on consumer acceptance of products. Good system design requires efficient information transfer to operators and supervisors, in terms of management information reports and data presentation on operator consoles.

The paper machine just described is an example of an important general class of problems called *multivariable control*.[10,11] Figure 1.13 shows the simplest (two-variable) example of this class in general form. Multivariable processes have multiple inputs and outputs that exhibit cross-coupling such that changes in an input (manipulated) variable such as M_1 cause responses in both the "intended" output C_1 but also in the "unintended" output C_2. If this cross-coupling is not too strong and/or performance need not be optimum, controls using individual feedback loops as in Fig. 1.13 may be acceptable. Sometimes, however, the individual controllers "fight" each other excessively. Then we may need to use a multivariable control theory that deals with the process cross-coupling by designing a cross-coupled controller (M_1 determined by *both* C_1 and C_2, M_2 similarly) to obtain improved performance.

Logic controls[12] are an important part of many control schemes, either as accessories to continuous systems or as complete entities in their own right. We here define logic controls as those that operate with "binary" (on/off) process signals rather than smoothly varying ones. (When a digital computer is interfaced with continuous processes, using A/D and D/A converters, the resulting system is *not* usually termed a logic-control system (even though the computer's internal workings are binary) because the *process* signals are not of an on/off nature.) The most extensive use of logic control is probably found in the area called manufacturing automation, where it is used to coordinate the functions of various machining, forming, assembly, and inspection operations into an integrated manufacturing process. For example, a cylinder block may be moving down a conveyor line. When it reaches a machining center, its presence is detected by a proximity sensor. This event causes the conveyor to momentarily stop while the block is pushed off the conveyor and onto a machine table by air cylinders.

---

[10]H. H. Rosenbrock, *Computer Aided Control System Design,* Academic, London, 1974.

[11]F. G. Shinskey, *Controlling Multivariable Processes,* ISA, Research Triangle Park, North Carolina.

[12]E. C. Fitch and J. B. Surjaamadja, *Introduction to Fluid Logic,* Hemisphere, Washington, D.C., 1978.

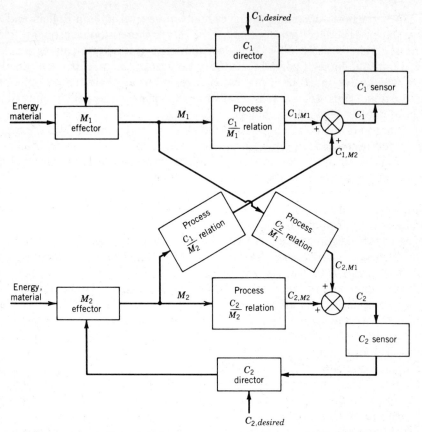

FIGURE 1.13   *Multivariable process controlled by individual feedback loops (Two-variable example).*

When the block is off the conveyor and on the machine table, the conveyor is restarted and other air cylinders come into action to clamp the block firmly in a desired position on the table. The table then indexes the block to the first machining position. When this position is reached, a boring tool is turned on and fed at rapid traverse to within 0.1 in. of cutting contact. At this point the feed rate is reduced to that desired for the actual boring and the tool is gradually fed in to the desired depth. When this depth is reached, the tool is backed away and retracted out of the hole. When the tool is clear, the table is indexed to the next position, where further machining operations are performed. This continues until all operations scheduled for this machine are complete, where-upon the cylinder block is unclamped, pushed back onto the conveyor (which has been momentarily stopped to receive it), and then transported on the conveyor to the next machining center.

The operation of this process can be thought of in terms of a sequence of interrelated logic statements.

IF the proximity sensor reports a block, THEN stop the conveyor.

IF the conveyor is stopped, AND a block is present, THEN actuate the air cylinders.

A relatively small number of basic logic functions (AND, OR, NOT, etc.) together with timing elements can be used to build up any required logic control system. Figure 1.14 shows how some basic functions can be implemented using electromechanical relays (switches). Electronic (no moving parts), pneumatic, and hydraulic logic devices are also available in a variety of forms. The *programmable logic controller,* a microprocessor-based general-purpose device provides a "menu" of basic operations that can be configured by programming to create a logic control system for any application. This "software" approach to logic control has many advantages, and although originally applied only to large systems has recently become economic even for small ones.

Logic controls are also used in important "accessory" functions on many continuous control systems. This refers to vital operations such as automatic start-up/shut-down procedures, emergency and protective alarming, selection of control modes based on process conditions, and interlocking of related functions. Recently, programmable logic controllers have been built with dual microprocessor systems that allow them to handle *both* the continuous and logical aspects of system control, providing a single machine for complete system operation.

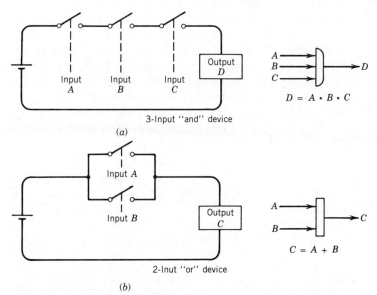

FIGURE 1.14 *Basic logic functions.*

A perusal of the library shelves devoted to the control systems field will reveal chapter headings and entire texts devoted to *nonlinear*,[13] *optimal*, and *adaptive* control. Much preliminary design of control systems can be successfully carried out using linear differential equations with constant coefficients as the mathematical model. Nonlinear effects introduce real mathematical difficulties but must sometimes be dealt with, mainly in two contexts. Once a preliminary design has been "roughed out" using rapid linear design methods that make clear important relations between system parameters and performance, it may be necessary to augment the linear model with one or more nonlinear effects that are not intentional features of the design but are known to be present in the real hardware. More often than not, these nonlinear effects are *undesirable* deviations from ideal behavior; things such as backlash in gears, stick/slip friction, and amplifier saturation. Second, we sometimes *intentionally* introduce certain nonlinear effects such as on–off control, variable gain, and self-tuning controllers, to improve system performance. Nonlinear control theory presents various approaches for dealing with both aspects of nonlinearity.

Good engineering design in any field has always striven for the "optimum" design, however in recent years formal mathematical approaches have received much attention, the *optimal control*[14] field being one of the most active. Here the controlled process is precisely defined in mathematical terms and we also must formulate a specific "cost" or "payoff" function which is to be minimized or maximized to achieve "best" system operation. One approach requires the designer to choose the *form* of the control law but leaves numerical values of control parameters open to choice. The optimal control theory then leads us to a parameter combination that results in the "best" system. A more general (and complicated) approach leaves both the form and numerical values of the control law open to determination by the optimalization procedure. Although such mathematical synthesis techniques have some appeal since they appear to avoid trial-and-error methods and minimize the importance of designer judgement and experience, the assumptions needed are often so unrealistic that the results become no more than general guidelines. Thus, although optimal control theories have made useful contributions to a few difficult control applications, the majority of practical designs are carried out with little or no consideration of these methods.

Adaptive control systems[15] can take a number of forms, but as the name suggests, they are intended to provide improved performance by adapting themselves to changing conditions. An "ordinary" feedback system exhibits a considerable tolerance for changes, thus special adaptive features are only needed for extreme situations. Whereas changes in commands, disturbances, and system parameters may all be of interest, parameter changes in the controlled process

---

[13]G. J. Thaler and M. P. Pastel, *Analysis and Design of Non-linear Feedback Control Systems,* McGraw–Hill, New York, 1962.
[14]A. E. Bryson and Y. C. Ho, *Applied Optimal Control,* Blaisdell, Waltham, Massachusetts, 1969.
[15]E. Mishkin and L. Braun, *Adaptive Control Systems,* McGraw–Hill, New York, 1961.

can be a major concern. Flight control systems for vehicles that operate over wide ranges of altitude and flight speed are a good example, since the vehicle's natural frequency may vary over a 10-to-1 range while damping ratio changes by 20 to 1. A controller designed for any one flight condition would be grossly inadequate at others. Combined air/space vehicles such as the space shuttle are extreme cases since aerodynamic control effectors (rudder, elevator, etc.) which are used in the atmosphere are totally useless in airless space and must be replaced by gas jets and the like, thus both vehicle characteristics and control hardware vary radically. The most common successful approach to adaptive control is a preprogrammed or scheduled scheme in which the system measures the environmental factors that are causing the changes in the controlled system's parameters and continuously adjusts the controller parameters to accommodate the current situation. In aircraft, the vehicle's natural frequency, damping ratio, and the like depend mainly on altitude and Mach number and can be calculated with adequate accuracy from measurements of these two quantities. It may then be possible to design a controller of a given form that changes its parameter numerical values (in response to altitude and Mach number data) so as to give good performance over the entire flight envelope. In extreme cases, logic controls that switch the *form* of the controller when radical process changes occur may be necessary. So-called *self-adaptive* systems which deal with the same problem but require no environmental measurements are also possible and can take a variety of forms.

Man-machine control systems generally use a human operator as one of the components in a feedback control loop so as to combine the best features of human and machine capabilities. The area in which the engineering of such systems has been furthest developed is probably that of flight-vehicle piloting, in which the human operator's behavior[16] is actually mathematically described with numerical values of gains, time-constants, and dead-times, so that the overall man-machine system may be rationally analyzed and designed. The area of *biological control systems*[17] employs mainly feedback system theory to understand and explain biological phenomena ranging from an organism's internal functions, such as body temperature regulation and posture control, to large scale effects, such as population dynamics of animal species and response of ecological systems. The *econometric models* used for predicting the behavior of economic systems[18] are also generally based on feedback principles.

A final comment in this section addresses the terminology *classical control theory* and *modern control theory,* phrases regularly encountered in reading recent texts and papers. Classical control theory was largely developed by engineers actually working on practical control system design and, by about 1960, comprised a large body of use-tested knowledge covering linear, nonlinear,

---

[16]E. O. Doebelin, *System Modeling and Response,* Chap. 13, Wiley, New York, 1980.

[17]J. H. Milsum, *Biological Control System Analysis,* McGraw–Hill, New York, 1966.

[18]A. Tustin, *Mechanism of Economic Systems,* Harvard University Press, Cambridge, Massachusetts, 1954.

sampled-data and digital systems, working with either deterministic or random inputs. This theory, combined with computer simulation and experimental development, provided an efficient framework for the design of feedback controls in all areas of application. In my opinion, this statement still holds today for 95% or more of the applications, even though 20 years of intense research by mathematicians and engineers has resulted in a vast body of so-called modern control theory, based on "state-space" methods, which was at one time proposed as a replacement for the classical methods. We cannot here pursue all the reasons for this situation but I will offer the following quotation[19] from a recent text devoted to control research.

> *After more than fifteen years have been spent developing the simplest and most thoroughly understood problem of modern control, the "linear regulator theory," practicing engineers still have no assurance that the theory will yield an acceptable solution to a particular problem, even if they are convinced that the quadratic performance index is quite appropriate. This is due to that fundamental deficiency of the theory which requires absolute fidelity of the mathematical model.*

Another, more recent view:[20]

> *Control theory made substantial progress in the period 1955–1970. Very little of this theory has, however, made its way into existing computer-control systems.*

Although research going beyond "classical" methods is necessary and desirable in all fields, new design methods must prove themselves in actual practice before they can displace well-accepted techniques, particularly if the new methods are couched in rather abstract mathematical terms. Except for a small number of quite specialized systems, modern control *theory* has really not been necessary for "modern" control *applications*. Practicing control designers often find that new *hardware* developments such as microprocessors and electro-optic measurement schemes have much greater impact on their designs than do the latest advances in control theory. Since this text is addressed to practical design rather than control research, the emphasis will be on classical methods. Although, 20 years ago, there might have been some risk involved in presenting "old" methods when "new and improved" ones seemed to be on the horizon, history has by now verified that classical methods may be augmented by "modern control theory" but will not be replaced by it.

We have now completed our overview of the control systems field, introducing much of the basic terminology and many of the fundamental ideas. To reem-

---

[19]R. E. Skelton and P. W. Likins, Techniques of Modeling and Model Error Compensation in Linear Regulator Problems, in *Control and Dynamic Systems*, Vol. 14, C. T. Leondes, ed., Academic, New York, 1978.
[20]K. J. Åström and B. Wittenmark, *Computer Controlled Systems*, Prentice-Hall, 1984, p. 7.

phasize the breadth of the area and to collect our thoughts on this diverse presentation it may be useful to list the major topics covered, in outline form:

1. Open-loop control, [including disturbance/command compensation (feed-forward)].
2. Closed-loop (feedback) control.
3. Servomechanisms/process control.
4. Cascade control.
5. Industrial robots.
6. Self-operated controllers.
7. Digital-computer control (centralized, distributed, microprocessor).
8. Sampled-data systems.
9. Multilevel hierarchical control strategy.
10. Logic control, programmable controllers.
11. Nonlinear, optimal, and adaptive control.
12. Man–machine/biological control.
13. Modern/classical control theory.

1.3 BASIC BENEFITS OF FEEDBACK CONTROL

Although a detailed exploration of the possibilities and problems of feedback control will occupy most of the rest of this text, some essential characteristics can be illustrated quite easily, and give a helpful preview of forthcoming developments. Figure 1.15a shows a simple open-loop system for controlling a first-order process. (Readers unfamiliar with the transfer function concept used in Fig. 1.15 will find it developed in Chapter 2; the analysis below requires only elementary differential equation background.) Controlled variable C is related to its desired value V and process disturbance U_p by

$$\left(\tau_p \frac{dC}{dt} + C\right) = (K_A K_C K_p)V + (K_N K_p)U_p \qquad (1.1)$$

where

$$\tau_p \triangleq \text{process time constant}$$
$$K_C \triangleq \text{controller sensitivity}$$
$$K_A \triangleq \text{sensitivity of reference input element}$$
$$K_N K_p \triangleq \text{process sensitivity to } U_p$$
$$K_p \triangleq \text{process sensitivity to } M$$

The symbol $\triangleq$ means "equal by definition"

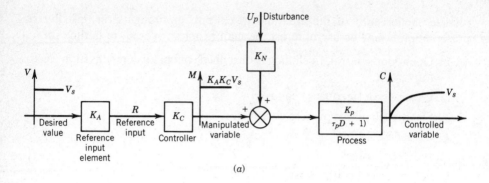

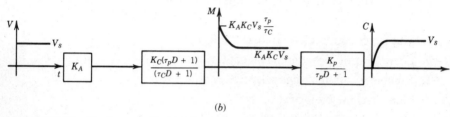

(b)

FIGURE 1.15 *Basic, and command-compensated open-loop systems.*

We choose K_A such that $K_A K_C K_p = 1.0$, giving $C \equiv V$ for any steady V if there is no disturbance. A step input V_s of V gives (for $U_p = 0$)

$$C = K_A K_C K_p (1 - e^{-(t/\tau_p)}) V_S \qquad (1.2)$$

whereas a step U_{ps} ($V = 0$) gives

$$C = K_N K_p (1 - e^{-(t/\tau_p)}) U_{ps} \qquad (1.3)$$

Equation 1.2 shows that any change in $K_A K_C K_p$ from the design value of 1.0 results in a directly proportional error in C, whereas a disturbance U_{ps} causes an additional error $K_N K_p U_{ps}$. Although the disturbance compensation (feedforward) schemes described in the previous section can reduce the error due to U_{ps}, that due to changes in $K_A K_C K_p$ is not amenable to such improvement.

Should faster response to V be desired, command compensation as in Fig. 1.15b may be possible. Using the operator notation $D \triangleq d/dt$ we can write

$$(\tau_C D + 1)M = K_A K_C (\tau_p D + 1)V \qquad (1.4)$$

$$(\tau_p D + 1)C = K_p M \qquad (1.5)$$

$$(\tau_C D + 1) \frac{(\tau_p D + 1)C}{K_p} = K_A K_C (\tau_p D + 1)V \qquad (1.6)$$

$$(\tau_C D + 1)C = K_A K_C K_p V \qquad (1.7)$$

Since the compensator time constant τ_C can be made much smaller than the process time constant τ_p, the response of C to V as given by Eq. 1.7 is now much faster than in Eq. 1.1. This is, of course, accomplished by the initial "over-response" seen in M. If we try for too much speedup, this high peak causes saturation nonlinearity in the controller and/or process and the predicted speedup is not realized.

To compare the behavior of this open-loop approach with feedback methods, we consider the system of Fig. 1.16. The system differential equation is

$$\left(\frac{\tau_p}{1 + K_C K_p K_H} D + 1\right) C = \frac{K_A K_C K_p}{1 + K_C K_p K_H} V + \frac{K_N K_p}{1 + K_C K_p K_H} U_p$$
$$- \frac{K_p K_C}{1 + K_C K_p K_H} U_s \qquad (1.8)$$

System linearity allows separate consideration of the three inputs V, U_p, and U_s, so let us take V as a step input V_s, with $U_p = U_s = 0$, giving

$$C = \frac{K_A K_C K_p}{1 + K_C K_p K_H} (1 - e^{-(t/\tau_{cl})}) V_s \qquad (1.9)$$

where

$$\tau_{cl} \triangleq \frac{\tau_p}{1 + K_C K_p K_H} \triangleq \text{closed-loop system time constant} \qquad (1.10)$$

In every feedback system, the static sensitivity (also called steady-state gain) between signals E and B is the single most important design parameter and is given the name *loop gain* and symbol $K(K = K_C K_p K_H$ in our example). Generally, all aspects of control system performance (steady-state accuracy, speed of response, etc.) improve when K is made larger, so large K is a basic design goal. (There is, however, always an upper limit on K, beyond which system stability suffers.) Equation 1.10 clearly shows that the time constant τ_{cl} which

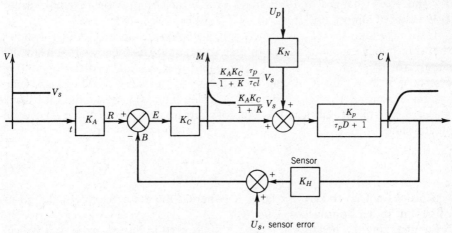

FIGURE 1.16 *Response of feedback system to step command.*

governs the speed of response of controlled variable C can be made much smaller (faster) than τ_p if $K = K_C K_p K_H$ is made large compared with 1.0. Thus if *process* design has brought τ_p to its minimum feasible value, feedback control allows significant *further* improvements in the speed of response of C to V, without *any* changes in the process itself. This capability for overcoming apparent limitations in basic hardware performance is one of the major contributions of feedback.

There is, of course, nothing "magical" about the speedup of C's response even though the process itself is "as slow as before." The manipulated variable M simply "overresponds" initially, as can be seen from its differential equation

$$(\tau_{cl}D + 1)M = \frac{K_A K_C}{1 + K_C K_p K_H} (\tau_p D + 1)V \qquad (1.11)$$

which may be compared with the open-loop command compensation of Eq. 1.4 ($\tau_{cl} < \tau_p$ is analogous to $\tau_c < \tau_p$). Thus command compensation and feedback achieve similar results (the initial overresponse of process input M) but by quite different means. However, they share the basic limitation on the degree of improvement possible, that is, the saturation nonlinearity caused by attempting excessive peaking in M.

Turning to steady-state behavior, Eq. 1.9 shows that if we choose $K_A = (1 + K_C K_p K_H)/K_C K_p$, then $C \equiv V_s$ in steady state. In actual practice, K_A and K_H are normally made equal and $K >> 1$, so that $K_A K_C K_p/(1 + K_C K_p K_H) \approx 1$ and C is very nearly equal to V_s in steady state. More importantly, however, if $K >> 1$, changes in K_C and/or K_p now have much less effect on accuracy than in an open-loop system. For example, if $K_A = K_H = 1$ and $K_C K_p = 100$, $C = 0.9901 V_s$. A 10% change in $K_C K_p$ to 110 would give $C = 0.9910 V_s$, only a 0.1% change; whereas an open-loop system would show a 10% change in the C/V relation. Thus if we can make loop gain high enough, system accuracy becomes very insensitive to changes in hardware parameter values other than K_A and K_H. (Changes of 10% in K_A or K_H for the previous example cause C/V errors of almost 10%.)

If we now take $V = U_s = 0$ and apply a step disturbance U_{ps}, Eq 1.8 shows that C responds exponentially (with time constant τ_{cl}) and levels off at a steady-state error of $[K_N K_p/(K + 1)]U_{ps}$, compared with $K_N K_p U_{ps}$ for an open-loop system. For loop gain $K = 100$, the error is 101 times less for the closed-loop system. Finally, if $V = U_p = 0$ but U_s is a step input U_{ss}, we see that feedback is *not* successful in rejecting this type of "disturbance" since Eq. 1.8 gives in steady state

$$C = - \frac{K_p K_c}{1 + K_p K_C K_H} U_{ss} \qquad (1.12)$$

which does not go to zero for large K (whereas the error due to U_p did go to zero), but rather approaches $- U_{ss}K_H$.

Summarizing the benefits of feedback discovered in this example, but typical in general of feedback systems with high loop gain, we can state that such systems:

1. Cause the controlled variable to accurately follow the desired value.
2. Greatly reduce the effect on the controlled variable of all external disturbances except those associated with the sensor.
3. Are tolerant of variations (due to wear, aging, environmental effects, etc.) in hardware parameters other than those of the sensor and reference input elements.
4. Can give a closed-loop response speed much greater than that of the components from which they are constructed.

1.4 THE ACCURACY–STABILITY TRADEOFF IN FEEDBACK SYSTEMS

I have mentioned several times that all feedback systems can become unstable if improperly designed (loop gain too high) and I now wish to give a brief qualitative discussion of this instability phenomenon. In essence, instability results from an improper balance between loop gain and system dynamic lags. If all the components of our system responded instantaneously, we could employ arbitrarily high gain values (shown to be very beneficial in the previous section) with no concern for instability. Instantaneous response is unfortunately *impossible* in the real world since it requires a system to go from one energy level to another in zero time, implying power supplies of *infinite* power. Thus, in any real-world component there is some kind of dynamic lagging behavior between input and output; we can make a sudden change in the input voltage to an electric oven but we will have to wait a while before the oven temperature comes up. For those readers with some system dynamics background, lagging behavior is characterized quantitatively by the time constants (τ's) of first order and the natural frequencies (ω_n's) of second-order components. Fast response (small τ's, large ω_n's) of components is desirable in feedback control systems; it results in greater accuracy since larger loop gain is allowed.

Why does the combination of excessive loop gain with excessive lags always result in feedback system instability? Consider a feedback system as in Fig. 1.3, with a fixed desired value and in equilibrium, with the controlled variable at the desired value. If a process disturbance occurs, the ensuing deviation of the controlled variable will cause a correction to be applied but it will be delayed by the cumulative lags of sensor, director, effector, and process. Eventually, however, the trend of the controlled variable caused by the disturbance *will* be reversed by the opposition of the process manipulated input, returning the controlled variable toward the desired value. Now if loop gain is high, a *strong* correction is applied and the controlled variable overshoots the desired value, causing a reversal in the algebraic sign of the system error (difference between desired value and controlled variable). Unfortunately, because of system lags, a reversal of *correction* does not occur immediately and the process manipulated input (acting on "old" information) is now actually driving the controlled variable in the *same* direction as it is already going, rather than opposing its excursions, leading to a larger deviation. Eventually the reversed error does cause a reversed correction but by then the controlled variable has *also* reversed and the correction is again in the wrong direction. The controlled variable is thus driven alternately in opposite directions and does not settle to an equilibrium

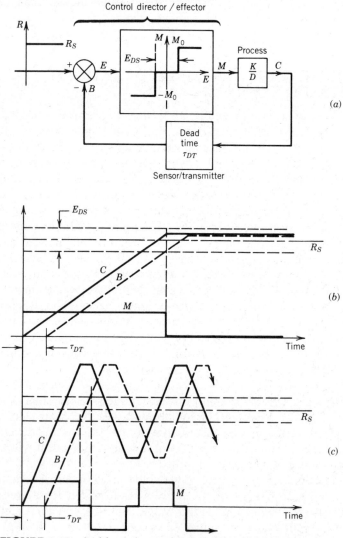

FIGURE 1.17 *Stable and unstable behavior of feedback system.*

condition. This oscillatory state is called *instability* and (except for certain classes of systems) is unacceptable as control system behavior.

A specific example of the general behavior just described may be useful. In Fig. 1.17 the liquid level C in a tank is manipulated by controlling the volume inflow rate M by means of a three-position on/off controller. The transfer function K/D between M and C represents the conservation of volume relation between inflow rate and liquid level

$$\frac{dC}{dt} = \frac{1}{\text{tank area}} M \triangleq KM$$

$$DC = KM \tag{1.13}$$

$$\frac{C}{M}(D) = \frac{K}{D}$$

The pump that manipulates M is shut off ($M = 0$) if the error E between desired tank level R and measured level B is less in absolute value than the *error dead space* $E_{DS}/2$. When the error exceeds these limits, the pump adds or removes liquid at a rate M_0. Our liquid-level sensor is assumed to measure C perfectly but there is a data-transmission delay of τ_{DT} seconds in sending this information to the controller. That is, the signal B is identical in form to C but is delayed by τ_{DT} seconds, a behavior given the name *dead time*.

Since instability can be triggered by either or both command and disturbance inputs, in this example we apply a step command input R_S and examine system response. Note that "loop gain" (the "strength" of the corrective effort) in this system depends on both M_0 and K. In Fig. 1.17b, modest values of M_0 and K give a relatively slow, but stable, response of C. If specifications require a faster response, M_0 and/or K may be increased, but in Fig. 1.17c the designer has gone too far with this, causing instability. Figure 1.17c also clearly shows how (because of time lags) "correction" M acts in a direction to increase (rather than reduce) the excursions of C during large parts of the cycle, a general condition for instability discussed earlier.

1.5 CONTROL SYSTEM DESIGN PROCEDURES

Just as for any other engineered product or service, the overall design of feedback control systems proceeds in systematic fashion through a sequence of steps as in Fig. 1.18. Rather than discussing this general process we will consider those aspects peculiar to feedback controls, which apply to the portions of the overall design process so designated on Fig. 1.18. Due to the closed-loop nature of the system configuration, the input of each component is affected by all the other components, requiring a design/analysis approach that considers the entire system simultaneously. Fundamentally, the system is described by differential equations, the most general and highly developed theory being based on ordinary

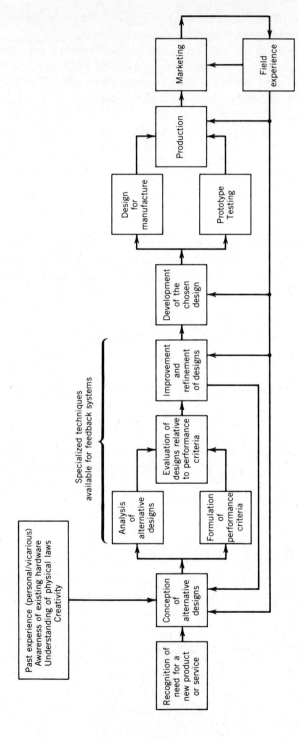

FIGURE 1.18 *Product design process outline.*

linear equations with constant coefficients, in which time is the independent variable.

Before one can write and solve the system differential equations to obtain system performance data, one must first *conceive* a system configuration and operating principle. New systems generally utilize combinations and/or variations of well-known principles, thus a designer must be familiar with this background (which is presented in the remainder of this text). In addition to familiarity with basic principles, practical designers must also maintain up-to-date knowledge of available hardware in their field of application. Trade journals, where manufacturers present advertisements and technical papers explaining the latest developments, are a major source of such information. Although new hardware is continually being developed, there are many classes of equipment that have been in wide use for decades and that show no signs of disappearing. Since the philosophy of this text attaches great importance to the benefits obtained from a concurrent treatment of theory and practice, hardware of such enduring nature will also be discussed at some length.

Once a tentative system concept has been developed through combination of basic principles and hardware familiarity, components must be "sized" to meet the needs of the specific application. For example, in the tracer lathe of Fig. 1.2c, use of a servovalve and hydraulic actuator represents the designer's choice (from several available alternatives) of a particular *form* of drive system; however, further decisions must be made as to the *size* of valve and actuator. These are determined from consideration of specifications such as the desired maxima of actuator force and speed required in the machining operation. Tentative choices of the form and "size" of *all* system components can be made in this way, and once this phase is complete, the form of the components' differential equations (and the numerical values of coefficients in these equations) become available, the simultaneous set of all these equations becoming the overall system description. Actually, at this stage, *one* numerical value, the loop gain, is left open to choice. Usually, loop gain is composed of several "bits and pieces" of gain, one for each component, so its adjustment to some desired value presents no practical difficulty. The "proper" value for loop gain is the one that meets the system accuracy specifications while maintaining adequate stability margins. A "brute force" approach to finding this value would be to repeatedly solve the system differential equation for the controlled variable (with the desired command/disturbance inputs applied) for a range of gain values and select the best. Computer-aided-design (CAD) techniques (such as the digital simulation language we will soon be using) can make such a trial-and-error technique quite cost effective. However, classical control theory provides more direct (and usually one-step) methods for solving this most basic system design problem of gain setting. The two most successful practical methods, *frequency response* and *root locus,* will be explained later in this text.

Although every feedback system must have its gain set, there is no guarantee that the optimum gain for the *initial* choice of components will meet all system specifications. When this failure occurs, changes *other* than loop gain must be tried. These include:

    a. Improved component dynamics (τ's, ω_n's, etc.)

    b. Addition of dynamic compensation elements.

    c. Change of controller type (on–off, proportional, integral, etc.).

    d. Totally new system concept (change from electric drive to hydraulic drive, for example).

Fortunately, the frequency-response and root locus design methods include techniques that provide guidance on the type and amount of change needed to help satisfy a particular unmet specification. The modified system must then again have its gain set, and it is hoped that sufficient iterations of this modify/ set gain/evaluate performance procedure lead to an acceptable design. Both the frequency-response and root locus approaches were originally implemented as "manual" graphical procedures, but they are now available as software packages for rapid computer-aided design on many computer systems. In fact, the control systems field was the leader in developing CAD procedures, which are now spreading rapidly to other fields.

BIBLIOGRAPHY

Books

1. O. Mayr, *The Origins of Feedback Control,* MIT Press, Cambridge, Massachusetts, 1970.

2. H. Chestnut and R. W. Mayer, *Servomechanisms and Regulating System Design,* Vol. 1, 1951, Vol. 2, 1955, John Wiley, New York.

3. F. G. Shinskey, *Process Control Systems,* McGraw–Hill, New York, 1979.

4. G. F. Franklin and J. D. Powell, *Digital Control of Dynamic Systems,* Addison–Wesley, Reading, Massachusetts, 1980.

5. H. H. Rosenbrock, *Computer-Aided Control System Design,* Academic Press, London, 1974.

6. E. C. Fitch and J. B. Surjaamadja, *Introduction to Fluid Logic,* Hemisphere Publ. Co., Washington, D.C., 1978.

7. G. J. Thaler and M. P. Pastel, *Analysis and Design of Nonlinear Feedback Control Systems,* McGraw–Hill, New York, 1962.

8. A. E. Bryson and Y. C. Ho, *Applied Optimal Control,* Blaisdell, Waltham, Mass., 1969.

9. E. Mishkin, et al., *Adaptive Control Systems,* McGraw–Hill, New York, 1961.

10. J. H. Milsum, *Biological Control System Analysis,* McGraw–Hill, New York, 1966.

11. J. H. Laning and R. H. Battin, *Random Processes in Automatic Control,* McGraw–Hill, New York, 1956.

12. H. S. Tsien, *Engineering Cybernetics,* McGraw–Hill, New York, 1954.

13. C. L. Smith, *Digital Computer Process Control,* International Textbook, Scranton, Pennsylvania, 1972.

14. Truxal, J. G., *Automatic Feedback Control System Synthesis*, McGraw–Hill, New York, 1955.

15. P. S. Buckley, *Techniques of Process Control*, John Wiley, New York, 1964.

16. Raven, F. H., *Automatic Control Engineering*, McGraw–Hill, New York, 1978.

17. R. S. Pressman and J. E. Williams, *Numerical Control and Computer-Aided Manufacturing*, John Wiley, New York, 1977.

18. R. Isermann, *Digital Control Systems*, Springer–Verlag, Berlin, Heidelberg, New York, 1981.

Periodicals

1. *Automatic Control*. English translation of Russian journal. Faraday Press, New York.

2. *Automation and Remote Control*. English translation of Russian Journal. Plenum Publications, New York.

3. *Automatica*. Pergamon Press, Oxford.

4. *Automatisierung*. In German.

5. *Automatisme*. In French.

6. *Canadian Controls and Instrumentation*.

7. *Control Engineering*.

8. Jour. of Dynamic Systems and Control. ASME Transactions.

9. IEEE Transactions:
   Automatic Control
   Industrial Electronics and Control Instrumentation
   Industry Applications
   Instrumentation and Measurement

10. *In. Tech.* Instrument Society of America (ISA).

11. *Trans. of ISA*.

12. *Jour. of the Instrumentation Control Assoc.* In Japanese.

13. *Trans. of the Soc. of Instrument and Control Engineers.* In Japanese; English abstract.

14. *Papers of Tech. Meet. on Auto. Control.* Inst. of Elect. Engineers, Japan. In Japanese.

15. *Trans. of the Inst. for Meas. and Control.* London.

PROBLEMS

1.1 Identify and explain the action of manipulated and disturbance inputs for the following processes:

   a. Ship steering.

   b. Automobile steering.

   c. Aircraft Mach number ("speed") control.

   d. Electric generator speed control in utility power station.

   e. Temperature control in automobile air conditioning system.

   f. Liquid-level control in steam boiler.

1.2 Choose a technical device or process familiar to you and trace its historical "control development" in a manner similar to that of Fig. 1.2.

1.3 In the system of Fig. 1.17, discuss the significance of the error dead space (E_{DS}) with regard to steady-state errors and stability.

1.4 Extend the concept of disturbance compensation (Fig. 1.4) to *multiple* disturbances by drawing and explaining the block diagram of such a system. Discuss the relative advantages of this scheme as compared with a feedback system with multiple disturbances.

1.5 A person carrying a full cup of coffee will often spill less if the cup is not watched. Assuming this to be true, discuss the phenomenon from a feedback-system viewpoint.

1.6 Draw block diagrams similar to that of Fig. 1.3 for the following feedback systems, clearly identifying the input energy/material, process, disturbances, controlled variable, measuring device, control director and effector, and desired value:

    a. A human operator steering a car that is coasting (no engine power) down a hill on a winding road

    b. The regulation of production by the manager of a manufacturing plant

    c. The operation of the law of supply and demand

    d. The tracer lathe of Fig. 1.2c

    e. The disk memory servo of Fig. 1.6

    f. The self-operated boiler control of Fig. 1.9

CHAPTER 2
Component Modeling

2.1 GENERALIZED BLOCK DIAGRAM OF A FEEDBACK SYSTEM

This chapter serves two primary functions, the presentation of some basic material from the system dynamics area, and familiarization with certain classes of control system hardware. For those (such as the author's students) who are already prepared in the system dynamics area, the former material will be review; however the hardware examples will contain considerable new and useful information. In a control systems text, presentation of basic system dynamics concepts must, for reasons of space, be limited to essentials, so the reader interested in more detail is referred to the literature.[1] To motivate the presentation and place it in a control systems context we begin by discussing the generalized feedback system diagram of Fig. 2.1.

The earlier block diagrams have by choice been of a general functional nature; it is now appropriate to begin using the working *operational block diagrams* necessary for actual system design and analysis. These use the transfer function concept, which allows the block diagram to communicate the numerical details

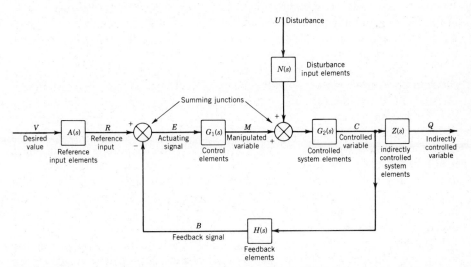

FIGURE 2.1   *Generalized operational block diagram of feedback control system.*

[1]E. O. Doebelin, *System Dynamics,* Merrill, Columbus, Ohio, 1972; E. O. Doebelin, *System Modeling and Response,* Wiley, New York, 1980.

of component and system behavior. Because I wish to introduce specific types of hardware in this chapter, Fig. 2.1 allows a useful classification since it identifies the basic functional components from which all feedback systems are built. It also represents the results of the nomenclature standardization efforts of the several engineering societies (ASME, IEEE, etc.) to ease communication in the control systems field by defining a standard set of symbols and word definitions. Note that two types of quantities require definition: *signals* and *systems*. Signals are the physical variables (voltage, pressure, temperature, etc.) that "flow" from one system (component) to another; whereas systems are the hardware components that perform the necessary operations. System descriptions in Fig. 2.1 consist of the transfer functions $A(s)$, $G_1(s)$, and the like, which are a shorthand graphic means of stating the component's differential equation.

To understand Fig. 2.1 we begin with the signal V, the *desired value* (of the controlled variable). This signal has the same units as the *controlled variable* C but may or may not exist as an actual physical quantity. For example, in a home heating system, V is the desired room temperature but this is not an actual measurable temperature, it exists only "mathematically" as the numbers (°F or °C) printed on the thermostat dial. In fact, the actual physical input to the control system is a mechanical rotation (angular degrees) of the dial which we manually enter as desired. This rotation is an example of a *reference input* R; that V and R need not be the same quantity is one reason the standard diagram provides the transfer function $A(s)$, *the reference input element*. In our example $A(s)$ is just a simple conversion factor, say, 5 angular degrees per °F. The reference input element can, when necessary, perform a more sophisticated function, such as the command compensation of Fig. 1.5 or noise filtering. In noise filtering, V might be a physical voltage, but "contaminated" by high frequency electrical noise. Then $A(s)$ might be a low-pass filter that passes the true desired value but screens out the spurious noise.

The *summing junction* symbol represents the comparison ($E = R - B$) of the reference input with the feedback signal B. The feedback element $H(s)$ is often a sensor for measuring C, but since its functions sometimes include more than simple measurement, the name *feedback element* (rather than *sensing element*) was chosen by the designers of the standard nomenclature. Use of symbol E for the *actuating signal* is a carry-over from earlier nonstandard usage in which it was called the error signal. Since system error is logically defined as ($V - C$), *actuating signal* is a more appropriate term for E, since $E = V - C$ only if $A(s) = H(s) = 1.0$, sometimes true but often not true. Note that the summing junction cannot change the units of R and B, thus R, B, and E *always* have exactly the same dimensions.

Functions of both the control director and control effector are included in $G_1(s)$, the standard symbol for the *control elements*. Standard terminology for the process being controlled, $G_2(s)$, is *controlled system elements*. Since controlled variable C is influenced by both *manipulated variable M* and *disturbance U* (U stands for "unwanted"), and since the effect on C of U and M would in general be different, the path from U to C is provided with $N(s)$, the *disturbance input elements*, allowing completely independent specification of C/U and C/M

relationships. Of course, the disturbance input elements are not system "components" intentionally added by the designer to allow the disturbance entry to the system but rather the necessary modeling of an unavoidable effect of U on C.

To provide for those situations where the quantity Q that we really wish to control cannot be measured, the path from C to Q through the *indirectly controlled system elements* $Z(s)$ is provided. This path is "open loop," thus we depend on $Z(s)$ to remain fixed if we are to accomplish accurate control of Q. Many numerically controlled machine tools fit this pattern. Piece dimension (Q) is the variable of prime interest but is often difficult to measure, so we instead measure tool-slide position (C). So long as backlash, elasticity, and tool wear are kept within allowable tolerances, $Z(s) \approx 1.0$ and accurate parts are produced.

Figure 2.1 defines the basic types of signals and components necessary for description of any feedback control system; however it must be adapted to the needs of each specific design. For example, disturbances may enter the system at several locations, not just at the process. This is easily accommodated by providing suitably located summing junctions, and defining disturbances U_1, U_2, U_3, and so forth with corresponding disturbance input elements N_1, N_2, N_3. Finally, when we deal with a specific application, rather than the abstract generality of Fig. 2.1, it is preferable to use the standard signal symbols as *subscripts* on symbols which relate more directly to the physical variables involved. For instance, in our earlier home heating example we would call the desired temperature T_V and the controlled variable T_C.

2.2 LAPLACE TRANSFORM METHODS, LINEARIZATION

A basic mathematical model used in many areas of engineering analysis is the ordinary linear differential equation with constant coefficients.

$$
a_n \frac{d^n q_o}{dt^n} + a_{n-1} \frac{d^{n-1} q_o}{dt^{n-1}} + \cdots + a_1 \frac{dq_o}{dt} + a_0 q_o =
$$

$$
b_m \frac{d^m q_i}{dt^m} + b_{m-1} \frac{d^{m-1} q_i}{dt^{m-1}} + \cdots + b_1 \frac{dq_i}{dt} + b_0 q_i \tag{2.1}
$$

where

$$q_o \triangleq \text{output (response) variable of physical system}$$

$$q_i \triangleq \text{input (excitation) variable of physical system}$$

$$t \triangleq \text{time}$$

$$a_n, b_m \triangleq \text{physical parameters of system}$$

Use of this model is widespread because straightforward analytical solutions are available no matter how high the order n of the equation, allowing treatment of linear systems of arbitrary complexity. Since every real-world system has some nonlinearity, such a math model is always an approximation to reality, but of course *no* math model ever exactly represents nature. Many real-world nonlinearities involve a "smooth" curvelinear relation between an independent variable x and a dependent variable y.

$$y = f(x) \tag{2.2}$$

A linear approximation to the curve, accurate in the neighborhood of a selected operating point x_0, is the tangent line to the curve at that point. This approximation is given conveniently by the first two terms of the Taylor series expansion of $f(x)$.

$$y = f(x_0) + \frac{df}{dx}\bigg|_{x=x_0} (x - x_0) + \frac{d^2f}{dx^2}\bigg|_{x=x_0} \frac{(x - x_0)^2}{2!} + \cdots \tag{2.3}$$

The linear approximation is

$$y \approx f(x_0) + \frac{df}{dx}\bigg|_{x=x_0} (x - x_0) \tag{2.4}$$

For example, in liquid-level control systems similar to that of Fig. 1.17, when the tank is not prismatic, a nonlinear volume/height relation exists and causes a nonlinear system differential equation. For a conical tank of height H and top radius R we would have

$$V = \frac{\pi R^2}{3H^2} h^3 \tag{2.5}$$

$$V \approx \frac{\pi R^2 h_0^3}{3H^2} + \frac{\pi R^2 h_0^2}{H^2} (h - h_0) \tag{2.6}$$

If we use the linearized approximate relation of Eq. 2.6 between V and liquid level h in our system differential equation it will remain linear. When a dependent variable y is related nonlinearly to several independent variables x_1, x_2, x_3, etc. according to

$$y = f(x_1, x_2, x_3, \ldots) \tag{2.7}$$

we may linearize using the multivariable form of the Taylor series

$$y \approx f(x_{10}, x_{20}, x_{30}, \ldots) + \frac{\partial f}{\partial x_1}\bigg|_{x_{10}, x_{20}, x_{30}, \ldots} (x_1 - x_{10}) + \frac{\partial f}{\partial x_2}\bigg|_{x_{10}, x_{20}, x_{30}, \ldots} (x_2 - x_{20})$$
$$+ \frac{\partial f}{\partial x_3}\bigg|_{x_{10}, x_{20}, x_{30}, \ldots} (x_3 - x_{30}) + \cdots \tag{2.8}$$

For example, in a ported gas-filled piston/cylinder where gas mass, temperature, and volume are all changing, the perfect gas law gives for the pressure p

$$p = \frac{RTM}{V} \tag{2.9}$$

whereas the linearized approximation would be

$$p \approx \frac{RT_0 M_0}{V_0} + \frac{RM_0}{V_0} (T - T_0) + \frac{RT_0}{V_0} (M - M_0) - \frac{RM_0 T_0}{V_0^2} (V - V_0) \quad (2.10)$$

Although the main purpose of this section is to present an introductory treatment of Laplace transform techniques, a brief review of the classical operator method for solving linear differential equations with constant coefficients will be useful. In Eq. 2.1, when input $q_i(t)$ is specified, the right-hand side becomes a known function of time, $f(t)$. The solution is a three-step procedure:

1. Find the complementary (homogeneous) solution q_{oc} for the equation with $f(t) = 0$.
2. Find the particular solution q_{op} with $f(t)$ present.
3. Get the complete solution $q_o = q_{oc} + q_{op}$ and evaluate the constants of integration by applying known initial conditions.

To find q_{oc}, rewrite Eq. 2.1 (with right-hand side zero) using the operator notation $D \triangleq d/dt$:

$$a_n D^n q_o + a_{n-1} D^{n-1} q_o + \cdots + a_1 D q_o + a_0 q_o = 0 \quad (2.11)$$

Once in operator form, the equation may be treated as if it were algebraic, so "factor out" q_o to get

$$(a_n D^n + a_{n-1} D^{n-1} + \cdots + a_1 D + a_0) q_o = 0$$

and then write the *system characteristic equation* as

$$a_n D^n + a_{n-1} D^{n-1} + \cdots + a_1 D + a_0 = 0 \quad (2.12)$$

We treat this as an algebraic equation in the unknown D and solve for the n roots $s_1, s_2, \ldots, s_n$. For $n \leq 4$, formulas (such as the quadratic formula, for $n = 2$) are available which allow solution for the roots in "letter form" if the a's are given as letters. If $n > 4$, it has been proven in algebra that no formulas exist and the roots can be found *only* if the a's are given as numbers. Also, the methods for root finding are now numerical algorithms which do not always work and always give approximate answers. Fortunately these algorithms almost always work in practical problems, give very accurate answers, and are available as easily used "canned" programs in most computer libraries.

Since root finding today is a rapid computerized operation, we assume all the roots are available and state "rules" (proven in differential equation courses, here assumed) that allow one to immediately write down q_{oc}. For a real, unrepeated root s_1 we get

$$q_{oc} = c_1 e^{s_1 t} \quad (2.13)$$

whereas a real root s_2 repeated m times gives

$$q_{oc} = c_0 e^{s_2 t} + c_1 t e^{s_2 t} + c_2 t^2 e^{s_2 t} + \cdots + c_m t^m e^{s_2 t} \quad (2.14)$$

When the a's in Eq. 2.12 are real numbers (the usual case), then any complex roots that might appear *always* come in pairs $a \pm ib$, for which the solution is

$$q_{oc} = ce^{at}\sin(bt + \phi) \tag{2.15}$$

For repeated root pairs $a \pm ib$, $a \pm ib$, and so forth, the solution takes the form

$$q_{oc} = c_0 e^{at} \sin(bt + \phi_0) + c_1 te^{at} \sin(bt + \phi_1) + \cdots \tag{2.16}$$

similar to the result for repeated real roots. The c's and ϕ's in all the complementary solutions given here are constants of integration whose numerical value cannot be found until the last step of the three-step solution process. As a numerical example, if a computer root finder has presented us with the following list of roots for a sixth-order equation:

$$-2.31, \ +1.71, \ -1.4 \pm i3.7, \ -5.3 \pm i1.61$$

we immediately write down the solution as

$$q_{oc} = c_1 e^{-2.31t} + c_2 e^{1.71t} + c_3 e^{-1.4t} \sin(3.7t + \phi_1) \\ + c_4 e^{-5.3t} \sin(1.61t + \phi_2) \tag{2.17}$$

Although the complementary solution can in principle *always* be found by this method, since the particular solution takes into account the "forcing function" $f(t)$ (right-hand side of equation), methods for getting the particular solution depend on the form of $f(t)$ and, in fact, for some $f(t)$'s no solution is known. Fortunately the *method of undetermined coefficients* provides a simple method of getting particular solutions for most $f(t)$'s of practical interest. To check whether this approach will work, differentiate $f(t)$ over and over. If repeated differentiation ultimately leads to zeros (e.g., $3 + 4t + 8t^3$), or else to repetition of a finite number of different time functions (e.g., $3 \cos(2t) - 8 \sin(7t)$), then the method will work. The particular solution will then be a sum of terms made up of each *different* type of function found in $f(t)$ and all its derivatives, each term multiplied by an unknown constant (undetermined coefficient). This solution is substituted into Eq. 2.1, making it an identity. One then gathers like terms on each side, equates their coefficients, and obtains a set of simultaneous algebraic equations that can be solved for all the undetermined coefficients. Having q_{oc} (has n unknown constants in it) and q_{op} (has *no* unknown constants), we add them to get the complete solution and then apply known initial conditions to find the n unknown constants.

Most devices and systems are modeled not by a single differential equation but rather by a simultaneous set of equations. Whether we use classical operator or Laplace transform methods, such sets are handled by reducing them to a single equation in the desired unknown. The classical operator method does this by treating the D-operator form of the equations as algebraic and then using any valid technique for solving n algebraic equations in n unknowns, such as substitution/elimination or determinants. Determinants are the most systematic method and are almost necessary for large numbers of equations. A simple example illustrates the procedure.

$$2\frac{dx_1}{dt} + 3x_1 - x_2 = 2t \tag{2.18}$$

$$x_1 - \frac{dx_2}{dt} - x_2 = 0 \tag{2.19}$$

$$(2D + 3)x_1 + (-1)x_2 = 2t \tag{2.20}$$

$$(1)x_1 + (-D - 1)x_2 = 0 \tag{2.21}$$

$$x_1 = \frac{\begin{vmatrix} 2t & -1 \\ 0 & -D-1 \end{vmatrix}}{\begin{vmatrix} 2D+3 & -1 \\ 1 & -D-1 \end{vmatrix}} = \frac{-2-2t}{-2D^2-5D-2} \tag{2.22}$$

$$(2D^2 + 5D + 2)x_1 = 2 + 2t \tag{2.23}$$

$$x_2 = \frac{\begin{vmatrix} 2D+3 & 2t \\ 1 & 0 \end{vmatrix}}{\begin{vmatrix} 2D+3 & -1 \\ 1 & -D-1 \end{vmatrix}} = \frac{-2t}{-2D^2-5D-2} \tag{2.24}$$

$$(2D^2 + 5D + 2)x_2 = 2t \tag{2.25}$$

The individual equations 2.23, 2.25 are solved by the procedures outlined earlier.

Although we will shortly be discussing Laplace *transfer functions,* the concept is also used with D operators. If we have reduced a set of simultaneous equations down to a single equation with a single output q_{o1} and several inputs q_{i1}, q_{i2}, etc. (or if there was only one such equation to begin with) we can write

$$(a_n D^n + \cdots + a_1 D + a_0)q_{o1} = (b_m D^m + \cdots + b_1 D + b_0)q_{i1} \tag{2.26}$$
$$+ (c_k D^k + \cdots + c_1 D + c_0)q_{i2} + \cdots$$

The *operational transfer function* $(q_{o1}/q_{i1})(D)$ relating output q_{o1} to input q_{i1} is defined by setting all other inputs in Eq. 2.26 to zero (superposition theorem allows this) and forming the ratio output/input

$$\frac{q_{o1}}{q_{i1}}(D) \triangleq \frac{b_m D^m + \cdots + b_1 D + b_0}{a_n D^n + \cdots + a_1 D + a_0} \tag{2.27}$$

Similarly,

$$\frac{q_{o1}}{q_{i2}}(D) \triangleq \frac{c_k D^k + \cdots + c_1 D + c_0}{a_n D^n + \cdots + a_1 D + a_0} \tag{2.28}$$

A major use of such transfer functions is in block diagrams such as Fig. 2.1. (We shall see shortly that the Laplace transfer functions (q_o/q_i) (s) have exactly the same form as the D-operator version except s is used in place of D.) Figure 2.2 shows how Eq. 2.26 through 2.28 are represented in block diagram form using transfer functions.

Although an introductory treatment of control systems (suitable for a required course that emphasizes principles) can be readily presented without use of Laplace transform methods, certain advanced analysis methods are most easily developed through their use. I now present, without proof, some basic Laplace transform techniques that will be useful later in the text. Numerous references[2] are available for the reader who requires more application details or mathematical background. One use of Laplace transforms is as an alternative method for solving linear differential equations with constant coefficients. Although this method will not solve any equations that cannot also be solved by the classical operator method, it presents certain advantages for some kinds of problems.

1. Separate steps to find the complementary solution, particular solution, and constants of integration are not used. The complete solution, including initial conditions, is obtained at once.

2. There is never any question about *which* initial conditions are needed; the solution process automatically introduces the correct ones. For sets of simultaneous equations, the "natural" initial conditions (those physically known) are all that is needed, whereas the classical operator method requires that we derive mathematically some *additional* initial conditions. Also, initial conditions in the classical method are evaluated at $t = 0^+$, a time just *after* the input is applied. For some kinds of systems and inputs,

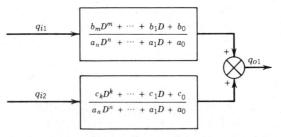

FIGURE 2.2 *Use of operational transfer functions in block diagrams.*

[2]Doebelin, *System Modeling and Response;* M. F. Gardner and J. L. Barnes, *Transients in Linear Systems,* Wiley, New York, 1954; W. Kaplan, *Operational Methods for Linear Systems,* Addison–Wesley, Reading, Massachusetts, 1962.

these $t = 0^+$ conditions are *not* the same as those *before* the input is applied, so extra work is required to find them. The transform method uses conditions *before* the input is applied; these are generally physically known and are often zero, simplifying the work.

3. For inputs that cannot be described by a single formula for their entire course, but must be defined over segments of time, the classical method requires a piecewise solution with tedious matching of final conditions of one piece and initial conditions of the next. The Laplace transform method handles such discontinuous inputs very neatly.

All the theorems and techniques of the Laplace transform derive from the fundamental definition for the direct Laplace transform $F(s)$ of the time function $f(t)$.

$$\mathscr{L}\,[f(t)] \triangleq F(s) \triangleq \int_0^\infty f(t)e^{-st}dt \qquad t > 0$$

$$(2.29)$$

$$s \triangleq \text{a complex variable} \triangleq \sigma + i\omega$$

The integral of Eq. 2.29 cannot be evaluated for all $f(t)$'s but when it can, it establishes a unique pair of functions, $f(t)$ in the time domain and its companion $F(s)$ in the s domain. It is conventional to use capital letters for s functions and lower case for t functions. Since comprehensive tables of such Laplace transform pairs have been published, it is rarely necessary for a transform user to actually work out Eq. 2.29. Table 2.1 is a brief table adequate for purposes of this text. When we use Laplace transforms to solve differential equations, we must transform entire equations, not just isolated $f(t)$ functions, so several theorems necessary for this will now be stated, without proof.

TABLE 2.1

	$F(s)$	$f(t)$
1	s	$\delta'(t)$, first derivative of unit impulse
2	1	$\delta(t)$, unit impulse
3	$1/s$	1, unit step, $u(t)$
4	$1/s^2$	t
5	$\dfrac{1}{s^n}$	$\dfrac{1}{(n-1)!}t^{n-1}$
6	$\dfrac{1}{s+a}$	e^{-at}
7	$\dfrac{1}{s(s+a)}$	$\dfrac{1-e^{-at}}{a}$

TABLE 2.1   *(Cont.)*

	$F(s)$	$f(t)$
8	$\dfrac{1}{s^2(s+a)}$	$\dfrac{e^{-at}+at-1}{a^2}$
9	$\dfrac{s+a_0}{s^2(s+a)}$	$\dfrac{a_0-a}{a^2}e^{-at}+\dfrac{a_0}{a}t+\dfrac{a-a_0}{a^2}$
10	$\dfrac{s^2+a_1 s+a_0}{s^2(s+a)}$	$\dfrac{a^2-a_1 a+a_0}{a^2}e^{-at}+\dfrac{a_0}{a}t+\dfrac{a_1 a-a_0}{a^2}$
11	$\dfrac{1}{(s+a)(s+b)}$	$\dfrac{e^{-at}-e^{-bt}}{b-a}$
12	$\dfrac{s+c}{(s+a)(s+b)}$	$\dfrac{(c-a)e^{-at}-(c-b)e^{-bt}}{b-a}$
13	$\dfrac{1}{s(s+a)(s+b)}$	$\dfrac{1}{ab}+\dfrac{be^{-at}-ae^{-bt}}{ab(a-b)}$
14	$\dfrac{s+c}{s(s+a)(s+b)}$	$\dfrac{c}{ab}+\dfrac{c-a}{a(a-b)}e^{-at}+\dfrac{c-b}{b(b-a)}e^{-bt}$
15	$\dfrac{s^2+a_1 s+a_0}{s(s+a)(s+b)}$	$\dfrac{a_0}{ab}+\dfrac{a^2-a_1 a+a_0}{a(a-b)}e^{-at}$ $-\dfrac{b^2-a_1 b+a_0}{b(b-a)}e^{-bt}$
16	$\dfrac{1}{s^2(s+a)(s+b)}$	$\dfrac{t}{ab}-\dfrac{a+b}{(ab)^2}+\dfrac{1}{a-b}\left(\dfrac{e^{-bt}}{b^2}-\dfrac{e^{-at}}{a^2}\right)$
17	$\dfrac{s+a_0}{s^2(s+a)(s+b)}$	$\dfrac{a_0-a}{a^2(b-a)}e^{-at}+\dfrac{a_0-b}{b^2(a-b)}e^{-bt}$ $+\dfrac{a_0}{ab}t+\dfrac{ab-a_0(a+b)}{(ab)^2}$
18	$\dfrac{s^2+a_1 s+a_0}{s^2(s+a)(s+b)}$	$\dfrac{a^2-a_1 a+a_0}{a^2(b-a)}e^{-at}$ $+\dfrac{b^2-a_1 b+a_0}{b^2(a-b)}e^{-bt}$ $+\dfrac{a_0}{ab}t+\dfrac{a_1 ab-a_0(a+b)}{(ab)^2}$
19	$\dfrac{1}{(s+a)(s+b)(s+c)}$	$\dfrac{e^{-at}}{(b-a)(c-a)}+\dfrac{e^{-bt}}{(a-b)(c-b)}$ $+\dfrac{e^{-ct}}{(a-c)(b-c)}$

	$F(s)$	$f(t)$
20	$\dfrac{s + a_0}{(s + a)(s + b)(s + c)}$	$\dfrac{a_0 - a}{(b - a)(c - a)} e^{-at}$
		$+ \dfrac{a_0 - b}{(a - b)(c - b)} e^{-bt}$
		$+ \dfrac{a_0 - c}{(a - c)(b - c)} e^{-ct}$
21	$\dfrac{s^2 + a_1 s + a_0}{s(s + a)(s + b)(s + c)}$	$\dfrac{a^2 - a_0 a + a_0}{(b - a)(c - a)} e^{-at}$
		$+ \dfrac{b^2 - a_1 b + a_0}{(a - b)(c - b)} e^{-bt}$
		$+ \dfrac{c^2 - a_1 c + a_0}{(a - c)(b - c)} e^{-ct}$
22	$\dfrac{1}{s^2 + a^2}$	$\dfrac{\sin at}{a}$
23	$\dfrac{s}{s^2 + a^2}$	$\cos at$
24	$\dfrac{1}{(s + a)^2 + b^2}$	$\dfrac{1}{b} e^{-at} \sin bt$
25	$\dfrac{s + a_0}{(s + a)^2 + b^2}$	$\dfrac{1}{b} [(a_0 - a)^2 + b^2]^{1/2} e^{-at} \sin(bt + \phi)$
		$\phi \triangleq \tan^{-1} \dfrac{b}{a_0 - a}$
26	$\dfrac{1}{s[(s + a)^2 + b^2]}$	$\dfrac{1}{b_0^2} + \dfrac{1}{bb_0} e^{-at} \sin(bt - \phi)$
		$\phi \triangleq \tan^{-1} \dfrac{b}{-a}, \ b_0 \triangleq \sqrt{a^2 + b^2}$
27	$\dfrac{s + a_0}{s[(s + a)^2 + b^2]}$	$\dfrac{a_0}{a^2 + b} + \dfrac{1}{b\sqrt{a^2 + b^2}}$
		$\times [(a_0 - a)^2 + b^2]^{1/2} e^{-at} \sin(bt + \phi)$
		$\phi \triangleq \tan^{-1} \dfrac{b}{a_0 - a} - \tan^{-1} \dfrac{b}{-a}$
28	$\dfrac{1}{s^2[(s + a)^2 + b^2]}$	$\dfrac{1}{a^2 + b^2} \left[t - \dfrac{2a}{a^2 + b^2} + \dfrac{1}{b} e^{-at} \sin(bt - \phi) \right]$
		$\phi \triangleq 2 \tan^{-1} \dfrac{b}{-a}$

TABLE 2.1   *(Cont.)*

F(s)	f(t)
29 $\dfrac{1}{(s + c)\,[(s + a)^2 + b^2]}$	$\dfrac{e^{-ct}}{(c - a)^2 + b^2}$
	$+\; \dfrac{1}{b[(c - a)^2 + b^2]^{1/2}}\, e^{-at}\,\sin(bt - \phi)$
	$\phi \triangleq \tan^{-1} \dfrac{b}{c - a}$
30 $\dfrac{1}{s(s + c)\,[(s + a)^2 + b^2]}$	$\dfrac{1}{c\sqrt{a^2 + b^2}} - \dfrac{1}{c[(a - c)^2 + b^2]}\, e^{-ct}$
	$+\; \dfrac{1}{b\sqrt{a^2 + b^2}\,[(c - a)^2 + b^2]^{1/2}}\, e^{-at}$
	$\sin(bt - \phi)$
	$\phi \triangleq \tan^{-1} \dfrac{b}{-a} + \tan^{-1} \dfrac{b}{c - a}$

Linearity Theorem

$$\mathcal{L}[a_1 f_1(t) + a_2 f_2(t)] = \mathcal{L}\,[a_1 f_1(t)] + \mathcal{L}\,[a_2 f_2(t)] = a_1 F_1(s) + a_2 F_2(s) \quad (2.30)$$

This theorem says we may transform an entire equation by adding the transforms of the individual terms. Also, the transform of a constant (a_1, a_2) times $f(t)$ is just the constant times the transform of $f(t)$.

Differentiation Theorem

$$\mathcal{L}\left[\frac{df}{dt}\right] = sF(s) - f(0) \quad (2.31)$$

$$\mathcal{L}\left[\frac{d^2 f}{dt^2}\right] = s^2 F(s) - sf(0) - \frac{df}{dt}(0) \quad (2.32)$$

$$\mathcal{L}\left[\frac{d^n f}{dt^n}\right] = s^n F(s) - s^{n-1}f(0) - s^{n-2}\frac{df}{dt}(0) - \cdots - \frac{d^{n-1}f}{dt^{n-1}}(0) \quad (2.33)$$

This theorem allows one to transform a derivative of any order and automatically inserts the necessary initial conditions into the solution process. That is, $f(0)$, $(df/dt)(0)$, and the like are the initial values of $f(t)$ and its derivatives, evaluated numerically at a time instant just *before* the driving input is applied.

Integration Theorem

$$\mathcal{L}[\textstyle\int f(t)dt] = \frac{F(s)}{s} + \frac{f^{(-1)}(0)}{s} \quad (2.34)$$

where $f^{(-1)}(0)$ is the initial value of $\int f(t)dt$. For example, if $f(t)$ were the *velocity* in a mechanical motion problem, $f^{(-1)}(0)$ would be the numerical value of the *displacement* just *before* the system input was applied. For higher order integrals

$$\mathscr{L}[f^{(-n)}(t)] = \frac{F(s)}{s^n} + \sum_{k=1}^{n} \frac{f^{(-k)}(0)}{s^{n-k+1}} \tag{2.35}$$

where

$$f^{(-n)}(t) \triangleq \int \cdots \int f(t)(dt)^n \text{ and } f^{(-0)}(t) \triangleq f(t) \tag{2.36}$$

A few more useful theorems remain to be given, but we now have enough to solve many differential equations. For example, suppose we have

$$5\frac{dy}{dt} + 10y = 10 \qquad y(0) = 0 \tag{2.37}$$

Using linearity and differentiation theorems and the transform table.

$$\mathscr{L}\left[5\frac{dy}{dt} + 10y\right] = \mathscr{L}[10] \tag{2.38}$$

$$5[sY(s) - y(0)] + 10Y(s) = 10/s \tag{2.39}$$

Whereas in the classical operator method we say "treat the *D*-operator equations *as if* they were algebraic," Eq. 2.39 *really is algebraic* and is solved for $Y(s)$ easily.

$$Y(s) = \frac{10}{s(5s + 10)} = 2\left[\frac{1}{s(s + 2)}\right] \tag{2.40}$$

Since we are really interested in $y(t)$, we use entry 7 of our table to transform $Y(s)$ back to the t domain (this is called the *inverse* transform)

$$y(t) = 1 - e^{-2t} \tag{2.41}$$

For the simultaneous equations 2.18, 2.19, we would have

$$2[sX_1 - x_1(0)] + 3X_1 - X_2 = \frac{2}{s^2} \tag{2.42}$$

$$X_1 - sX_2 + x_2(0) - X_2 = 0 \tag{2.43}$$

(note that for compactness we now write X_1 rather than $X_1(s)$ since the capital letter *implies* a function of s.) Equations 2.42 and 2.43 *are* algebraic equations and can be solved for either X_1 or X_2 using any valid algebraic technique, such as using determinants. In this case, substitution and elimination may be quicker.

$$X_1 = (s + 1) X_2 - x_2(0) \tag{2.44}$$

$$2\{s[(s + 1)X_2 - x_2(0)] - x_1(0)\} + 3[(s + 1)X_2 - x_2(0)] - X_2 = \frac{2}{s^2} \tag{2.45}$$

$$X_2 = \frac{1}{s^2 (s + 2) (s + 0.5)} + \frac{sx_2(0) + 3x_2(0) + x_1(0)}{2(s + 2) (s + 0.5)} \tag{2.46}$$

Note that to find x_2 from Eq. 2.46 requires knowledge only of the "natural" initial conditions $x_1(0)$, $x_2(0)$; whereas solving for x_2 from Eq. 2.25 (by the classical method) requires that we *derive* a value for $\dot{x}_2(0)$. Taking $x_1(0) = x_2(0) = 0$ in Eq. 2.46 and using Table Entry 16 we get

$$x_2(t) = t - 2.5 + 2.67e^{-0.5t} - 0.167e^{-2t} \tag{2.47}$$

When all initial conditions are taken as zero, when we Laplace transform a set of simultaneous equations relating outputs q_{o1}, q_{o2}, . . . to inputs q_{i1}, q_{i2}, . . . and combine these algebraically to get a single equation in q_{o1}, this equation will look exactly like Eq. 2.26 except that all D's will be replaced by s's. The definition of the *Laplace transfer function* $(Q_{o1}/Q_{i1})(s)$ is the ratio of $Q_{o1}(s)$ to $Q_{i1}(s)$ when all initial conditions are taken to be zero and all other inputs (q_{i2}, q_{i3}, etc.) are set to zero. Thus Laplace transfer functions are identical in form to the D-operator versions of Eq. 2.27, 2.28 except that all D's are replaced by s's.

$$\frac{Q_{o1}}{Q_{i1}} (s) = \frac{b_m s^m + \cdots + b_1 s + b_0}{a_n s^n + \cdots + a_1 s + a_0} \tag{2.48}$$

$$\frac{Q_{o1}}{Q_{i2}} (s) = \frac{c_k s^k + \cdots + c_1 s + c_0}{a_n s^n + \cdots + a_1 s + a_0} \tag{2.49}$$

The example of Fig. 1.17 briefly introduced the system dynamic element called dead time (also referred to as transport lag or discrete delay). The Laplace transform provides a theorem useful for such elements and for dealing efficiently with discontinuous inputs. This theorem uses delayed unit step functions, so we first define (see Fig. 2.3a) the unit step function $u(t)$ as

$$u(t) = 1.0 \text{ for } t > 0 \tag{2.50}$$
$$u(t) = 0 \quad \text{for } t < 0$$

and the delayed unit step function $u(t - a)$ as

$$u(t - a) = 1.0 \text{ for } t > a \tag{2.51}$$
$$u(t - a) = 0 \quad \text{for } t < a$$

Figure 2.3b defines the behavior of a dead-time element ($\tau_{DT} \triangleq$ dead time); whereas the *delay theorem* states that

$$\mathscr{L}[f(t - a)u(t - a)] = e^{-as}F(s) \tag{2.52}$$

Note that multiplying $f(t - a)$ by $u(t - a)$ "turns off" (multiplies by zero) the f function for all $t < a$. From Eq. 2.52 and Fig. 2.3b it is clear that the Laplace transfer function of a dead-time element is given by

$$\frac{Q_o}{Q_i} (s) = e^{-\tau_{DT}s} \tag{2.53}$$

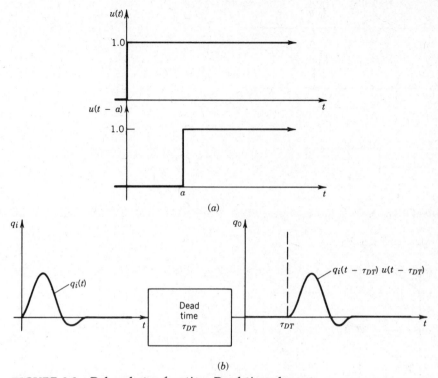

(a)

(b)

FIGURE 2.3 *Delayed step function. Dead-time element.*

Figure 2.4 shows another major type of application for the delay theorem. Problems of this type (discontinuous input) can be solved by the classical operator method but require a tedious piece-wise procedure. Using Laplace transform, our first step is to write $e_i(t)$ in a form that facilitates application of the delay theorem.

$$e_i(t) = tu(t) - (t - 1)u(t - 1) - u(t - 1) \qquad (2.54)$$

$$E_i(s) = \frac{1}{s^2} - \frac{e^{-s}}{s^2} - \frac{e^{-s}}{s} \qquad (2.55)$$

For zero initial conditions

$$E_o(s) = \left(\frac{1/RC}{s + 1/RC}\right)\left[\frac{1}{s^2} - \frac{e^{-s}}{s^2} - \frac{e^{-s}}{s}\right] \qquad (2.56)$$

$$E_o(s) = \frac{1/RC}{s^2(s + 1/RC)} - \frac{(1/RC)e^{-s}}{s^2(s + 1/RC)} - \frac{(1/RC)e^{-s}}{s(s + 1/RC)} \qquad (2.57)$$

Using Table Entry 8 (taking $RC = 1$) and the delay theorem;

$$e_o(t) = (e^{-t} + t - 1) - u(t - 1)[e^{-(t-1)} \\ + (t - 1) - 1] - u(t - 1)[1 - e^{-(t-1)}] \qquad (2.58)$$

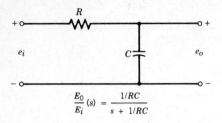

$$\frac{E_0}{E_i}(s) = \frac{1/RC}{s + 1/RC}$$

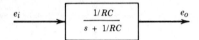

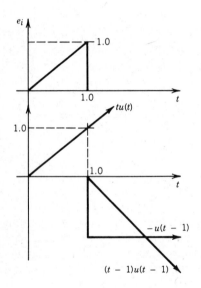

FIGURE 2.4 *Use of delay theorem for discontinuous input.*

In graphing $e_o(t)$, note that the terms multiplied by $u(t-1)$ contribute *nothing* to the graph for $t < 1$.

Initial Value Theorem and Final Value Theorem

If we have found $Q_o(s)$ and wish only to find $q_o(\infty)$, the final "steady-state" value of $q_o(t)$, this can be quickly done *without* doing the complete inverse transform by use of the *final value theorem*

$$\lim_{t \to \infty} f(t) = \lim_{s \to 0} sF(s) \tag{2.59}$$

if the system and input are such that the output approaches a constant value as $t \to \infty$. This theorem is not of great utility; its results can generally be obtained as quickly by classical methods, because the particular solution in such cases is obvious by inspection. The *initial value theorem*

$$\lim_{t \to 0^+} f(t) = \lim_{s \to \infty} sF(s) \qquad (2.60)$$

is occasionally useful for finding the value of $f(t)$ just *after* ($t = 0^+$) the input has been applied. In getting the $F(s)$ needed to apply Eq. 2.60, our usual definition of initial conditions as those *before* the input is applied must be used.

A concept of wide use in advanced system dynamics[3] is that of the *impulse function*; we develop it here only to the point needed for this text. We define $\delta(t)$, the unit impulse function, as

$$\delta(t) = 0 \qquad t \neq 0$$
$$\int_{-\epsilon}^{+\epsilon} \delta(t)dt = 1 \qquad \epsilon > 0 \qquad (2.61)$$

Note that this definition, unlike those for more familiar functions, does not allow one to graph $\delta(t)$ versus t. (In fact, because of $\delta(t)$'s peculiar behavior, mathematicians refuse to call it a function and have developed a whole branch of mathematics called the theory of distributions to put these methods on a rigorous foundation. Engineers, the original inventors of the concept, however, generally use impulse function techniques in a formal way for practical applications.) To get some appreciation of the nature of the impulse function, various approximating functions such as $p(t)$ (see Fig. 2.5) are useful.

$$\delta(t) = \lim_{b \to 0} p(t) \qquad (2.62)$$

As $b \to 0$, the height of $p(t)$ approaches infinity, the width approaches zero, but the area is always equal to one, satisfying the definition 2.61. Another useful feature of $\delta(t)$ is its relation to the widely used step function $u(t)$. In Fig. 2.5, approximating functions $u_a(t)$ and $\delta_a(t)$ for $u(t)$ and $\delta(t)$ show that the step function is the integral of the impulse function, or, conversely, the impulse function is the derivative of the step function. Note that "conventional" calculus does not allow differentiation of discontinuous functions such as the step, showing the utility of impulse function concepts in extending our analysis capabilities. When we multiply $\delta(t)$ by some number, giving, say, $3.82\,\delta(t)$ or $-1.37\,\delta(t)$, we increase the "strength" of the impulse, but "strength" now means *area*, not height as it does for "ordinary" functions. The Laplace transform of $\delta(t)$ exists and can be found either from definitions 2.29 and 2.61 or from the differentiation theorem.

$$\mathcal{L}[\delta(t)] = \mathcal{L}\left[\frac{du}{dt}\right] = sU(s) = s\frac{1}{s} = 1.0 \qquad (2.63)$$

Further "peculiar" but useful functions can be defined by repeatedly differentiating $\delta(t)$; however, we here have no need for them so conclude our treatment.

---

[3]Doebelin, *System Modeling and Response*, Chap. 3.

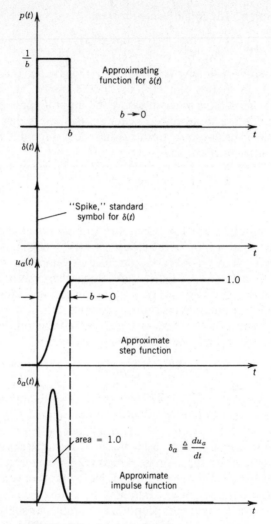

FIGURE 2.5 *Impulse function.*

Although a comprehensive transform table will allow inverse transformation of many practical problems, we will occasionally need the *partial-fraction expansion method* to handle general cases. We assume that the function $F(s)$ to be inverse transformed is of the form

$$F(s) = \frac{N(s)}{D(s)} = \frac{B_p s^p + B_{p-1} s^{p-1} + \cdots + B_1 s + B_0}{s^n + A_{n-1} s^{n-1} + \cdots + A_1 s + A_0} \qquad (2.64)$$

Values of s which make $F(s)$ equal to zero are called *zeros* of $F(s)$, while values of s that make $F(s)$ equal to infinity are called *poles* of $F(s)$. The system characteristic equation is $D(s) = 0$, and we must first find its n roots. (Since this "factoring" is needed even when we use the transform table, root finding is an unavoidable step in the solution process, whether we use classical or transform methods.) We now also assume $n > p$ in Eq. 2.64 and that $D(s) = 0$ has no repeated roots. (Exceptions to these assumptions are rare; their treatment is covered in the literature.[4]) Once $D(s)$ is factored we can write

$$F(s) = \frac{N(s)}{D(s)} = \frac{N(s)}{(s - s_1)(s - s_2) \cdots (s - s_n)} \qquad (2.65)$$

which, from algebra, we know can be written as

$$\frac{N(s)}{D(s)} = \frac{K_1}{s - s_1} + \frac{K_2}{s - s_2} + \cdots + \frac{K_k}{s - s_k} + \cdots + \frac{K_n}{s - s_n} \qquad (2.66)$$

where the K's are real or complex numbers to be found, but which do not contain s. Once the K's are found, inverse transformation of the simple individual terms is easy. To find K_k, a typical term, multiply Eq. 2.66 through by $(s - s_k)$ and then set $s = s_k$ to make K_k "stand alone."

$$\frac{N(s)(s - s_k)}{(s - s_1) \cdots (s - s_k) \cdots (s - s_n)} = \frac{K_1(s - s_k)}{(s - s_1)} + \cdots + K_k$$

$$+ \cdots + \frac{K_n(s - s_k)}{s - s_n} \qquad (2.67)$$

$$K_k = \frac{N(s)}{(s - s_1) \cdots (\quad) \cdots (s - s_n)} \bigg|_{s = s_k} \qquad (2.68)$$

A numerical example illustrates the procedure.

$$F(s) = \frac{1}{s(s + 1)(s + 2)(s + 1 + i1)(s + 1 - i1)(s + 2 + i2)(s + 2 - i2)} \qquad (2.69)$$

$$= \frac{K_1}{s} + \frac{K_2}{s + 1} + \frac{K_3}{s + 2} + \frac{K_4}{s + 1 - i1} + \frac{K_5}{s + 1 + i1}$$

$$(2.70)$$

$$+ \frac{K_6}{s + 2 - i2} + \frac{K_7}{s + 2 + i2}$$

$$K_1 = \frac{1}{(1)(2)(2)(8)} = 0.0313 \qquad K_2 = \frac{1}{(-1)(1)(1)(5)} = -0.2$$

$$K_3 = \frac{1}{(-2)(-1)(2)(4)} = 0.0625$$

$$K_4 = \frac{1}{(-1 + i)(i)(1 + i)(i2)(1 + i3)(1 - i)} = 0.056 \underline{/-26.5°}$$

$$K_6 = \frac{1}{(-2 + i2)(-1 + i2)(i2)(-1 + i3)(-1 + i1)(i4)} = 0.0041 \underline{/-8.2°}$$

[4]Ibid., pp. 39–47.

Note that for complex root pairs it is necessary only to work out the $K(K_4, K_6)$ that is associated with the *negative* imaginary part, the companion $K(K_5, K_6)$ can be shown always to be the complex conjugate (same magnitude, negative angle). Furthermore, the inverse transform for *both* complex factors of any root pair $a \pm ib$ can be shown to be

$$f_{\text{pair}}(t) = 2|K|e^{at}\sin(bt + \underline{/K} + 90°) \qquad (2.71)$$

Using Table Entries 3 and 6 for the real roots and Eq. 2.71 for the complex root pairs, we quickly obtain

$$f(t) = 0.0313 - 0.2e^{-t} + 0.0625e^{-2t} + 0.112e^{-t}\sin(t + 63.5°)$$
$$+ 0.0082e^{-2t}\sin(2t + 81.8°) \qquad (2.72)$$

2.3 SYSTEM FREQUENCY RESPONSE AND FREQUENCY SPECTRUM METHODS

The response of systems to sinusoidal inputs, called *frequency response*, is the basis of many important analysis and design methods for dynamic systems of all kinds (including control systems) and is covered in detail in system dynamics texts.[5] Space limits us to a brief summary treatment. In the general model of Eq. 2.1, when $q_i = q_{i0} \sin \omega t$, the right-hand terms can be combined into a single term $A \sin(\omega t + \alpha)$. The method of undetermined coefficients then allows us to get the particular solution for q_o, which will always be of the form $q_{o0} \sin(\omega t + \phi)$. For a *stable* system, the complementary solution will always disappear as time goes by since the terms will be of form ce^{-at} or $ce^{-at}(\sin bt + \theta)$, with "a" a positive number. By definition, system frequency response refers to the "sinusoidal steady state" that obtains *after* the transient complementary solution dies out, thus the complementary solution is of *no* interest and is not calculated. Our interest is only in the *amplitude ratio* q_{o0}/q_{i0} and the *phase angle* (ϕ) of output with respect to input in the sinusoidal steady state. Both amplitude ratio and phase angle change when we use a different frequency ω, and our desired results are graphs of these two quantities against frequency, the so-called *frequency-response curves*.

Although the frequency response curves may be obtained by getting the particular solution to the differential equation for many different values of ω, the *sinusoidal transfer function* $(Q_o/Q_i)(i\omega)$ provides a much better method. It can be shown that if we replace s by $i\omega(i \triangleq \sqrt{-1}, \omega \triangleq$ sinusoidal frequency, radian/time) in any Laplace transfer functions $(Q_o/Q_i)(s)$, we get a complex number $M\underline{/\phi}$, where M will be the amplitude ratio q_{o0}/q_{i0} and ϕ will be the angle by which the q_o sine wave leads the q_i sine wave. (Negative ϕ means q_o

---

[5]Ibid., Chap. 3.

lags q_i). For example, a first-order system such as the circuit of Fig. 2.4 would give

$$\frac{E_o}{E_i}(s) = \frac{K}{\tau s + 1}$$ (2.73)

$$K \triangleq \text{steady-state gain} = 1.0 \frac{\text{volt}}{\text{volt}} \quad \tau \triangleq RC \triangleq \text{time constant, } s$$

$$\frac{E_o}{E_i}(i\omega) = \frac{K}{i\omega\tau + 1} = \frac{K\underline{/0°}}{\sqrt{(\omega\tau)^2 + 1^2} \underline{/\tan^{-1}\omega\tau}}$$

$$= \frac{K}{\sqrt{(\omega\tau)^2 + 1}} \underline{/-\tan^{-1}\omega\tau} = M\underline{/\phi}$$ (2.74)

Equation 2.74 allows us to plot the frequency-response curves of Fig. 2.6. Note that at low frequency, e_o's amplitude is nearly the same as e_i's and the two sine waves are nearly in phase; whereas at high frequency, e_o becomes smaller and smaller, with $\phi \rightarrow -90°$. These curves make clear that this circuit is a *low-pass filter,* allowing low-frequency signals to get through essentially unchanged, while strongly attenuating high-frequency signals. The frequency-response curves are useful tools in the design and use of such filters.

Since a general case for sinusoidal transfer functions can be easily set up (see Eq. 2.48), general-purpose computer programs for computing and/or graphing

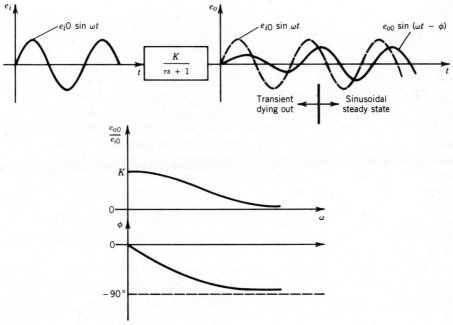

FIGURE 2.6 *System frequency response.*

frequency response are widely available in computer libraries. These programs are quick and easy to use since basically one need only specify numerical values for the "*a*" and "*b*" coefficients and specify increments and ranges for frequency ω. If a convenient high-level engineering language such as SPEAKEASY[6] is available, one can get results almost as quickly without having a "canned" frequency-response program to use. For example, if we need frequency-response graphs over $1 < \omega < 1000$ for

$$\frac{Q_o}{Q_i}(s) = \frac{0.001s^2 + 0.205s + 1}{6 \times 10^{-6}s^3 + 0.00166s^2 + 0.116s + 1} \qquad (2.75)$$

we could quickly type in the following SPEAKEASY program at a graphics terminal.

```
DOMAIN COMPLEX       allows use of complex numbers
W1=GRID(1,10,1)      sets up array of frequency values going from 1 to 10
                       in increments of 1
W2=GRID(20,100,10)     more frequency values
W3=GRID(200,1000,20)     more frequency values
W=ARRAY(W1,W2,W3)     assembles all frequency values into a single array
S=W*1I     sets s equal to iω
NUM=.001*S*S+.205*S+1     computes numerator of sinusoidal transfer
                          function for all ω values
DEN=6E-6*S**3+.00166*S*S+.116*S+1     computes denominator
STF=NUM/DEN     computes sinusoidal transfer function
AR=ABS(STF)     computes amplitude ratio
PHI=PHASE(STF)     computes phase angle
GRAPHICS(TEK4010)     calls graphics package
GRAPH(AR:W); OVERLAY; GRAPH(PHI:W)     plots frequency-response
                                       graphs
```

At the design stage, frequency response must be computed from theoretical models; however when actual hardware is available, frequency-response curves can be obtained by experimental measurement. Such measurements may be used to verify theoretical predictions or, in those cases where no adequate theory exists, may constitute the only model available for the device. Testing of this type is so common that commercial test equipment designed to carry out the experiments rapidly and accurately is widely available. Details on the equipment and techniques used to develop mathematical models by experimental testing are available in the literature.[7]

When one has the frequency-response curves for any system and is given a specific sinusoidal input, say $37.1 \sin(28t + 18°)$, it is an easy calculation to get the sinusoidal output. What is not obvious, but is extremely important, is that

---

[6]Speakeasy Computing Corp., 222 W. Adams Street, Chicago, Illinois 60606.
[7]Doebelin, *System Modeling and Response,* Chap. 6.

the frequency-response curves are really a *complete* description of the system's dynamic behavior and allow one to compute the response for *any* input, not just sine waves. Every dynamic signal has a *frequency spectrum* and if we can compute this spectrum and properly combine it with the system's frequency response, we can calculate the system time response. The details of this procedure depend on the nature of the input signal; is it periodic, transient, or random? For periodic signals (those that repeat themselves over and over in a definite cycle), *Fourier series*[8] is the mathematical tool needed to solve the response problem. Although a single sine wave is an adequate model of some real-world input signals, the general periodic signal fits many more practical situations. This is mainly because many engineering calculations involve machinery running at essentially constant speed under steady-state operating conditions. Under such conditions, every physical variable (pressure, vibratory motion, flow rate, temperature, etc.) associated with the machine goes through a repetitive cycle of some kind.

Figure 2.7 shows a periodic function $q_i(t)$ of arbitrary wave form. It can be shown that such functions can be represented by an infinite series of terms called the Fourier series.

$$q_i(t) = \frac{a_o}{T} + \frac{2}{T} \sum_{n=1}^{\infty} \left[a_n \cos\left(\frac{2\pi n}{T} t\right) + b_n \sin\left(\frac{2\pi n}{T} t\right) \right] \qquad (2.76)$$

$$a_n \triangleq \int_{-(T/2)}^{T/2} q_i(t) \cos\left(\frac{2\pi n}{T} t\right) dt \qquad (2.77)$$

$$b_n \triangleq \int_{-(T/2)}^{T/2} q_i(t) \sin\left(\frac{2\pi n}{T} t\right) dt \qquad (2.78)$$

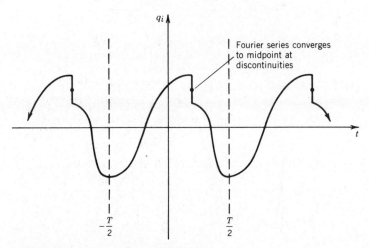

FIGURE 2.7   *General periodic function.*

---

[8]Ibid., pp. 92–99.

The square wave of Fig. 2.8a is a simple example.

$$\frac{a_0}{T} = \text{average value} = \frac{\int_{-0.01}^{0} -0.5 \, dt + \int_{0}^{+0.01} 1.5 \, dt}{0.02} = 0.5$$

$$a_n = \int_{-.01}^{0} -0.5 \cos \frac{2n\pi t}{0.02} \, dt + \int_{0}^{+0.01} 1.5 \cos \frac{2n\pi t}{0.02} \, dt = 0$$

$$b_n = \int_{-0.01}^{0} -0.5 \sin \frac{2n\pi t}{0.02} \, dt + \int_{0}^{+0.01} 1.5 \sin \frac{2n\pi t}{0.02} \, dt = \frac{1 - \cos(n\pi)}{50n\pi}$$

Equation 2.76 then gives

$$q_i(t) = 0.5 + \frac{4}{\pi} \sin(100\pi t) + \frac{4}{3\pi} \sin(300\pi t) + \frac{4}{5\pi} \sin(500\pi t) + \cdots \quad (2.79)$$

The term for $n = 1$ is called the fundamental or first harmonic and always has the same frequency as the repetition rate of the original periodic wave form (50

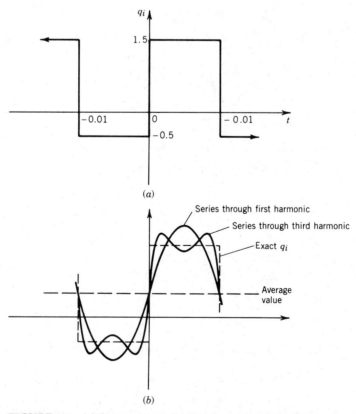

(a)

(b)

FIGURE 2.8 (a) *Square wave.* (b) *Fourier series curve fit*

Hz in our example); whereas $n = 2, 3, \ldots$ gives the second, third, and so forth harmonic frequencies as integer multiples of the first. Our square-wave example has only the first, third, fifth, and so forth harmonics. Figure 2.8*b* shows how the curve fit improves going from the first to the third harmonic; more terms (harmonics) in the series give an even better fit, an infinite number give a "perfect" fit.

For a signal of arbitrary periodic shape (rather than the simple and symmetrical square wave), the Fourier series will generally include *all* the harmonics and both sine and cosine terms. We can combine the sine and cosine terms using $A \cos \omega t + B \sin \omega t \equiv C \sin(\omega t + \alpha)$, where $C \triangleq \sqrt{A^2 + B^2}$ and $\alpha \triangleq \tan^{-1}(A/B)$, thus Eq. 2.76 can always be put in the form

$$q_i(t) = A_{i0} + A_{i1} \sin(\omega_1 t + \alpha_1) + A_{i2} \sin(2\omega_1 t + \alpha_2) + \cdots \quad (2.80)$$

A graphical display of the amplitudes (A_{ik}) and phase angles (α_k) of the sine waves in Eq. 2.80 versus frequency is called the *frequency spectrum* of q_i (see Fig. 2.9*a*; Fig. 2.9*b* for our square wave example). If a periodic q_i is applied as

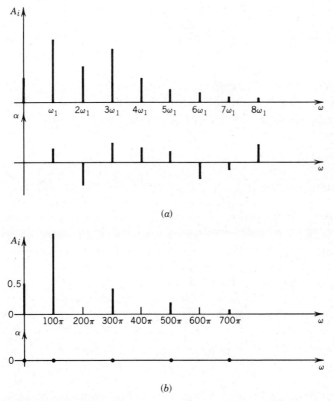

FIGURE 2.9   *Discrete spectrum of general periodic function and square wave example.*

input to a system with sinusoidal transfer function $W(i\omega)$, after transients have died out, the output q_o will be in a periodic steady state given by

$$q_o(t) = A_{o0} + A_{o1}\sin(\omega_1 t + \theta_1) + A_{o2}\sin(2\omega_1 t + \theta_2) + \cdots \quad (2.81)$$

where

$$A_{ok} \triangleq A_{ik}|W(i\omega_k)| \qquad \theta_k \triangleq \alpha_k + \underline{/W(i\omega_k)} \quad (2.82)$$

Equation 2.81 follows directly from the superposition theorem and the definition of a sinusoidal transfer function. Although the Fourier series generally has an infinite number of terms, results of practical accuracy can always be obtained with a finite (sometimes quite small) number of terms since *both* A_{ik} and $|W(i\omega_k)|$ tend toward zero for real-world signals and systems as $\omega \to \infty$, making A_{ok} more and more negligible for higher frquencies. This general result is extremely useful in deciding on the proper mathematical model to use for a component in a particular application. The input signals (commands and disturbances in control systems) of each specific application will have their own characteristic frequency spectra, showing which ranges of frequency are important and which are not. A familiar example is high fidelity music reproduction systems. The components of such systems need to perform well in the frequency range of 20 to 20,000 Hz; frequencies outside this range are of no interest since they are inaudible to human beings. Once we can estimate the frequency range of interest, we can then choose the simplest mathematical model whose frequency response adequately corresponds with actual behavior over this range.

Although the Fourier series allows us to get the frequency spectra of periodic signals, for transient signals we need the Fourier transform[9] $Q_i(i\omega)$, which formally corresponds to the Laplace transform of Eq. 2.29 when we let $s = i\omega$.

$$Q_i(i\omega) \triangleq \int_0^\infty q_i(t)e^{-i\omega t}\,dt = \int_0^\infty q_i(t)\cos \omega t\,dt - i \int_0^\infty q_i(t)\sin \omega t\,dt \quad (2.83)$$

The function $Q_i(i\omega)$ gives the frequency spectrum of the transient $q_i(t)$ and will in general be a smooth *continuous spectrum* rather than the "spikes" *(discrete spectrum)* of the Fourier series. To get the transient response of a system with sinusoidal transfer function $W(i\omega)$ to a transient $q_i(t)$, we again multiply the input spectrum $Q_i(i\omega)$ by $W(i\omega)$ to get $Q_o(i\omega)$ (similar in principle to Eq. 2.81 and 2.82), but now $Q_o(i\omega)$ is a continuous function and we must do an inverse Fourier transform[10] to obtain $q_o(t)$. The important point for our present discussion, however, is that the input frequency spectrum $Q_i(i\omega)$ will, analogously to a Fourier series, have ranges of frequencies where it is significant and ranges where it is negligible, again allowing us to select component models such that their frequency response is adequate *only where needed*.

---

[9]Ibid., pp. 102–111.
[10]Ibid., pp. 102–111.

The rectangular pulse of Fig. 2.10 is a transient that makes a convenient example. In getting $Q_i(i\omega)$ it is often easier to get $Q_i(s)$ and *then* let $s = i\omega$ rather than to use Eq. 2.83 directly.

$$q_i(t) = Au(t) - Au(t - T)u(t - T) \tag{2.84}$$

$$Q_i(s) = \frac{A}{s} - \frac{Ae^{-Ts}}{s} \tag{2.85}$$

$$Q_i(i\omega) = \frac{A}{i\omega}(1 - \cos \omega T + i \sin \omega T)$$

$$= \frac{\sqrt{2}A}{\omega}\sqrt{1 - \cos \omega T} \bigg/ \tan^{-1}\frac{\cos \omega T - 1}{\sin \omega T} \tag{2.86}$$

If we were to actually calculate $q_o(t)$ by inverse transforming $Q_i(i\omega)W(i\omega)$, both the magnitude and phase of Eq. 2.86 would be important. However, if our interest is only in deciding on the range of frequencies where $Q_i(i\omega)$ is non-negligible, we need consider only the magnitude. We see that for ω's beyond about $6\pi/T$ or $8\pi/T$, $|Q_i(i\omega)|$ is very small relative to its largest values. Thus, for example, if a system is subjected to rectangular pulses with $T \approx 0.01s$, our

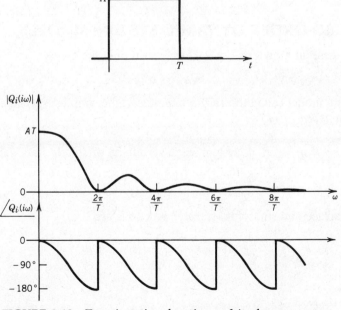

FIGURE 2.10 *Transient time function and its frequency spectrum.*

system model's frequency response need be "correct" only to about $\omega = 800\pi$ rad/s. Note that as T becomes smaller, the frequency range of importance becomes larger, a characteristic of all transients, not just the simple rectangular pulse.

For random signals[11] the appropriate frequency spectrum quantity is called the *power spectral density*. Again, our system model need have accurate frequency response only in frequency ranges where this quantity is nonnegligible. The frequency spectra of real-world periodic, transient, or random signals are today often quickly and easily calculated by an instrument called a Fast Fourier Transform (FFT) Analyzer.[12] The physical pressure, temperature, force, and the like are transduced to a proportional voltage by suitable measurement instrumentation, sent into the analyzer, digitized, and processed by an FFT algorithm to produce the desired spectral measurement in a fraction of a second. Two-channel analyzers can simultaneously process the input $q_i(t)$ and output $q_o(t)$ of a system component to experimentally find its sinusoidal transfer function from

$$\frac{Q_o}{Q_i}(i\omega) = \frac{Q_o(i\omega)}{Q_i(i\omega)} \qquad (2.87)$$

This can be done with swept-sine-wave, transient, or random-input signals, depending on the needs of the particular test. The analyzers provide only tables of numerical values (or graphs), but various curve-fitting procedures[13] can be used to get analytical transfer functions if these are needed.

2.4   ZERO-ORDER DYNAMIC SYSTEM MODELS

If, in the general model of Eq. 2.1, we retain only the terms

$$a_0 q_o = b_0 q_i \qquad (2.88)$$

we create a model called the *zero-order system,* the simplest special case of Eq. 2.1. Its *standard form* is

$$q_o = K q_i \qquad K \triangleq \text{system static sensitivity or steady-state gain} \qquad (2.89)$$

$$\triangleq \frac{b_o}{a_o}$$

and its Laplace and sinusoidal transfer functions are

$$\frac{Q_o}{Q_i}(s) = K \qquad \frac{Q_o}{Q_i}(i\omega) = K\underline{/0^\circ} \qquad (2.90)$$

[11]Ibid., pp. 111–135.
[12]Ibid., p. 243.
[13]Ibid., pp. 244–251.

Figure 2.11 shows its step response and frequency response. This model is "dynamically perfect" since, no matter what form the input might take, the output duplicates its shape exactly. If the inputs to a component have a frequency spectrum with cutoff frequency ω_{co} as in Fig. 2.12, then a zero-order model is an adequate representation of its behavior. (Cutoff frequency is defined as the frequency beyond which the magnitude of the signal's spectrum is judged negligible.)

A component invariably modeled as zero order is the potentiometer[14] of Fig. 2.13. When we desire a manually set reference input in the form of a voltage,

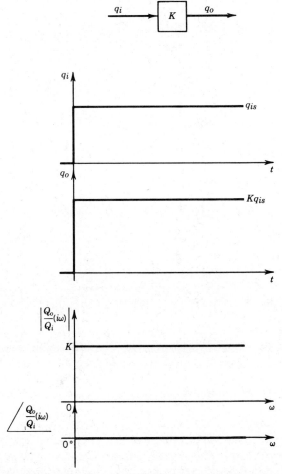

FIGURE 2.11   *Zero-order system: step and frequency response.*

[14]E. O. Doebelin, *Measurement Systems,* 3rd Ed., McGraw–Hill, New York, 1983, pp. 218–225.

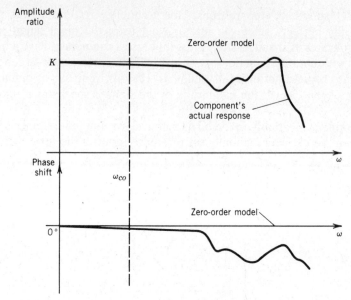

FIGURE 2.12 *Validation of zero-order model.*

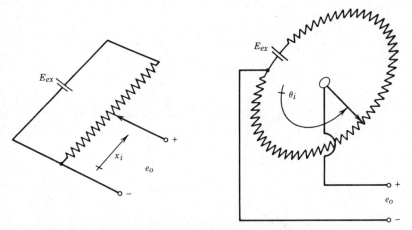

FIGURE 2.13 *Translational and rotational potentiometers.*

a potentiometer is often used as reference input element since adjustment of a calibrated dial attached to the θ_i shaft allows us to select any voltage from zero to E_{ex}. Potentiometers are also used as sensors for controlled variables of rotary or translational displacement. In both types of applications, a zero-order model is appropriate because voltage e_o follows motion θ_i or X_i almost instantly since circuit parasitic inductance and capacitance are very small. When electronic amplifiers are used as components in systems that are not completely electronic, but rather have some mechanical moving parts, pneumatic or hydraulic devices, and/or thermal dynamic effects, the amplifiers are easily made much faster than the other system components and are thus generally modeled as zero order. Note that the very same amplifier properly modeled as zero order in an electromechanical system might need to be modeled as first or second-order (or higher) if it were used in an *all* electronic system, where *all* the components might be very fast. That is, when we neglect some dynamic effects that are truly present, it must be *relative* to those of slower system components that dominate system response.

Devices for comparing reference input R with feedback signal B (Fig. 2.1) are also often treated as zero order. In Fig. 2.14a, 2.14b, the gear differential and summing link, respectively, used to compare mechanical rotational or translational displacements in some servo systems, are inherently zero order if elasticity is neglected since the inputs are motions (rather than forces) and thus the output motion is *kinematically* (rather than dynamically) related to the inputs. Many pneumatic process control systems use the opposed-metal-bellows arrangement of Fig. 2.14c to compare pressures. The pressures p_R and p_B often vary quite slowly, making inertial effects negligible; whereas use of flexure pivots (rather than rolling or sliding bearings) in the mechanisms makes friction negligible. This leaves only elastic effects, which give a zero-order relation between displacement X_E and pressure difference $(p_R - p_B)$. The potentiometer bridge (Fig. 2.14d) and wheatstone bridge (Fig. 2.14e) have negligible capacitive and inductive effects, making them essentially resistive, and thus zero-order, devices. Analysis of the summing amplifier of Fig. 2.14f requires the usual op-amp assumptions (current into op-amp is zero, summing-junction voltage is zero, op-amp has instantaneous response).

$$\frac{e_B - 0}{R} + \frac{-e_R - 0}{R} = \frac{0 - e_o}{R} \tag{2.91}$$

$$e_o = e_R - e_B \tag{2.92}$$

Although not instantaneous, the op-amp dynamics are usually negligible in systems that are not all electronic, making the zero-order model of Eq. 2.92 valid.

2.5 FIRST-ORDER DYNAMIC SYSTEM MODELS

The zero-order model does not always adequately represent a device's behavior over the frequency range of interest; the next more complicated model available from Eq. 2.1 is the first-order.

$$a_1 \frac{dq_o}{dt} + a_0 q_o = b_1 \frac{dq_i}{dt} + b_0 q_i \tag{2.93}$$

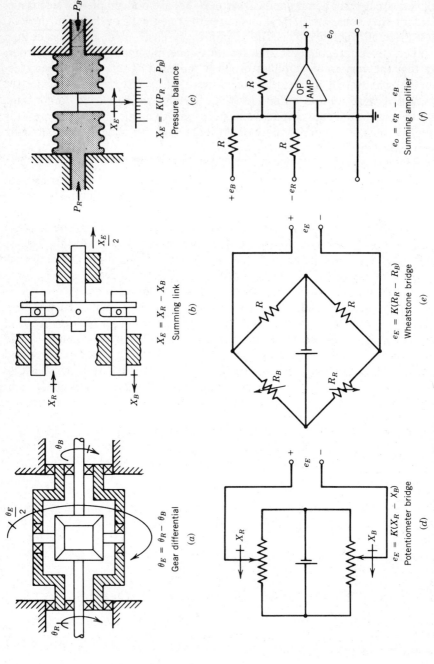

(c) Pressure balance
$$X_E = K(P_R - P_B)$$

(f) Summing amplifier
$$e_o = e_R - e_B$$

(b) Summing link
$$X_E = X_R - X_B$$

(e) Wheatstone bridge
$$e_E = K(R_R - R_B)$$

(a) Gear differential
$$\theta_E = \theta_R - \theta_B$$

(d) Potentiometer bridge
$$e_E = K(X_R - X_B)$$

FIGURE 2.14 *Comparators/error detectors.*

whose most common version is

$$a_1 \frac{dq_o}{dt} + a_0 q_o = b_0 q_i \tag{2.94}$$

with standard form

$$\tau \frac{dq_o}{dt} + q_o = K q_i \tag{2.95}$$

where

$$\tau \overset{\Delta}{=} \text{time constant} \overset{\Delta}{=} \frac{a_1}{a_0} \quad K \overset{\Delta}{=} \text{steady-state gain} \tag{2.96}$$

The Laplace and sinusoidal transfer functions are

$$\frac{Q_o}{Q_i}(s) = \frac{K}{\tau s + 1} \quad \frac{Q_o}{Q_i}(i\omega) = \frac{K}{\sqrt{(\omega\tau)^2 + 1}} \underline{/-\tan^{-1}\omega\tau} \tag{2.97}$$

whereas the response to a step input q_{is} is easily found to be

$$q_o = K q_{is}(1 - e^{-(t/\tau)}) \tag{2.98}$$

From the step and frequency response graphs of Fig. 2.15 it is clear that system speed of response is governed entirely by the time constant τ, large values of τ giving the sluggish response that creates stability problems in high-gain feedback systems; whereas systems with $\tau \to 0$ approach the instantaneous response of the zero-order model.

I introduce at this point a logarithmic technique of plotting frequency-response curves that is widely used in system dynamics. The curves are plotted on semilog graph paper using the logarithmic horizontal axis for frequency (three cycles is often sufficient). Amplitude ratio is plotted on the vertical (linear) scale as a decibel (dB) value, according to

$$\text{dB} \overset{\Delta}{=} 20 \log_{10}(\text{amplitude ratio}) \tag{2.99}$$

whereas phase angle is plotted directly on the linear vertical scale. For first-order terms of the form $1/(i\omega\tau + 1)$ and $(i\omega\tau + 1)$, it can be shown that the amplitude ratio in dB approaches straight-line asymptotes at low and high frequencies, allowing quick manual plotting. At low frequencies the straight-line asymptote is the horizontal 0 dB line; whereas at high frequencies the asymptote slopes downward $[1/(i\omega\tau + 1)]$ or upward $(i\omega\tau + 1)$ at 20 dB/decade (6 dB/octave). A decade is any 10 to 1 frequency change; whereas an octave is a 2 to 1 change. These two straight lines intersect at $\omega_b \overset{\Delta}{=} 1/\tau$, the so-called breakpoint frequency. To plot the amplitude ratio curve quickly for any first-order system, first locate ω_b and then draw in the two asymptotes. The exact curve is then obtained by correcting the asymptotes 3 dB at the breakpoint and 1 dB at frequencies ± 1 octave from ω_b. Phase-angle curves are asymptotic to $0°$ at low frequency and $\pm 90°$ at high, with accurate intermediate points obtained from the table of Fig. 2.16, which illustrates the entire procedure. This form of

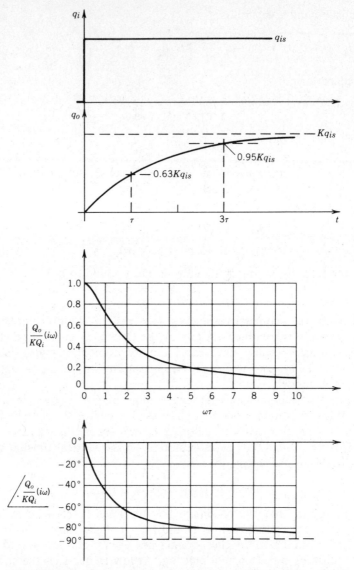

FIGURE 2.15 *Step and frequency response of first-order model.*

frequency-response graphing is applicable to linear dynamic systems in general (not just first-order systems) and has the following advantages:

1. Rapid manual graphing is possible.
2. Wide ranges of amplitude ratio and frequency are conveniently displayed.
3. Amplitude ratio exhibits straight-line-asymptote regions of definite slope. These are helpful in identifying model type from experimental data.

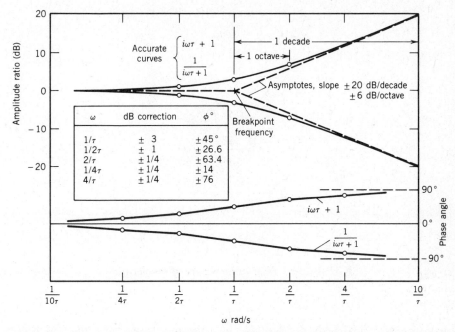

FIGURE 2.16 *Logarithmic ("Bode") plotting of first-order frequency response.*

4. Complex transfer functions are easily plotted and understood as graphical sums of simple (zero-order, first-order, second-order, etc.) basic systems since the dB (logarithmic) technique changes multiplication into addition.

Although the number of control system components adequately modeled as zero order is modest, first-order models are used for many sensors, controllers, actuators, and processes. Electrical temperature sensors such as thermocouples, resistance temperature detectors (RTDs), and thermistors, when not enclosed in protective wells, can be modeled with a single thermal capacitance and resistance as in Fig. 2.17. Conservation of energy gives

Energy in − energy out = additional energy stored

$$\frac{T_f - T_s}{R_t}\, dt - 0 = C_t\, dT_s \qquad (2.100)$$

$$\tau \frac{dT_s}{dt} + T_s = T_f \qquad \tau \triangleq R_t C_t \qquad (2.101)$$

where

$$R_t \triangleq \text{thermal resistance} \triangleq \frac{1}{(\text{film coefficient})(\text{surface area})}$$

$$C_t \triangleq \text{thermal capacitance} \triangleq (\text{sensor mass})(\text{specific heat})$$

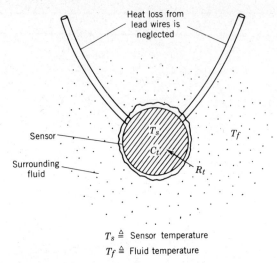

$T_s \triangleq$ Sensor temperature

$T_f \triangleq$ Fluid temperature

FIGURE 2.17 *First-order model of temperature sensor.*

Since the electrical response (voltage in a thermocouple, resistance in RTDs and thermistors) follows T_s instantly, the thermal time constant of Eq. 2.101 gives the complete dynamics of such sensors.

Many pneumatic and hydraulic components make use of the nozzle/flapper device shown in Fig. 2.18 (pneumatic version) to convert small motions to related pressures with extremely high sensitivity. For flapper displacement x_i at a fixed operating point x_{i0}, output pressure p_o becomes steady at p_{o0}, a value between fixed supply pressure p_s and atmospheric. Small perturbations x_{ip} away from x_{i0} will now cause corresponding perturbations p_{op} away from from p_{o0} and we wish to model the p_{op}/x_{ip} dynamic relation. For the volume V containing p_o, the perfect gas law gives

$$p_o = \frac{RT}{V} M \qquad (2.102)$$

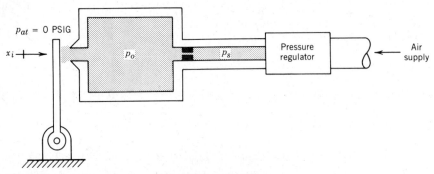

FIGURE 2.18 *Pneumatic nozzle-flapper amplifier.*

Industrial pneumatic devices of this type work with small pressure ranges ($3 \leq p_o \leq 15$ psig) and very small flow rates, thus air temperature tends to stay near room temperature. Furthermore, T in Eq. 2.102 is absolute, thus even a 20°F change in temperature from 70°F would be only about a 4% (20/530) change in absolute temperature, thus we assume constant T with small error. Then Eq. 2.102 can be differentiated to get

$$\frac{dp_o}{dt} = \frac{RT}{V}\frac{dM}{dt} \tag{2.103}$$

where RT/V is a constant. The rate of change of mass, dM/dt, is the difference between mass inflow rate from supply pressure p_s to chamber pressure p_o and nozzle mass outflow rate from p_o to p_{at}. Under our assumptions, nozzle flow rate is a function of two independent variables, x_i and p_o. This nonlinear function can be estimated from fluid mechanics theory but we here choose to go directly to its linearized approximation.

$$G_n \approx G_{n0} + \left.\frac{\partial G_n}{\partial x_i}\right|_{\substack{x_i = x_{i0} \\ p_o = p_{o0}}} (x_i - x_{i0}) + \left.\frac{\partial G_n}{\partial p_o}\right|_{\substack{x_i = x_{i0} \\ p_o = p_{o0}}} (p_o - p_{o0}) \tag{2.104}$$

$$G_n \approx G_{n0} + K_{nx}x_{ip} + K_{np}p_{op} \tag{2.105}$$

where G_{n0} is the operating-point value of G_n. For the inflow rate G_s we could again obtain a nonlinear function of p_o only from fluid mechanics; its linearized version would be

$$G_s \approx G_{s0} + \left.\frac{\partial G_s}{\partial p_o}\right|_{p_o = p_{o0}} (p_o - p_{o0}) \tag{2.106}$$

$$G_s \approx G_{s0} - K_{sf}p_{op} \tag{2.107}$$

where the operating-point value G_{s0} must equal G_{n0} since a steady $p_o = p_{o0}$ requires equality of inflow and outflow. Equation 2.103 now gives

$$\frac{dp_{op}}{dt} = \frac{dp_o}{dt} = \frac{RT}{V}(-K_{sf}p_{op} - K_{nx}\,x_{ip} - K_{np}p_{0p}) \tag{2.108}$$

$$(\tau D + 1)p_{op} = Kx_{ip}$$

where

$$\tau \triangleq \frac{V}{RT(K_{sf} + K_{np})} \qquad K \triangleq \frac{K_{nx}}{K_{sf} + K_{np}} \tag{2.109}$$

Although fluid mechanics theory is needed to estimate values of τ and K before a device is built, experimental testing of existing devices can obtain more accurate values and check the validity of the linearizing approximations. Step-function testing and static calibration can obtain data similar to that in Fig. 2.19a, 2.19b quite easily. Since the human eye is inadequate to judge whether the p_o versus t curve of Fig. 2.19b is truly exponential, the semilog replot of

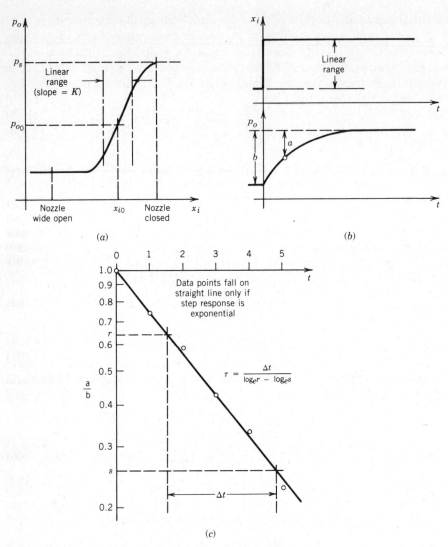

FIGURE 2.19   *Experimental modeling of nozzle-flapper.*

this data as in Fig. 2.19c is used to determine whether response is sufficiently close to the theoretical first-order model, and, if so, to obtain a better numerical value of τ than one could get from the 63.2% point on Fig. 2.19b. This experimental/graphical procedure is, of course, useful for validating any first-order device, not just the nozzle-flapper of this example. Frequency-response experiments, with results plotted as in Fig. 2.16, can also be used to verify (or refute) first-order behavior and obtain numerical values of K and τ.

Figure 2.20 shows mechanical spring/damper systems useful as so-called "lead" and "lag" controllers. In Fig. 2.20a, applying Newton's law to a fictitious

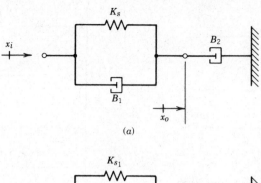

(a)

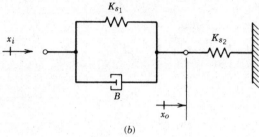

(b)

FIGURE 2.20 *Mechanical lag and lead controllers.*

mass M located at x_o gives

$$\Sigma F = M\ddot{x}_o$$

$$K_s(x_i - x_o) + B_1(\dot{x}_i - \dot{x}_o) - B_2\dot{x}_o = M\ddot{x}_o = 0 \qquad (2.110)$$

$$\frac{x_o}{x_i}(s) = \frac{\tau_1 s + 1}{\tau_2 s + 1} \qquad (2.111)$$

where

$$\tau_2 \triangleq \frac{B_1 + B_2}{K_s} \qquad \tau_1 \triangleq \frac{B_1}{K_s} \qquad (2.112)$$

Since $\tau_2 > \tau_1$, the frequency response of this system has a lagging phase angle at all frequencies. We will see later that its amplitude ratio characteristic gives a useful control action called "lag compensation." Fig. 2.21a gives the step and frequency response. For the system of Fig. 2.20b

$$K_{s1}(x_i - x_o) + B(\dot{x}_i - \dot{x}_o) - K_{s2}x_o = M\ddot{x}_o = 0 \qquad (2.113)$$

$$\frac{x_o}{x_i}(s) = K\frac{\tau_1 s + 1}{\tau_2 s + 1}$$

where

$$\tau_2 \triangleq \frac{B}{K_{s1} + K_{s2}} \qquad \tau_1 \triangleq \frac{B}{K_{s1}} \qquad K \triangleq \frac{K_{s1}}{K_{s1} + K_{s2}} = \frac{\tau_2}{\tau_1} \qquad (2.114)$$

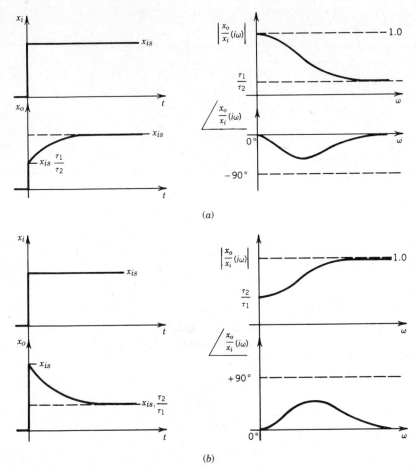

FIGURE 2.21 *Step and frequency response.* (a) *lag controller.* (b) *lead controller.*

Here, $\tau_2 < \tau_1$, and the frequency response exhibits a leading phase angle, a control characteristic useful for another class of practical problems (see Fig. 2.21b for dynamic response curves). From the initial "overresponse" for a step input, we can see that such a controller is useful, for example, in implementing dynamic compensation for process lags as in Fig. 1.5.

Electrical versions of lag and lead controllers can be realized in either passive (no internal power source) form (Fig. 2.22a) or the active (op-amp) type (Fig. 2.22b). The op-amp circuit can give either lag or lead behavior, depending on numerical values, can provide a wider practical range of τ values, and because of its internal power source, can supply more output energy to its load than it takes from the device supplying e_i, giving a desirable buffering effect. Using the

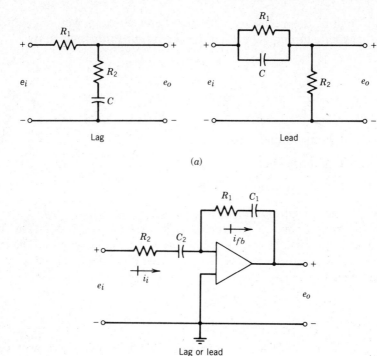

FIGURE 2.22   *Electrical lag and lead controllers.*

definition of *operational impedance,* $Z(D) \triangleq (e/i)(D)$ we can write

$$i_i = \frac{e_i - 0}{Z_i(D)} = i_{fb} = \frac{0 - e_o}{Z_{fb}(D)} \qquad (2.115)$$

$$\frac{e_i}{R_2 + 1/C_2 D} = \frac{-e_o}{R_1 + 1/C_1 D} \qquad (2.116)$$

$$\frac{e_o}{e_i}(s) = \frac{-(C_2/C_1)\,(R_1 C_1 s + 1)}{R_2 C_2 s + 1} = K\frac{\tau_1 s + 1}{\tau_2 s + 1} \qquad (2.117)$$

Note that we can get $|K| > 1$ (amplification) irrespective of the relative magnitudes of τ_1 and τ_2, a desirable feature not obtained with the passive circuits.

Electric drives using dc motors are widely used in servomechanisms. Up to about 10 hp (8 kW), machines with permanent-magnet (PM) fields are popular, and control is achieved by manipulating the armature current. For larger machines, a wound field is necessary, requiring an electrical power supply for the field. The magnetic torque T_m exerted on the motor rotor is given by

$$T_m = K_t i_a \quad \text{PM field} \tag{2.118}$$

$$T_m = K_{tf} i_f i_a \quad \text{wound field} \tag{2.119}$$

where

$$K_t, K_{tf} \overset{\Delta}{=} \text{motor torque constants}$$

$$i_f \overset{\Delta}{=} \text{motor field current}$$

$$i_a \overset{\Delta}{=} \text{motor armature current}$$

For a wound field, the proportionality between T_m and i_f exists only for i_f values below the saturation point of the iron in the field; beyond this point, increases in i_f cause no further increase in torque, giving nonlinear behavior. If the motor is rotating at speed ω, a "back emf" e_b (generator effect) exists across the armature, given by

$$e_b = K_e \omega \quad \text{PM field} \tag{2.120}$$

$$e_b = K_{ef} i_f \omega \quad \text{wound field} \tag{2.121}$$

where K_e, $K_{ef} \overset{\Delta}{=}$ motor back-emf constants. The field circuit of a wound-field machine is usually modeled as inductance L_f and resistance R_f; whereas the armature circuits of both PM and wound-field machines usually can be modeled as pure resistance R_a (brush resistance plus armature resistance), since armature inductance is often negligible. Mechanically, the motor rotor and attached mechanical load can often be treated as inertia and viscous friction. All the electrical and mechanical constants mentioned can be estimated from theory, but more accurate values are available from experimental testing once the machine has been built.

When the field (either PM or wound) is constant, we get the armature-controlled motor of Fig. 2.23, giving

$$e_i - i_a R_a - K_e \omega - 0 \tag{2.122}$$

$$K_t i_a - B\omega = J\dot{\omega} \tag{2.123}$$

which leads to

$$\frac{\omega}{e_i}(s) = \frac{K}{\tau s + 1} \tag{2.124}$$

where

$$K \overset{\Delta}{=} \frac{K_t}{BR_a + K_t K_e} \qquad \tau \overset{\Delta}{=} \frac{JR_a}{BR_a + K_t K_e} \tag{2.125}$$

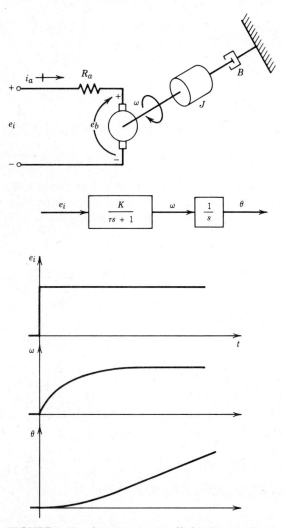

FIGURE 2.23 *Armature-controlled DC motor/load.*

When the motor is used to control shaft position θ rather than speed ω we have

$$\frac{\theta}{e_i}(s = \frac{K}{s(\tau s + 1)} \tag{2.126}$$

We shall see later that the integrating effect $1/s$ in Eq. 2.126 gives a feedback-type position control system better steady-state error behavior than the corresponding speed-control system. Short stroke translational motors ("voice coil actuators") have been available for some time. Long stroke motors appeared only recently as standard components and provide a design alternative to rotary motors with lead screw or rack-and-pinion mechanisms. Translational and rotational dc motor dynamics have the same form.

2.6   SECOND-ORDER DYNAMIC SYSTEM MODELS

When a physical system exhibits a natural *oscillatory* behavior, a first-order model (or even a cascade of several first-order models connected output-of-first to input-of-second, etc.) cannot provide the desired response. The simplest linear system that does provide this possibility is obtained from Eq. 2.1 as

$$a_2 \frac{d^2 q_o}{dt^2} + a_1 \frac{dq_o}{dt} + a_0 q_o = b_2 \frac{d^2 q_i}{dt^2} + b_1 \frac{dq_i}{dt} + b_0 q_i \qquad (2.127)$$

The most basic and common version of this general case is

$$a_2 \frac{d^2 q_o}{dt^2} + a_1 \frac{dq_o}{dt} + a_0 q_o = b_0 q_i \qquad (2.128)$$

whose standard form is

$$\frac{1}{\omega_n^2} \frac{d^2 q_o}{dt^2} + \frac{2\zeta}{\omega_n} \frac{dq_o}{dt} + q_o = K \, q_i \qquad (2.129)$$

$$\omega_n \triangleq \text{undamped natural frequency} \triangleq \sqrt{a_0/a_2}, \text{ rad/time} \qquad (2.130)$$

$$\zeta \triangleq \text{damping ratio} \triangleq a_1/2\sqrt{a_2 a_0} \qquad (2.131)$$

$$K \triangleq \text{steady-state gain} \triangleq b_0/a_0 \qquad (2.132)$$

Laplace and sinusoidal transfer functions are

$$\frac{Q_o}{Q_i} (s) = \frac{K}{\dfrac{s^2}{\omega_n^2} + \dfrac{2\zeta s}{\omega_n} + 1} \qquad (2.133)$$

$$\frac{Q_o}{Q_i} (i\omega) = \frac{K}{\sqrt{\left[1 - \left(\dfrac{\omega}{\omega_n}\right)^2\right]^2 + \dfrac{4\zeta^2 \omega^2}{\omega_n^2}}} \Big/ \tan^{-1} \frac{2\zeta}{(\omega/\omega_n) - (\omega_n/\omega)} \qquad (2.134)$$

For the step response (see Fig. 2.24), three different forms of solution (underdamped, critically damped, and overdamped), corresponding to $\zeta < 1$, $\zeta = 1$, and $\zeta > 1$, are found because the two roots of the characteristic equation may be complex, real and equal, or real and unequal.

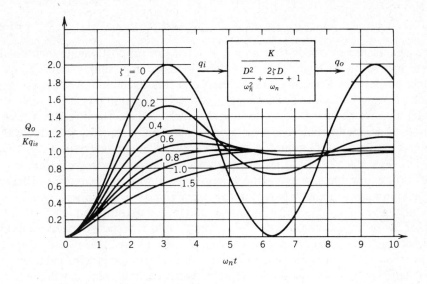

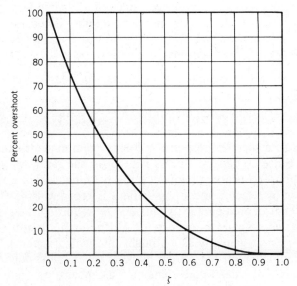

FIGURE 2.24 *Step response of second-order systems.*

$$q_o = Kq_{is}\left[1 - \frac{1}{\sqrt{1-\zeta^2}}e^{-\zeta\omega_n t}\sin\left(\omega_n\sqrt{1-\zeta^2}\,t + \sin^{-1}\sqrt{1-\zeta^2}\right)\right] \quad \zeta < 1 \quad (2.135)$$

$$q_o = Kq_{is}\left[1 - \frac{\zeta + \sqrt{\zeta^2 - 1}}{2\sqrt{\zeta^2 - 1}}e^{(-\zeta + \sqrt{\zeta^2 - 1})\omega_n t}\right.$$

$$\left. + \frac{\zeta - \sqrt{\zeta^2 - 1}}{2\sqrt{\zeta^2 - 1}}e^{(-\zeta - \sqrt{\zeta^2 - 1})\omega_n t}\right] \quad \zeta > 1 \qquad (2.136)$$

$$q_o = Kq_{is}[1 - (1 + \omega_n t)e^{-\omega_n t}] \quad \zeta = 1 \tag{2.137}$$

The significance of steady-state gain K as a scale or proportionality factor is the same here as in all linear system models. Increase in K simply scales up q_o in direct proportion. Damping ratio ζ clearly controls oscillation; $\zeta < 1$ is required for oscillatory behavior. The *undamped* case ($\zeta = 0$) is not physically realizable (total absence of energy loss effects such as friction, electrical resistance, etc.) but gives us, mathematically, a sustained oscillation at frequency ω_n. Note that natural oscillations of damped systems ($\zeta \neq 0$) are at *damped natural frequency* $\omega_{n,d} \triangleq \omega_n\sqrt{1 - \zeta^2}$, *not* at ω_n. In hardware design, an *optimum value* of $\zeta \approx 0.64$ is often used to give maximum response speed without excessive oscillation. (This optimum is actually based on *frequency response,* which we discuss next, but also gives desirable step response.) Undamped natural frequency ω_n is the major factor in response speed. For a given ζ, in fact, response speed is directly proportional to ω_n, as can be seen from Eq. 2.135 through 2.137 where ω_n *always* appears as $\omega_n t$, showing that the same stage of the response is achieved in exactly one half the time if ω_n is doubled. Thus, when second-order components are used in a feedback system design, large values of ω_n (small lags) are desirable since they allow use of larger loop gain before stability limits are encountered.

For frequency response (Fig. 2.25), all values of ζ are covered by a single formula (Eq. 2.134). A resonant peak occurs for $\zeta < 0.707$.

$$\text{peak frequency} \triangleq \omega_p = \omega_n\sqrt{1 - 2\zeta^2} \tag{2.138}$$

$$\text{peak amplitude ratio} = \frac{K}{2\zeta\sqrt{1 - \zeta^2}} \tag{2.139}$$

The role of ω_n as a speed-of-response factor is again evident, from Eq. 2.134 and Fig. 2.25. Since ω_n always appears with ω (as the ratio ω/ω_n), if a system has a "flat" response to a certain frequency, this flat range will be exactly doubled if ω_n is increased by a factor of two, assuming ζ is kept the same. The optimum value of $\zeta = 0.64$, quoted earlier, is obtained by requiring the system to have the widest possible range of frequency for which amplitude ratio is "flat" and phase angle varies most nearly as a straight line with frequency. When these two requirements are met, the system behaves (over the stipulated frequency range) essentially like a dead-time element. This means that input signals pass through it with shape undistorted but with a time delay. Such behavior is especially desirable in sensors but it also is a good compromise for many general applications. When $\zeta > 1$, we can always factor the second-order term into two first-order terms $(\tau_1 s + 1)(\tau_2 s + 1)$ which are generally easier to plot, using the logarithmic scheme discussed earlier. For underdamped systems, this is not possible and the second-order form must be plotted directly. There are again two straight-line asymptotes, which intersect at a breakpoint exactly at $\omega = \omega_n$. The high-frequency asymptote has a slope of -40 dB/decade (-12 dB/octave), twice that of a first-order system. Corrections to the asymptotes now depend on ζ, requiring use of a standard chart, Fig. 2.26, which also gives the phase-angle curves.

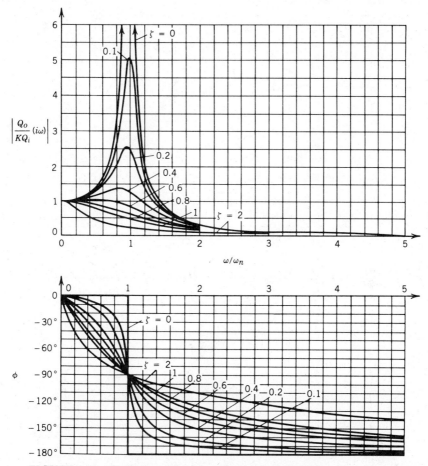

FIGURE 2.25 *Frequency response of second-order systems.*

For existing components, we can again run experimental step-function and/ or frequency-response tests to verify second-order behavior and get numbers for K, ζ, and ω_n. If $\zeta < 1.0$, step tests give us ζ from percent overshoot (Fig. 2.24) and $\omega_{n,d}$ from counting and timing cycles, allowing calculation of ω_n. Over-damped second-order systems may be visually hard to tell from first-order by their step responses. (In fact, for heavy damping, a first-order model may be sufficiently accurate.) Frequency-response testing takes more effort but gives more conclusive results since a -40 dB/decade slope at high frequencies and $\phi \rightarrow -180°$ cannot possibly come from a first-order system.

Our first example is the float-type tank level sensor (Fig. 2.27) used in some liquid-level control systems. The index mark for float motion x is attached to the float at the "waterline" when level h and x are both steady. Float buoyant force then varies nonlinearly with immersion $(h - x)$, as in Fig. 2.27, and is equal to float weight W_f when $(h - x)$ is zero. The damper is included in the

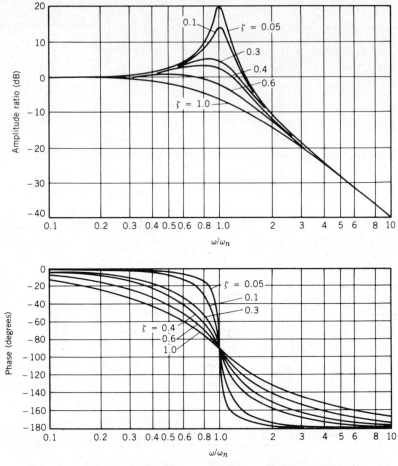

FIGURE 2.26   *Logarithmic plotting of second-order frequency response.*

model to represent both unavoidable friction effects and possible intentional damping. We assume the electrical displacement transducer (such as an LVDT) exerts negligible force on the float rod. By restricting float motion to the linear range L_f, we get from Newton's law

$$[W_f + A_f\gamma(h - x)] - W_f - B\dot{x} = M_f\ddot{x} \qquad (2.140)$$

$$\left[\frac{D^2}{\omega_n^2} + \frac{2\zeta D}{\omega_n} + 1\right]x = Kh \qquad (2.141)$$

Where

$$\omega_n \triangleq \sqrt{\frac{A_f\gamma}{M_f}} \text{ rad/sec} \qquad \zeta \triangleq \frac{B}{2\sqrt{A_fM_f\gamma}} \qquad K \triangleq 1.0 \text{ cm/cm} \qquad (2.142)$$

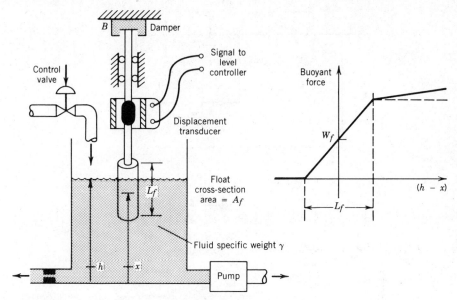

FIGURE 2.27 *Float-type tank level sensors.*

To measure rapid changes in h accurately, ω_n must be sufficiently large. Since γ is not a design variable, we strive for large values of A_f/M_f, suggesting the use of *hollow* floats. If the frequency spectrum of the h variations expected in actual operation permits, our model can, of course, be simplified to first- or even zero-order. If the tank has a very large diameter and if the possible inflow and outflow rates are modest, h *cannot* change rapidly, and using a simpler model may be justifiable.

Controllers whose transfer function is a ratio of two second-order polynomials are often useful in control system design. When the controller is to be of the electronic analog type, the op-amp configuration of Fig. 2.28 is quite versatile. It was used, for example, to control an inertial guidance platform[15] that used air-bearing gyroscopes. Such gyros have practically no friction (damping) and require artificial stabilization, provided by a feedback control scheme. Application of standard op-amp circuit analysis techniques leads to

$$\frac{e_o}{e_i}(s) = \frac{(AB + AC\tau_2 + \tau_1\tau_2)s^2 + (AC + \tau_1 + \tau_2)s + 1}{\tau_1\tau_2 s^2 + (\tau_1 + \tau_2)s + 1} \quad (2.143)$$

where

$$A \triangleq R_2C_1 \quad B \triangleq R_4C_2 \quad C \triangleq R_6/R_5$$

$$\tau_1 \triangleq R_1C_1 \quad \tau_2 \triangleq R_3C_2$$

[15]J. L. Dooley, Active Compensation Networks Stabilize Guidance Platform, *Electromechanical Design*, April 1965, pp. 36–42.

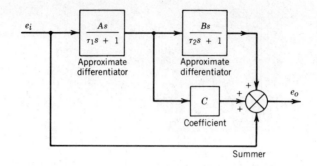

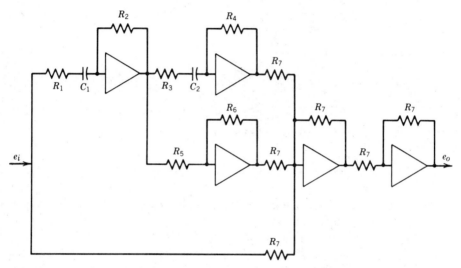

FIGURE 2.28 *Op-amp electronic controller.*

The referenced example used R's $\le 100,000$ Ω and C's ≤ 0.03 μF to implement

$$\frac{e_o}{e_i}(s) = \frac{421(s^2 + 40s + 1300)}{(s + 740)^2} \tag{2.144}$$

(The limitation on C kept capacitor physical size small enough to allow mounting of the network directly on the gyro gimbal, a packaging advantage. Resistance limitations are generally set by op-amp performance characteristics and noise pickup problems.) Since the frequency response of the complete system (including the platform mechanical dynamics) extends to only about 40 rad/s, we see that the denominator of Eq. 2.144 has little dynamic effect, making the controller act essentially as a combination of proportional, first derivative, and second derivative effects. We shall see later that such derivative control modes are powerful stabilizing influences in feedback systems.

Many servomechanisms involve mechanical systems that can be modeled as

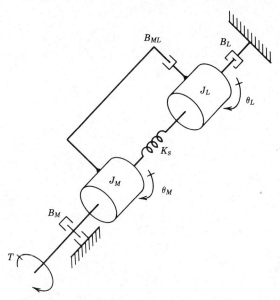

FIGURE 2.29 *Mechanical resonant/antiresonant system.*

in Fig. 2.29. The inertia J_M represents the rotating member of an electric or hydraulic motor that applies the torque T; whereas J_L is the driven load. In low-power servos the shafts can often be treated as having negligible elasticity; however high-power systems may require modeling the elasticity with K_s as in Fig. 2.29. Damping B_{ML} represents frictional losses in the shaft. Dampers B_M and B_L may also be included in some models; we here neglect them. Application of Newton's law to each inertia in turn gives

$$T - B_{ML}(s\theta_M - s\theta_L) - K_s(\theta_M - \theta_L) = J_M s^2 \theta_M \qquad (2.145)$$

$$B_{ML}(s\theta_M - s\theta_L) + K_s(\theta_M - \theta_L) = J_L s^2 \theta_L \qquad (2.146)$$

Manipulation leads to

$$\frac{\theta_L}{T}(s) = \frac{K(\tau s + 1)}{s^2\left[\dfrac{s^2}{\omega_R^2} + \dfrac{2\zeta_R s}{\omega_R} + 1\right]} \qquad (2.147)$$

where

$$K \triangleq 1/(J_M + J_L) \qquad \tau \triangleq B_{ML}/K_s$$

$$\omega_R \triangleq \sqrt{\frac{K_s(J_M + J_L)}{J_M J_L}} \qquad \zeta_R \triangleq \frac{B_{ML}}{2\sqrt{K_s J_M J_L/(J_M + J_L)}} \qquad (2.148)$$

also

$$\frac{\theta_M}{T}(s) = \frac{K\left[\dfrac{s^2}{\omega_{AR}^2} + \dfrac{2\zeta_{AR}}{\omega_{AR}}s + 1\right]}{s^2\left[\dfrac{s^2}{\omega_R^2} + \dfrac{2\zeta_R}{\omega_{AR}}s + 1\right]} \qquad (2.149)$$

where

$$\omega_{AR} \triangleq \sqrt{\frac{K_s}{J_L}} \qquad \zeta_{AR} \triangleq \frac{B_{ML}}{2\sqrt{K_s J_L}} \qquad (2.150)$$

The transfer function

$$\frac{\ddot{\theta}_M}{KT}(s) = \frac{\left[\dfrac{s^2}{\omega_{AR}^2} + \dfrac{2\zeta_{AR}s}{\omega_{AR}} + 1\right]}{\left[\dfrac{s^2}{\omega_R^2} + \dfrac{2\zeta_R s}{\omega_R} + 1\right]} \qquad (2.151)$$

exhibits both *resonance* and *antiresonance* phenomena. The most extreme examples of these effects occur if $B_{ML} = 0$. This frictionless case is not physically possible but is approached by some practical systems. Taking $\zeta_{AR} = \zeta_R = 0$ and noting that $\omega_R > \omega_{AR}$, we can sketch the frequency response of Eq. 2.151 as in Fig. 2.30. At the antiresonant frequency ω_{AR}, the numerator quadratic becomes precisely zero, thus torque applied at this frequency produces *no* motion ($\ddot{\theta}_M$, $\dot{\theta}_M$, or θ_M) of J_M. This type of behavior is also called a *notch filter*

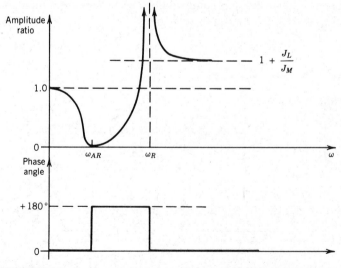

FIGURE 2.30 *Frequency response of resonant/antiresonant system.*

effect. At the resonant frequency ω_R, the denominator quadratic is zero, making the amplitude ratio infinite and giving huge motions for even the tiniest applied torque. Servos that involve such resonant/antiresonant behavior require special control techniques to achieve optimum performance. We will develop some of these methods in later chapters.

Although electric drives such as that of Fig. 2.23 are in wide use, hydraulic actuation is preferable in certain circumstances. Figure 2.31a indicates that when *both* high speed (frequency response) and high power are required, hydraulic systems may be mandatory or desirable, with 2.31b showing typical industrial applications and 2.31c displaying some aerospace/military uses. In addition to speed and power considerations, many additional factors bear on the electric/ hydraulic decision for those applications where both are technically feasible. Electric power is more readily available, cleaner and quieter, and easier to transmit, but may create electrical interference with low-level data signals and can cause overheating problems at low speeds. Hydraulic drives tend to be stiffer with respect to load disturbances; the flowing fluid is a convenient cooling medium; intermittent and stalled operation without damage is possible; but oil leakage and flammability and fluid contamination may pose problems.

Although a wide variety of detailed hydraulic control schemes are in use, a useful overall classification is that of *pump-controlled* versus *valve-controlled* systems. Pump-controlled systems are usually relatively high power (above 10 or 20 hp) applications, where efficiency is economically significant and response speed requirements are modest (less than 10 Hz frequency response). Valve-controlled systems are faster but are generally quite inefficient. For a low-power system, inefficiency has little economic impact. For fast high-power systems where speed specifications can be met *only* by valve control, the economic cost of low efficiency must be accepted.

We consider first a pump-controlled system ("hydrostatic drive") shown in simplified form in Fig. 2.32a. A variable-displacement positive displacement pump is driven at constant speed by some source of mechanical power. The displacement (volume of oil passed through pump per radian of shaft rotation) is adjustable by a stroke control lever (while pump is running) from negative maximum, through zero, to positive maximum. In a servo system the pump stroke would be controlled to produce just the fluid power needed to drive the load; if no load motion is requested the pump "idles" at zero displacement. This production of fluid power only when needed is the basis of the good efficiency. A fixed-displacement rotary motor (or hydraulic cylinder, if translatory motion is desired) is connected to the pump as shown, and drives some useful mechanical load. Thus the load can be driven in forward or reverse at a range of speeds from zero to maximum by controlling the pump's displacement.

To overcome the problems of leakage and cavitation possible in a practical system, the idealized version of Fig. 2.32a must be augmented with the replenishing and supercharging subsystem of Fig. 2.32b. There a small fixed displacement ("charge") pump and relief valve establish a base pressure of a few hundred psi throughout the system before the main pump is activated, continuously

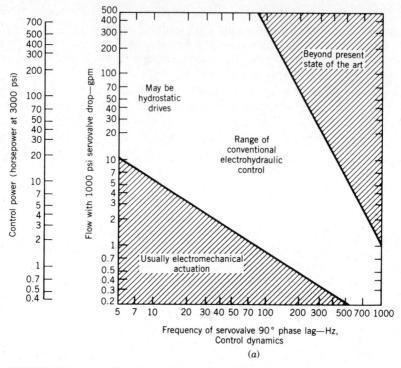

FIGURE 2.31 *Applications of servo hydraulics.* **(a)** *Electric vs. hydraulic systems.* **(b)** *Industrial application* **(c)** *Aerospace/military applications.*
Source: Moog Inc., Catalog 801–678, East Aurora, New York, 1982.

making up any leakage that may occur. When the main pump is stroked, fluid is drawn from the upstream line, tending to reduce its pressure. Without the supercharging system, this pressure could drop to fluid vapor pressure, causing cavitation with its attendant noise and wear problems. The charge pump system prevents this by maintaining the low-pressure line at the set base pressure; while the check valve on the high-pressure side allows this pressure to build up as needed to drive the load.

To design servo systems that use the arrangement of Fig. 2.32*b* for actuation we need the transfer functions relating motor velocity ω_m to pump stroke ϕ_i and disturbing load torque T_l. Application of conservation of volume, together with Newton's law, will yield the desired results. For the high-pressure side of the system we can write: pump gross flow rate-motor/pump leakage flow minus compressibility "flow" equals motor displacement flow. Taking all variables as dynamic perturbations away from an operating point

$$\omega_p d_p \phi_i - K_{lpm} p_1 - \frac{V}{M_B}\frac{dp_1}{dt} = d_m \omega_m \qquad (2.152)$$

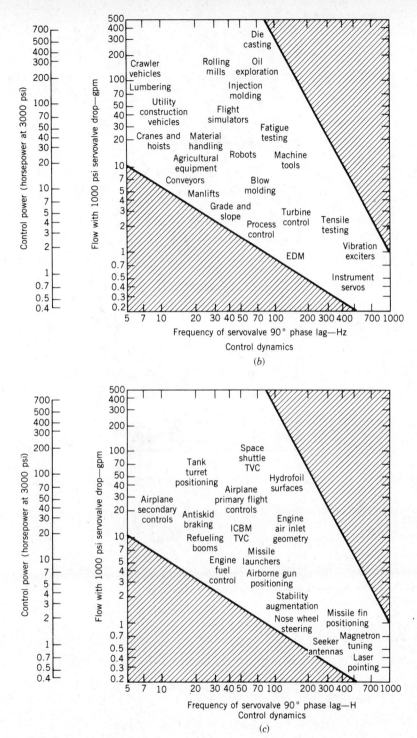

FIGURE 2.31 (Cont.)

97

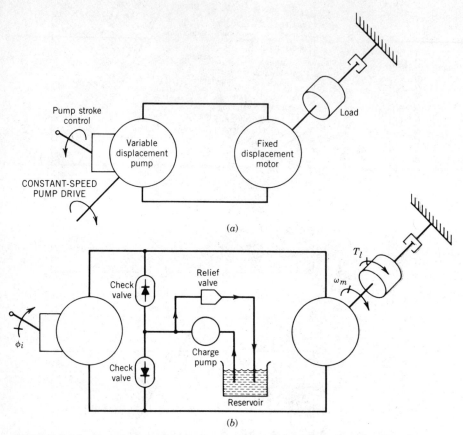

FIGURE 2.32   *Pump-controlled hydraulic system. (a) Simplified. (b) Practical.*

where

$$\omega_p \triangleq \text{steady pump speed, rad/s}$$

$$d_p \triangleq \text{pump maximum displacement, in}^3/\text{radian of shaft rotation}$$

$$\phi_i \triangleq \text{pump normalized stroke, } -1.0 \leq \phi_i \leq 1.0$$

$$K_{lpm} \triangleq \text{combined pump/motor leakage coefficient, (in}^3/\text{s)/psi}$$

(Leakage rate is really $K_{lpm}(p_1 - p_{\text{base}})$ but p_{base} has no perturbation.)

$$p_1 \triangleq \text{perturbation in high-side pressure, above base pressure}$$

$$V \triangleq \text{volume exposed to pressure } p_1 \text{ (one side of pump, one line, one side of motor)}$$

$$M_B \triangleq \text{fluid effective bulk modulus (working value of } 100,000 \text{ psi used to estimate effects of air bubbles and component elasticity}[16])$$

$d_m \triangleq$ fixed motor displacement, in^3/rad

$\omega_m \triangleq$ motor speed, rad/s

Newton's law at the motor shaft gives

$$T_l + d_m p_1 - B\omega_m = J \frac{d\omega_m}{dt} \qquad (2.153)$$

where B and J are, respectively, the viscous damping and inertia at the motor shaft, and motor torque is $d_m p_1$[17].

Solving for $\omega_m(s)$ gives

$$\omega_m(s) = \frac{K_\phi \, \phi_i(s) + K_T(\tau s + 1)T_l(s)}{\dfrac{s^2}{\omega_n^2} + \dfrac{2\zeta s}{\omega_n} + 1} \qquad (2.154)$$

where

$$K_\phi \triangleq \frac{\omega_p d_p d_m}{d_m^2 + K_{lpm}B} \quad \frac{\text{rad/s}}{\% \text{ stroke}} \qquad K_T \triangleq \frac{K_{lpm}}{d_m^2 + K_{lpm}B} \quad \frac{\text{rad/s}}{\text{in.-lb}_f} \qquad (2.155)$$

$$\omega_n \triangleq \sqrt{\frac{M_B(d_m^2 + K_{lpm}B)}{JV}} \quad \text{rad/s} \qquad \tau \triangleq \frac{V}{M_B K_{lpm}} \quad \text{s} \qquad (2.156)$$

$$\zeta \triangleq \frac{K_{lpm}JM_B + VB}{2} \sqrt{\frac{1}{VJM_B(d_m^2 + K_{lpm}B)}} \qquad (2.157)$$

Figure 2.33 shows construction details of a practical system. The pump displacement control shown there uses manual input (control handle) to a small valve-controlled servo system, and mechanical feedback from the pump swashplate angle. Either or both the input and feedback functions can be performed electrically in electrohydraulic systems of this type.

We now study in more detail *valve-controlled* hydraulic systems such as that used to stroke the pump in many pump-controlled systems and for the direct control of load motion in myriad other applications. Whereas in pump-controlled systems the fluid power supply must be included in the system model, analysis of valve-controlled systems can proceed without consideration of power supply details if one *assumes* the existence of a power source of constant supply pressure p_s, irrespective of flow demand. Power supplies that approximate this behavior are available in several different forms that trade off complexity, cost, efficiency, and static/dynamic pressure regulation accuracy. Figure 2.34a shows the simplest

---

[16]Doebelin, "System Dynamics.", p. 187.
[17]Ibid., p. 228.

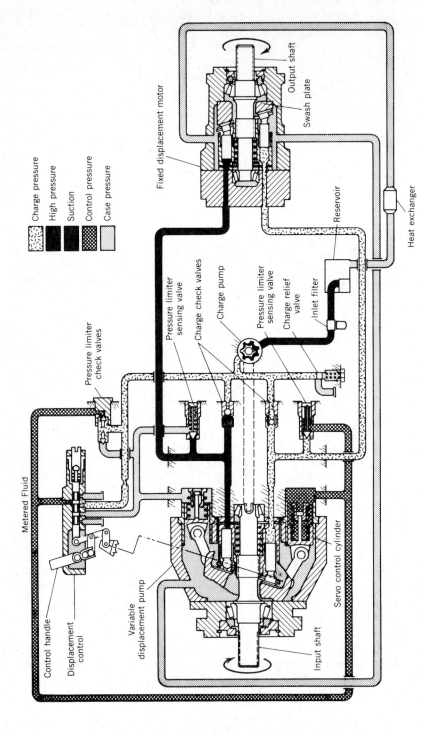

FIGURE 2.33 *Constructional details of pump-controlled hydraulics.*
Source: Sundstrand Corp., Bulletin 9779, Ames, Iowa, 1982.

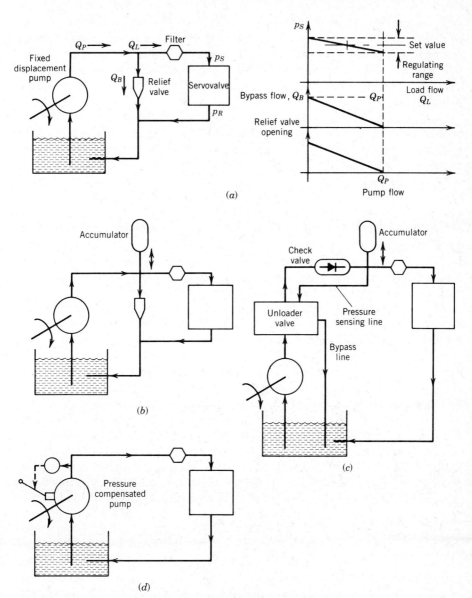

FIGURE 2.34 *Power supplies for valve-controlled hydraulics.*

supply, the *relief-valve* type. The spring-loaded relief valve is completely shut until pressure reaches the low end of the regulating range, whereupon it opens sufficiently to bypass any pump flow not required by the servovalve. The fluid power (volume flow rate times relief valve pressure drop) of the bypassed flow is completely lost and converted to heat, which may require installation of an oil cooling system. When the servo system requires no flow, *all* the pump-generated power is converted to heat, giving zero efficiency. Supply pressure p_s varies by about $\pm 3\%$ over the regulating range for steady conditions and response to transient flow requirements is quite fast; thus the standard assumption of constant p_s used in servo-system analysis is usually reasonable. Pump size (gal/min) must be chosen to match the largest anticipated servo-system demand.

Since servo-system flow demands are by nature intermittent, addition of an *accumulator*[18] in Fig. 2.34b allows use of a smaller pump since part of the required transient flow is now supplied by the accumulator, which is recharged during servo idle periods. The smaller pump wastes less energy during periods when all the flow is bypassed, increasing efficiency and perhaps allowing deletion of the heat exchanger. Pump pressure pulsations (related to pump speed and number of cylinders) are also reduced by the accumulator. Further increase in efficiency is provided by the *unloader-valve* type of supply of Fig. 2.34c. Here an accumulator is mandatory since the unloader valve disconnects the pump from the load when the pressure-sensing line reports that set pressure has been reached. Although the pump continues to pump at full flow, little energy is used since the check valve isolates the pump from system pressure and the unloader valve provides a bypass to the reservoir at nearly zero gauge pressure. Unloader valves have a regulating range similar to relief valves, the pump cycling between load and unload to maintain system pressure near the set point. Figure 2.34d shows a *pressure-compensated pump*[19] type of supply. Here a variable-displacement pump uses a pressure sensing scheme to adjust its displacement so as to keep system pressure near the set value and supply *only* the pump flow needed by the load, giving high efficiency.

Having briefly described several means of maintaining p_s nearly constant, we now proceed to analyze the valve-controlled servo of Fig. 2.35a (shown in a neutral position) under this assumption. (Figure 2.35b shows a sophisticated application of valve-controlled servos where the motions of six actuators are coordinated to provide roll, pitch, and yaw rotary motions plus x,y,z, translation in a flight simulator. Here the valves are electrically positioned and motion feedback is also electrical.) If a positive step motion command x_{Vs} is given by moving the valve spool to the right (this requires only a few ounces or pounds of force), the left valve port is opened to p_S (say 1000 psi) whereas the right is opened to p_R, which is slightly above atmospheric pressure but which we approximate as zero gauge pressure in our model. Cylinder pressure p_{cl} thus rises (while p_{cr} falls), producing an accelerating force on mass M that moves with displacement x_C. (Mass M includes *all* parts that move with x_C.) Since the valve *housing* is attached to the load (mechanical feedback), as x_C approaches the commanded value x_{Vs}, the valve ports are closing, and M decelerates and finally comes to rest with $x_C = x_{Vs}$. This *hydromechanical servo system* acts as a power

[18]E. O. Doebelin, *System Dynamics*, Merrill, Columbus, Ohio, 1972, pp. 185–192.
[19]E. O. Doebelin, *System Modeling and Response*, Wiley, New York, 1980, pp. 400–405.

amplifier since the power (force × velocity) applied with the motion x_V is much (perhaps 20,000 times) smaller than that developed at the load.

The servovalve, a precise and expensive component intolerant of fluid contamination, is the heart of such systems and is available in several forms[20] such as nozzle flapper and jet pipe, but the spool type of Fig. 2.35 is most common. Most spool-type servovalves are intended to be the zero lap form of Fig. 2.36 since the "flow gain" (slope of Fig. 2.36 graphs) is then most constant; however manufacturing tolerances, elastic/thermal expansion, wear, and the like dictate

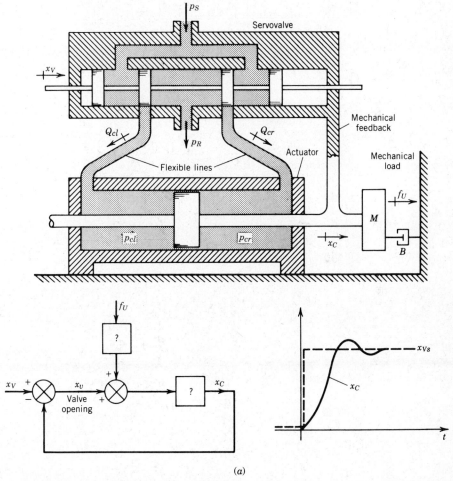

(a)

FIGURE 2.35   *Valve-controlled servo system.*

[20]H. E. Merritt, *Hydraulic Control Systems,* Wiley, New York, 1967, Chap. 7.

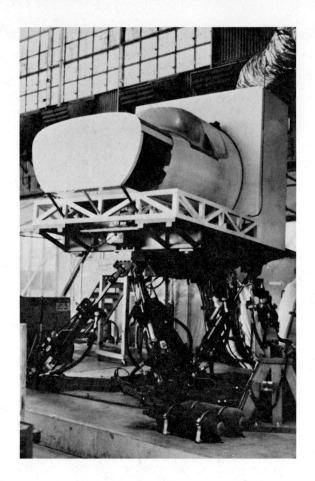

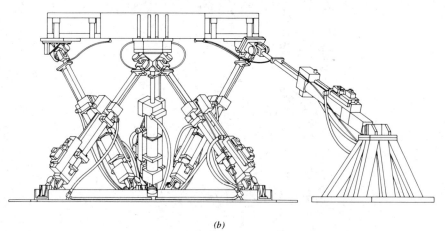

(b)

FIGURE 2.35 (Cont.)
Source: McFadden Systems Inc., South Gate, California.

that a real valve will always exhibit either underlap or overlap behavior. Since the full stroke of a typical servovalve may be only 0.030 in, the overlap/underlap effects are usually very small, perhaps on the order of 0.0001 in (not the exaggerated dimensions necessary for clarity in Fig. 2.36). Underlap has the disadvantage of there being power waste due to leakage even when the valve is "closed," but the advantage of immediate flow response for the slightest spool motion. In fact, the flow gain near null is *twice* that of a zero-lap valve since *two* flow orifices are changing area. Overlapped valves have less null leakage (theoretically zero) but the dead space for small spool motions is quite undesirable for precise load motion control. Since all three forms of valve exhibit some leakage in their neutral positions, the actuator pressures p_{cl} and p_{cr} will each come to $p_S/2$ at the servo rest condition (f_U assumed zero), since only then would leakage from p_S to, say, p_{cl} be exactly equal to that from p_{cl} to p_R ($p_R \approx 0$), giving no net flow to the actuator. Thus for the system of Fig. 2.35, with

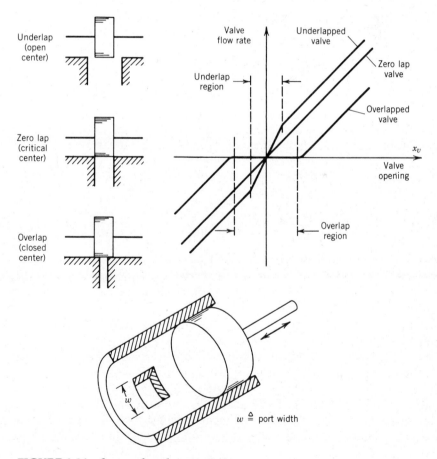

FIGURE 2.36 *Servovalve characteristics.*

$p_S = 1000$ psi, p_{cl} and p_{cr} would both be 500 psi when the valve is in neutral and $f_U = 0$.

We will now develop both nonlinear and linearized models for valve-controlled servos, and introduce the use of the digital simulation language CSMP, a powerful analysis and design tool for dynamic systems of all kinds. We choose to consider a valve with underlap $x_u = 0.0001$ in and port width $w = 0.5$ in. Using standard fluid mechanics formulas for steady liquid flow through sharp-edged orifices, the volume flow rate Q_{cl} to the left end of the cylinder is

$$Q_{cl} = C_d w(x_u + x_v) \sqrt{\frac{2(p_S - p_{cl})}{\rho}} - C_d w(x_u - x_v) \sqrt{\frac{2(p_{cl} - 0)}{\rho}} \quad (2.158)$$

We will take the orifice discharge coefficient C_d to be 0.6 and fluid mass density $\rho = 7.8 \times 10^{-5}$ lb$_f$-sec^2/in.^{4} Similarly, for the right end of the cylinder

$$Q_{cr} = -C_d w(x_u + x_v) \sqrt{\frac{2(p_{cr} - 0)}{\rho}} + C_d w(x_u - x_v) \sqrt{\frac{2(p_S - p_{cr})}{\rho}} \quad (2.159)$$

Just as in the pump-controlled system, the volume flow rate Q_{cl} is "used up" in supplying compressibility, leakage, and actuator displacement flows

$$Q_{cl} - \frac{(V_{lo} + A_p x_c)}{M_B} \frac{dp_{cl}}{dt} - K_{pl}(p_{cl} - p_{cr}) = A_p \frac{dx_C}{dt} \quad (2.160)$$

Piston leakage coefficient K_{pl} is included, even though translational actuators usually have piston rings that are essentially leak free, since our analysis also holds for rotary motors (which *do* have leakage) and since we sometimes pipe in an *intentional* "leak" in the form of a needle valve between the ends of the cylinder to get a damping effect. The compressed volume $V_{lo} + A_p x_c$ includes the line from valve to actuator, plus the left end of the actuator, whose piston area is A_p. For the actuator's right side

$$Q_{cr} - \frac{(V_{ro} - A_p x_C)}{M_B} \frac{dp_{cr}}{dt} + K_{pl}(p_{cl} - p_{cr}) = -A_p \frac{dx_C}{dt} \quad (2.161)$$

Newton's law for the mass M gives

$$(p_{cl} - p_{cr})A_p - B \frac{dx_C}{dt} + f_U = M \frac{d^2 x_C}{dt^2} \quad (2.162)$$

whereas the kinematic relation of x_V, x_C, and x_v provides the final equation $x_v = x_V - x_C$.

When Eq. 2.158 and 2.159 are substituted into 2.160 and 2.161, x_v is replaced by $x_V - x_C$, and Eq. 2.162 is included, we get a complete set of three equations in the unknowns x_C, p_{cl}, and p_{cr}. Unfortunately, due to the square roots and products of variables in Eq. 2.158 and 2.159 and products of variables in Eq. 2.160 and 2.161 this equation set is nonlinear and unsolvable analytically. Analog or digital simulation handles such equations easily but requires use of specific numerical values for all parameters, initial conditions, and inputs, so general solutions are not possible. We will shortly develop a digital simulation and get some actual numerical results.

If we restrict our analysis to small perturbations around a chosen operating point, a linearized approximate model may be obtained that provides many useful results. Valve flow equations, such as Eq. 2.158 and 2.159, can be thought of in general as relations between a dependent variable (flow rate) and two independent variables (spool motion and cylinder pressure), and can thus be linearized about any desired operating point using our usual Taylor series expansion

$$Q_v \approx Q_{v,0} + \frac{\partial Q_v}{\partial x_v}\bigg|_{\substack{x_{v,p} \\ \text{operating point}}} + \frac{\partial Q_v}{\partial p_c}\bigg|_{\substack{p_{c,p} \\ \text{operating point}}} \tag{2.163}$$

The partial derivatives, evaluated numerically at the operating point, can be estimated theoretically from equations such as Eq. 2.158 and 2.159 or, more accurately, from experimental tests once the valve has been built. Using the definitions

$$\text{flow gain} \triangleq C_x \triangleq \frac{\partial Q_v}{\partial x_v}\bigg|_{\text{operating point}} \qquad \text{pressure coefficient} \triangleq C_p \triangleq -\frac{\partial Q_v}{\partial p_c}\bigg|_{\text{operating point}} \tag{2.164}$$

Eq. 2.163 becomes

$$Q_v \approx Q_{v,0} + C_x x_{v,p} - C_p p_{c,p} \tag{2.165}$$

If we take $Q_{v,0} = 0$ and assume the numerical values of C_x and C_p are equal for the Q_{cl} and Q_{cr} equations (correct assumptions for commonly used operating points), we can write

$$Q_{cl} \approx C_x x_{v,p} - C_p p_{cl,p} \qquad Q_{cr} \approx -C_x x_{v,p} - C_p p_{cr,p} \tag{2.166}$$

To linearize Eqs. 2.160 and 2.161 we need only take the volumes $V_{l0} + A_p x_C$ and $V_{r0} - A_p x_C$ to be constant at $V_{l0} = V_{r0}$, a good approximation for small changes in x_C. Our linearized equation set then becomes

$$(C_x x_{v,p} - C_p p_{cl,p}) - \frac{V_{l0}}{M_B}\frac{dp_{cl,p}}{dt} - K_{pl}(p_{cl,p} - p_{cr,p}) = A_p \frac{dx_{C,p}}{dt} \tag{2.167}$$

$$(-C_x x_{v,p} - C_p p_{cr,p}) - \frac{V_{l0}}{M_B}\frac{dp_{cr,p}}{dt} + K_{pl}(p_{cl,p} - p_{cr,p}) = -A_p \frac{dx_{C,p}}{dt} \tag{2.168}$$

$$(p_{cl,p} - p_{cr,p})A_p - B\frac{dx_{C,p}}{dt} + f_{U,p} = M\frac{d^2 x_{C,p}}{dt^2} \tag{2.169}$$

We could now get six transfer functions relating outputs x_{Cp}, $p_{cl,p}$, and $p_{cr,p}$ to inputs $x_{v,p}$ and $f_{U,p}$. Our main interest is in $x_{C,p}$, so (using determinants, or substitution and elimination) we obtain

$$\frac{\dot{x}_{C,p}}{x_{v,p}}(s) = \frac{K_x}{\dfrac{s^2}{\omega_n^2} + \dfrac{2\zeta s}{\omega_n} + 1} \tag{2.170}$$

$$\frac{\dot{x}_{C,p}}{f_{U,p}}(s) = \frac{K_f(\tau s + 1)}{\dfrac{s^2}{\omega_n^2} + \dfrac{2\zeta s}{\omega_n} + 1} \tag{2.171}$$

where

$$K_x \triangleq \frac{2C_x A_p}{2A_p^2 + B(C_p + 2K_{pl})} \frac{\text{in/s}}{\text{in}} \qquad K_f \triangleq \frac{1}{\dfrac{2A_p^2}{2K_{pl} + C_p} + B} \frac{\text{in/s}}{\text{lb}_f} \tag{2.172}$$

$$\omega_n \triangleq \sqrt{\frac{M_B[2Ap^2 + B(C_p + 2K_{pl})]}{MV_{l0}}} \qquad \tau \triangleq \frac{V_{l0}}{M_B(C_p + 2K_{pl})} \tag{2.173}$$

$$\zeta \triangleq \frac{B + \left(\dfrac{2M_B M}{V_{l0}}\right)K_{pl} + \left(\dfrac{M_B M}{V_{l0}}\right)C_p}{2\sqrt{\dfrac{M_B M}{V_{l0}}[2A_p^2 + B(C_p + 2K_{pl})]}} \tag{2.174}$$

These transfer functions are useful for *any* application that uses a servovalve, actuator, and load, not just our specific hydromechanical servo of Fig. 2.35.

Our linearized model is extremely useful since it relates important response characteristics such as speed, gain, and damping to basic system design parameters such as mass and piston area through our familiarity with basic second-order systems. Its main drawback, uncertainty about its accuracy, can ultimately be resolved only by experimental testing; however comparison of linearized-model predictions with digital simulation of the more physically correct nonlinear equations can establish an improved level of confidence without the time and expense of experimentation. Since powerful and easy-to-use digital simulation languages have been in use for over twenty years and are now widely available, I wish to introduce this computer-aided-design capability early in the text and show its application to many aspects of control system design. Fortunately, these digital simulation languages[21] are quite easy to understand and we will simply present and explain the programs needed for a given application rather than try to teach the reader the language. For those who wish to study the languages in more depth, the CSMP (Continuous System Modeling Program, IBM Corp.),

---

[21]F. E. Cellier, Simulation Software: Today and Tomorrow, in *Simulation in Engineering Sciences,* J. Burger and Y. Jarny (Eds.), Elsevier, 1983, pp. 3–19.

which we will use, is explained in various references.[22] Two other popular languages, ACSL (Advanced Continuous Simulation Language) and CSSL (Continuous System Simulation Language) are described in detail in their user's manuals.[23,24] Fortunately, all continuous system simulation languages have much in common. Having learned one, a user can pick up any of the others very quickly. Thus our use of the specific language CSMP really introduces the reader to this whole *class* of languages. Finally, the languages ACSL and CSSL, scaled down to the limits of personal computers, are starting to become available.

We will now give a CSMP III program that will study both the nonlinear and linearized hydraulic servo models so that we can judge the validity of the linearization technique for this application.

```
TITLE NONLINEAR AND LINEARIZED HYRAULIC SERVO
INIT
PARAM PS=1000.      1000 psi supply pressure
PARAM AA=1.0,BB=0.0     selects small (0.002 in.) step input
PARAM CD=0.6     discharge coefficient
PARAM W=0.5     valve port width = 0.5 in.
PARAM XU=.0001     underlap of 0.0001 in.
PARAM RHO=7.8E-5     fluid density
PARAM AP=2.0     piston area of 2.0 in.²
PARAM BM=100000.     bulk modulus 100,000 psi
PARAM M=.03     mass 0.03 (lb_f-s²)/in
PARAM KPL=.001     leakage coefficient 0.001 (in.³/s)/psi
PARAM B=100.     viscous damping 100. lb_f/(in./s)
PARAM VLO=4.0 ⎫ compressed volumes at operating
PARAM VRO=4.0 ⎭ point are both 4.0 in.³
PARAM KSP=0.0     no spring in mechanical load
INCON PCLO=500. ⎫ initial conditions (pressures) are
INCON PCRO=500. ⎭ both 500 psi
INCON XPO=0.0 ⎫ initial conditions (displacement and velocity) of
INCON VPO=0.0 ⎭ mechanical load are both zero
      H1=CD*W*SQRT(2./RHO)     constant for nonlinear model
      KP=0 ⎫ valve coefficients
      KX=CD*W*SQRT(PS/RHO) ⎭ for linearized model
DYNAM
*     NONLINEAR MODEL
      SW1=XV+XU ⎫ switching effects
                ⎬ to model
      SHUT1=FCNSW(SW1,0.0,0.0,1.0) ⎭ underlapped valve
```

Here:

$PARAM$ AP=2.0 — piston area of 2.0 in.²

---

[22]Doebelin, *System Dynamics; System Modeling and Response;* F. H. Speckhart and W. L. Green, *A Guide to Using CSMP,* Prentice–Hall, Englewood Cliffs, N.J., 1976; Document SH19-7001-2, IBM Corp.

[23]Mitchell and Gauthier Assoc's., Concord, Mass.

[24]Simulation Services, Chatsworth, CA.

```
SW2 = XV - XU                                              ⎫ switching effects
                                                           ⎪ to model
SHUT2 = FCNSW(SW2,1.0,0.0,0.0)                             ⎭ underlapped valve
QCL = H1*(XU+XV)*SQRT(PS-PCL)*SHUT1-H1*(XU-XV)*
SQRT(PCL)*SHUT2
QCR = -H1*(XU+XV)*SQRT(PCR)*SHUT1+H1*(XU-XV)*
SQRT(PS-PCR)*SHUT2
PCLDOT = (QCL-KPL*(PCL-PCR)-AP*VP)*(BM/
(VL0+AP*XP))
PCL = INTGRL(PCL0,PCLDOT)          integrate pressure rate to get pres-
                                   sure
PCRDOT = (QCR+KPL*(PCL-PCR)+AP*VP)*(BM/
(VR0-AP*XP))
PCR = INTGRL(PCR0, PCRDOT)
ACCP = (AP*(PCL-PCR)+FI-B*VP-KSP*XP)/M
VP = INTGRL(VP0,ACCP)         integrate acceleration to get velocity
XP = INTGRL(XP0,VP)        integrate velocity to get displacement
FI = 0.0       take disturbing force as zero
XI = AA*.002*STEP(.003)+BB*.02*STEP(.003)         desired
                                                  value is
                                                  0.002 in.
                                                  step at t =
                                                  0.003s, or
                                                  0.02 in,
                                                  depending
                                                  on values
                                                  of AA and
                                                  BB
```

```
XV = XI - XP       mechanical feedback
*    LINEARIZED MODEL
QCLL = KX*XVL-KP*PCLL                  ⎫ linearized valve
QCRL = -KX*XVL-KP*PCRL                 ⎭ flow model
PCLLDT = (QCLL-KPL*(PCLL-PCRL)-AP*VPL)*(BM/VL0)
PCLL = INTGRL(0.0,PCLLDT)      integrate pressure rate
PCRLDT = (QCRL+KPL*(PCLL-PCRL)+AP*VPL)*(BM/VL0)
PCRL = INTGRL(0.0,PCRLDT)      integrate pressure rate
ACCPL = (AP*(PCLL-PCRL)+FI-B*VPL-KSP*XPL)/M
VPL = INTGRL(VP0,ACCPL)        integrate acceleration
XPL = INTGRL(XP0,VPL)       integrate velocity
XVL = XI - XPL       mechanical feedback
PCRLT = PCRL+PCR0       add pressure perturbation to operating point
PCLLT = PCLL+PCL0       add pressure perturbation to operating point
TIMER FINTIM = .03,DELT = .00001,OUTDEL = .0001
OUTPUT TIME,XP,XPL
PAGE GROUP,XYPLOT,HEIGHT = 5.5,WIDTH = 6.5
OUTPUT TIME,QCL,QCLL
```

```
PAGE GROUP,XYPLOT,HEIGHT=5.5,WIDTH=6.5
OUTPUT TIME,PCL,PCLLT
PAGE GROUP,XYPLOT,HEIGHT=5.5,WIDTH=6.5
END
PARAM AA=0.0,BB=1.0    reruns entire problem using a 0.02 rather than
                       0.002 in. step input
END
```

We note first that both the nonlinear and linearized models are run *simultaneously* (not sequentially) simply by using different symbols (e.g., XP, XPL) for the same physical variable in the two models. Statements sandwiched between the cards labelled INIT and DYNAM, such as KX = . . . used to compute valve coefficients, are computed only once at $t = 0$ but the results are available whenever needed later in the program. Cards labeled PARAM give numerical values of system parameters, whereas INCON does the same for initial conditions. To automatically rerun the entire program for several (up to 50) values of a single parameter, merely write, for example, PARAM M=(.03,.05,.07,.09) to see the effect of varying the mass. Cards after DYNAM state and solve the system differential equations, using the same scheme as analog simulation. That is, we must first give an expression that allows calculation, using FORTRAN plus a versatile menu of special CSMP statements, of the highest derivative of each unknown. Then using the INTGRL statement, which invokes a numerical integration algorithm, we integrate the highest derivative successively to generate all the lower derivatives and finally the unknown itself.

The SHUT1 and SHUT2 statements use the CSMP function switch operator called FCNSW to model the nonlinear valve orifice area variation as spool position XV goes from zero through XU, the underlap. Figure 2.37 defines the operation of FCNSW. The independent variable in CSMP is called TIME unless renamed by the user. Other variables may be assigned names consisting of one to six

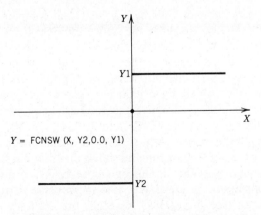

Y = FCNSW (X, Y2,0.0, Y1)

FIGURE 2.37 *CSMP function switch operator.*

alphanumeric characters, but the first character must be alphabetic. The sequence of most CSMP statements (for example, *all* those between DYNAM and TIMER) is not critical since a sorting algorithm puts them in proper order. Statements QCL= and QCR= implement Eq.'s 2.158 and 2.159 using FORTRAN plus FCNSW, whereas PCLDOT= gets the highest derivative of p_{cl}, from Eq. 2.160 (note that actuator displacement is named XP, velocity VP). The INTGRL statement (whose general form is INTGRL (initial value, quantity to be integrated)) then gets us p_{cl}, and a similar procedure is used for p_{cr}. Then ACCP= states the load acceleration from Eq. 2.162, except we have added a spring of stiffness KSP to the load (a PARAM card set KSP=0.0, so we do agree with Eq. 2.162). Two integrations then get us VP and XP. Finally XI= applies a step input of size 0.002 at TIME=0.003 (note XI is the X_V of Fig. 2.35). The general form of the CSMP step input is STEP (TS), which causes a step of size 1.0 to occur at time TS. Delaying the step by 3 ms in our problem allows us to clearly see the initial state of the system. This completes the simulation of the nonlinear model, and those who are unfamiliar with CSMP (or similar languages) should already begin to appreciate the ease with which we can study complex physical effects using such tools.

Proceeding now to the linearized model, we should first state that if we only wanted the response of $x_{C,p}$ to a step input of $x_{v,p}$ as given by Eq. 2.170, we could do this with a *single* CSMP statement, CMPXPL, which models second-order systems. Since we desire information also on pressures, flow rates, and the like, we use the more detailed approach shown, starting with the flow equations QCLL= and QCRL= based on Eq. 2.166. The remaining equations should be self-explanatory. Every CSMP program needs a TIMER card to give problem duration FINTIM, computing increment DELT (FINTIM divided by a few thousand, not critical) and graphing increment OUTDEL. Since the nonlinear and linearized models run simultaneously, they share the same TIMER card. To get graphical output (printer plots) of up to five named variables A, B, C, D, E plotted all on one graph against TIME, just write OUTPUT A, B, C, D, E. For higher quality electrostatic plotter graphs, our program uses the PAGE GROUP, XYPLOT statement, PAGE GROUP forcing both variables on each graph to be plotted to the same scale, giving a good visual comparison of nonlinear/ linearized models. Finally, we rerun the entire problem for a larger step input (0.020 in.) to see the effect on linearization accuracy, since linearized models generally get worse as we depart further from the operating point. Note that this requires only that, after the END card, we supply those parameter values (in this case AA, BB) which have changed. Unchanged ones need not be repeated, and another END card signals the end of the new run. This use of END cards and new parameter values (TIMER and OUTPUT can also be changed) can be applied as many times as we wish, to explore variations in the problem.

Figure 2.38*a* for the 0.002-in. command shows that motion, flow, and pressure are all more oscillatory in the nonlinear (more realistic) model than in the linearized approximation. Interestingly, just the reverse is true for the larger (0.02 in.) step input (see Fig. 2.38*b*). Overall, the accuracy of the linearized model seems quite acceptable for both sizes of input.

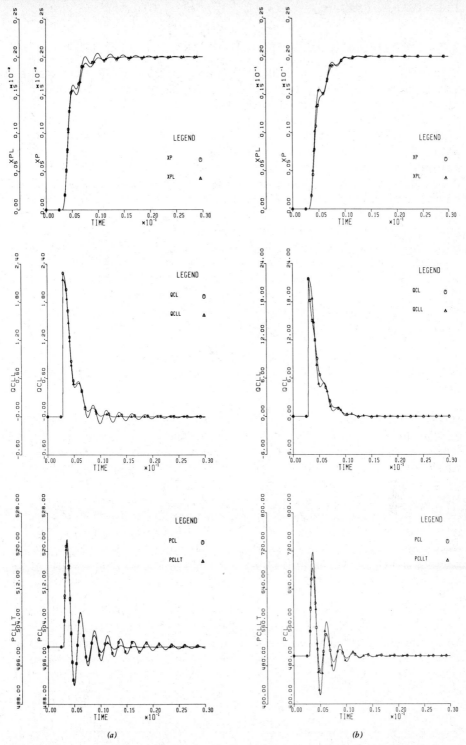

FIGURE 2.38 *Digital simulation results for nonlinear and linearized hydraulic servo (a) 0.002-in. input. (b) 0.02-in. input.*

2.7   DEAD-TIME EFFECTS

Section 2.2 introduced some mathematical aspects of dead-time elements; this section gives some practical applications in modeling devices that appear in control systems. Many materials such as metals, plastics, and rubber are manufactured in sheet form using rolling processes as in Fig. 2.39a. The control effector changes sheet thickness by adjusting the separation between the rolls; however the sheet thickness sensor must often (because of space limitations or environmental factors) be located some distance "downstream." This causes a dead-time between actual changes in thickness and their measurement. A similar effect occurs in the relation between temperatures T_{o1} and T_{i2} in the process vessels connected by a pipeline in Fig. 2.39b, and in the mass flow rates of granular material carried by the conveyor belt in Fig. 2.39c. Note that, depending on conditions, dynamic effects *in addition* to the dead-time effect may be needed for accurate modeling of some of these systems. For example, T_{i2} may exhibit a temperature loss due to heat transfer, in addition to a delay due to dead-time. Also, the numerical value of dead-time need not be fixed; the conveyor belt, for example, need not run at constant speed.

Figures 2.39d and 2.39e illustrate dead-time effects associated with wave propagation phenomena. The pneumatic signal transmission system (common in industrial process controls) suffers from a dead-time of about 1 s for every 300m of tubing, since the "speed of sound" (propagation velocity of pressure disturbances) is about 300 m/s. A volume at the receiving end may result in dynamics that are a combination of dead-time and first-order effects, as shown. When such pneumatic delays become excessive, electrical cable signal transmission (propagation velocity $\approx$ 186,000 miles/s) may be necessary. Even such "speed of light" transmission can cause nonnegligible delays in the remote control of a lunar robot vehicle directed from Earth (Fig. 2.39e), since the time between a steering command and the visual picture (relayed to Earth from a vehicle-mounted TV camera) of the results of that command is about 2.6 s. Considering dead-time effects is also useful in modeling the dynamics of vehicles such as the train car of Fig. 2.39f, where the track profile inputs at the rear wheels are similar to those at the front wheels, but are delayed. Finally, the growing application of digital computers in control systems requires the use of dead times to model the computational delay inherent in such systems (see Fig. 2.39g).

When dead-time elements are embedded in feedback loops, the simple analysis methods of Section 2.2 do not usually provide analytical solutions; in fact, exact solutions are often not available. We then turn to approximations, frequency-response methods, or computer simulation. The frequency response of a dead time can be found exactly from its Laplace transfer function (Eq. 2.53) as

$$\frac{Q_o}{Q_i}(i\omega) = e^{-\tau_{DT} i\omega} = \cos(\omega\tau_{DT}) - i\sin(\omega\tau_{DT}) = 1\underline{/-\omega\tau_{DT}} \quad (2.175)$$

and graphed as in Fig. 2.40. This result can also be obtained directly from $q_o(t) = q_i(t - \tau_{DT}) = q_{io}\sin[\omega(t - \tau_{DT})]$. Note that the phase lag increases

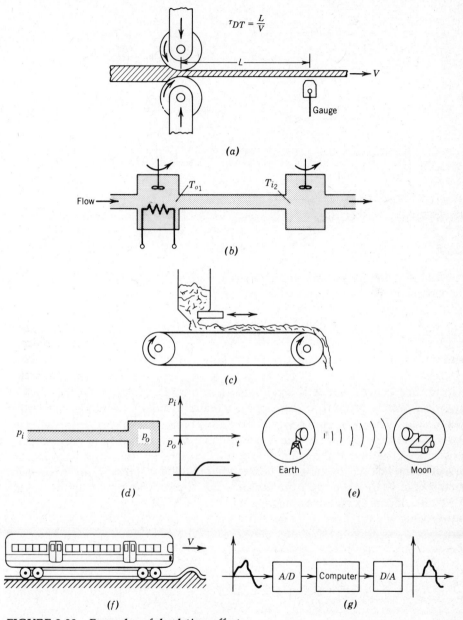

FIGURE 2.39 *Examples of dead-time effects.*

without bound, whereas first-order, second-order, and so forth systems always have an asymptote, such as $-90°$ or $-180°$. We will see later that this unbounded phase angle aggravates stability problems in feedback systems with dead times. Whereas frequency-response design methods for feedback systems treat dead times exactly, differential equation methods (including the Laplace transform) usually require that we approximate the dead time.

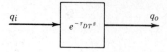

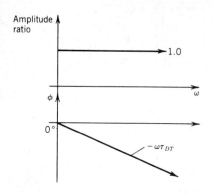

FIGURE 2.40 Dead-time frequency response.

The simplest dead-time approximation can be developed by the graphical approach of Fig. 2.41, from which we see

$$q_o(t) \approx q_i(t) - \tau_{DT}\frac{dq_i}{dt} \qquad Q_o(s) \approx (1 - \tau_{DT}s)Q_i(s) \qquad (2.176)$$

(This result can also be obtained by taking the first two terms of the Taylor series expansion of $e^{-\tau_{DT}s}$.) The accuracy of this approximation depends on the dead time being sufficiently small *relative* to the rate of change of the slope of q_i. If q_i were a ramp (*constant* slope), the approximation would be *perfect* for *any* value of τ_{DT}. When the slope of q_i varies rapidly, only small τ_{DT}'s will give a good approximation. A frequency-response viewpoint gives a more general accuracy criterion; if the amplitude ratio and phase of the approximation are "sufficiently close" to the exact curves of Fig. 2.40 for the range of frequencies present in q_i, then the approximation is valid. The Pade' approximants[25] provide a family of approximations of increasing accuracy (and complexity), the simplest two being

$$\frac{Q_o}{Q_i}(s) = \frac{2 - \tau_{DT}s}{2 + \tau_{DT}s} \qquad (2.177)$$

$$\frac{Q_o}{Q_i}(s) = \frac{2 - \tau_{DT}s + (\tau_{DT}s)^2/6}{2 + \tau_{DT}s + (\tau_{DT}s)^2/6} \qquad (2.178)$$

Analog computer simulation of dead times often implements one of the above approximations and thus gives no accuracy improvement over corresponding analytical calculations. In some cases, use of a tape recorder as an electromechanical dead-time simulator is accurate and practical. Here, the analog computer voltage for q_i is sent to the recorder write head whereas q_o is taken from the read head; head-to-head spacing and tape speed determining the dead time.

[25]J. G. Truxal, *Automatic Feedback Control System Synthesis*, McGraw–Hill, New York, 1955, pp. 548–551.

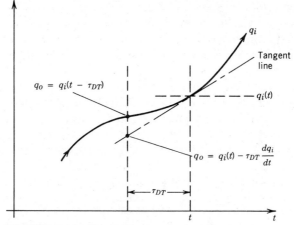

FIGURE 2.41 *Dead-time approximations.*

Digital simulation, which has largely replaced analog as a general simulation tool, has particular advantages in dead-time problems and would be the preferred method today. Here the simulation is essentially perfect since we merely ask the computer to "remember" what was happening τ_{DT} seconds ago at the input and recreate it now at the output, a task easily accomplished with digital memory. In the CSMP language introduced earlier, simulation of a fixed dead time requires only a single statement

$$Y = \mathrm{DELAY(N,DT,X)} \tag{2.179}$$

where

$X \triangleq$ name of dead-time input variable

$Y \triangleq$ name of dead-time output variable

$DT \triangleq$ numerical value of dead time, say 5.0 s

$N \triangleq$ integer number of computing increments (DELT's) in DT, say 50 if DELT=0.1 and DT=5.0.

Exact analytical methods usually are not available for *variable* dead times; however digital simulation provides an accurate and convenient approach. Although the actual algorithm is somewhat more complicated than for a fixed dead time, it is still invoked by a single simple statement, called PIPE[26] in the CSMP language, for example.

2.8 DISTRIBUTED-PARAMETER MODELS

At the macroscopic level, physical phenomena are most correctly described by distributed-parameter (partial differential equation) mathematical models. The lumped-parameter (ordinary differential equation) models of this chapter are

---

[26]Doebelin, *System Modeling and Response*, pp. 193–201.

really approximations to these more exact representations. These approximations are almost universally used in control system design because the mathematics is greatly simplified and the accuracy is generally adequate. Since distributed-parameter models are only occasionally encountered, we give only a brief overview.

When the tubing or hose connecting the servovalve to the actuator in hydraulic servosystems such as that of Fig. 2.35 is sufficiently short, we can neglect pipeline dynamics and assume the same pressure and flow rate at both ends of the pipe, as was done in that analysis. For longer runs of pipe such assumptions are not correct and one must take into account wave propagation effects, using partial differential equations. A configuration useful for demonstrating pipeline dynamics is that of Fig. 2.42, where a pipe of length L is terminated in a flow resistance R_f. Distributed-parameter modeling[27] (wave equation) leads to the transfer function

$$\frac{P(L,s)}{P(0,s)} = \frac{1}{\cosh(T_p s) + \dfrac{Z_0}{AR_f} \sinh(T_p s)} \tag{2.180}$$

where

$T_p \triangleq$ propagation time $\triangleq L\sqrt{K_c \rho_0}$ = time for pressure disturbance to travel distance L

$Z_0 \triangleq$ characteristic impedance $\triangleq \sqrt{\rho_0/K_c}$

$A \triangleq$ pipe flow area

$R_f \triangleq$ fluid resistance $\triangleq$ pressure drop/volume flow rate

$\rho_0 \triangleq$ fluid mass density

$K_c \triangleq$ line compliance $\triangleq 1/M_B + 2r_0/(Et_w)$

$M_B \triangleq$ fluid bulk modulus

$E \triangleq$ pipe material modulus of elasticity

$t_w \triangleq$ pipe wall thickness

$r_0 \triangleq$ mean radius of pipe (assumed thin walled)

If the model of Eq. 2.180 is included in a larger system model and inputs are applied, one finds that the solution for the system outputs is almost always frustrated because the transcendental functions (cosh, sinh, etc.) do not allow analytical inverse transformation. This is true not only for this example but for partial differential equation models in general. Also, simulation methods such as CSMP cannot be directly applied. These analytical roadblocks inhibit wide

---

[27]Ibid., Chap. 9.

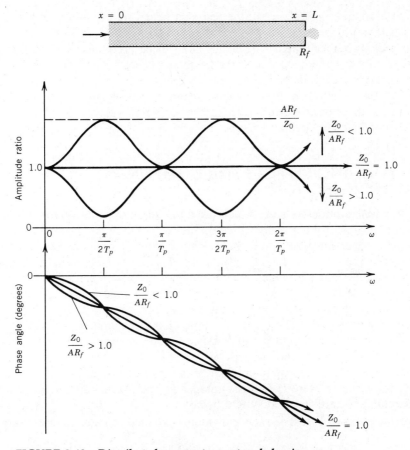

FIGURE 2.42 *Distributed-parameter system behavior.*

use of distributed-parameter models even though their accuracy is superior to lumped representations.

Although time-domain solutions are difficult or impossible, frequency-response methods provide effective analysis tools. Simple substitution of $i\omega$ for s gives for Eq. 2.180

$$\frac{P(L,i\omega)}{P(0,i\omega)} = M\underline{/\phi}$$

(2.181)

$$= \frac{1}{\sqrt{[\cos(\omega T_p)]^2 + [(Z_0/AR_f)\sin(\omega T_p)]^2}} \underline{/-\tan^{-1}[(Z_0/AR_f)\tan(\omega T_p)]}$$

which graphs as in Fig. 2.42. For $Z_0/AR_f = 1.0$ we have a pure dead-time response, thus pressure disturbances at $x = 0$ are reproduced exactly at $x = L$, but T_p seconds later. For $Z_0/AR_f > 1.0$, peaks and valleys occur, but the peaks have an amplitude ratio of 1.0. For $Z_0/AR_f < 1.0$, true resonances occur with an infinite number of natural frequencies given by $\omega = n\pi/2T_p$, $n = 1,3,5, \ldots$.

For a closed-end pipe ($R_f = \infty$), the resonant peaks are of infinite height, since there is then no damping.

To judge whether pipeline dynamics need to be accounted for in a specific application requires knowledge of the frequency spectrum of system excitation. If this spectrum extends only to a frequency ω_{co} and if $P(L,i\omega)/P(0,i\omega)$ is close to $1.0\underline{/0°}$ over this range, then pipeline dynamics can be ignored. Since in Fig. 2.42, T_p is directly proportional to L, the longer the pipeline, the lower the frequency at which pipeline dynamics become significant.

2.9 LOADING EFFECTS FOR COMPONENT INTERCONNECTION

Transfer function methods are widely used in system dynamic analysis for representing chains of interconnected components by multiplying together the individual transfer functions to get the overall transfer function of the chain. This widespread practice assumes, however, that *loading effects* at the component interfaces are negligible, which may or may not be true. It can be shown[28] that the (unloaded) transfer function is an *incomplete* component description and that to properly account for interconnection effects one must know *three* component characteristics. These are: the unloaded transfer function of the "upstream" component, the output impedance of the "upstream" component, and the input impedance of the "downstream" component. Only when the ratio of output impedance over input impedance is small compared to 1.0 (over the frequency range of interest) does the unloaded transfer function give an accurate description of interconnected system behavior.

For the tanks of Fig. 2.43a the individual unloaded transfer functions are

$$\frac{q_{o1}}{q_{i1}}(s) = \frac{1}{\dfrac{A_{T1}R_{f1}}{\gamma}s + 1} \qquad \frac{q_{o2}}{q_{i2}}(s) = \frac{1}{\dfrac{A_{T2}R_{f2}}{\gamma}s + 1} \qquad (2.182)$$

and when interconnected in the *nonloading* fashion of Fig. 2.43b, the simple product of these transfer functions gives the overall relation correctly as

$$\frac{q_{o2}}{q_{i1}}(s) = \frac{1}{\dfrac{A_{T1}A_{T2}R_{f1}R_{f2}}{\gamma^2}s^2 + \left(\dfrac{A_{T1}R_{f1} + A_{T2}R_{f2}}{\gamma}\right)s + 1} \qquad (2.183)$$

When connected in the manner of Fig. 2.43c, loading *is* present; Eq. 2.183 is *not* correct; and analysis of the complete system from basic principles gives

$$\frac{q_{o2}}{q_{i1}}(s) = \frac{1}{\dfrac{A_{T1}A_{T2}R_{f1}R_{f2}}{\gamma^2}s^2 + \left(\dfrac{A_{T1}R_{f1} + A_{T1}R_{f2} + A_{T2}R_{f2}}{\gamma}\right)s + 1} \qquad (2.184)$$

---

[28]Ibid., Chap. 7.

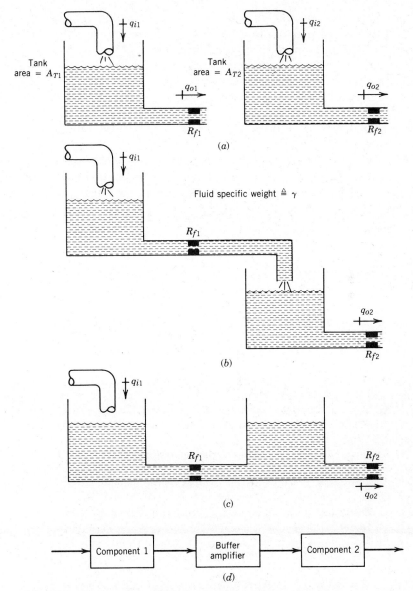

FIGURE 2.43 *Loading effects in component interconnection.*

Note that if $A_{T1}R_{f2}$ is sufficiently small (relative to $A_{T1}R_{f1}$ and $A_{T2}R_{f2}$), then Eq. 2.183 is a good approximation to Eq. 2.184 since the loading effect is small.

In general, loading effects occur because when analyzing an isolated component (one with no other component connected at its output), we assume no power is being withdrawn at this output location. When we later decide to attach another component to the output of the first, this second component *does* withdraw some power, violating our earlier assumption and thereby invalidating the analysis (transfer function) based on this assumption. One way to reduce loading effects to negligible proportions is to interpose *buffer amplifiers* between the components (see Fig. 2.43*d*). These devices, which may be electrical, pneumatic, hydraulic, or the like to suit the physical nature of the connected components, have their own internal power supplies and can thus supply power to a downstream component *without* withdrawing this power from the upstream component. That is, such amplifiers have a high input impedance and a low output impedance.

When we model chains of components by simple multiplication of their individual transfer functions, later in the text, the assumption will always be that loading effects are either not present, have been proven negligible, or have been made negligible by use of buffer amplifiers.

BIBLIOGRAPHY

Books

1. W. R. Ahrendt and C. J. Savant, Jr., *Servomechanism Practice*, McGraw–Hill, New York, 1960.

2. B. W. Andersen, *The Analysis and Design Of Pneumatic Systems*, Wiley, New York, 1967.

3. N. H. Beachley and H. L. Harrison, *Introduction to Dynamic Systems Analysis*, Harper and Row, New York, 1978.

4. R. H. Cannon, *Dynamics of Physical Systems*, McGraw–Hill, New York, 1967.

5. C. M. Close and D. K. Frederick, *Modeling and Analysis of Dynamic Systems*, Houghton Mifflin, Boston, 1978.

6. I. Cochin, *Analysis and Design of Dynamic Systems*, Harper and Row, New York, 1977.

7. D. M. Considine, *Process Instruments and Controls Handbook*, McGraw–Hill, New York, 1960.

8. E. O. Doebelin, *System Dynamics*, C. E. Merrill, Columbus, Ohio, 1972.

9. E. O. Doebelin, *System Modeling and Response*, Wiley, New York, 1980.

10. E. O. Doebelin, *Measurement Systems, 3rd Ed.*, McGraw–Hill, New York, 1983.

11. J. E. Gibson, and F. B. Tuteur, *Control System Components*, McGraw–Hill, New York, 1958.

12. M. F. Gardner and J. L. Barnes, *Transients in Linear Systems*, Wiley, New York, 1954.

13. C. M. Haberman, *Engineering Systems Analysis*, C. E. Merrill, Columbus, O., 1965.

14. D. C. Karnopp and R. C. Rosenberg, *System Dynamics*, Wiley, New York, 1975.

15. H. E. Merritt, *Hydraulic Control Systems*, Wiley, New York, 1967.

16. K. Ogata, *System Dynamics*, Prentice-Hall, Englewood Cliffs, N.J., 1978.

17. R. Oldenburger, *Mathematical Engineering Analysis*, Macmillan, New York, 1950.

18. J. B. Reswick and C. K. Taft, *Introduction to Dynamic Systems*, Prentice-Hall, Englewood Cliffs, N.J., 1967.

19. J. L. Shearer, A. T. Murphy and H. H. Richardson, *Introduction to System Dynamics*, Addison-Wesley, Reading, Mass., 1967.
20. J. G. Truxal, *Control Engineers Handbook*, McGraw–Hill, New York, 1960.
21. A. Tustin, *Direct Current Machines for Control Systems*, Macmillan, New York, 1952.

PROBLEMS

2.1 Using Fig. P2.1 as a guide, make up operational block diagrams for the systems of:

a. Fig. 1.2c.   b. Fig. 1.6.

c. Fig. 1.7.   d. Fig. 1.9.

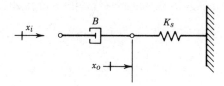

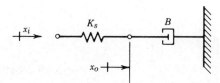

FIGURE P2.1.

2.2 Linearize the following expressions about a general operating point:

a. Heat radiation $Q = KT^4$ (K = constant).

b. Pendulum torque $T = WR \sin \theta$ (W, R are constants).

c. Orifice flow $Q = K\sqrt{p_1 - p_2}$ (K = constant).

d. Electrostatic force $F = K/r^2$ (K = contant).

2.3 A control system has the characteristic equation

$$D^3 + 2D^2 + D + K = 0$$

where K is the system loop gain. Using the SPEAKEASY (or other available) polynomial root finder, find the characteristic equation roots for K values from 0 to 3, going by increments of 0.1. What is the particular significance of the K value of 2.0?

2.4 Get the result of Eq. 2.58 by classical (rather than transform) methods, graph it and compare with Eq. 2.58.

2.5 Treating the nozzle flow as an orifice flow and using standard fluid mechanics results for orifice flows, get analytical expressions for K_{nx} and K_{np} in Eq. 2.105.

2.6 Calculate the step input responses of the systems of:

    a. Fig. 2.20*a*   **b.** Fig. 2.20*b*.

2.7 Derive transfer functions and compute step responses of the systems of Fig. P2.1. Also sketch frequency response curves.

2.8 Derive $(e_o/e_i)(s)$ for the system of Fig. P2.2.

2.9 Derive $(e_o/e_i)(s)$ for each of the systems of Fig. 2.22*a*.

2.10 For the system of Fig. 2.23, derive $(i_a/e_i)(s)$ and $(e_b/e_i)(s)$.

2.11 For the field-controlled dc motor of Fig. P2.3, derive $(\omega_o/e_i)(s)$ and $(\omega_o/T_i)(s)$.

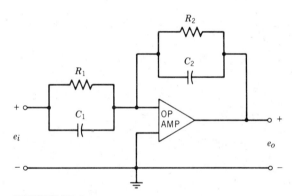

FIGURE P2.2.

2.12 Derive a first-order model for the float system of Fig. 2.27 by making suitable simplifying assumptions.

2.13 Derive Eq. 2.143 and choose a set of R's and C's that meet the restrictions stated there and give the numerical values of Eq. 2.144.

2.14 Plot frequency response curves for Eq. 2.144.

2.15 Derive $(\theta_L/T)(s)$ for the system of Fig. 2.29 with B_L and B_M present.

2.16 Derive the transfer function $(\theta_L/\theta_M)(s)$ for the system of Fig. 2.29 with B_L and B_M present.

2.17 Derive $(p_1/\phi_i)(s)$ and $(p_1/T_l)(s)$ for the system of Fig. 2.32.

2.18 Derive $(p_{cl,p}/x_{v,p})(s)$ and $(p_{cl,p}/f_{u,p})(s)$ for the system of Eq. 2.167 through 2.169.

2.19 Derive $(p_{cr,p}/x_{v,p})(s)$ and $(p_{cr,p}/f_{u,p})(s)$ for the system of Eq. 2.16 through 2.169.

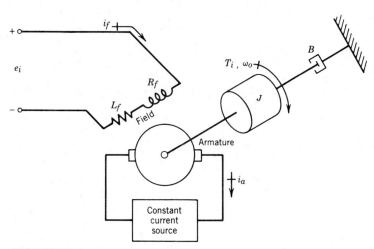

FIGURE P2.3.

2.20 In K_x and ω_n of Eq. 2.172 and 2.173, the term $B(C_p + 2K_{p1})$ is often negligible relative to $2A_p^2$. Also the volume V_{t0} may often be approximated as $A_pL/2$ if tubing volume is negligible relative to cylinder volume ($L \triangleq$ piston total stroke). Make these simplifications in K_x and ω_n and then discuss the trade off between steady-state gain and speed of response in terms of the design variable A_p.

2.21 Add a spring K_s between M and the base in Fig. 2.35. Derive $(x_{c,p}/x_{v,p})(s)$ and $(x_{c,p}/f_{u,p})(s)$ for the linearized model.

2.22 Repeat Problem 2.21 for $p_{cl,p}$ and $p_{cr,p}$ as outputs.

2.23 Using CSMP (or other available simulation language) write a program similar to that of the text, except that the valve is overlapped rather than underlapped; for the system of Fig. 2.35.

2.24 Compare the frequency response of dead-time approximations 2.177 and 2.178 with the exact curves. Also use CSMP (or other available simulation language) to compare the step and ramp responses.

2.25 Hydraulic fluid with $M_B = 100,000$ psi and $\rho_0 = 7.5 \times 10^{-5}$ lb$_f$ $-$ sec^2/in.^{4}

flows in ¾ in. diameter steel tubing of 0.040 in. wall thickness and 10 ft length. System excitation is known to be below 10 Hz. The terminating flow resistance varies widely and is not known. Estimate, using conservative calculations, whether line dynamics need to be considered.

2.26 Derive Eq. 2.182 and 2.184.

CHAPTER 3
Component Sizing for the Basic System

The subject of this chapter is not addressed by most control system texts since it involves practical considerations that are difficult to treat comprehensively outside the actual design environment of industrial practice. I recognize these difficulties, but because I feel avoidance of this area leaves a disturbing gap in discussions of the control system design process, a brief treatment will be presented.

3.1 ESTIMATION OF CRITICAL VALUES OF SYSTEM COMMAND AND DISTURBANCE INPUTS

When we speak of component sizing, our concern is mainly with the *power-handling* devices that comprise the control effectors and the controlled process itself. These must be properly sized to accommodate the maximum energy, force, and the like, demands of system operation. "Sizing" of the *information-handling* hardware found in reference-input, sensing, and control-director sections of the system is usually a relatively straightforward process of choosing suitable full-scale ranges, resolutions, and so forth. For example, if a servo-mechanism must position a rotary load over 180° of travel with 0.5° resolution, certain classes of rotary displacement sensors are applicable. Choice of a specific type and model involve practical considerations such as cost, weight, space, reliability, power supply, and environmental factors.

To rationally select the power-handling components, one must study the forms and magnitudes of the system command and disturbance inputs, since these will determine the power demands on these components. The specific details of commands and disturbances are often somewhat nebulous so one must usually deal with conservative estimates that allow sufficient safety factors for the uncertainties involved. A useful approach often is to define a *representative* command or disturbance of specific mathematical form, and base calculation on this assumed form, even though the actual input will never be exactly like this. If the representative input is judiciously chosen and safety factors properly applied, this is a practical approach. An example[1] of this approach is found in a servo-

---

[1]H. Chestnut and R. W. Mayer, *Servomechanisms and Regulating System Design,* Vol. II; Wiley, New York, 1955, p. 43.

mechanism whose task is to track a flying target. Of the infinite variety of target trajectories possible, we select as representative a target moving at constant velocity V along a straight-line path at a constant altitude Z_0 above a horizontal plane (see Fig. 3.1). We chose time $t = 0$ when the target is at its closest approach, with range R_0 and horizontal distance X_0. Two servosystems, azimuth and elevation, would be required to keep an antenna, gun, missile launcher, etc. pointed at the target. The desired values of the controlled variables of these systems would be, respectively, the azimuth angle A and elevation angle E; thus we are interested in describing mathematically the variation of these angles. Since the usual procedure is to consider the two axes of control separately in preliminary design, we focus on, say, the azimuth axis to get

$$A = \tan^{-1}\frac{Vt}{X_o} = \tan^{-1}at \qquad a \triangleq \frac{V}{X_o} \tag{3.1}$$

Although the elevation axis would probably experience a torque load related to elevation angle (due to gravity loads), there probably would be no azimuth torque related to azimuth angle. However frictional and inertial azimuth torques would be present and these depend on azimuth velocity and acceleration, so we need to calculate these.

$$\dot{A} = a\cos^2 A \qquad \ddot{A} = -a^2(\sin 2A)(\cos^2 A) \tag{3.2}$$

To estimate the maximum torque required of the azimuth drive motor we need estimates of the drive friction and inertia. We also need to decide on a "worst case" combination of the various types of load. The graphs of Fig. 3.2 show that the peaks of velocity and acceleration do not occur simultaneously; however, they do occur quite close in time, so it would not be unreasonable to

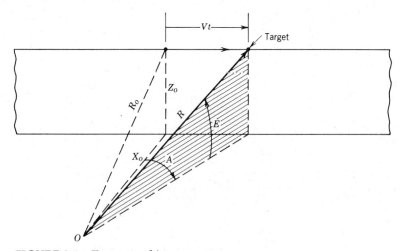

FIGURE 3.1   *Target tracking geometry.*

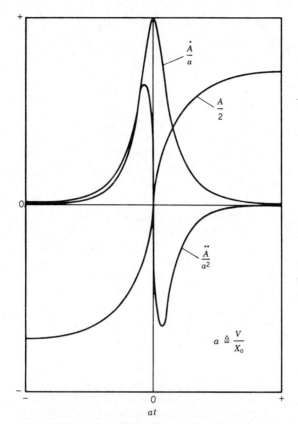

FIGURE 3.2 *Nondimensional azimuth angle, rate and acceleration for target tracking servo.*

treat them as simultaneous for a conservative design criterion. Note also that the graphs are nondimensionalized by dividing by "a" and a^2; so to get actual numerical values, we must decide on the maximum value of V/X_0 that we wish the system to be able to deal with. So far we have considered input commands only; disturbance torques must also be taken into account. For an antenna, wind gust torques are a possibility; whereas a missile launcher might experience disturbance torques due to rocket motor thrust at launch. If the timewise variation of all the torques during a typical "duty cycle" is available, a graph of total torque versus time can be used to obtain the maximum instantaneous torque needed for motor sizing. If such details are not available, the assumption that all maximum values occur simultaneously can be used as a conservative design criterion.

Total torque = dry friction torque + disturbance torque
+ (maximum velocity)(viscous friction coefficient)   (3.3)
+ (maximum acceleration)(system inertia)

Some other types of example may be helpful in getting a general appreciation of component sizing considerations. For a contouring machine tool (such as a lathe), typical parts to be machined may be defined. The shape of these parts and the possible machining speeds and feeds can be used to calculate the transverse position, velocity, and acceleration of the tool slide, analogous to the target motion analysis performed earlier. Disturbing forces such as the cutting force can be estimated from machining data. In sizing a furnace for a building temperature control system, weather records can provide data on the coldest winter days. This establishes a maximum ΔT across the building walls and allows estimation of the maximum heat loss that the furnace must supply. Industrial furnaces (heat treating, etc.) often chose the furnace rating at twice the heat loss when operating at design temperature. For on–off control this gives a duty cycle of 50% on and 50% off, resulting in a desirable symmetrical temperature/time graph. Of course if a system specification requires a very fast heat-up time from a cold start, this two-to-one overdesign may not be satisfactory and a larger heater may be needed.

3.2   GEARED SYSTEMS, OPTIMUM GEAR RATIOS, AND MOTOR SELECTION

Servomechanisms may be direct drive (motor coupled directly to load) or geared systems. For geared systems, choice of the motor also involves a choice of gear ratio. Numerous studies[2] of such questions are recorded in the literature. We first derive a useful result called "the equivalent dynamic system for a gear train." In Fig. 3.3 we show a single mesh of gears and assume no backlash or elasticity in either gear teeth or shafts. These assumptions are reasonable in many cases for high-quality, well-adjusted gearboxes that do not exhibit torsional vibration problems. Applying Newton's law to each shaft in turn we get

$$T_1 - B_1\omega_1 + T_g = J_1\dot{\omega}_1 \tag{3.4}$$

$$T_2 - B_2\omega_2 - nT_g = J_2\dot{\omega}_2 \tag{3.5}$$

where

$n \overset{\Delta}{=}$ gear ratio $= \omega_1/\omega_2$

$T_g \overset{\Delta}{=}$ torque exerted on gear one by gear two

Suppose our main interest is in the number two shaft (perhaps its angle is the controlled variable in a feedback system). We can reduce the two equations to

[2]W. R. Ahrendt and C. J. Savant, *Servomechanism Practice*, McGraw–Hill, New York, 1960, p. 30; J. E. Gibson and F. B. Tuteur, *Control System Components*, McGraw–Hill, New York, 1958, p. 320.

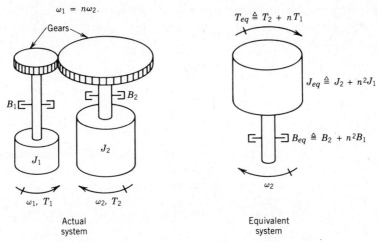

FIGURE 3.3 *Equivalent dynamic system for a gear train.*

a single equation in ω_2 and thereby define the equivalent dynamic system as follows.

$$(T_2 + nT_1) - (B_2 + n^2B_1)\omega_2 = (J_2 + n^2J_1)\dot{\omega}_2 \qquad (3.6)$$

$$T_{eq} - B_{eq}\omega_2 = J_{eq}\dot{\omega}_2 \qquad (3.7)$$

This is also called "*referring* the torque, damping, and inertia of shaft one to shaft two." We see that torques refer as the first power of the gear ratio; whereas damping and inertia refer as n^2. This result can be used for gear trains of any number of meshes by selecting the shaft of interest (can be at either end of the train, or interior) and repeatedly applying the rules just derived to finally end up with a single-shaft equivalent system. This is quicker and less subject to error than each time writing and solving the entire set of simultaneous equations; so we recommend this shortcut procedure whenever it is applicable. An important specific consequence of these results is that *motor inertia is often the dominant inertia in a servo system*. This can be seen by a numerical example where $J_{motor} = 1$, $J_{load} = 100$, and $n = 100$. Physically, the load appears 100 times "larger" than the motor, but because of the high (but not unusual) gear ratio, the motor's inertial *effect* is $1 \times 10,000$ or *100 times* larger than the load. Thus measures to "lighten" the load inertia (drill holes, etc.) are misplaced; we should really be striving for a lower inertia *motor*.

When frictional and other load effects are negligible and inertia is dominant, an optimum gear ratio that maximizes load shaft acceleration for a given input torque exists and may be found as follows. The equivalent system has the same acceleration as the load shaft and we can write

$$nT_1 = (J_2 + n^2J_1)\dot{\omega}_2 \qquad (3.8)$$

Since the torque effect on acceleration increases with n, whereas the inertial effect decreases acceleration as n^2, an optimum n should exist, which we find by standard calculus methods.

$$\frac{d\dot{\omega}_2}{dn} = \frac{(J_2 + n^2J_1)T_1 - (2nJ_1)nT_1}{(J_2 + n^2J_1)^2} = 0 \text{ for a maximum} \qquad (3.9)$$

$$n_{opt} = \sqrt{\frac{J_2}{J_1}} \qquad (3.10)$$

Thus if $J_2 = 25J_1$, a gear ratio of 5 will give the load shaft its greatest possible accleration for a given input torque. Note that this rule makes the referred inertia of shaft one just equal to the actual inertia of shaft two. A similar analysis[3] for a system with only viscous friction and inertia on both shafts shows that to obtain a specified load velocity ω_L and load acceleration $\dot{\omega}_L$ with minimum input torque requires a gear ratio

$$n_{opt} = \sqrt{\frac{B_2\omega_L + J_2\dot{\omega}_L}{B_1\omega_L + J_1\dot{\omega}_L}} \qquad (3.11)$$

When friction, load, and acceleration torques are all significant, gear ratio selection is less straightforward. Consider the linear speed/torque curve (typical of an armature-controlled dc motor with fixed field) in Fig. 3.4. Since motor

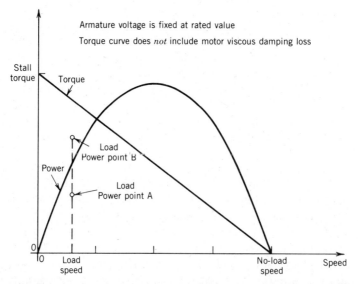

FIGURE 3.4 *Performance characteristics for armature-controlled dc motor.*

---

[3]J. G. Truxal, Ed., Control Engineers' Handbook, McGraw–Hill, New York, 1958, Sec. 13 p. 6.

power is the product of speed and torque, we get a parabolic speed/power curve whose maximum occurs at one half the no-load speed. One possible design criterion requires that we operate the motor at this maximum power point when it is supplying the greatest demand of the load. If we use Eq. 3.3 to compute the maximum power demand at maximum velocity, it is unlikely that this maximum speed will coincide with the peak power speed of the motor; thus a direct drive (gear ratio = 1.0) would not satisfy our requirement to operate at the peak power point. Fig. 3.4 shows two load power points at the same maximum load speed. If maximum load power corresponded to point *A*, a direct drive would be possible since the motor has more power than needed at this speed. The "excess" power simply means that more acceleration is available than is required. If instead our calculations had given point *B*, direct drive would not be possible since the load requires more power than the motor can supply at that speed. However, this load power is less than the *peak* motor power, so suitable gearing can reconcile the supply/demand mismatch. Since one finds both direct-drive and geared systems in practical use, it is clear that the one scheme is not always preferable to the other. Direct drive is favored because of reduced backlash and compliance and longer motor life due to lower speed; whereas the advantages of geared drives include smoother motor operation at higher speeds, possibly higher torsional natural frequency, and the lower cost of smaller motors. Of course, if a sufficiently large motor is simply not available, then a geared drive is a necessity.

As a numerical example, consider a drive in which maximum load is estimated as 30 N-m dry frictional torque, 40 N-m load torque, and load maximum speed and acceleration are 5 rad/s and 100 rad/s^2, respectively. Load viscous friction is 2 N-m/(rad/s) and load inertia is 0.2 N-m/(rad/s^2). Neglecting motor friction and inertia since the motor has not yet been chosen, we get

$$\text{Maximum power} = 5[30 + 40 + 5(2.0) + 100(0.2)] = 500 \text{ W} \quad (3.12)$$

Having decided on a certain family of dc motors on the basis of price, availability, and size, we scan the specification sheets and find the standard motor closest to our requirements has a peak power of 750 W at 100 rad/s, a stall torque of 15 N-m and a no-load speed of 200 rad/s (see Fig. 3.4). A quick check shows that direct drive is not possible with this motor since the required torque at 5 rad/s is 100 N-m, whereas the motor has available only 15 N-m at stall. Since a geared drive will be required, note that because the motor provides *no* torque beyond 200 rad/s, the maximum possible gear ratio will be less than 200/5 = 40. To see whether we can design for the peak power point (100 rad/s), let us check a gear ratio of 100/5 = 20. We now need the motor damping and inertia, which we get from the motor specification sheet as 0.01 N-m/(rad/s) and 0.0001 N-m/(rad/s^2). The required drive torque is

$$T_{\text{req'd}} = 30 + 40 + 5[(0.01 \times 400) + 2] + 100[(0.0001 \times 400) + 0.2] \quad (3.13)$$
$$= 124 \text{ N-m}$$

whereas the available motor torque is $7.5 \times 20 = 150$ N-m; so some extra torque is available and a gear ratio of 20 would work, as would a range of n's in the neighborhood of 20. The reader should check $n = 5$ and $n = 30$ to see that these ratios do *not* meet requirements.

Further details of gear-ratio selection, such as the inertia of the gears themselves (neglected in our analysis above), minimization of this inertia, and optimum selection of the number of stages in the gear train (e.g., 20 to 1 cannot be obtained in a single mesh for spur gears) will not be pursued here but are treated in the literature.[4] Thermal problems[5] related to adequate power dissipation in the motor and the associated power amplifier also need to be considered. This has some interaction with motor selection since a given motor is often available with several different windings (armature resistance), and a winding that most closely matches the power amplifier output stage must be selected.[6]

3.3 VALVE-CONTROLLED HYDRAULIC ACTUATORS; MAXIMUM POWER TRANSFER

In valve-controlled hydraulic servo systems, component sizing involves mainly the proper choice of servovalve and actuator. Choice of net piston area in translatory actuators and displacement (in.^{3}/rad) in rotary actuators is related to requirements of maximum force/torque and hydraulic natural frequency. Supply pressure is usually fixed somewhat arbitrarily by availability of standard lines of components rated at "conventional" pressures, such as 1000 psi for many industrial systems and 3000 psi for aircraft systems, though, of course, there are exceptions to these rules. Actuators sized to meet maximum load requirements may or may not be adequate in terms of hydraulic natural frequency (see Eq. 2.173). This will become apparent only during the servo-system phase of system design since the closed-loop system response speed depends critically on the open-loop component dynamics such as hydraulic natural frequency. If an actuator sized for maximum load does not allow a sufficiently high closed-loop speed of response, a larger actuator must be used; it will then be overdesigned in terms of maximum load, but this is generally acceptable.

The *maximum power transfer theorem* is sometimes used to assist in making sizing decisions. Consider the system of Fig. 3.5 with p_S fixed, $p_R = 0$, no leakage or mechanical friction, and a load force F which is allowed to move at different steady velocities V. If $V \equiv 0$ (piston clamped) and the valve is opened, $p_1 = p_S$, $p_2 = 0$, and we get the maximum force $F_{max} = A_p p_S$; but *no* fluid power is converted into useful mechanical power and $P_L = 0$. Conversely, if

[4]Ibid., Sect. 13-5 to 13-11.

[5]S. L. Tako, Understanding DC Servoamplifiers, Machine Design, May 10, May 24, June 7, 1979; H. Chestnut and R. W. Mayer, *Servomechanisms* pp. 89–118.

[6]S. L. Tako, Servoamplifiers; Direct Drive Servo Design Handbook, Inland Motor Corp., Radford, Va., 1964.

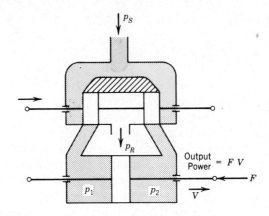

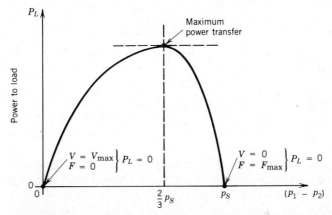

FIGURE 3.5 *Maximum power transfer in valve-controlled actuators.*

$F \equiv 0$, velocity V is at its maximum value but now $p_1 = p_2 = 0$, and again no useful output power results. Between these two extreme conditions of zero-power output there *must* be a point of maximum fluid-to-mechanical power conversion, which can be found by standard calculus maximization methods. The volume flow rate Q through both servovalve ports is equal and given by

$$Q = K\sqrt{p_s - p_1} = K\sqrt{p_2 - 0} = A_p V \qquad (3.14)$$

whereas

$$A_p(p_1 - p_2) = F \qquad (3.15)$$

The mechanical output power is given by

$$P_L = FV = [(p_1 - p_2)A_p]\left[\frac{K}{A_p}\sqrt{p_s - p_1}\right] \qquad (3.16)$$

Substituting $p_2 = p_S - p_1$ and setting $dP_L/dp_1 = 0$ leads to

$$p_1 = \frac{5}{6}p_S \qquad p_2 = \frac{1}{6}p_S \qquad p_1 - p_2 = \frac{2}{3}p_S \qquad (3.17)$$

We thus have the rule that to maximize power transfer, the load pressure drop should be two thirds the supply pressure.

Because of the unsteady nature of servo-system duty cycles, it is impossible to maintain $p_1 - p_2 = (2/3)p_S$ at all times, so the maximum power transfer theorem is often used only at the maximum load condition as an aid in selecting actuator size. (One of my graduate students has studied[7] a power supply of the pressure compensated variable-displacement pump type in which a load Δp sensor continuously adjusts p_S to maintain the two-thirds relation "most of the time", with significant energy savings.) If we have estimated the maximum load F_m from an equation similar to Eq. 3.3, then actuator area can be calculated as

$$A_p = \frac{3F_m}{2p_S}$$

Once actuator size is fixed, an estimate of maximum required actuator velocity allows sizing of the servovalve. It is recommended[8] that load pressure drop never exceed $(2/3)p_S$ so that valve flow gain, which depends on valve Δp, is not reduced below about 60% of the no-load value. That is, the servovalve is non-linear and we wish to keep changes in its linearized gain within the safety margins ("gain margin") usually provided in servo-system design. Having calculated the maximum actuator flow rate and the valve pressure drop corresponding to that condition, one can choose a valve size such that the wide-open valve will pass at least this much flow rate at that pressure drop. As a check, other known system operating conditions of flow rate and load pressure drop should be calculated to make sure they fall within the acceptable operating region of Fig. 3.6.

3.4   PROCESS-CONTROL-VALVE SIZING; TECHNICAL AND ECONOMIC CONSIDERATIONS

A very large percentage of the process systems used to control temperature, pressure, flow rate, liquid level, and the like use some sort of valve as the control effector (final control element), thus proper choice of these valves is an extremely common and important task which has received considerable attention

---

[7]A. Pery, A Theoretical and Experimental Study of Hydraulic Power Supplies Using Pressure-Compensated Pumps, Their Influence on Servosystem Dynamic Response and Their Utilization in Energy-Saving Configurations, Ph.D. Diss., Ohio State University, Mechanical Engineering Department, 1983.

[8]H. E. Merritt, *Hydraulic Control Systems,* Wiley, New York, 1967, p. 227.

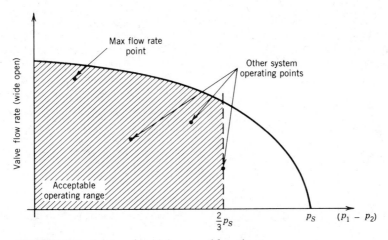

FIGURE 3.6   *Servovalve sizing considerations.*

in the literature.[9] Technical considerations include choice of characteristic to counteract process nonlinearities, prevention of cavitation, and acoustic noise problems. Since a large valve can easily cost several thousand dollars, and the energy loss due to valve pressure drop is a significant ongoing cost, economic factors must also be carefully considered.

The term *valve characteristic* refers to the graphical relation between the valve flow rate and the valve stem motion (translation or rotation) when valve pressure drop is kept constant. *Installed characteristic* refers to the relation between valve flow rate and valve stem motion for the complete system (valve and piping) when the overall system pressure drop is kept constant. Installed characteristic is the pertinent factor in determining the effect of a valve on overall control system behavior. If all components (including the process) in a feedback control system are essentially linear, then the installed characteristic of the valve should also be linear, since then system performance (stability, response speed, etc.) will be uniform over all system operating conditions. If other system components (usually the process) exhibit significant nonlinearities, then we can choose a valve characteristic that is the inverse of the process nonlinearity, thereby linearizing the overall system and gaining the benefits of uniform performance.

Methods for obtaining desired valve (and thereby installed) characteristics include the design of the valve itself, use of a characterizing valve positioner (often a small pneumatic/mechanical feedback system using properly shaped cams to achieve any desired characteristic), or in computer controlled systems, using a software approach that uses an "ordinary" valve but computes a proper valve input signal to obtain the desired characteristic. Simplicity and low cost often favor the use of valves specifically designed to give certain commonly

---

[9]*ISA Handbook of Control Valves,* 2nd Ed., ISA, Research Triangle Park, NC, 1976.

desired characteristics. The most common characteristics used are square-root, linear, equal-percentage (exponential), and hyperbolic. Of these, the linear and equal percentage are available as inherent valve characteristics from many manufacturers, for example[10] the popular rotary-type valves in ball (equal-percentage) and butterfly (linear) forms. Figure 3.7 illustrates some of these concepts. Proper choice of the type of characteristic requires detailed analysis[11] of the specific system behavior. For example,[12] for flow control between two sources of pressure, if an orifice-type flowmeter (nonlinear) is used for measurement,

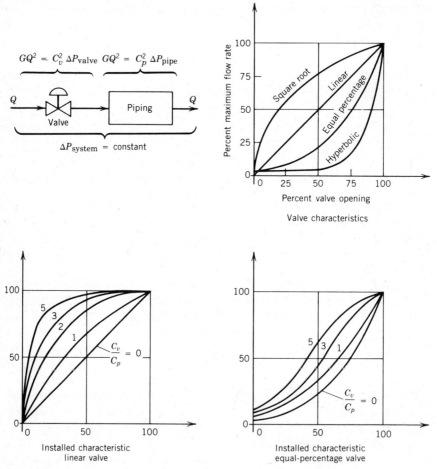

FIGURE 3.7   *Control valve characteristics and piping effects.*

[10]*Control Valve Tech. Bull. 284*, Jamesbury Corp., Worcester, Mass., 1976.
[11]J. E. Valstar, Selecting Control Valve Characteristics, *Cont. Eng.*, Mar. 1959, pp. 103–108; How Piping Influences Valve Performance, *Cont. Eng.*, Apr. 1959., pp. 123–125; P. S. Buckley, *Techniques of Process Control*, Wiley, New York, 1964, p. 152.
[12]Buckley, "Process Control," p. 152.

and if piping pressure drop is negligible compared to valve drop, then a square-root type of valve characteristic is indicated. If piping ΔP *is* significant, then a linear valve is better. For a linear flowmeter, negligible pipe ΔP requires a linear valve; whereas significant piping ΔP indicates an equal-percentage type.

In addition to the type of characteristic, it is also necessary to choose the actual size of the valve. Industry standards[13] and most manufacturers' handbooks specify size in terms of a factor called C_v, the valve sizing coefficient. From a fluid mechanics viewpoint the valve is treated as an orifice with various correction factors for deviations between theory and reality. The simplest situation is turbulent, noncavitating liquid flow in a valve whose diameter is the same as the pipe in which it is inserted (control valves are often less than line size). This situation corresponds to the one under which C_v was experimentally defined for the particular valve and the valve flow equation thus takes its simplest form

$$q = C_v \sqrt{\frac{\Delta P}{G}} \qquad (3.18)$$

where

$q \overset{\Delta}{=}$ volume flow rate, gal/min

$\Delta P \overset{\Delta}{=}$ pressure drop across valve, psi

$G \overset{\Delta}{=} \dfrac{\text{density of flowing fluid at flowing temperature}}{\text{density of water at } 60°F}$

Equation 3.18 is taken as a mathematical model and the constant C_v found by experimental measurements of q, ΔP, and G for each specific valve in the manufacturer's catalog, thus the catalog can tabulate the C_v values. Note that for turbulent, noncavitating liquid flow, the only significant fluid property is the density and that C_v values have an uncertainty of a few percent. That is, Eq. 3.18, like any math model of a real-world process, is not perfect; thus over a (turbulent) range of q values, C_v is not really perfectly constant. (To define the valve characteristic, C_v values can be measured at, say, every 10% of valve opening from 10 to 100% open and these values are tabulated in manufacturer's catalogs.) The usual situation in valve sizing is that q, ΔP, and G are known from basic process calculations and one can then easily compute the needed C_v from Eq. 3.18 and chose a suitable valve. Usually, minimum, "normal," and maximum flow conditions are specified and the chosen valve must be capable of handling all these while maintaining its particular characteristic (linear, equal-percentage, etc.). This is called *rangeability*.[14] A ball valve might typically have

---

[13]*Instrument Soc. of America*, Std. S39.1, Control Valve Sizing Equations for Incompressible Fluids; S39.2, Control Valve Capacity Test Procedure for Incompressible Fluids; S39.3, Control Valve Sizing Equations for Compressible Fluids; S39.4, Control Valve Capacity Test Procedures for Compressible Fluids.

[14]J. E. Valstar, Sizing Valves for Maximum Flow and Control Rangeability, *Cont. Eng.*, May 1959, pp. 120–122.

a rangeability of 50 to 1; that is; it follows the equal-percentage characteristic reasonably well from about 2% open to 100% open. Normal flow rate is generally known from steady-state processing conditions; whereas maximum flow rate depends mainly on the deviation from normal flow needed to counteract a disturbance.

When actual application conditions differ from the simplest example just presented, Eq. 3.18 requires modification; however a systematic approach[15] is still possible. Conditions that can be accomodated include valve size different from line size, choked flow, transitional flow, laminar flow, two-phase flow, compressible flow, vaporizing, cavitating, and flashing liquids, and dry or saturated steam flow. Driskell gives a complete discussion with numerical examples. Valve sizing is also sometimes affected by acoustic noise considerations since valves can become strong noise sources under some flow conditions. Methods for estimating acoustic noise generation at the valve sizing stage of design are available.[16]

Although control valves are usually sized on the technical factors that have just been briefly reviewed, consideration of *economics* in terms of first cost versus operating cost may provide additional useful insights (helpful in reaching final sizing decisions) which are increasingly significant as energy costs rise. Our discussion follows the analysis of Schmidt.[17] Basically, a small valve costs less initially but requires a higher ΔP to pass the required flow rate. The fluid power dissipated into heat is the product of valve ΔP and volume flow rate and represents an ongoing operating cost. A larger valve has less ΔP and thus less operating cost, but costs more initially. Conceivably, there exists an optimum valve size that minimizes the total cost; and we now show a procedure for choosing such a valve.

A standard formula[18] for the annual cost C_A of a piece of equipment is

$$C_A = (C_i - C_s)\left[\frac{I(1 + I)^m}{(1 + I)^m - 1}\right] + C_s I + C_o \qquad (3.19)$$

where

$C_i \triangleq$ initial cost (purchase price + installation cost)

$C_s \triangleq$ salvage value

$I \triangleq$ annual interest rate

$m \triangleq$ number of years useful life

$C_o \triangleq$ annual operating expense

[15]L. Driskell, Control Valve Sizing with ISA Formulas, *Instr. Tech.*, July, 1974, pp. 33–48.

[16]*Control Valve Tech. Bull. 284*, Jamesbury Corp., Worcester, Mass., 1976; *ISA Handbook of Control Valves*, 2nd Ed., Chap. 6, Part 3, ISA, Research Triangle Park, N.C., 1976.

[17]W. J. Schmidt, Control Valves: First Cost Versus Operating Cost, *Cont. Eng.*, May 1961, pp. 109–112.

[18]E. L. Grand and W. G. Ireson, "Principals of Engineering Economy," Ronald Press, New York, 1970, p. 74.

Our objective is to choose a valve that minimizes C_A. For our control valve example we will take $C_s = 0$, $m = 10$ yr, $I = 8\%$, and installation cost to be 15% of purchase price C_p. The bracketed expression in Eq. 3.19 is called the capital recovery factor R, 0.149 in our case. Equation 3.19 can then be written as

$$C_A = RC_p(1.15) + C_o \tag{3.20}$$

We now need to express both C_p and C_o in terms of valve "size" C_v. To find C_p pricing data were gathered from six valve manufacturers for valves of a chosen type in sizes from 1 to 12 in. pipe diameter. Plotting valve price versus valve C_v gives a reasonably smooth curve which can be curve-fitted with various forms of analytical functions. Schmidt found the following function to give a good fit to the empirical valve cost data.

$$C_p = 60.2\, C_v^{0.374} + 2.41\, C_v^{0.732} + 0.0340\, C_v^{1.492} \tag{3.21}$$

Operating cost C_o is the cost of power to pump the process fluid through the control valve only. We assume a combination of control valve and piping as in Fig. 3.7 to obtain an expression for C_o. For the normal flow rate Q_n, the valve is partially open and we have $Q_n^2 = (kC_v)^2 \dfrac{\Delta p_v}{G}$, $0 \le k \le 1.0$.

$$Q_n^2 \left[\frac{1}{(kC_v)^2} + \frac{1}{C_p^2} \right] = \frac{\Delta p_{vn} + \Delta p_{pn}}{G} = \frac{\Delta p_s}{G} \tag{3.22}$$

For the valve wide open, flow is Q_{max}, $k = 1.0$ and

$$Q_{max}^2 \left[\frac{1}{C_v^2} + \frac{1}{C_p^2} \right] = \frac{\Delta p_{v\,max} + \Delta p_{p\,max}}{G} = \frac{\Delta p_s}{G} \tag{3.23}$$

From Eq. 3.22 and 3.23 we can get

$$\Delta p_{vn} = \left(\frac{Q_{max}}{Q_n} \right)^2 \left[1 + \left(\frac{C_p}{C_v} \right)^2 \right] \Delta p_{pn} - \Delta p_{pn} \tag{3.24}$$

The fluid power expended in pumping fluid through the control valve is

$$Q_n\, \Delta p_{vn} = \frac{Q_{max}^2}{Q_n} \Delta p_{pn} - Q_n\Delta p_{pn} + \left(\frac{Q_{max}^2}{Q_n} C_p^2 \Delta p_{pn} \right) \frac{1}{C_v^2} \tag{3.25}$$

Since the units of Eq. 3.25 are those (gpm, psi) used in the conventional C_v definition, we convert to kilowatts and also introduce a pump efficiency E_p and motor efficiency E_m.

$$\text{motor kilowatts} = \frac{Q_n\Delta p_{vn}}{2298\, E_p E_n} \tag{3.26}$$

Typically $E_m = 0.90$ and $E_p = 0.75$. Assuming a 90% plant-utilization factor (7920 h/yr) and a power cost of \$0.01/kW h (note that Schmidt's prices are from 1959) we get

$$\text{valve yearly operating cost} = C_o = 0.0511\, Q_n\Delta p_{vn}. \tag{3.27}$$

We have related all parts of Eq. 3.20 to C_v and can now differentiate it with respect to C_v and carry out our minimization of C_A.

$$\frac{dC_A}{dC_v} = 0 = 0.171 \left[22.5C_v^{-0.626} + 1.765C_v^{-0.268} + 0.051C_v^{0.492} \right]$$

$$+ (0.0511) \left(-\frac{2}{C_v^3} \right) GQ_nQ_{max}^2$$

(3.28)

This can be put in the form

$$\frac{C_v^3}{0.598} \left[22.5C_v^{-0.626} + 1.765C_v^{-0.268} + 0.051C_v^{0.492} \right] = GQ_nQ_{max}^2 \qquad (3.29)$$

For a given example we would know G, Q_n, and Q_{max} and would want to find the C_v which satisfies Eq. 3.29. This is most easily accomplished by graphing the left side of Eq. 3.29 against C_v, as in Fig. 3.8. For a given value of GQ_nQ_{max} we can then immediately pick off the graph the valve (C_v value) which minimizes the annual cost. Schmidt carries through a numerical example for which the optimum valve found by this method has an initial cost of \$2820 (annual cost

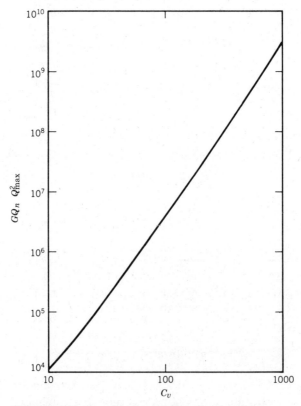

FIGURE 3.8   *Graphical solution of valve sizing equation.*

$2184) whereas a valve sized by "conventional" (noneconomic) methods cost $2090 initially but had a $2677 cost per year, indicating savings of $493/yr or $4930 over the 10-yr life of the valve.

BIBLIOGRAPHY

Journals

1. J. M. Jutila, Control Valves: An Industry at Full Throttle, *In. Tech.*, Nov. 1979, p. 7.

2. D. G. Wolter, Control Valve Selection: A Practical Guide, *Instr. Tech.*, Oct. 1977, pp. 55–63.

3. P. Wing, Plain Talk on Valve Rangeability, *Instr. Tech.*, Apr. 1978, pp. 53–57.

4. P. Wing, Determining and Using the Control Valve Pressure Recovery Factor, *In. Tech.*, Aug. 1979, pp. 55–59.

5. G. L. Roth, Factors in Selecting Valves for Compressible Flow, *Cont. Eng.*, Dec. 1955, pp. 46–53.

6. R. E. Gooch, The Energy Dimension in Control Valve Actuation, *Actuator Syst.*, Nov. 1980, pp. 26–32.

7. M. Dobrowolski, Guide to Selecting Rotary Control Valves, *In. Tech.*, Dec. 1981, pp. 35–39.

8. M. Adams, Control Valve Dynamics, *Instr. Tech.*, July 1977, pp. 51–56.

9. O. P. Lovett, Jr., Valve Sizing Equations for Computers, *ISA Jour.*, Vol. 6, No. 11, Nov. 1959, pp. 48–53.

10. H. M. Paynter, Characterizing Valves through Feedback, *Cont. Eng.*, Oct. 1954, pp. 47–49.

11. Q. V. Koecher, Characterized Valve Actuators, *Instr. Tech.*, Mar. 1977, pp. 47–52.

12. D. Stevens, Performance Features of Hydraulic Valve Operators, *Actuator Syst.*, Aug. 1981, pp. 22–26.

13. W. G. Holzbock, Factors in Selecting Valve Actuators for Electronic Control Systems, *ISA Jour.*, Sept. 1956.

14. C. D. Close, Valve Actuators Tie Precision to Power, *Cont. Eng.*, Sep. 1955, pp. 97–104.

15. W. H. Brand and E. F. Holben, Electric Actuators: Examination and Evaluation, *ISA Jour.*, Aug. 1957, pp. 326–331.

PROBLEMS

3.1 Derive Eq. 3.2.

3.2 Use the equivalent dynamic system concept to define single-shaft equivalent systems whose motion is identical with that of:

    a. Shaft 1.

    b. Shaft 3.

    c. Shaft 4.

in the geared system of Fig. P3.1.

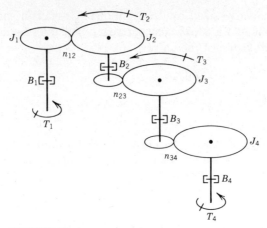

FIGURE P3.1.

3.3 Derive Eq. 3.11.

3.4 For the example of Eq. 3.13, define the complete range of acceptable gear ratios by plotting (on the same graph) curves of required torque and available torque against gear ratio.

3.5 For the example of Eq. 3.13, Eq. 3.11 does not exactly apply since torques other than viscous and inertial are present. Use Eq. 3.11 anyway to estimate a useable gear ratio and comment on the results.

3.6 For the system of Fig. 3.5, maximum power transfer is *not* the same as maximum efficiency. Derive a formula for efficiency and discuss this result in relation to maximum power transfer.

3.7 Develop a digital computer program for economic valve sizing based on Eq. 3.29. (Those having access to the SPEAKEASY engineering computer language will find this easy using the ZEROS (F:X) procedure.)

CHAPTER 4
Logic Controls

4.1 DEFINITION OF LOGIC CONTROL

A definition of *logic controls* that adequately describes most applications is that they are controls that work with one-bit binary signals. That is, the system needs only to know that a signal is absent or present, its *exact* size is not important. The one-bit limitation is intended to exclude the large and important field of digital computer control (covered in a later chapter) since, conventionally, computer control also uses binary signals, but usually with many bits. The type of application and the analysis methods are quite different for logic controls and digital computer controls, which is why we make the distinction. I should also point out, however, that the most important logic control device, the so-called *programmable logic controller* (PLC or PC) uses computers (microprocessors) to accomplish its functions, but the type of computer and how it is used are quite different from the conventional "digital computer control system." For most readers, these distinctions between "logic control" and "digital computer control" will become clear only when both areas are studied further.

4.2 OVERVIEW OF LOGIC-CONTROL FORMS AND APPLICATIONS

Logic control systems can involve both *combinational* and *sequential* aspects. Combinational effects are implemented by a proper interconnection of basic logic elements such as AND, OR, NOT, so as to produce a desired output or outputs when a certain combination of present inputs exists. Sequential effects use logic elements together with memory elements (flip-flops, counters, timers, etc.), to ensure that a chain of events occurs in some desired sequence. The present status of outputs depends both on the past and present status of inputs. Although the feedback of discrete events at the output can sometimes influence input action in logic systems, I prefer to treat logic controls separately from the main focus of the text, feedback control.

Manufacturing automation is a name often used for one of the largest areas of logic control application. For example, a cylinder block may be moving down a conveyor line. When it reaches a machining center, its presence is detected by a proximity sensor of some kind. This event causes the conveyor to momentarily stop while the block is pushed off the conveyor and onto a machine table by air cylinders. When the block is off the conveyor and on the machine table, the conveyor is restarted and other air cylinders come into action to clamp the block firmly in a desired position on the table. The table then indexes the block to the first machining position. When this position is reached (and sensed by limit switches) a boring tool is turned on and fed at rapid traverse to within 1 mm of cutting contact. Now the feed rate is reduced to that desired for the actual boring and the tool is gradually fed to the desired depth. When this depth is reached, the tool is backed away and retracted out of the hole. When the tool is clear, the table is indexed to the next position, where further machining operations are performed. This continues until all operations scheduled for this machine are complete, whereupon the cylinder block is unclamped. A robot with a suitably designed gripping hand now grasps the part and moving through a preprogrammed set of motions, places it in a gaging machine where several critical measurements are to be made.

We can see that the operation of such an automatic process may be thought of in terms of a sequence of interrelated logic statements, similar to those used in computer programming. For example

IF the proximity sensor reports a block, THEN stop the conveyor.

IF the conveyor is stopped, AND a block is present, THEN actuate the air cylinders.

Note also that, as mentioned earlier, the signals used are of the "either/or" variety, a signal is either present or absent; thus one binary bit of information is sufficient to describe the state of any individual signal. Industrial applications of logic control fall mainly into two types, those in which the control system is entirely (or almost entirely) based on logic principles and those that are mainly of a continuous feedback nature and use a "relatively small" amount of logic in auxiliary functions such as start-up/shut-down, safety interlocks and overrides, and mode switching. Programmable controllers, originally intended for "100%" logic systems, have in recent years added the capability of conventional feedback control, making them very popular since one controller can now handle in an integrated way *all* aspects of operation of a practical system that includes both types of control problems. General purpose digital computers could also handle such situations but they have not proven to be as popular as the industrially hardened (rugged and reliable) and easily programmable PLC's.

We have so far (and rightly) emphasized the role of PLC's in logic control. Originally cost-effective for only large-scale systems, small versions are now available that compete for tasks that could alternatively be handled with as few as 10 discrete logic devices or relays. However, some applications of small and medium scale are accomplished by approaches other than those using PLC's. If electrical control seems the most reasonable, the traditional approach is to use relay logic, and this electromechanical (relays having moving parts) type of design is still the best for some types of applications. Before the era of PLC's, relays were, in fact, the major type of system and this historical development explains why the most modern, microprocessor-based PLC's still are usually programmed according to relay ladder diagrams. This feature has been responsible for much of the widespread and rapid acceptance of PLC's, since personnel trained in relay systems were very comfortable with PLC programming because it was made "user-friendly" by setting it up to emulate a relay system. That is, the computer was forced to "learn" the already-familiar human language rather than making the humans learn a new and strange computer language.

If electrical control *without* moving parts is desired, all the necessary elements are available as individual solid-state electronic devices. A more convenient approach in many cases may be a *binary logic module*.[1] These are single circuit cards that offer a *selection* ("menu") of uncommitted logic elements that can be easily interconnected with each other and interfaced directly with the "outside world," such as an analog controller. A typical menu for such a device might be

four AND gates	three set–reset flip-flops
four OR gates	three one-shot multivibrators
four inverters (NOT)	two time delays
	one power-up timer

We will later see an example of the use of a single such module to implement a push button controlled manual shut-down and reset as well as an externally operated switch for automatic shut-down under alarm conditions. Larger applications could utilize several of these modules. The reader may think that a microprocessor might also become a suitable choice as system size increased; however, we are here considering "one-of-kind" applications where a control engineer wants to modify some plant hardware for a specific system or a few such systems. Here the expense and time associated with microprocessor system development could not be written off because of the small numbers of systems produced. If a large number of systems is involved, then a microprocessor becomes economically feasible but *we don't develop it ourselves,* we buy it ready-made as a PLC.

Leaving the realm of electricity completely, we now consider pneumatic and hydraulic hardware for implementing logic systems; both have been available for many years. Though they may not seem as glamorous as microprocessors, the continued profitable existence of companies specializing in this hardware gives evidence that they continue to find useful and economic application. Although the "no moving parts to wear or jam" philosophy and the blinding speed of electronics are important features of this technology, pneumatics/hydraulics also offers advantages which might be decisive in certain applications. In the cylinder-block machining process described earlier (and in many others) the speed of electronics is not really needed since the actuators (air or hydraulic cylinders, solenoids, etc.) *do* have moving parts and they limit the speed of the overall system *far* below the capabilities of the electronic devices. Thus, as long as pneumatic or hydraulic logic is faster than the actuators (and it generally is) logic speed is not a real limitation. The most popular pneumatic logic systems are designed to work at the same pressures (about 100 psi) as the actuators; thus the "brains" and the "muscles" of the system need no special interfacing

---

[1] J. Hampel, Using Digital Logic in Analog Control Systems, *In. Tech.*, Dec. 1979, pp. 39–41.

hardware. Electronic logic controls are very low power and need interfacing to high-power electrical actuators such as motors and solenoids. If an electronic logic system controls pneumatic/hydraulic actuators (often the case) then even more extensive interfacing is needed. Finally, certain specific advantages, such as a lack of spark hazards for combustible/explosive environments, may result in a choice of pneumatics over electronics. The control engineer, whether, by training, a mechanical or an electrical engineer, must thus keep an open mind when making such decisions, become familiar with *all* the alternatives, and only then make an informed choice based on all the factors rather than on prejudices.

4.3   THE BASIC LOGIC ELEMENTS

Although specialized functions are useful in certain situations, most logic control systems may be implemented with the three basic elements AND, OR, and NOT, or versions of these. An AND device may have any number of inputs and one output. To turn the output on, *all* the inputs must be on. This function is most easily visualized in terms of the switch arrangement of Fig. 4.1*a*. The *truth table* of Fig. 4.1*b* is another useful tool for this and other elements or systems. Various standardized symbols are available for drawing circuits; Fig. 4.1*c* shows conventional two-input ANDs of electronic and pneumatic type. When hardware is available only for two-input devices (such as the moving-part pneumatic logic we will shortly discuss), one can use the *cascaded* or *fan-in* arrangement of Fig. 4.1*d* to achieve AND functions with any number of inputs. The relay *ladder network* representation of Fig. 4.1*e* is important for two reasons. First, it is the scheme actually used for logic systems that use electromechanical relays as the hardware; and second, most electronic, microprocessor-based PLC's (which *do not* actually use relays) employ this type of display to program the machine. Finally, Fig. 4.1*f* shows the *Boolean algebra* expression for a three-input AND. Boolean algebra is sometimes used in analysis and design of logic systems and we will give a few more details shortly. The expression $A \cdot B \cdot C = D$ is read "A and B and C equals D."

Figure 4.2 gives similar details for the logical OR function and should be self-explanatory. The Boolean expression $A + B + C = D$ is read "A or B or C equals D." The NOT function is defined as having a single input and a single output. The output is "on" if the input is "off" and vice versa. Using electrical switches, the NOT function is obtained as in Fig. 4.3 simply by *defining* the open position of the switch as "on" and the closed position as "off." The Boolean expression $\overline{A} = B$ is read "not A equals B," and the overbar is used in general for applying a NOT function to any symbol or expression. Two other logic functions in common use that can be related to the three basic functions just presented are the NAND ("not and") and NOR ("not or"). For two inputs A and B, if C is to be a NAND output we would have $\overline{A \cdot B} = C$, whereas a NOR output would be $\overline{A + B} = C$. Figure 4.4 shows some symbols.

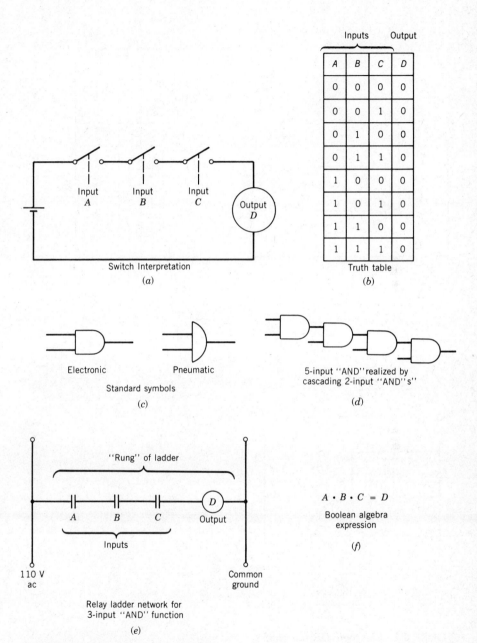

Inputs   Output

A	B	C	D
0	0	0	0
0	0	1	0
0	1	0	0
0	1	1	0
1	0	0	0
1	0	1	0
1	1	0	0
1	1	1	0

Switch Interpretation
(a)

Truth table
(b)

Electronic   Pneumatic
Standard symbols
(c)

5-input "AND" realized by
cascading 2-input "AND"s
(d)

"Rung" of ladder
A   B   C
Inputs
D
Output

110 V
ac
Common
ground

Relay ladder network for
3-input "AND" function
(e)

$A \cdot B \cdot C = D$

Boolean algebra
expression

(f)

FIGURE 4.1   *The logic AND function.*

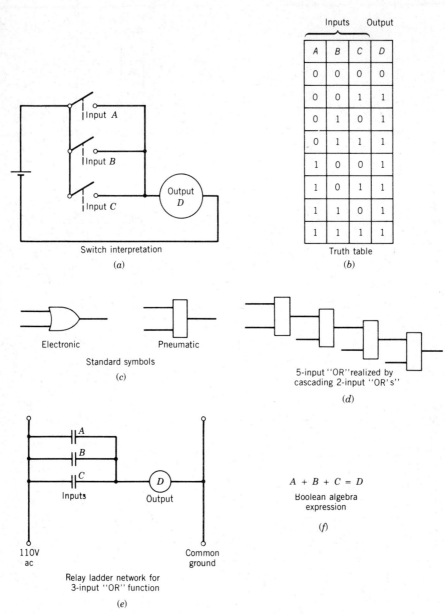

FIGURE 4.2   *The logic OR function.*

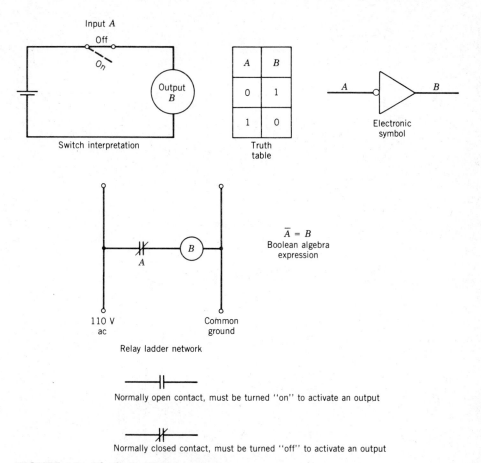

FIGURE 4.3 *The logic NOT function.*

When electromechanical ("moving parts") switches or relays are used to implement logic operations, the OFF condition is represented by a zero voltage and the ON condition is represented by the power supply voltage, perhaps 24 V dc or 110 V ac. For solid state electronic elements (no moving parts), the OFF condition might typically be the range 0 to 4.5 V, whereas ON would be any voltage between 7.5 and 12.0 V. Figures 4.5, 4.6, and 4.7 show the construction and explain the operation of one manufacturer's pneumatic logic elements of the OR, NOT, and AND types (other useful functions are also available). These devices are particularly convenient since the operating pressure is the same standard "shop air" range (60 to 150 psig) as is used for the power system components such as air cylinders, thus special interfacing devices are usually unnecessary. Although, of course, not approaching the switching speed of electronic logic, these pneumatic elements do respond in 5 to 10 ms, making them fast enough for many industrial applications.

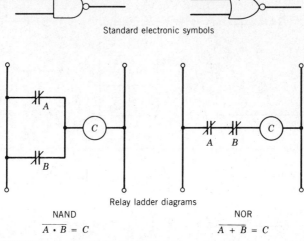

Standard electronic symbols

Relay ladder diagrams

NAND   NOR

$$\overline{A \cdot B} = C \qquad\qquad \overline{A + B} = C$$

FIGURE 4.4   *NAND and NOR logic functions.*

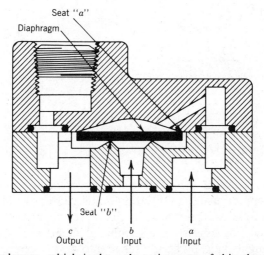

The flexible diaphragm, which is the only active part of this element, touches both seats if the Inputs are OFF. Pressure from either Input "a" or "b" will force the diaphragm from its respective seat, thereby pressurizing Output "c" (ON).

Flow is possible only from

Input "a" → Output "c"
Input "b" → Output "c"
Output "c" → Input "b", but only if Input "a" is OFF

FIGURE 4.5   *Pneumatic OR device.*
Source: ARO Corp., Bryan, Ohio.

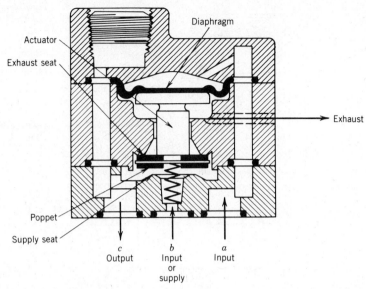

If Input "a" is OFF (discharged), supply air from port "b" can pressurize Output "c" (ON). Exhaust seat is closed.

If Input "a" is ON (pressurized), the diaphragm forces actuator and poppet downward, which opens Output "c" to Exhaust and closes supply seat. Therefore the Output is OFF (discharged).

Ports "a" and "b" may both serve as signal inputs, simultaneously. However, only "b" alone supplies flow to output "c".

This element operates with a snap action, provided a supply is present at port "b" prior to port "a" being pressurized. Element actuation occurs when pressure at port "a" approximates 65% of pressure of port "b".

FIGURE 4.6 *Pneumatic NOT device.*
Source: ARO Corp., Bryan, Ohio.

4.4 ELEMENTS OF BOOLEAN ALGEBRA

Many practical logic control systems are developed entirely by "common-sense" approaches without benefit of mathematical techniques; however a brief exposure to some elementary aspects of Boolean algebra may be useful. More details are available in the literature.[2] Our illustrations will use the standard pneumatic symbols since we wish to shortly give a pneumatic application example. With AND and OR expressions, parentheses, connectives, and factoring follow the same rules as for ordinary algebra, treating the dot (·) as the multiplication sign and the plus (+) as the addition sign. The number of parenthesis sets *(including a bracket over the whole expression)* is equal to the number of

---

[2]E. C. Fitch and J. B. Surjaatmadja, *Introduction to Fluid Logic*, McGraw–Hill, New York, 1978.

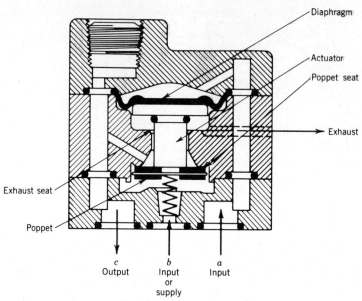

If both Inputs "a" and "b" are OFF (discharged), the Output "c" is connected to Exhaust (OFF).

If Input "b" alone is ON, the poppet is forced against poppet seat, thereby closing it, which leaves the Output connected to Exhaust (OFF).

If Input "a" alone is ON (pressurized), the diaphragm forces actuator and poppet downward, thereby closing exhaust seat. However, Output "c" is still OFF, because it is now connected to Input "b" which is OFF.

If Inputs "a" and "b" are ON (pressurized), the diaphragm forces the poppet down. Exhaust seat is held closed and poppet seat is open, thereby making connection Input "b" → Output "c". Therefore the output is ON.

Ports "a" and "b" may both serve as signal inputs, simultaneously. However, only "b" alone supplies flow to output "c".

This element operates with a snap action, provided a supply is present at port "b" prior to port "a" being pressurized. Element actuation occurs when pressure at port "a" approximates 65% of pressure of port "b".

FIGURE 4.7   *Pneumatic AND device.*
Source: ARO Corp., Bryan, Ohio.

logic functions. Figure 4.8 illustrates this rule and shows how factoring can be used to reduce the amount of hardware required to implement a given function.

Another useful rule (sometimes called DeMorgan's theorem) says that a NOT bar can be taken off a whole expression and placed over each term individually by changing the connectives from AND (·) to OR (+); see Fig. 4.9a. Furthermore, a NOT bar can be taken off a whole expression and placed over each term individually by changing the connectives from OR to AND; see Fig. 4.9b. Both manipulations are reversible (from right to left) and apply to any number

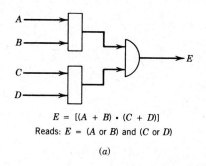

$E = [(A + B) \cdot (C + D)]$

Reads: $E = (A \text{ or } B) \text{ and } (C \text{ or } D)$

(a)

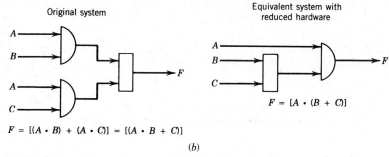

Original system

Equivalent system with reduced hardware

$F = [(A \cdot B) + (A \cdot C)] = [(A \cdot B + C)]$

$F = [A \cdot (B + C)]$

(b)

FIGURE 4.8 *Connectives, parentheses, and factoring.*

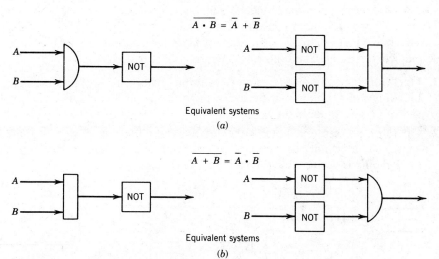

$\overline{A \cdot B} = \overline{A} + \overline{B}$

Equivalent systems

(a)

$\overline{A + B} = \overline{A} \cdot \overline{B}$

Equivalent systems

(b)

FIGURE 4.9 *DeMorgan's theorem.*

of terms. However the connectives to be manipulated in one step must be either all AND ($\cdot$) or all OR ($+$). For example

$$\overline{(A \cdot B)} + \overline{C} = \overline{(\overline{A} + \overline{B})} + \overline{C} = \overline{A} + \overline{B} + \overline{C} = \overline{A \cdot B \cdot C} \qquad (4.1)$$

If AND and OR functions are mixed in an expression, several manipulations may be needed to arrive at the simplest form (see Fig. 4.10).

Elimination of *redundancy* in a circuit reduces the needed number of logic elements and can often be facilitated with Boolean algebra. Some useful redundancy theorems are

$$(A + A) = A$$
$$(A \cdot A) = A$$
$$(A + A \cdot A) = A$$
$$\overline{\overline{A}} = A$$
$$\overline{\overline{A + B + C}} = A + B + C$$

Equal terms with connectives AND, OR in a Boolean expression are unnecessary repetitions. Two NOT's cancel each other for individual terms or whole expressions.

Figure 4.11 shows how redundancy elimination together with DeMorgan's theorem can reduce a system with four logic elements to one with none.

4.5   A SIMPLE EXAMPLE OF LOGIC CONTROL

Figure 4.12 shows an automatic machine for piercing holes in tubing. The machine is designed to handle two different lengths of tubing; the longer pieces are to have two holes pierced horizontally; whereas the shorter pieces have one hole pierced vertically. The following operating cycle is desired.

1. Operator manually loads one piece of tubing (short or long) into piercing block, pushing tubing in as far as it will go. Long pieces actuate a sensing switch G, short pieces do not.

2. Operator presses and holds either of two start buttons A and B. This causes an air cylinder to clamp the tubing firmly in vee-blocks. Occurrence of clamping is verified by a sensing switch C.

$$\overline{(A + B)} \cdot \overline{C} = \overline{A} \cdot \overline{B} \cdot \overline{C} = \overline{A + B} + \overline{C}$$

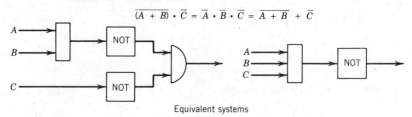

Equivalent systems

FIGURE 4.10   *Hardware simplification.*

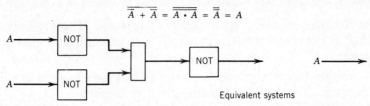

$$\overline{\overline{A} + \overline{A}} = \overline{\overline{A} \cdot \overline{A}} = \overline{\overline{A}} = A$$

Equivalent systems

FIGURE 4.11   *Simplification through redundancy and deMorgan's theorem.*

3. If clamping is verified and operator is still holding in A or B, the vertical piercing cylinder is actuated if the tube is a short one, the two horizontal cylinders are actuated if the tube is long.

4. Operator visually notes that piercing is completed and releases buttons A, B. Piercing cylinders retract. Completion of retraction is sensed by switches D, E, and F. When D, E, and F signal completion and A and B are also still released, the clamping cylinder is retracted; operator unloads workpiece and inserts a new one.

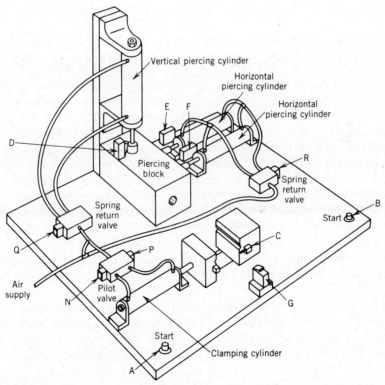

FIGURE 4.12   *Tube-piercing machine.*

Equipment available for actuating the various cylinders is as follows.

1. The clamping cylinder is controlled by a pilot valve. If port N is pressurized and P is not, the cylinder extends. If P is pressurized and N is not, it retracts.

2. The vertical piercing cylinder is controlled by a spring-return valve. If port Q is pressurized, cylinder extends; if not, it retracts.

3. The horizontal piercing cylinders are both controlled (in parallel) by a single spring-return valve. If R is pressurized, cylinders both extend; if not, they retract.

As a first step in system design, we use an OR device as in Fig. 4.13a to implement the logic equation $A + B = N$, which clamps the workpiece if either A or B are pressurized. Horizontal piercing is implemented with the equation $(A + B) \cdot C \cdot G = R$, as in Fig. 4.13$b$. Vertical piercing uses the equation $(A + B) \cdot C \cdot \overline{G} = Q$; Fig. 4.13$c$. The operator at this point visually notes the completion of piercing and releases A and B. Finally, to unclamp the workpiece (pressurize P) we use the equation $\overline{(A + B)} \cdot E \cdot F \cdot D = P$ to ensure that piercing cylinders have been retracted (Fig. 4.13d). The total system diagram (Fig. 4.13e) is slightly simpler that the sum of its parts since the signals (A + B) and (A + B) $\cdot$ C are used in several places.

4.6 CONVERTING A SIMPLE RELAY SYSTEM TO SOLID-STATE ELECTRONICS USING A BINARY LOGIC MODULE

In section 4.2 I mentioned binary logic modules; I now show a simple example of their use, following Hampel[3] closely. Figure 4.14 shows a simple interlock and shut-down circuit in relay-ladder-diagram form. The load is energized through contact 2CR1, which is operated by relay coil CR1. The load can be started if the reset button (normally open) is depressed, AND if the shut-down button (normally closed) is NOT depressed, AND if the auto-shut-down contact (normally closed) is closed. Under these conditions, coil CR1 is energized. This changes the state of load contact 2CR1 and lets it perform the desired control function. Energizing coil CR1 also closes contact 1CR1 so that the load remains operative when the (spring-loaded) reset button is released. (This is called a latching function). For shut-down, the manual push button may be depressed OR the auto-shut-down contact may be opened (by some emergency condition). Coil CR1 is then deenergized, interrupting power to the load and opening contact 1CR1.

Figure 4.15 is the first step in converting the relay system to the logic module implementation. The input devices are the same as for the relay system except that they are wired as contact closures to ground rather than as current carrying switches. The latching function of the relay network, which uses contact 1CR1

---

[3]J. Hampel, *Digital Logic*, pp. 39–41.

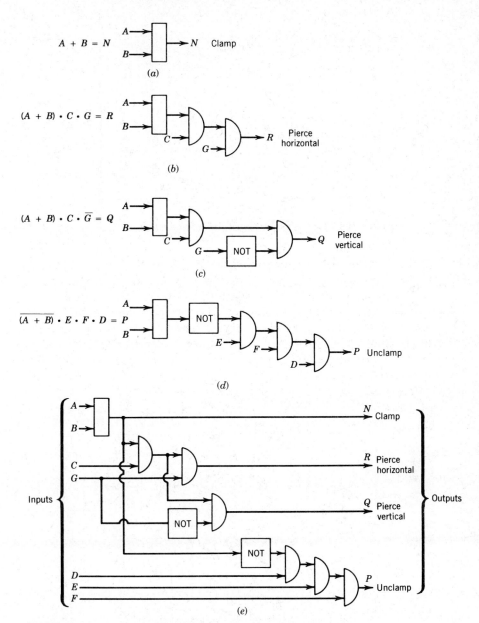

$A + B = N$ N Clamp

(a)

$(A + B) \cdot C \cdot G = R$ R Pierce horizontal

(b)

$(A + B) \cdot C \cdot \overline{G} = Q$ Q Pierce vertical

(c)

$\overline{(A + B)} \cdot E \cdot F \cdot D = P$ P Unclamp

(d)

Inputs   Outputs

(e)

FIGURE 4.13   *Logic diagram for tube piercing machine.*

to memorize the pushing of the reset button, is implemented in solid-state logic with a set–reset flip-flop. A logic 1 signal is momentarily developed when the reset button is depressed and grounds the set input of the flip-flop; this causes the flip-flop output to assume a logic 1 state. The manual and auto-shut-down inputs are normally grounded; so the output of each NOT element (and consequently of the OR gate) is logic 0. Should either the auto OR the manual

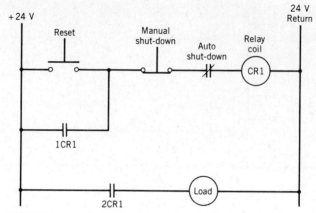

FIGURE 4.14 *Ladder network for relay system.*

shut-down contact operate and unground the associated line, the output of the OR gate becomes logic 1. This causes the flip-flop to change state, producing a logic 0 output and causing the solid state switch to disconnect the load.

In the relay system, a failure and subsequent restoration of power would cause the loop to be shut down until reset by the operator. This function is now achieved by using a power-up timer, which produces a momentary logic 1 pulse when energy is first applied. If the timer output is used as a reset input to the flip-flop, the module will be switched to logic 0 every time power is interrupted and then restored.

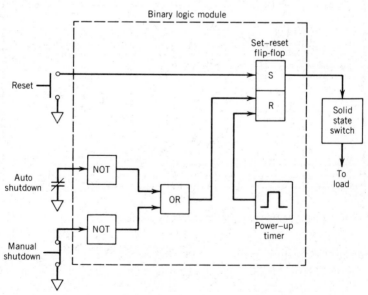

FIGURE 4.15 *Initial configuration of binary logic module.*

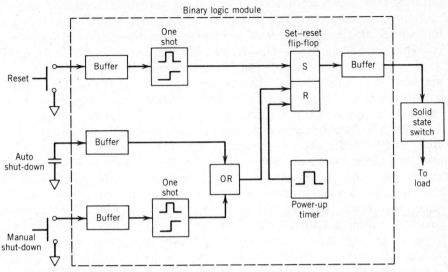

FIGURE 4.16 *Final configuration of binary logic module.*

The diagrams of Figs. 4.14 and 4.15 are functionally equivalent; however, Fig. 4.15 can be simplified to lower the component count. If the auto and manual shut-down contacts are changed from normally closed to normally open, both NOTs can be eliminated. The normal input to the OR through the open contact is logic 0. An auto or manual shut-down command applies a logic 1 to the OR input. Some further improvements are incorporated in the final version of Fig. 4.16. Input and output buffers minimize effects of transients and overvoltages. Also, problems of contact bounce and short contact closures are avoided using one-shot multivibrators. These condition the pushbutton signals into sharp, clean pulses for setting or resetting the flip-flop.

In this example, the entire function of the relay interlock and shut-down system is performed on one binary logic module; in fact three fourths of the circuits on the module are unused and available to implement other logic functions. It takes only 10 wire connections to "program" the general-purpose module for this specific application.

4.7 PROGRAMMABLE LOGIC CONTROLLER

As mentioned previously, the programmable logic controller (PLC or PC) is by far the most important commercially available logic control device. Its development began in 1968 when General Motors Corp. solicited the electronics industry for an alternative to the electromechanical relay logic systems used to control machining, materials handling, and assembly in automobile manufacture. Their relay systems had become very large and complex and required a great amount of rewiring and even outright scrapping when changes in manufacturing processes were needed. This became particularly critical during the yearly model changeovers typical of the auto industry. A number of manufacturers responded and the software-based programmable controller was born. It could duplicate all the logic control functions of relay networks, but implemented them in software (a program), just like a computer. This meant that process revisions or complete model changeovers did not now require scrapping or tedious rewiring of relay systems, but instead needed only relatively quick and easy changes in the PLC's program.

In 1969, PLC's were first actually used in the auto industry and by 1971 their utility in other industries was recognized and applied. By 1973 "smart" PLC's using arithmetic operations, printer control, matrix operations, CRT interface, and the like began to appear. The capability to control "continuous" processes involving such things as pressure, temperature, and flow feedback control loops using proportional, integral, and derivative control was added in about 1975, greatly increasing potential applications, since one machine could now be used to handle the logical and the continuous feedback aspects of system control in an integrated fashion. About 1976, the coordination of several individual PLC's in a factory into an integrated computer-managed overall manufacturing system began. To complement these advances into operations of larger and larger scale, very small PLC's (based on microprocessors) were now developed to economically replace systems of only 8 or 10 relays. Systems now range from those costing several hundred dollars and replacing 8 or 10 relays to $50,000 items replacing hundreds of relays and a minicomputer.

Although understanding the internal details of PLC's is neither necessary to use them nor appropriate for this text, Fig. 4.17 shows a general overview of the basic elements. All programmable controllers contain a central processing unit (CPU), memory (the established program), power supply, input/output (I/0) modules, and a programming device for programming, editing, and troubleshooting. The CPU, upon receiving instructions from the memory (together with feedback on the status of the input/output devices), generates commands to the outputs by means of the output module. These commands control the output elements on a machine or process. Devices such as relay coils, solenoid valves, indicator lamps, and motor starters are typical loads to be controlled. The machine or process input devices transmit status signals to the input modules, which in turn generate logic signals for use by the CPU. In this way the CPU monitors elements such as push buttons, selector switches, and relay contacts on a machine or process.

The programming device is often detachable and shared among several PLC's since it is needed only when a program is being developed or altered. Once an operating program is in memory, the programmer can be removed. Although the structure of Fig. 4.17 is strongly reminiscent of general purpose computers, recall that although the PLC is a computerlike device, it is industrially hardened for the factory-floor environment and the programming device speaks a high-level langauge (usually relay-ladder-network diagrams) making it easy to use by noncomputer specialists. These features have really been the key to its suc-

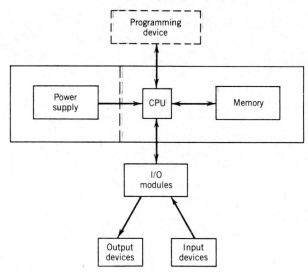

FIGURE 4.17 *Programmable controller architecture.*
Source: A look at Programmable Controllers, Gould, Inc.,
Modicon Div., Andover, Massachusetts.

cess. The larger PLC's are usually supplied with a CRT display for the programmer that presents a graphic picture whose appearance closely resembles a
traditional relay-ladder-network diagram. One can thus "draw" the control
diagram one element at a time and actually see the element connections being
established on the CRT display. A keyboard that has a key for each needed
logic operation is used to enter the ladder diagram into the CRT display, one
element at a time. When the CRT display shows the desired final diagram, the
PLC has been programmed and can be used to control the system. The PLC
provides a variety of input/output modules to which the external input and output
devices can be connected by ordinary hand wiring.

Although I do not intend to present comprehensive performance specifications, I will mention the concept of *scan time.* A PLC simulates the action of
relays in a serial rather than parallel fashion. In an actual relay network, many
relays can be reacting simultaneously, whereas a PLC must "scan" through the
entire network to determine the status of all elements before it can perform its
programmed function and reset all the outputs to their new updated values.
The time required for this is called scan time and might typically be 20 ms for
a 2K memory. This is a limitation on response speed that must be recognized.
If a 10-in.-long part is to be photoelectrically detected when moving on a 100
in./sec conveyor belt, we have at most 100 ms to "see" it and a 20-ms scan time
is adequate. A 1-in.-long part would be in view only 10 ms and could be missed
by a 20-ms scan time. When PLC's include capability for analog process control
loops, a similar scanning limitation will exist. Typically, a PLC might take 500

ms to scan eight analog control loops; so each loop might be sampled only twice per second. If the process variables are fast responding, such sampling delays are intolerable. These limitations are being removed by employing several microprocessors, each optimized for its own control task, within a single PLC. A recent development[4] of this sort allows PLC logic control to be integrated with high-speed, servo-motion controls, which require even higher response speed than process control loops. The motion controller combines a typical motor, tachometer, and encoder servo system with a high-speed microprocessor that is in turn interfaced with the PLC, as in Fig. 4.18. Although the PLC is limited by its rather slow scantime, the motion control processor can implement complete displacement, velocity, and acceleration commands at high speed.

Although it is not appropriate here to teach the reader the programming details for any particular manufacturer's PLC, I do wish to show a fairly comprehensive example ladder diagram and program listing for real-world application to give the reader some appreciation for their nature. We use a case study[5] of a woodworking machine used in a furniture factory.[6] In this double-end trim and bore machine, a wood piece is manually inserted and the operator depresses a cycle start switch. The piece is clamped in place and both ends are

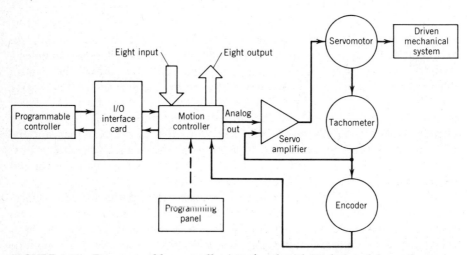

FIGURE 4.18   *Programmable controller interfaced with High-Speed Servo System.*

---

[4]Servo Controller Adds Speed, Accuracy to Programmable Controller, *Comp. Design,* Dec. 1981, p. 32; B. R. Beadle, Almost CNC-Controlling Machine Motion with Smart PC's, *Prod. Eng.,* Jul. 1982, pp. 47–50; R. B. Noha, Servo-Oriented PC Controls Flying Cutoff Saw, *Hy. and Pneu.,* Nov. 1983, pp. 103–106.
[5]ECD 2100 A, Texas Instruments Inc., Industrial Controls, Johnson City, Tennessee.
[6]Colonial Wood Products, Elizabethton, Tennessee.

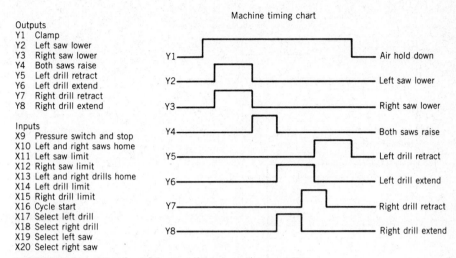

Outputs
Y1 Clamp
Y2 Left saw lower
Y3 Right saw lower
Y4 Both saws raise
Y5 Left drill retract
Y6 Left drill extend
Y7 Right drill retract
Y8 Right drill extend

Inputs
X9 Pressure switch and stop
X10 Left and right saws home
X11 Left saw limit
X12 Right saw limit
X13 Left and right drills home
X14 Left drill limit
X15 Right drill limit
X16 Cycle start
X17 Select left drill
X18 Select right drill
X19 Select left saw
X20 Select right saw

Machine timing chart

Y1 — Air hold down
Y2 — Left saw lower
Y3 — Right saw lower
Y4 — Both saws raise
Y5 — Left drill retract
Y6 — Left drill extend
Y7 — Right drill retract
Y8 — Right drill extend

FIGURE 4.19 *Woodworking machine input/output list.*

trimmed by lowering two circular saws. After the saws return to their upper position, two drills bore both ends to a preset depth. When the drills retract, the wood piece is released so it can be removed. It is also possible to select right, left, or both saws and right, left, or both drills. Sequencing provides that saws and drills are not actuated simultaneously. A self-check feature shuts down operations should input switches fail in a closed position.

Figure 4.19 lists inputs and outputs and gives a machine timing chart (sequencing is handled by a drum timer feature in the PLC which simulates the behavior of a rotating mechanical drum timer but is a keyboard programmable software device just like the logic portions of the PLC). A complete relay ladder diagram with some explanatory notes is given in Fig. 4.20. This was used as a work sheet to actually program the PLC. Figure 4.21 shows the actual program used to implement the functions of Fig. 4.20. Even though we have not explained the programming language, comparison of Figs. 4.20 and 4.21 together with noting some obvious logiclike program statements (STR Y4, AND YS, AND Y7, etc.) should convince the reader that the act of programming the machine while looking at the ladder diagram is probably not too difficult.

As a concluding comment we should note that PLC's have not totally replaced relays; some systems are still best done entirely as relays; whereas many PLC systems use relays in important adjunct roles. The following table[7] summarizes the pertinent selection criteria.

---

[7]C. C. Jones and L. T. Thompson, The Role of Relays in Machine Control, *Prod. Eng.,* Mar. 1982, pp. 64–66.

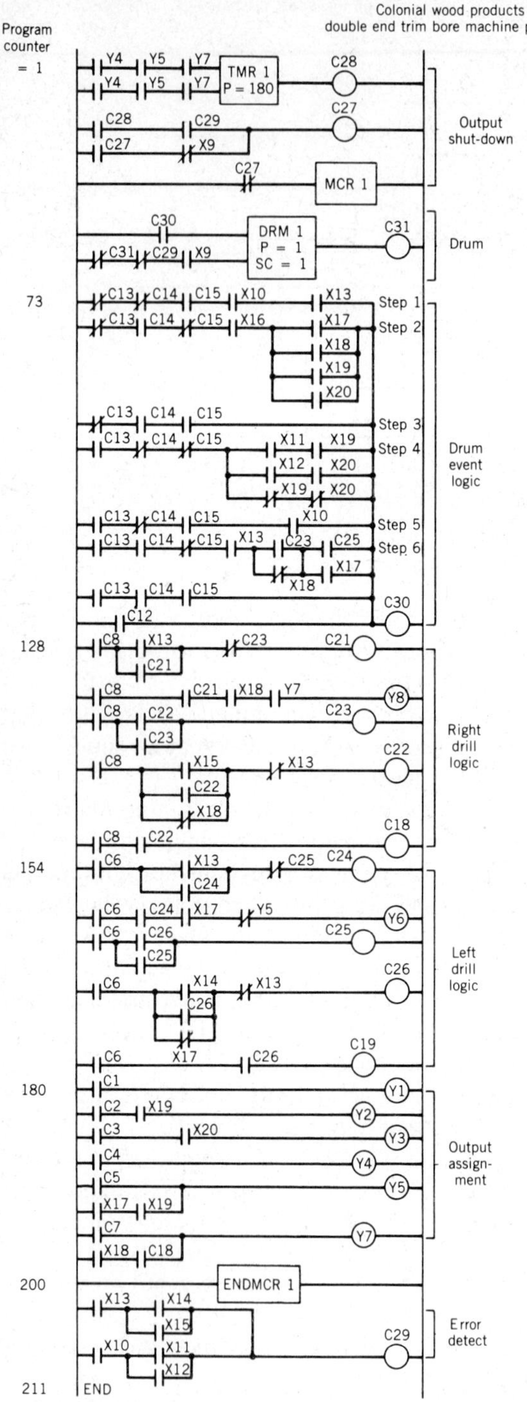

Timer 1 is activated when outputs Y4, Y5, and Y7 are all on simultaneously as in the error condition. If all stay on for 3s the timer activates C28. If C29 (the error detect relay) or X9 (the pressure switch) is closed when C28 energizes, then C27 latches on. This energizes the MCR to disable all the machine outputs.

The top line into the drum contains C30 contact which allows the use of the event logic to step the drum. The bottom line contains C31 contact which allows automatic reset of the drum upon error detection, and X9 contact which allows reset of the drum in the event of pressure loss or stop signal presence.

The drum event logic section contains contacts from the drum and input devices to step the drum upon completion of certain events. Each step in the drum is coded with the step number in the outputs C12, C13, C14, and C15 (see drum timer programming form). These contacts are used to allow only one of the parallel lines in the event logic section to be completed at a time. When the drum is on Step 1 C12 = OFF, C13 = OFF, C14 = ON, and X10 and X13 are closed. Thus for the drum to progress to Step 2 both drills must be at home base (X13) and both saws must be at home base (X10). Similar events must be completed to progress through the remaining steps.

Since each drill must operate independently but simultaneously, additional logic must be added to sequence these drills. If the right drill is not called for by the operator (X18 is not closed) then the logic prevents the right drill from being extended. If it is called for and the drum is on Step 6 (C8 is closed), then the drill extend output Y8 is energized. When the drill limit is reached (X15 closed) C22 latches on, which turns on C18, which turns on Y7 (see output assignment section). When the drill returns to home base both the drill extend Y8 and drill return Y7 outputs are disabled.

Identical logic exists for the left drill sequence. When both drills have completed their sequence, the drum steps to Step 16 and resets to Step 1.

The output section takes the outputs of the drum C1 to C8 and combines them with other contacts to activate the PC outputs. For example, the left saw output (Y2) will be energized when C2 is energized only if the left saw is called for (X19 = ON).

The error detection section checks for limit switch malfuctions if X13 is closed, indicating that both drums are at home base, and if either X14 or X15 is closed, indicating that a drill is at its outer limit, then any error is flagged on C29.

Other errors could be selected and incorporated into this circuit.

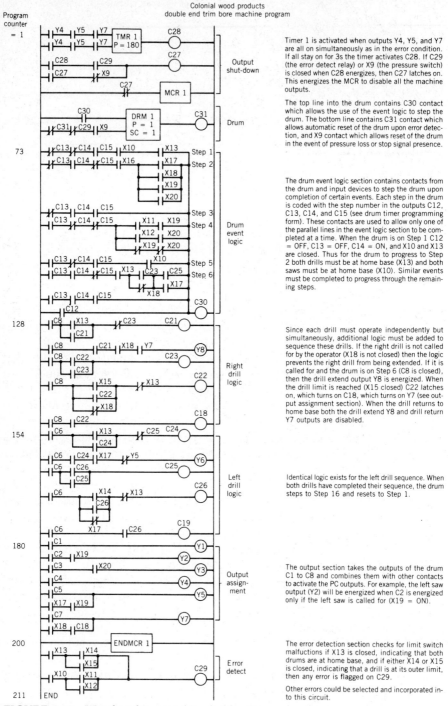

FIGURE 4.20 ***Woodworking machine ladder diagram.***

Relays Alone	PC's Alone	Both
a. Loads above 3 amp and little need for logic	a. Light loads and complex logic	a. Redundant safety
b. Applications with only one or two relays	b. Systems requiring PC functions not found in relays; e.g., data handling	b. Heavy loads and complex logic
c. Systems requiring many load relays where addition of PC is uneconomical	c. Systems with extensive logic needs where redesign can reduce loads to PC rating (2 to 3 amps)	c. Applications requiring 100% shut-off (*must* use relays)
d. Systems with infrequent cycling and relatively simple logic	d. Systems suitable for relays or PC where space is limited	d. System where line loss weakens signal through PC output
e Systems suitable for relays or PC where space is no problem		

Ladder Diagram Program Code

Shown here is the ladder diagram code of the Colonial Wood Products program which can be easily entered via TI's Model 510 hand-held programmer.

```
STR  Y4      AND X13      AND  C15     AND STR      AND N C25    STR  C3
AND  Y5      STR N C13    AND  X10     AND N C23    OUT  C24     AND  X20
AND  Y7      AND C14      STR  C13     OUT  C21     STR  C6      OUT  Y3
STR  Y4      AND N C15    AND  C14     STR  C8      AND  C24     STR  C4
AND  Y5      AND X16      AND N C15    AND  C21     AND  X17     OUT  Y4
AND  Y7      STR X17      AND  X13     AND  X18     AND N Y5     STR  C5
TMR  1       OR  X18      STR  C23     AND N Y7     OUT  Y6      STR  X17
OUT  C28     OR  X19      OR N X18     OUT  Y8      STR  C6      AND  C19
STR  C28     OR  X20      STR  C25     STR  C8      STR  C26     OR  STR
OR   C27     AND STR      OR N X17     STR  C22     OR   C25     OUT  Y5
STR  C29     STR N C13    AND STR      OR   C23     AND STR      STR  C7
OR N X9      AND C14      AND STR      AND STR      OUT  C25     STR  X18
AND  STR     AND C15      STR  C13     OUT  C23     STR  C6      AND  C18
OUT  C27     STR C13      AND  C14     STR  C8      STR  X14     OR  STR
OUT  C27     AND N C14    AND  C15     STR  X15     OR   C26     OUT  Y7
STR N C27    AND N C15    STR  C12     OR   C22     OR N X17     END  M1
MCR  1       STR X11      OR  STR      OR N X18     AND STR      STR  X13
STR  C30     AND X19      OR  STR      AND STR      AND N X13    STR  X14
STR N C31    STR X12      OR  STR      AND N X13    OUT  C26     OR  X15
AND N C29    AND X20      OR  STR      OUT  C22     STR  C6      AND  STR
AND  X9      STR N X19    OR  STR      STR  C8      AND  C26     STR  X10
DRM  1       AND N X20    OR  STR      AND  C22     OUT  C19     STR  X11
OUT  C31     OR  STR      OR  STR      OUT  C18     STR  C1      OR  X12
STR N C13    OR  STR      OUT C30      STR  C6      OUT  Y1      AND  STR
AND N C14    AND STR      STR  C8      STR  X13     STR  C2      OR  STR
AND  C15     STR C13      STR  X13     OR   C24     AND  X19     OUT  C29
AND  X10     AND N C14    OR  C21      AND STR      OUT  Y2      END
```

FIGURE 4.21 *Woodworking machine program listing.*

BIBLIOGRAPHY

Books

1. E. C. Fitch and J. B. Surjaatmadja, *Introduction to Fluid Logic,* McGraw–Hill, New York, 1978.
2. E. L. Holbrook, The *How* of Pneumatic Logic Control, Clippard Inst. Lab., Cincinnati, Oh.

Journals

1. W. G. Holzbock, Designing with Moving Parts Logic, *Hydraulics and Pneumatics,* Sep. 1979, pp. 100–104.

Drum Timer Programming Form

The drum timer programming form is used to facilitate entry of the drum elements for Colonial's program

		C	C	C	C	C	C	C	C	C	C	C	C	C	C	C
Drum 01 PRESET = 01		0	0	0	0	0	0	0	0	0	1	1	1	1	1	1
SCN/CNT = 01		1	2	3	4	5	6	7	8	9	0	1	2	3	4	5
STEP	CNT/ STP*															
1	00001	1	0	0	1	1	0	1	0	0	0	0	0	0	0	1
2	00001	0	0	0	0	0	0	0	0	0	0	0	0	0	1	0
3	00060	1	0	0	0	0	0	0	0	0	0	0	0	0	1	1
4	00001	1	1	1	0	0	0	0	0	0	0	0	0	1	0	0
5	00001	1	0	0	1	0	0	0	0	0	0	0	0	1	0	1
6	00001	1	0	0	0	0	1	0	1	0	0	0	0	1	1	0
7	00000	0	0	0	0	0	0	0	0	0	0	0	0	1	1	1
8	00000	0	0	0	0	0	0	0	0	0	0	0	1	0	0	0
9	00000	0	0	0	0	0	0	0	0	0	0	0	1	0	0	1
10	00000	0	0	0	0	0	0	0	0	0	0	0	1	0	1	0
11	00000	0	0	0	0	0	0	0	0	0	0	0	1	0	1	1
12	00000	0	0	0	0	0	0	0	0	0	0	0	1	1	0	0
13	00000	0	0	0	0	0	0	0	0	0	0	0	1	1	0	1
14	00000	0	0	0	0	0	0	0	0	0	0	0	1	1	1	0
15	00000	0	0	0	0	0	0	0	0	0	0	0	1	1	1	1
16	00000	0	0	0	0	0	0	0	0	0	0	0	1	1	1	1

*Counts/step indicates length of time out puts are in the prescribed state for each step.
1—indicates output will be ON during step.
0—indicates output will be OFF.

FIGURE 4.21 (Cont.)

2. Air Logic and Limit Valves Insure Welding Accuracy, *Hydraulics and Pneumatics,* Jul. 1978, p. 22.

3. L. Teschler, Update on Electronic Logic, *Machine Design,* Aug. 20, 1981, pp. 63–67.

4. W. P. Bowdry, Air Logic Earns Its Bread, *Hydraulic and Pneumatics,* Jun. 1980, pp. 53–56.

5. D. J. Sternaman, MPL Automates Transfer Impression Operation, *Hydraulics and Pneumatics,* Apr. 1980, pp. 67, 102.

6. J. S. Stecki and O. A. Reddecliffe, Designing Complex Control Systems with the Computer, *Hydraulics and Pneumatics,* Dec. 1979, p. 54, Jan. 1980, p. 107.

7. B. E. McCord, Simple Checklists Compare Cost of Air Versus Electrical Logic, *Machine Design,* Nov. 8, 1979, pp. 108–112.

8. S. C. Tsai, A Simplified Method of Logic Design of Pneumatic Sequential Control Circuits, ASME Paper 75-WA/Aut-22, 1975.

9. L. L. Boulden, Moving-Part Pneumatic Logic, *Machine Design,* March 4, 1971, pp. 78–84.

10. C. Bert, Applying Air Logic to Process Machines, *Automation,* Jul. 1971, pp. 38–41.

11. J. Crain, Miniature Pneumatic Logic Control, *Automation,* Mar. 1970, pp. 105–107.

12. W. D. Ludwig, Special Pneumatic Circuit Provides Control of Machine Masses, *Automation,* Oct. 1958, pp. 82–86.

13. N. C. Persson, Programmable Controllers Today, *Actuator Syst.,* Aug. 1981, pp. 10–21.

14. D. Deltano, Programming Your PC, *Inst. and Cont. Syst.,* Jul. 1980, pp. 37–40.

15. J. Prioste and T. Balph, Relay to IC Conversion, *Machine Design,* May 28, 1970, pp. 148–152.

16. B. R. Rusch, The Fine Points of Programmable Control, *Machine Design,* May 22, 1980, pp. 88–91.

17. F. Gruner, Static Switching, *Machine Design,* Jan. 22, 1970, pp. 121–129.

18. Programmable Controllers—Microprocessors in Overalls, *Mech. Eng.,* Feb. 1982, pp. 26–31.

19. N. Andreiev, Programmable Logic Controllers—an Update, *Cont. Eng.,* Sept. 1972, pp. 45–47.

20. D. Henry, PLCs Reach for New Process Control Applications, *Instr. Tech.,* Mar. 1976, pp. 45–48.

21. C. Trott, Choosing a Programmable Controller, *Digital Design,* Nov. 1975, pp. 42–46.

22. F. R. Hollo, Programmable Controllers—Familiarity Breeds Success, *Automation,* Feb. 1974, pp. 68–72.

23. J. J. Rodriguez and H. L. Harrison, Safer "Software" for Programmable Controllers, *Machine Design,* Aug. 12, 1976, pp. 76–80.

24. J. A. Seibel and C. L. Aronson, Control of Machines with Programmable Controllers, ASME Paper 72-DE-50, 1972.

25. G. W. Younkin, Single Time-Shared Drive Positions Two Axes, *Cont. Eng.,* Aug. 1960, pp. 122–126.

26. N. E. Igla and E. R. Forman, Improve Memory Usage by Compressing Programmable Control Logic, *In. Tech.,* Nov. 1981, pp. 73–76.

27. B. D. Stanton, 5 Hybrid Circuits for Process Control, *Cont. Eng.,* Apr. 1967, pp. 65–68.

CHAPTER 5
System Performance Specifications

5.1 RELATIONS BETWEEN PRACTICAL ECONOMIC AND DETAILED TECHNICAL PERFORMANCE CRITERIA

Most of our discussions in this chapter will involve rather specific mathematical performance criteria whereas the ultimate success of a controlled process generally rests on economic considerations which are difficult to calculate. This rather nebulous connection between the technical criteria used for system design and the overall economic performance of the manufacturing unit is not peculiar to the control field and results in the need for much exercise of judgement and experience in decision making at the higher management levels. Although control system designers must be cognizant of these higher-level considerations they, of necessity, usually employ rather specific and relatively simple performance criteria when evaluating their designs.

5.2 BASIC CONSIDERATIONS

Recall the basic feedback system block diagram defined in Chapter 2 and shown in abridged form as Fig. 5.1. It is conventional to work with controlled variable C (rather than indirectly controlled variable Q) since Q and C are often equal, and when not, Q is easily related to C through $Z(s)$, which is outside the feedback loop (see Fig. 2.1). Since the control system objective is that C follow the desired value V and ignore disturbance U, technical performance criteria must have to do with how well these two objectives are attained. Clearly the performance

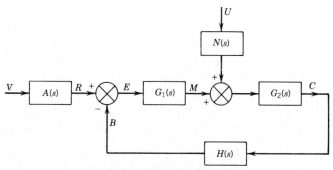

FIGURE 5.1 *General linear feedback system.*

depends both on the system characteristics and on the nature of V and U. We here again encounter practical difficulties, since the precise mathematical functions for V and U will not generally be known in practice. For example, in a building heating system, we may be quite clear that V will be a constant 70°F but who can predict the variation of the main disturbance (outdoor temperature)? In a contouring machine tool, who can know at the design stage all the different shapes of parts (desired value V) that the machine will need to produce during its lifetime? For a radar missile tracker how can we know the trajectories to be flown by all possible enemy targets?

This brief discussion makes clear that the random nature of many practical commands and disturbances makes difficult the development of performance criteria based on the actual V and U experienced by the real system. Although the theory[1] of control-system response to random input signals has been developed to a considerable degree, the methods are generally difficult to use and still sometimes require unrealistic restrictions on the detailed nature of the random inputs. It is thus much more common to base performance evaluation on system response to simple "standard" inputs such as steps, ramps, and sine waves. This approach has been successful for a number of reasons.

1. In many areas, experience with the actual performance of various classes of control systems has established a good correlation between the response of systems to standard inputs and the capability of the systems to accomplish their required tasks.

2. Design is much concerned with the *comparison* of competitive systems. This comparison can often be made nearly as well in terms of standard inputs as for real inputs.

3. The simplicity of form of the standard inputs facilitates mathematical analysis and experimental verification.

4. For linear systems with constant coefficients, theory shows that the response to a standard input of frequency content adequate to exercise all significant system dynamics can then be used to find mathematically the response to *any* form of input.

The standard performance criteria in common use may be classified as falling into the *time domain* or the *frequency domain*. Time-domain specifications have to do with the response to steps, ramps, parabolas, and the like; whereas frequency-domain specifications are concerned with certain characteristics of the system frequency response. Both types of specifications are often applied to the same system to ensure that certain behavior characteristics will be obtained. Any given specification will also usually include a tolerance band on each side of the nominal value to allow for system-to-system variability and variations within a given system due to environmental effects such as temper-

---

[1] J. H. Laning Jr. and R. H. Battin, *Random Processes in Automatic Control,* McGraw–Hill, New York, 1956.

ature changes. Since all other performance specifications are meaningless unless the system is absolutely stable, we *assume* absolute stability for the remainder of this chapter. Methods for determining absolute stability will be covered in detail in Chapter 6.

In considering the general nature of specifications, some particular examples are useful. In a contouring machine tool, if any one point on the contour lies outside the allowed dimensional tolerance band, the part will not be acceptable. Here the "instantaneous" correspondence of C to V would be of interest in specifications. In a house heating system a temperature steadily above the desired value causes not only continued discomfort of the occupants but also an economic loss since a higher than necessary heat loss is being supported by the furnace. Here the integral of error over time would be of interest. This would also be true in a thickness control for a rolling mill, since continued thickness error produces continuous scrap product.

5.3 Time-Domain Performance Specifications

For linear systems, the superposition principle allows us to consider response to commands apart from response to disturbances. If both occur simultaneously, the total response is just the superposition of the two individual responses. In nonlinear systems, such separate treatment with subsequent superposition is not valid. Figure 5.2 shows a typical closed-loop system response of C to a step of V when $U \equiv 0$. Both time-domain and frequency-domain design criteria generally are intended to specify one or the other of

1. Speed of response.
2. Relative stability.
3. Steady-state errors.

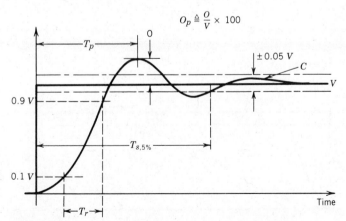

FIGURE 5.2 *Time-domain (transient) response criteria.*

Rise time T_r and *peak time T_p* are intended as speed-of-response criteria. Clearly, the smaller these values, the faster the system response. *Percent overshoot O_p* is used mainly for relative stability. Values in excess of about 30% may indicate that a system is dangerously close to absolute instability. Many systems are designed for 10 to 20% overshoot. No overshoot at all is sometimes necessary; but usually this needlessly penalizes speed of response. *Decay ratio,* the ratio of the second overshoot divided by the first, is a relative stability criterion often used in process control systems, the most common design value being ¼. *Settling time T_s,* the time to get *and stay* within a specified percentage (say 2%, 5%, or 10%) of V combines stability and speed-of-response aspects and is widely used. Note in Fig. 5.3 that system A (which has a much shorter rise time than system B) is, in a sense, not really faster than B since it does not *settle* in the neighborhood of the desired value until a later time.

When discussing steady-state errors, we should first note that certain math models of systems will predict, for given commands or disturbances, steady-state errors that are precisely zero, but no real system can achieve this perfection. Thus a realistic steady-state error specification will never require zero error, even though the system math model might predict this. Nonzero errors are always present because of nonlinearities, measurement uncertainties, and the like. For our usual linear, time-invariant math models, steady-state error is defined as follows. Set up the closed-loop system differential equation in which error $V - C$ is the unknown. Solution of this equation gives a complementary (transient) solution that always decays to zero for an absolutely stable system. The remaining (particular) solution is, by definition, the steady-state error, whether it is itself steady or time varying. That is, a steady-state error need not be a constant value.

The steady-state error E_{ss}, of course, depends on both the system and the input command or disturbance that causes the error. We will discover later a certain pattern of behavior as the input is made more difficult from the steady-state viewpoint. The "easiest" input is generally a step since it requires only a

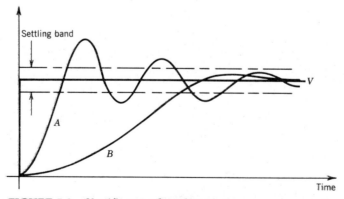

FIGURE 5.3 *Significance of settling time.*

steady response once the transient is over. Systems with integral control can theoretically give precisely zero E_{ss} for step inputs, as in Fig. 5.4a. The very same system subjected to a (more difficult) ramp input may now show the constant E_{ss} of Fig. 5.4b. The even more difficult parabola of Fig. 5.4c can cause an E_{ss} that increases linearly with respect to time. This type of pattern can be expected for both commands and disturbances in all linear systems, though details will vary.

When the system input is a disturbance $U(V \equiv 0.0)$ some of these criteria can still be applied, although others cannot. It is still possible to define a peak time T_p; however T_r, T_s, and O_p are all referenced to the step size V, which is now zero, thus they cannot be used (see Fig. 5.5). One possibility is to use the peak value C_p as a reference value to define a T_r and a T_s. To replace O_p as a stability specification one could use the decay ratio defined earlier or perhaps the number of cycles to damp the amplitude to, say, 10% of C_p. The smaller the number of cycles, the better the stability. The definition of steady-state errors still applies and we would again expect the same trend of worsening error as U changed from step to ramp to parabola, similar to the behavior in Fig. 5.4.

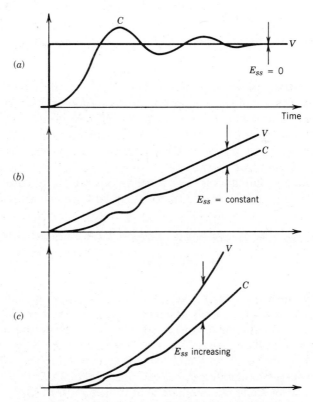

FIGURE 5.4   *Effect of command severity on steady-state error.*

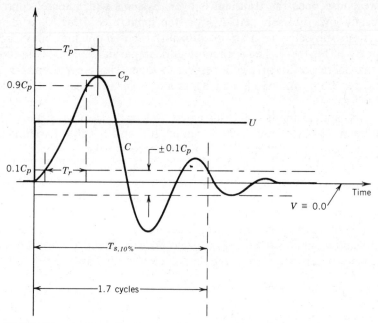

FIGURE 5.5 *Time-domain specifications for disturbance input.*

Although the types of specifications just described are most commonly used, the reader will occasionally encounter one form or another of *optimal perform-ance criteria,* so we now explore this area briefly. The concept in general is that, rather than quoting several specifications, each pertinent to an *individual* aspect of behavior (e.g., T_p or O_p), we attempt to formulate one overall criterion, which when minimized or maximized represents the optimum system. These criteria often involve integrating some function of system error over some time interval when the system is subjected to a standard command or disturbance such as a step input. A common example is the integral of absolute error (IAE) defined by

$$\text{IAE} \triangleq \int_0^\infty |V - C| dt \qquad (5.1)$$

One use for the IAE (or similar criteria) is to fix all system parameters except one (perhaps the loop gain), apply a step command, compute the IAE and vary the gain until a minimum value of IAE is found. This gain would then be used as the design value. This approach is not unreasonable since both high-gain systems (fast, but very oscillatory) and low-gain systems (sluggish) will give large IAE values, whereas moderate gain values give smaller IAE; thus a minimum IAE should exist for some intermediate gain value (see Fig. 5.6).

Another similar criterion is the ITAE (integral of time multiplied by absolute error) which exhibits the additional useful features that the initial large error

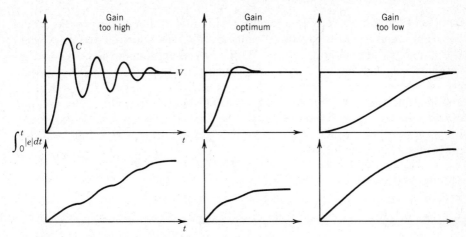

FIGURE 5.6 *The IAE optimal response criterion.*

(unavoidable for a step input) is not heavily weighted, whereas errors that persist are more heavily weighted.

$$\text{ITAE} \triangleq \int_0^\infty t\,|V - C|\,dt \tag{5.2}$$

Both the IAE and ITAE will go to infinity for system/input combinations that exhibit steady-state errors. Since this prevents definition of an optimum system, these criteria are usually then modified so that the integral runs only up to some finite time T, which usually is chosen to emphasize the transient period of operation. An optimal criterion much used for random inputs is the mean-square error $\overline{e^2}$.

$$\overline{e^2} \triangleq \frac{1}{T} \int_0^T |V - C|^2\,dt \tag{5.3}$$

where T is chosen to be long enough to get statistically reliable results.

Availability of a digital simulation language, such as the CSMP used in this text, makes the application of these optimal performance criteria quite convenient. As an example, consider the system of Fig. 5.7, where the process is

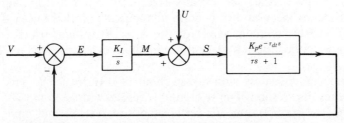

FIGURE 5.7 *Example of system used for optimal criteria simulation study.*

modeled as a first-order system together with a dead time. This process model has been found to fit the measured behavior of many industrial processes when K_p, τ, and τ_{dt} are suitably chosen. We use an integral controller K_I/s and investigate step inputs of V and U, for which this system has zero steady-state errors. We want to fix τ and τ_{dt} at chosen values, vary loop gain $K \triangleq K_I K_p$ over a range, find those K's for which IAE and ITAE exhibit minimum values, and observe the nature of system response when these optimum K values are used. For $\tau = 5$, $\tau_{dt} = 1.0$, $K_p = 1.0$ and a unit step input of $V(U \equiv 0.0)$, a CSMP program might be as follows.

```
PARAM TAU = 5.
PARAM K = (.5,3.,5.,8.,10.)
V = STEP (0.0)
U = 0.0
E = V-C
EK = K*E
M = INTGRL (0.0,EK)
EABS = ABS(E)
IAE = INTGRL (0.0, EABS)
EABST = TIME*EABS
ITAE = INTGRL (0.0, EABST)
S = U+M
C1 = REALPL (0.0,TAU,S)
C = DELAY (100, 1.0, C1)
TIMER FINTIM = 40., DELT = .01, OUTDEL = .8
OUTPUT C
PAGE MERGE
OUTPUT IAE
PAGE MERGE
OUTPUT ITAE
PAGE MERGE
END
```

   The results of this study were not very encouraging. In Fig. 5.8 we see first that both IAE and ITAE do not show large changes when K covers a wide range from systems that are somewhat sluggish to systems that are almost unstable. This makes location of a definite minimum somewhat uncertain. Also IAE gives a nearly minimum value for the $K = 10$ system, which is almost unstable, showing the criterion's lack of discrimination against such undesirable systems.
   A repetition of this study with a step input of disturbance U gave similar inconclusive results. Although more favorable results with IAE and ITAE have been reported in the literature for certain types of systems and inputs, there does not seem to be any evidence that these criteria are in any way universally useful. A more appropriate and successful application for these criteria seems to be as a means of selecting the best system from several *different types,* each

type having had its gain *already* adjusted to satisfy some conventional relative stability requirement, such as O_p or M_p (see Fig. 5.9). In the process control field, for example, it is common to design for quarter-amplitude damping (decay ratio = ¼). When the desired degree of relative stability has already been assured by this criterion, the use of a minimum value of, say, IAE to select the best among several competing designs becomes more practical. In fact, the integral of error itself (not its absolute value) can now be used as a measure of average error, since all competing systems have a similar degree of oscillation. This simpler (and linear) criterion has been quite successful in evaluating both the control difficulty of a process and the effectiveness of the control system.[2]

5.4 FREQUENCY-DOMAIN PERFORMANCE SPECIFICATIONS

We saw in Chapter 2 that if we know the frequency response of any linear system we can compute the system response to any form of input. It should

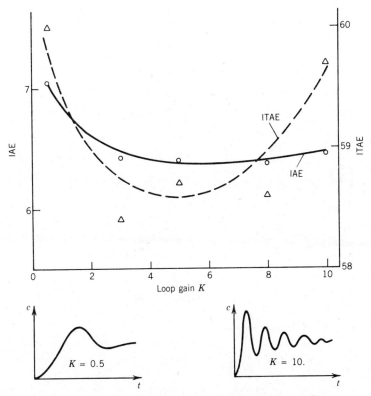

FIGURE 5.8 *Results of optimal criteria simulation study.*

[2]F. G. Shinskey, *Process-Control Systems*, 2nd Ed., McGraw–Hill, New York, 1979, pp. 86–87.

thus not be surprising that control system performance can be (and is) described in frequency-response terms. In Fig. 5.1 if we let V be a sine wave ($U \equiv 0.0$) and wait for transients to die out, every signal will be a sine wave of the same frequency. We can then speak of amplitude ratios and phase angles between various pairs of signals. The most important pair involves V and C since C is supposed to duplicate V and, ideally, $(C/V)(i\omega) \equiv 1.0$ for all frequencies. We can measure $(C/V)(i\omega)$ experimentally once a system has been built, and in fact such measurements are so common that special instruments called frequency-response analyzers are available to make them fast and accurate. At the design stage we must calculate $(C/V)(i\omega)$ analytically from

$$[A(i\omega)V - H(i\omega)C]G_1G_2(i\omega) = C \tag{5.4}$$

$$\frac{C}{V}(i\omega) = \frac{AG_1G_2(i\omega)}{1 + G_1G_2H(i\omega)} \tag{5.5}$$

A typical curve of $(C/V)(i\omega)$, the closed-loop frequency response, is shown in Fig. 5.9. The amplitude ratio and phase angle will approximate the ideal $1.0\underline{/0°}$ for some range of "low" frequencies but will deviate for higher frequen-

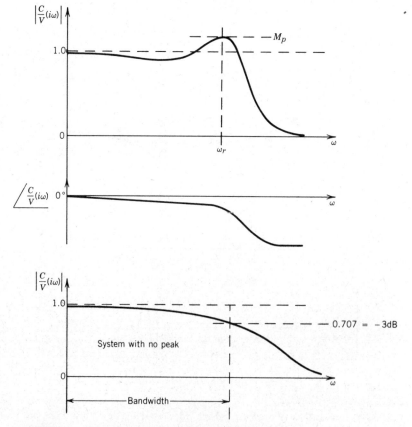

FIGURE 5.9   *Closed-loop frequency response criteria.*

cies. Many systems are designed to exhibit a resonant peak of controlled height in the range 1.2 to 1.4. The frequency ω_r (resonant frequency) at which this peak occurs is a speed of response criterion; the higher ω_r, the faster the system. The height M_p (peak amplitude ratio) of the peak is a relative stability criterion; the higher the peak, the poorer the relative stability. If no specific requirements are pushing the designer in one direction or the other, $M_p = 1.3$ is often used to give a good compromise between speed and stability. For systems that exhibit no peak (sometimes the case), the *bandwidth* is used for speed of response specification. Bandwidth is the frequency at which amplitude ratio has dropped to $1/\sqrt{2}$ times its zero-frequency value. It can, of course, be specified even if there is a peak.

With respect to disturbances, if we set $V \equiv 0.0$ and let U be a sine wave, we can measure or calculate $(C/U)(i\omega)$, which should of course ideally be zero for all frequencies. A real system cannot achieve this perfection but will behave typically as in Fig. 5.10. For low frequencies the feedback control action is effective in keeping C near the ideal value of zero. Just beyond this range, control becomes ineffective but the process is still capable of responding to U, giving the response shown. For still higher frequencies, control is still ineffective but now the process itself is too slow to react to U and response approaches zero again.

Finally I'll explain two open-loop performance criteria in common use to specify relative stability. The open-loop frequency response is defined as $(B/E)(i\omega)$; and we first note that it could be measured for an operating closed loop system as in Fig. 5.1 or, alternatively, one could *open the loop* by removing the summing junction at R, B, and E and just input a sine wave at E and measure the response at B. This latter procedure is clearly valid since $(B/E)(i\omega) = G_1 G_2 H(i\omega)$ in either case. Open-loop experimental testing has the advantage that open-loop systems are rarely absolutely unstable, thus there is little danger of starting up an untried apparatus and having destructive oscillations occur before it can be safely shut down. The utility of the open-loop frequency response rests on the *Nyquist stability criterion* which will be presented in detail in Chapter 6, but which we here use in simplified form to discuss the criteria *gain margin* and *phase margin*. To develop these concepts we must first plot $(B/E)(i\omega)$ in

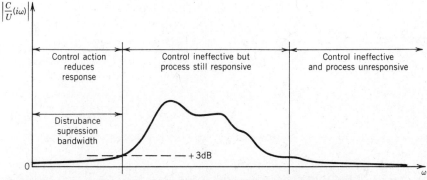

FIGURE 5.10 *Closed-loop frequency response to disturbance input.*

polar (rather than our usual rectangular) form as in Fig. 5.11. A *simplified* version of the Nyquist stability criterion states that if $|(B/E)(i\omega)| > 1.0$ at the point where $\underline{/(B/E)(i\omega)} = -180°$ then the *closed-loop* system obtained by connecting the usual summing junction at B and E will be absolutely unstable (see Fig. 5.12). Note that the point $1\underline{/-180°} = -1$ thus becomes a critical stability point on polar plots of $(B/E)(i\omega)$. Gain margin and phase margin are in the nature of safety factors such that $(B/E)(i\omega)$ stays far enough away from $1\underline{/-180°}$ on the safe (stable) side. In Fig. 5.13a we define gain margin as the multiplying factor by which the steady-state gain of $(B/E)(i\omega)$ could be increased (nothing else in $(B/E)(i\omega)$ being changed) so as to put the system on the edge of instability ($(B/E)(i\omega)$ passes exactly through the -1 point). This is called *marginal stability*. Phase margin is the number of degress of additional phase lag (*nothing* else being changed) required to create marginal stability. Figures 5.13b and 5.13c make clear that *both* a good gain margin and a good phase margin are needed; neither is sufficient by itself. General numerical design goals for gain margin and phase margin cannot be given since systems that satisfy other specific performance criteria may exhibit a wide range of gain and phase margins. It is possible, however, to give useful lower bounds; gain margin should usually exceed 2.5 and phase margin should exceed 30°.

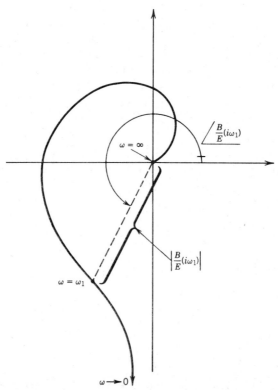

FIGURE 5.11 *Polar plot of open-loop frequency response.*

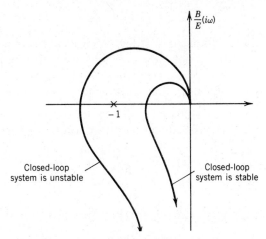

FIGURE 5.12 *Nyquist stability criterion.*

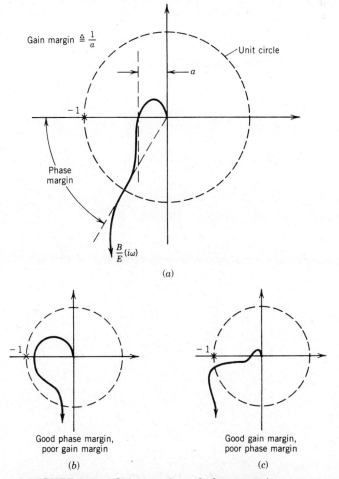

FIGURE 5.13 *Gain margin and phase margin.*

PROBLEMS

5.1 Run the digital simulation study of Fig. 5.8 for the case where $V \equiv 0.0$ and $U = 1.0$. Note that now not only loop gain K but the *individual* values K_I and K_p must be considered. Devise a plan for dealing with this complication, run the necessary simulations, and comment on the results.

5.2 Would you expect the results of Fig. 5.8 to be independent of the numerical values of τ and τ_{dt}? Devise simulations to explore this effect, run them, and comment on the results.

5.3 Repeat the study of Fig. 5.8 using, however, the criterion of Eq. 5.3, and comment on the results.

5.4 Repeat the study of Fig. 5.8 for the case where $\tau_{dt} \equiv 0.0$ and comment on the results.

5.5 Repeat Problem 1 for the case $\tau_{dt} \equiv 0.0$.

5.6 Using the optimum (based on IAE) value of loop gain found in Problem 4, compute and graph $(C/V)(i\omega)$ and comment on the results.

5.7 Repeat Problem 6 using the ITAE criterion.

5.8 In Problem 6, find the gain margin and phase margin.

CHAPTER 6
Absolute Stability Criteria

6.1 OVERVIEW

Of the various possible definitions of *absolute stability* we choose a relatively simple and practical one. If a system in equilibrium is momentarily excited by command and/or disturbance inputs and those inputs are then removed, the system must return to equilibrium if it is to be called absolutely stable. If action persists indefinitely after excitation is removed, the system is adjudged absolutely unstable. The analytical study of stability becomes a study of the stability of the solutions of the closed-loop system's differential equations and has received much attention from mathematicians. Results of practical use to engineers are mainly limited to linear systems with constant coefficients, where an exact and complete stability theory has been known for a long time. For nonlinear systems, the major useful result is an approximation technique called describing function theory, which we will discuss in a later chapter. Exact, general results for linear time-variant and nonlinear systems are nonexistant and those few results available for special cases are rarely of practical engineering use. Fortunately, the linear time-invariant theory is adequate for many practical systems, whereas the describing function approximation has a good record of success with many nonlinear systems. When neither of these approaches seems to work, we can always fall back on digital simulation, such as CSMP. Here no general results are possible but one can often explore enough typical inputs and system parameter values to gain a high degree of confidence in stability for any specific system.

6.2 LINEAR SYSTEMS WITH CONSTANT COEFFICIENTS

Here a complete and general stability theory has been available for many years, based on the locations in the complex plane of the roots of the closed-loop system characteristic equation, stable systems having all their roots in the left half plane. Although nowadays digital computer root-finding routines can find excellent approximations to the actual numerical roots of a specific polynomial quickly and easily, there is still need for general methods of determining the presence of unstable roots without actually finding their numerical values. Two such methods are in common use. One, the Routh stability criterion, works with the closed-loop characteristic equation in an algebraic fashion; whereas the second, the Nyquist stability criterion, is a graphical technique based on the open-loop frequency response polar plot. Both methods give the same result, a statement of the number (but *not* the specific numerical values) of unstable roots. This information is generally adequate for design purposes.

185

For the general feedback system of Fig. 2.1 the closed-loop system equation will always be of the form (using D operators)

$$\text{(polynomial in } D)Q = \text{(polynomial in } D)V + \text{(polynomial in } D)U$$

Our stability definition requires that when V and/or U are "turned on" and then later returned to zero, Q should return to zero. The solution for Q will always have two parts, the complementary (transient) solution and the particular (steady-state) solution. If V and U both eventually return to zero, then their particular solutions must also return to zero. Thus if system response is to persist after V and U have returned to zero, this response must come from the complementary solution alone. A stable system must thus have a complementary solution which eventually decays to zero. Recall that all terms of the complementary solution arise from the roots of the system characteristic equation. These roots are either real or complex, repeated or unrepeated. Since for a root s the corresponding complementary solution term has the form ce^{st}, we see that, if real, s must be negative, and if complex, s must have a negative real part, thus *stable systems must have all their roots in the negative half of the complex plane,* which is the basic stability criterion for linear, constant-coefficient systems. Roots falling on the imaginary axis are either zero or pairs of pure imaginaries. A single zero root gives a constant (nonzero) response; whereas repeated zero roots give ct, ct^2, and so forth, none of which go to zero. Pairs of pure imaginary roots give oscillations that neither die down nor build up in amplitude (the so called *marginal stability* case), and not decaying to zero they are not stable.

The previously mentioned stability criterion is of great practical use because many systems are nearly linear for suitably small excursions away from an equilibrium operating point. It does, however, suffer one fault and that is that it predicts excursions of infinite magnitude for unstable systems since the unstable terms of the complementary solution take the form, for example, $3e^{4t}$ or $0.68e^{7t}\sin(8t-18°)$. Since infinite motions, voltages, temperatures, and the like require infinite power supplies, no real-world system can conform to such a mathematical prediction, casting possible doubt on the validity of our linear stability criterion since it predicts an impossible occurence. What actually happens is that oscillations, if they are to occur, start small, under conditions favorable to and accurately predicted by the linear stability theory. They then start to grow, again following the exponential trend predicted by the linear model. Gradually, however, the amplitudes leave the region of accurate linearization, and the linearized model, together with all its mathematical predictions, loses validity. Since solutions of the now nonlinear equations are usually not possible analytically we now must rely on experience with real systems and/or nonlinear computer simulations when explaining what *really* happens as unstable oscillations build up. First, practical systems often include overrange alarms and safety shut-offs that automatically shut down operation when certain limits are exceeded. If such safety features are not provided, the system may destroy itself, again leading to a shut-down condition. If safe or destructive shut-down does not occur, the system usually goes into a *limit cycle oscillation*, an ongoing, nonsinusoidal oscillation of *fixed* amplitude. The wave form, frequency, and amplitude of limit cycles is governed by nonlinear math models that are usually analytically unsolvable; however the earlier mentioned describing-function approximation often gives good predictions of amplitude and frequency. We should also note that although limit-cycle instabilities are usually intolerable in practical systems, this is not always the case. Almost every residential heating or air conditioning system uses an on–off type of control that continually limit cycles; however this "unstable" behavior is accepted since the amplitude of the temperature fluctuations is small enough to be unobjectionable.

6.3 THE ROUTH STABILITY CRITERION

To use the Routh stability criterion we must have in hand the characteristic equation of the closed-loop system's differential equation. If the system includes any dead times these *must be approximated* with polynomial forms in D, since Routh's criterion requires the characteristic equation to be a polynomial in D. To get the system characteristic equation we can write directly from Fig. 2.1

$$\left\{\left[A(D)V - \frac{H}{Z}(D)Q\right]G_1(D) + N(D)U\right\}G_2(D) = \frac{1}{Z(D)}Q \qquad (6.1)$$

$$[1 + G_1G_2H(D)]Q = [AG_1G_2Z(D)]V + [NZG_2(D)]U \qquad (6.2)$$

In general, each individual transfer function could have both a numerator and denominator polynomial, allowing us to define

$$AG_1G_2Z(D) \triangleq \frac{G_{nV}(D)}{G_{dV}(D)} \qquad NZG_2(D) \triangleq \frac{G_{nU}(D)}{G_{dU}(D)} \qquad G_1G_2H(D) \triangleq \frac{G_n(D)}{G_d(D)} \qquad (6.3)$$

If Eq. 6.2 were actually to be written out as a differential equation, all denominator polynomials would have to be cleared out first, allowing us to write the system characteristic equation as

$$G_{dV}G_{dU}(G_n + G_d) = 0 \qquad (6.4)$$

The terms G_{dV} and G_{dU} are almost always themselves stable (no right-half plane roots) and when they are not stable it is generally obvious since these terms usually are already in factored form where unstable roots are apparent. For these reasons it is conventional to concentrate on the term $G_n + G_d$ which came from the original $1 + G_1G_2H$ term which describes the behavior of the feedback loop without including "outside" effects such as $A(D)$, $N(D)$, and $Z(D)$. When we proceed in this fashion we are really examining the stability behavior of the closed loop, rather than the entire system. Since any instabilities in the "outside the loop" elements are so rare and also usually obvious, this common procedure is reasonable.

Proceeding along these lines we will write a general form of the system characteristic equation as

$$1 + G_1G_2H(s) = G_n(s) + G_d(s) = a_n s^n + a_{n-1} s^{n-1} + \cdots + a_1 s + a_0 = 0 \quad (6.5)$$

We assume the term a_0 is nonzero; otherwise the characteristic equation has one or more zero roots which we easily detect and which do not correspond to stable systems. The Routh stability criterion was developed in the 1870s by E. J. Routh, a British dynamicist (the German A. Hurwitz developed a similar criterion about the same time). Its proof involves some delicate algebra and is rarely presented, even in advanced control texts, so I only explain its use. We first arrange the coefficients of Eq. 6.5 in two rows.

a_n	a_{n-2}	a_{n-4}	a_{n-6}
a_{n-1}	a_{n-3}	a_{n-5}	$\cdots$

Then form a third row

$$b_1 \qquad b_2 \qquad b_3 \qquad \cdots$$

by following the pattern

$$b_1 \triangleq \frac{a_{n-1}a_{n-2} - a_n a_{n-3}}{a_{n-1}} \qquad b_2 \triangleq \frac{a_{n-1}a_{n-4} - a_n a_{n-5}}{a_{n-1}}$$

$$b_3 = \frac{a_{n-1}a_{n-6} - a_n a_{n-7}}{a_{n-1}} \qquad \text{and so forth.}$$

When the third row has been completed, a fourth row is formed from the second and third in exactly the same fashion as the third was formed from the first and second. This is continued until no more rows and columns can be formed, giving a triangular sort of array. If the numbers become cumbersome, their size may be reduced by multiplying any row by any positive number. If one of the a's is zero, it is entered as a zero in the array.

An example illustrates the procedure.

$$D^5 + 2D^4 + 4D^2 + D + 1 = 0 \quad (6.6)$$

1	0	1
2	4	1
$\dfrac{0-4}{2}$	$\dfrac{2-1}{2}$	0
$\dfrac{-8-1}{-2}$	$\dfrac{-2-0}{-2}$	0
$\dfrac{\dfrac{9}{4}+2}{\dfrac{9}{2}}$	0	
1		

which reduces to

$$
\begin{array}{ccc}
1 & 0 & 1 \\[4pt]
2 & 4 & 1 \\[4pt]
-2 & \dfrac{1}{2} & \\[10pt]
\dfrac{9}{2} & 1 & \\[10pt]
\dfrac{17}{18} & & \\[10pt]
1 & &
\end{array}
$$

Although it is necessary to form the entire array, its evaluation depends always on only the first column. Routh's criterion says that the number of roots not in the left half plane is equal to the number of changes of algebraic sign in the first column. *Thus a stable system must exhibit no sign change in the first column.* Our example shows two such changes ($+2$ to -2, -2 to 9/2), thus two unstable roots exist; however we do not know their numerical values.

Although the complete Routh procedure as just explained gives a correct result in every case, two special situations are worth memorizing as shortcuts. It can be shown that if the original system characteristic equation (Eq. 6.5) *itself* shows any sign changes, there is really no point in carrying out the Routh procedure; the system will always be unstable. Similarly, if there are any "gaps" (zero coefficients) in the characteristic equation (such as the missing cubic coefficient in Eq. 6.6), the system is always unstable. These two rules are worth memorizing since they often alert the analyst (at an early stage) to outright mistakes in the mathematical model, since correct math models almost always have no gaps or sign changes. Note, of course, that a lack of gaps or sign changes is a necessary but *not* a sufficient condition for stability.

Although not of much practical significance, since they rarely occur in practical problems, two special cases are now mentioned since they can occur mathematically. The first is when a term in the first column is zero but the remaining terms in its row are not all zero, causing a division by zero when forming the next row. The second occurs if *all* the terms in the second or any further row are zero, giving the indeterminate from 0/0. Special tricks to handle both these problems are available.[1]

Until this point we have presented the Routh criterion as merely a means to answer yes–no questions concerning absolute stability. It is also often useful in developing design guidelines helpful in making trade-off choices among system physical parameters, as will be shown by the next example. In the electromechanical positioning servomechanism of Fig. 6.1 we can write down the closed-loop-system differential equation "at sight" from the given block diagram.

---

[1]E. O. Doebelin, *Dynamic Analysis and Feedback Control*, McGraw–Hill, New York, 1962, pp. 176–177.

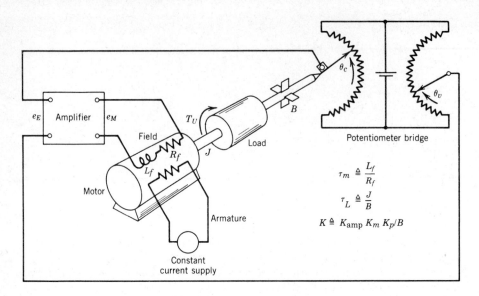

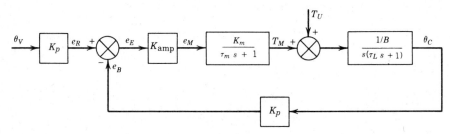

FIGURE 6.1   *Use of Routh criterion to develop design relations.*

$$\left[(\theta_V K_p - \theta_C K_p)\frac{K_{amp}K_m}{\tau_m s + 1} + T_U\right]\frac{1/B}{s(\tau_L s + 1)} = \theta_C \tag{6.7}$$

$$[\tau_m \tau_L s^3 + (\tau_m + \tau_L)s^2 + s + K]\theta_C = K\theta_V + (1/B)(\tau_m s + 1) T_U \tag{6.8}$$

Application of Routh's criterion shows that to ensure absolute stability we require

$$\text{loop gain } K < \frac{1}{\tau_m} + \frac{1}{\tau_L} \tag{6.9}$$

Note that the smaller the electric motor field lag τ_m and the mechanical inertia/ friction lag τ_L, the larger we can make loop gain before instability occurs. This confirms our oft-stated rule that slow response in components generally leads to undesirably low loop gain in the feedback system. Furthermore, if tentative numerical values for τ_m and τ_L have been selected (say $\tau_m = 0.05s$, $\tau_L = 0.15s$) we can combine Eq. 6.9 with the lower-bound suggestion for gain margin of 2.5 (given in Chapter 5) to establish a trial value for loop gain.

$$K_{\text{trial}} = \frac{\dfrac{1}{0.05} + \dfrac{1}{0.15}}{2.5} = 11.1 \qquad (6.10)$$

We can now easily set up a digital simulation (such as CSMP) of this system and quickly explore a range of gain values in the neighborhood of 11.1 to select a value meeting system specifications.

6.4 THE NYQUIST STABILITY CRITERION

Advantages of the Nyquist criterion (developed by H. Nyquist in 1932) over the Routh criterion include:

1. If some system components are modeled experimentally using frequency response measurements, these measurements can be used *directly* in the Nyquist criterion. The Routh criterion would first require the fitting of some analytical transfer function to the experimental data; this involves extra work and reduces accuracy since curve fitting procedures are never perfect.

2. Being a frequency response method, the Nyquist criterion handles dead times without approximation since the frequency response of a dead time is exactly known.

3. In addition to answering the question of absolute stability, Nyquist also gives some useful results on relative stability. Furthermore, the graphical plot used, $(B/E)(i\omega)$, keeps the effects of individual pieces of hardware more apparent (Routh tends to "scramble them up") making needed design changes more obvious.

The Nyquist criterion can be mathematically proven at various levels[2] of rigor and completeness. Here I choose only to explain its application and give a brief indication of its plausibility. To establish some confidence in its plausibility, consider a sinusoidal input to the open-loop configuration of Fig. 6.2a. Suppose that at some frequency, $(B/E)(i\omega) = 1\underline{/-180°}$. If we would then close the loop as in Fig. 6.2b, the signal $-B$ would now be exactly the same as the original E excitation sine wave and an *external* source for E would no longer be required; the closed loop system would maintain a steady *self-excited* oscillation of fixed amplitude, the situation we have earlier called marginal stability. It thus appears that if the open-loop curve $(B/E)(i\omega)$ for any system passes through the -1 point, the closed loop system will be marginally stable. This plausibility argument does not make clear what happens if the curve does not go exactly through -1. The complete answer to this and other questions requires a rigorous proof,

---

[2]Ibid. pp. 180–187, J. J. D'Azzo and C. H. Houpis, Feedback Control System Analysis and Synthesis, 2nd Ed., McGraw–Hill, New York, 1960, pp. 747–753.

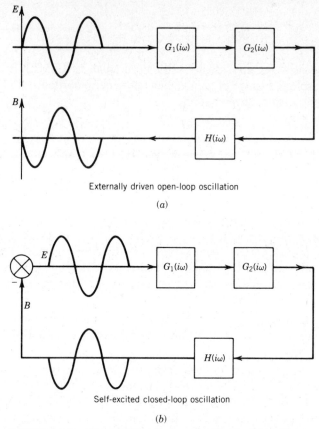

Externally driven open-loop oscillation

(a)

Self-excited closed-loop oscillation

(b)

FIGURE 6.2   *Plausibility demonstration for Nyquist criterion.*

which makes use of complex-variable theory and results in a criterion that gives exactly the same type of answer as the Routh criterion (i.e., the number of unstable closed-loop roots).

I now state the Nyquist criterion as a step-by-step procedure.

1. Make a polar plot of $(B/E)(i\omega)$ for $0 \leq \omega < \infty$, either analytically or by experimental test for an existing system.

2. Although negative ω's have no physical meaning, the mathematical criterion requires that we plot $(B/E)(-i\omega)$ on the same graph. Fortunately this is easy since $(B/E)(-i\omega)$ is just a reflection about the real (horizontal) axis of $(B/E)(+i\omega)$. If $(B/E)(i\omega)$ has no terms $(i\omega)^k$ (integrators) as multiplying factors in its denominator, Step 2 results in a closed curve with ω running from $-\infty$ to 0 to $+\infty$ as we go around the curve (see Fig. 6.3a.)

3. If $(B/E)(i\omega)$ has $(i\omega)^k$ as a multiplying factor in its denominator, the plots for $+\omega$ and $-\omega$ will "go off the paper" as $\omega \rightarrow 0$ and we will not get a single closed curve (see Fig. 6.3b). The rule for closing such plots says to

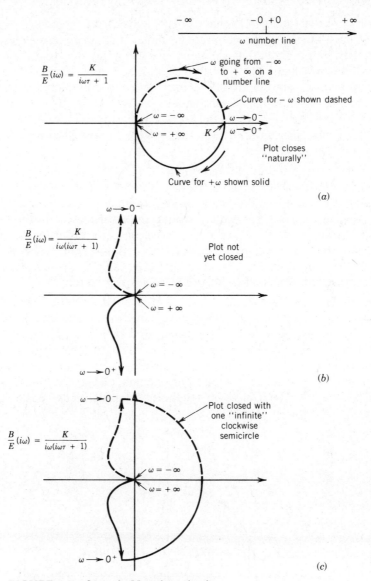

FIGURE 6.3 *Steps in Nyquist criterion.*

connect the "tail" of the curve at $\omega \to 0^-$ to the tail at $\omega \to 0^+$ by drawing k clockwise semicircles of "infinite" radius (see Fig. 6.3c). Application of this rule will always result in a single closed curve so that one can start at the $\omega = -\infty$ point and trace completely around the curve toward $\omega = 0^-$ and $\omega = 0^+$ and finally to $\omega = +\infty$, which will always be the same point (the origin) at which we started with $\omega = -\infty$.

4. We must next find the number N_p of poles of $G_1 G_2 H(s)$ that are in the right half of the complex plane. *This will almost always be zero* since these poles are the roots of the characteristic equation of the open-loop system and open-loop systems are rarely unstable. If the open-loop poles are not already factored and thus apparent, one can apply the Routh criterion to find out how many unstable ones there are, if any. If $G_1G_2H(i\omega)$ is not known analytically but rather by experimental measurements on an existing open-loop system, then it *must* have zero unstable roots or else we would never have been able to run the necessary experiments because the system would have been unstable. We thus generally have little trouble finding N_p and it is usually zero.

5. We now return to our plot of $(B/E)(i\omega)$, which has already been reflected and closed in earlier steps. Draw a vector whose tail is bound to the -1 point and whose head lies at the origin, where $\omega = -\infty$. Now let the head of this vector trace completely around the closed curve in the direction from $\omega = -\infty$ to 0^- to 0^+ to $+\infty$, returning to the starting point. Keep careful track of the total number of *net* rotations of this test vector about the -1 point, calling this number N_{p-z} and making it positive for counterclockwise rotations and negative for clockwise rotations.

6. In this final step we subtract N_{p-z} from N_p. This number will always be zero or a positive integer and will be equal to the number of unstable roots for the closed-loop system, the same kind of information given by the Routh criterion. Applying this procedure to the two examples of Fig. 6.3a and 6.3c, we get $N_p = N_{p-z} = 0$ for both cases, indicating the closed-loop systems to be absolutely stable. The example of Fig. 6.4 has $N_p = 0$ and $N_{p-z} = -2$, so its closed-loop system is unstable with two unstable roots.

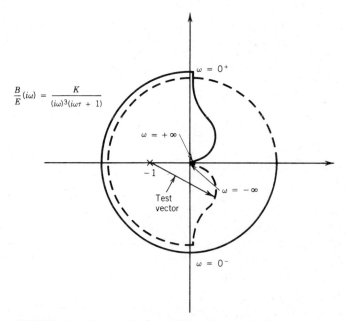

$$\frac{B}{E}(i\omega) = \frac{K}{(i\omega)^3(i\omega\tau + 1)}$$

FIGURE 6.4   *Example of unstable system.*

We have mentioned that the Nyquist criterion treats without approximations systems with dead times; Fig. 6.5 is an example. Note that since the open-loop system is stable, $N_p = 0$. With low loop gain the closed-loop system for Fig. 6.5a has $N_{p-z} = 0$ and is stable. Raising the gain as in Fig. 6.5b expands the spirals (caused by the dead time's ever-increasing phase lag) sufficiently to cause the test vector to now experience two net rotations, making $N_{p-z} = -2$ and causing closed-loop instability. Because the dead-time's phase lag behavior causes the $(B/E)(i\omega)$ curve to spiral unendingly into the origin, further gain increases expand more and more of these spirals out to the region beyond the -1 point, causing N_{p-z} to increase, indicating the presence of more and more unstable roots.

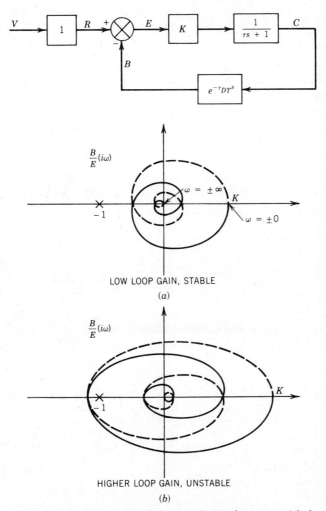

FIGURE 6.5 *Nyquist stability analysis of system with dead time.*

Another significant concept clearly illustrated by the Nyquist criterion is that of *conditional stability*. In all our stability discussions, starting, qualitatively, even in the first chapter, we have implied that increase in loop gain always leads to instability (at least in those system models capable of instability). We now wish to show that in some situations (which do correspond to real physical systems) a gain increase can make an unstable system stable and a gain decrease can make a stable system unstable, a phenomenon called conditional stability. Assume in Fig. 6.6 that $N_p = 0$. The Nyquist criterion shows the system of Fig. 6.6*a.* to be stable, yet when we *reduce* the gain as in Fig. 6.6*b* we get instability. A *further* gain reduction as in Fig. 6.6*c* does return us to stability however, and a gain increase as in Fig. 6.6*d* causes instability. We see that in systems of this type, stability is assured only for a certain *range* of loop gain; gains *either* lower or higher than this range result in instability. I present this example mainly to disabuse the reader of the notion that gain loss in a stable system never has adverse stability effects, a notion I had to this point encouraged by several examples. This conditional stability phenomenon is actually of some practical interest since a fair number of high-performance systems exhibit this general shape of Nyquist diagram. By itself this shape presents no problem, however

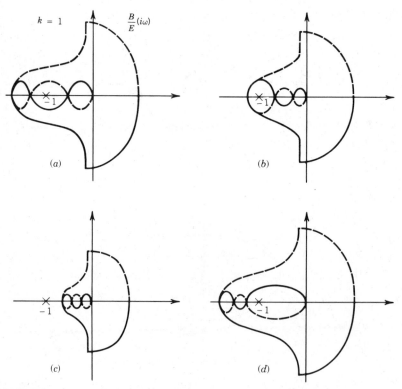

FIGURE 6.6 *Conditional stability.*

when it is combined with one or more system components that exhibit *saturation nonlinearity*, as in Fig. 6.7, instability can be triggered whenever commands and/or disturbances are large enough to cause significant saturation. Although saturation is a nonlinear effect, it can be approximated linearly as a gain reduction, the gain reduction becoming greater as saturation increases. With sufficient saturation, the linearized gain decrease can be sufficient to change the system from Fig. 6.6a to 6.6b, causing instability, as can be seen from Fig. 6.7.

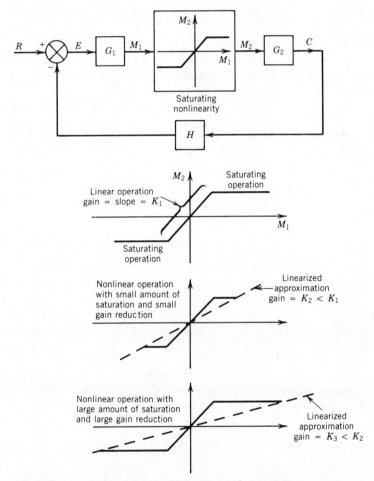

FIGURE 6.7   *Linearized approximation of saturating system.*

6.5 ROOT-LOCUS INTERPRETATION OF STABILITY

Our final explanation of the stability question uses the root locus concept. In later chapters, the root locus method will be developed as one of the major design tools of the control-system field. Our purposes here are served by a brief discussion of one aspect only. Consider, as an example, a system (such as that of Fig. 6.1) with open-loop transfer function

$$\frac{B}{E}(s) = \frac{K}{s(\tau_1 s + 1)(\tau_2 s + 1)} \tag{6.11}$$

The closed-loop characteristic equation would be

$$\tau_1\tau_2 s^3 + (\tau_1 + \tau_2)s^2 + s + K = 0 \tag{6.12}$$

Let us assume that τ_1 and τ_2 have been chosen and that we wish to explore the effect of varying loop gain K on system stability. For each value of K, Eq. 6.12 has three roots, which may be plotted in the complex plane (see Fig. 6.8). For $K = 0$ these roots are $-1/\tau_2$, $-1/\tau_1$ and, 0. For K a small number we find (perhaps using a digital computer root-finding routine) that we get a new root slightly to the left of $-1/\tau_1$, one slightly to the right of $-1/\tau_2$, and the third slightly to the left of 0. Since K can be varied in as small steps as we please, the roots also move in infinitesimal increments, tracing out continuous curves that are called the *root loci*. For low values of K we get three negative real roots, indicating a stable system with three exponentially decaying terms in its complementary solution. As K is increased the two smaller roots approach each other while the third heads toward $-\infty$. For a particular K value the two smaller roots merge, corresponding to the repeated roots of a quadratic factor of Eq.

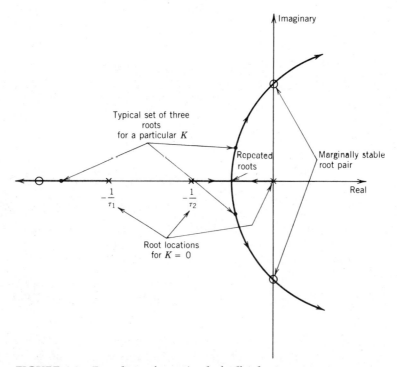

FIGURE 6.8   *Root locus for a simple feedback system.*

6.12, the so-called critically damped case. Further K increases cause these two roots to split off from the real axis and become a pair of complex roots $a \pm ib$, giving an underdamped (oscillatory) response term, while the third root proceeds as before toward $-\infty$, a stable response term. If K is made even larger, the complex root pair approaches the imaginary axis and for one specific K value we get a pair of pure imaginary roots, the *marginally stable* case, whose oscillation neither dies down nor builds up. Yet larger K's push the two roots into the right half plane, giving absolute instability. Figure 6.8 shows the complete root locus plot for our example. Every linear, time invariant feedback system has a root locus plot and these are extremely helpful in system design and analysis, though of course the detailed shapes of the curves will vary from one system to another. In using root locus methods later, we will see that both the "brute force" root-finding method using digital computer polynomial root finders, and also certain clever graphical approaches will be useful in establishing the actual shape of a particular root locus.

PROBLEMS

6.1 Derive Eq. 6.9.

6.2 Use digital simulation as suggested after Eq. 6.10 to choose a loop gain that gives θ_C a 30% overshoot for a step command θ_V. For the system so designed, find:

    a. Peak time T_p.

    b. 5% settling time.

    c. Steady-state error for $\theta_V = 0.5t$ radians (t in seconds).

    d. Phase margin and gain margin.

    e. ω_r and M_p for $(C/V)(i\omega)$.

6.3 In Problem 2, suppose we wish to maintain the same gain margin as obtained there but want to raise the gain so as to reduce the ramp steady-state error to one half its earlier value. If only τ_L and loop gain are adjustable, find the necessary value of τ_L.

6.4 Investigate the stability of these systems using both Routh and Nyquist criteria. For Routh analysis of dead times, approximations must be used; try all three (Eqs. 2.176 through 2.178).

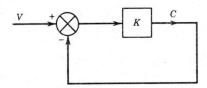

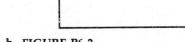

a. FIGURE P6.1.   **b. FIGURE P6.2.**

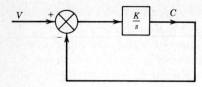

c. FIGURE P6.3.

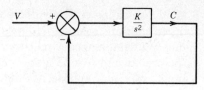

d. FIGURE P6.4.

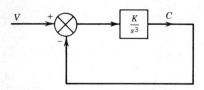

e. FIGURE P6.5.

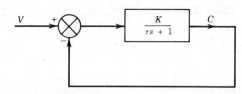

f. FIGURE P6.6.

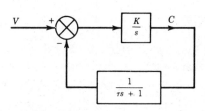

g. FIGURE P6.7.

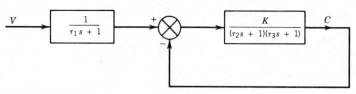

h. FIGURE P6.8.

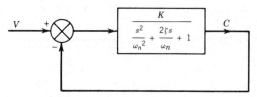

i. FIGURE P6.9.

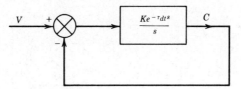

j. FIGURE P6.10.

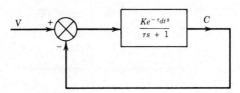

k. FIGURE P6.11.

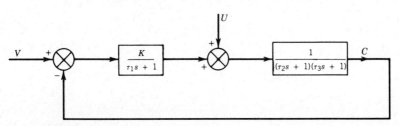

l. FIGURE P6.12.

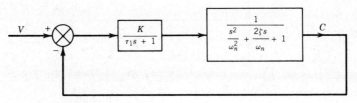

m. FIGURE P6.13.

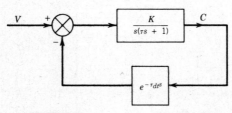

n. FIGURE P6.14.

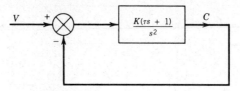

o. FIGURE P6.15.

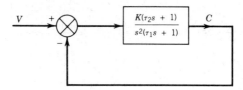

p. FIGURE P6.16.

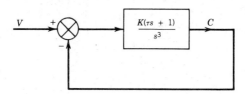

q. FIGURE P6.17.

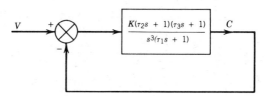

r. FIGURE P6.18.

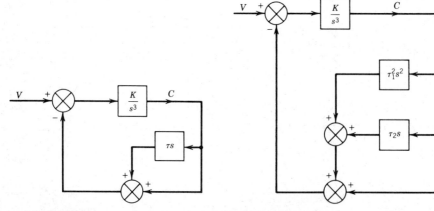

s. FIGURE P6.19.　　　　　　　**t. FIGURE P6.20.**

6.5 An experimental open-loop frequency response test was run on a system, giving the following measured results for $(B/E)(i\omega)$.

ω Rad/S	Amplitude Ratio	Phase Angle (Degrees)	ω	Amplitude Ratio	Phase Angle (Degrees)
5	24.1	−179	55	1.5	−135
10	11.8	−177	60	1.0	−130
15	9.9	−175	65	0.5	−132
20	6.1	−171	75	0.28	−170
30	4.0	−160	90	0.26	−180
40	3.1	−150	110	0.19	−210
50	1.9	−140	170	0.06	−260

Check for absolute stability, gain margin, and phase margin of the closed-loop system.

6.6 Using both Routh and Nyquist methods, find the loop gain needed to give each system a gain margin of 3. For this gain value, then find the phase margin.

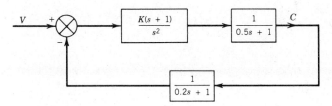

a. FIGURE P6.21.

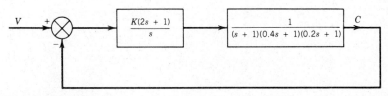

b. FIGURE P6.22.

6.7 Using digital simulation, check the stability of the nonlinear system of Fig. P6.23 for step commands of size $V = 0.5$ and for $V = 3.0$. It is suggested that the problem encountered can be solved by making τ_2 sufficiently small. Use digital simulation to see whether this suggestion works.

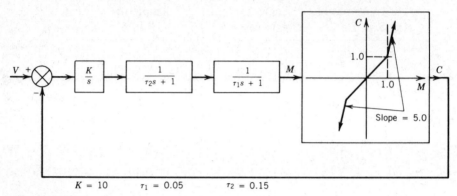

$$K = 10 \qquad \tau_1 = 0.05 \qquad \tau_2 = 0.15$$

FIGURE P6.23.

CHAPTER 7
Open-Loop Input-Compensated (Feedforward) Control

7.1 DISTURBANCE-COMPENSATED CONTROL

I introduced briefly in Chapter 1 the concept of input (disturbance and command) compensated control and now I wish to elaborate on these types of open-loop controls. Recall that these control modes can be used both by themselves and in combination with feedback schemes, the open-loop modes serving to deal with *specific* major disturbances or commands, the feedback mode taking care of all other sources of error, including those not anticipated by the designer.

When disturbance compensation, or disturbance feedforward as it is often called, is used *alone* to counteract a specific process disturbance, the diagram of Fig. 1.4 (as modified in Fig. 7.1) is appropriate. Process dynamics G_2 and G_N are considered given and unalterable, whereas disturbance compensation elements G_C and G_s and main controller G_1 are subject (within limits of practical realizability) to our design. From the relation

$$C(s) = (AG_1G_2)V(s) + (G_NG_2 - G_sG_CG_1)U(s) \qquad (7.1)$$

we see that to completely remove the influence of disturbance U (no matter what its nature) on controlled variable C we must make $G_NG_2 = G_sG_CG_1$. This ideal can, of course, not be achieved in practice for a number of reasons. The essential difficulty is that G_NG_2 and $G_sG_CG_1$ are neither perfectly known nor absolutely fixed. If, say, G_NG_2 is not described with mathematical exactness, then how can one design $G_sG_CG_1$ to be equal to it? Or, if both G_NG_2 and $G_sG_CG_1$ *were* perfectly known and equal at one instant in time, environmental effects of one kind or another (temperature, ageing, etc.) will surely cause

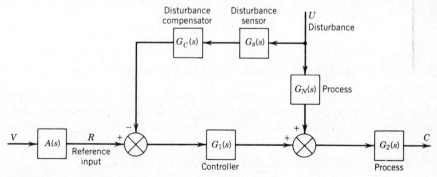

FIGURE 7.1 *System with disturbance feedforward.*

changes (perhaps large) as time goes by, again causing a mismatch of $G_N G_2$ with $G_s G_C G_1$. These difficulties may prevent achievement of perfection, but certainly do not prevent such control schemes from being practical methods of achieving improved control performance.

Although Fig. 7.1 (with its transfer functions) implies perfectly linear time-invariant system behavior, it should be clear that the basic concept of disturbance compensation is not so limited. So long as the differential equations (linear or not) describing signal propagation from U to C by the process path are identical with the equations for the compensating path through the sensor, compensator, and controller, the desired cancellation of the effects of U on C will be achieved.

The design of a disturbance compensation system thus consists essentially of mathematically modeling (as best possible) the C/U path through the process and then obtaining control apparatus (sensor, compensator, controller) such that the compensation path transmission characteristics closely match those of the process path. Since compensation can never be perfect, an important part of the design process is to estimate the expected deviations in the static and dynamic characteristics of both paths and study the influence of such changes on the control quality, that is, how well C follows V. Digital simulation (such as our CSMP) is an ideal method for carrying out such studies since it accommodates both linear and nonlinear systems and allows rapid evaluation of the various combinations of deviations from ideal in the process and compensation paths.

Since the hardware in the compensation path is "instrumentation-type" equipment of our own selection, changes in its behavior are often minor when compared with those encountered in the process path. Thus the maintenance of a close match between the two paths is most often compromised by changes in the process, and this problem has, in fact, often frustrated practical application of the feedforward principle. The common presence today of significant computer capacity in control systems may provide a means to overcome these types of difficulties. The computer is used to implement a *process identification scheme*[1] which, intermittently or on a continuous basis, makes sufficient measurements and calculations to identify the current characteristics of the process path. Once these characteristics are known, parameters in the compensation path can be adjusted so that the match between the two paths is again reestablished and improved control is maintained.

7.2 COMMAND-COMPENSATED CONTROL

The basic concept of Fig. 1.5 is made more specific in Fig. 7.2. In Fig. 7.2a the command compensation G_C is shown combined with a feedback mode of control, a scheme to be discussed in a later chapter. Here we are interested in command

---

[1]E. O. Doebelin, *System Modeling and Response*, Wiley, New York, 1980, Chap. 6.

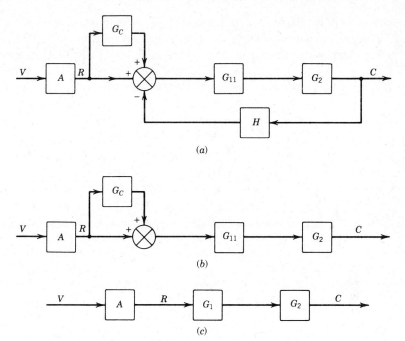

FIGURE 7.2 *Systems with command feedforward.*

compensation *alone*, thus Fig. 7.2*b* is appropriate. From this figure we see that

$$C(s) = [A(1 + G_C)G_{11}G_2]V(s) \qquad (7.2)$$

Since both G_C and G_{11} are items of controller hardware open to the designer's choice, the transfer function $A(1 + G_C)G_{11}$ gives no more design freedom than would G_{11} alone, thus the scheme reduces to Fig. 7.2*c*. That is, command compensation, used alone, can be implemented using just the basic controller G_1 suitably designed, G_C is not needed.

Since ideally we want $C \equiv V$, a "suitably designed" G_1 would be $1/G_2A$ since then $G_1G_2 \equiv 1.0$. This is the design principle of command compensation. As usual, the perfection implied by $C \equiv V$ is in practice unattainable; however significant improvements in control can often be achieved. The difficulty here is that the process dynamics G_2 are often "lagging" types, such as $K/(\tau s + 1)$, which means that for perfect compensation we need $G_1 = (\tau s + 1)/K$, a "leading" or differentiating type of response. Thinking now in terms of frequency response $(i\omega\tau + 1)/K$, such a leading type of hardware is impossible to construct in the real world since it implies response out to infinitely high frequencies. A *practical* goal for the controller design would be $G_1 = G_3/G_2A$, giving

$$\frac{C}{V}(i\omega) = \frac{G_3}{AG_2}(i\omega)AG_2(i\omega) = G_3(i\omega) \qquad (7.3)$$

The dynamics $G_3(i\omega)$ can easily be made to have a steady-state gain of 1.0 and can often be made to have a wider range of flat frequency response than did $G_2(i\omega)$, thus giving improved control, as shown in Fig. 7.3. For our earlier example where $G_2 = K/(\tau s + 1)$, we could use one of the lead controllers $(\tau s + 1)/K(\tau_C s + 1)$, discussed in Chapter 2, for G_1, giving (assuming $A(s) = 1.0$)

$$\frac{C}{V}(i\omega) = \frac{\tau s + 1}{K(\tau_C s + 1)}\left(\frac{K}{\tau s + 1}\right) = \frac{1}{\tau_C s + 1} \qquad (7.4)$$

If we can make τ_C significantly less than τ, then the C/V of Eq. 7.4 will be an improvement over G_2. Such improvements are often practical.

In addition to the earlier mentioned impossibility of perfect $(C \equiv V)$ response, this approach also suffers in practice from the unavoidable mismatch between the G_2 function in the controller and that in the process. Controller and (more likely) process dynamics will drift around, preventing the perfect "cancellation" of the numerator and denominator G_2's in Eq. 7.3 and a degradation of response. Fortunately, considerable static and dynamic mismatch can be tolerated and still result in significant net response improvement, making the approach practical except in cases of extreme mismatch. Again, digital simulation gives a convenient analysis method for the study of mismatch effects.

7.3 EXAMPLES OF COMMAND AND DISTURBANCE FEEDFORWARD

Steam generation equipment for process plants or electric power production often employs feedforward techniques at several points. In Fig. 7.4 the steam drum liquid-level control uses a combination of feedforward (disturbance compensation) and feedback. The main disturbance on liquid level is the rate at which steam is used by the various loads (three shown in Fig. 7.4). Since a steam-usage flowmeter is standard instrumentation provided for monitoring system operation, implementation of a disturbance compensation control mode that uses this same flow-rate signal is convenient. The feedback level control would normally compare the measured and desired levels and change the feedwater flow rate to keep the level constant. Sudden load (steam usage) changes would, however, cause considerable level fluctuations since correction is delayed by the lag between steam usage and drum level. By augmenting the feedback mode with disturbance measurement and compensation, the feedwater flow rate can be made to respond almost immediately to changes in steam usage, maintaining a better mass balance (feedwater in, steam out) for the steam drum. When measured and desired drum levels are equal, note that the feedwater flow rate setpoint is equal to the steam usage rate. Unavoidable errors and mismatches will of course keep the balance between feedwater inflow and steam outflow from being perfect; however any such mismatch will, over time, cause the measured level to gradually deviate from the desired level, biasing the

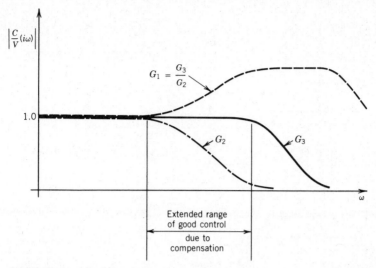

FIGURE 7.3 *Improvement in command-following due to feedforward compensation.*

feedwater set point in the proper direction to correct the error. Since the disturbance compensation takes care of most of the major disturbance, the feedback level control can be relatively low gain and slow acting, a desirable feature since the level measurement is often quite noisy due to sloshing in the steam drum.

In addition to maintaining a material (mass) balance, boiler control must also deal with the energy balance.[2] In Fig. 7.4, the major energy input is provided by fuel combustion, which is manipulated by changing the firing rate. When energy output (steam usage) increases, boiler pressure will drop unless energy input (firing rate) is increased. Modeling the loads as a single equivalent valve of "opening" C_v, the steam mass flow rate W_s can be approximated as[3]

$$W_s = C_v p \tag{7.5}$$

where C_v is not really the valve flow area (opening) but is proportional to it, and is taken as a measure of energy demand. The steam-flow-rate meter measures W_s as

$$W_s = k\sqrt{p\Delta p} \tag{7.6}$$

and, combining Eq. 7.5 and 7.6 we get

$$C_v = k\sqrt{\frac{\Delta p}{p}} \triangleq \text{"energy demand"} \tag{7.7}$$

[2]F. G. Shinskey, *Controlling Multivariable Processes*, ISA, Research Triangle Park, North Carolina, 1981, pp. 171–173; *Process Control Systems*, McGraw–Hill, New York, 1979, pp. 237–238;
[3]Shinskey, *Multivariable Processes*, pp. 171–173; *Process Control*, pp. 237–238.

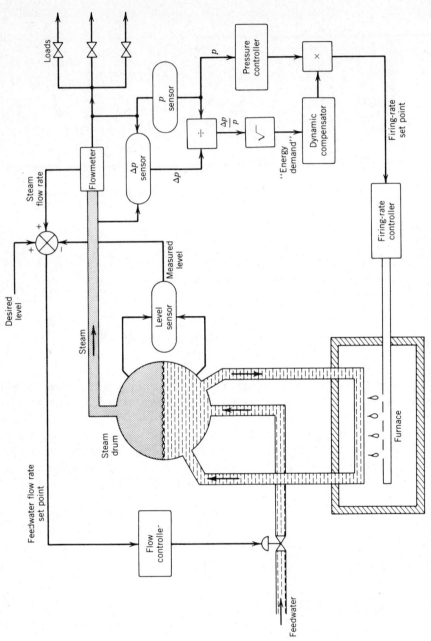

FIGURE 7.4 *Use of combined feedforward/feedback in steam generation system.*

If a user valve opens, note that pressure p drops and flow rate (and thus Δp) rises, thus Eq. 7.7 properly indicates an increase in demand. If flow rate alone (Eq. 7.6) had been used as an indication of demand, the increase in Δp and decrease in p would tend to offset each other, giving a false demand signal. In Fig. 7.4 the C_v disturbance compensation signal is used as a multiplying factor in a pressure feedback system. If energy demand is constant, the pressure-control feedback loop will maintain constant pressure by adjusting the fuel firing rate to combat all energy input disturbances such as changes in fuel heating value. The major output disturbance (steam usage) is measured and used to change firing rate in a feedforward manner, responding to load changes more quickly than possible with a pure pressure feedback system.

Heat exchangers, such as that in Fig. 7.5, often use disturbance compensation (feedforward) to achieve better control of liquid outlet temperature T_C by measuring and correcting for two major disturbances, cold liquid mass flow rate G_L and inlet temperature T_I. A steady-state energy balance gives

$$G_L c_L (T_C - T_I) = G_S H_S \tag{7.8}$$

where

$c_L \overset{\Delta}{=}$ liquid specific heat

$G_S \overset{\Delta}{=}$ steam mass flow rate

$H_S \overset{\Delta}{=}$ steam enthalpy change per unit mass

To make T_C equal to a desired value T_V, we must provide a steam flow rate G_S given by

$$G_S = \frac{G_L c_L (T_V - T_I)}{H_S} \tag{7.9}$$

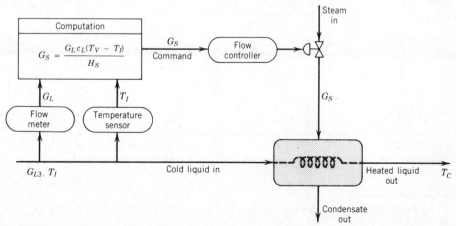

FIGURE 7.5   Heat exchanger control using disturbance compensation only.

Assuming T_V, c_L, and H_S to be known, measurement of G_L and T_I allows calculation of the needed G_S, which becomes the set point of a (closed-loop or open-loop) steam flow controller. As usual, modeling inaccuracies in Eq. 7.9 prevent perfect control of T_C. If specifications cannot be met, we can augment the feedforward control mode with feedback in several ways.[4]

    a. Add the computed steam flow signal G_S to the "error" (actuating) signal of a feedback loop on T_C and use this sum as the steam flow command.

    b. Use the actuating signal of a T_C feedback loop as a $\pm$ bias on the desired value T_V in the feedforward controller.

    c. Use the actuating signal of a T_C feedback loop as a $\pm$ bias on some other parameter (such as H_S) in the feedforward controller.

Figure 7.6 shows how method *b* might be implemented. If Eq. 7.8 is modified to include dynamic effects, all these control schemes may likewise be changed to achieve better control, since then both transient and steady-state behavior can be compensated. If an accurate analytical model is not available, good results are often obtainable by installing an adjustable lead-lag element $(\tau_1 s + 1)/(\tau_2 s + 1)$ in the feedforward control path and adjusting the values of τ_1 and τ_2 by trial and error to approximately compensate for the unknown process dynamics.

    Command feedforward has been used in various ways for servomechanism applications.[5] Figure 7.7 shows a combined feedforward/feedback system for

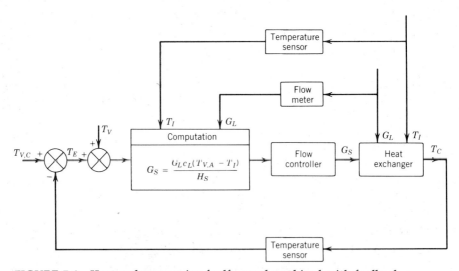

FIGURE 7.6 *Heat exchanger using feedforward combined with feedback.*

[4] C. L. Smith, *Digital Computer Process Control,* Intext Publ., Scranton, Pennsylvania, 1972, pp. 205–209.

[5] R. L. Duthie, Feed Forward can Improve Feedback Controls, *Cont. Eng.,* May 1959, pp. 136–140.

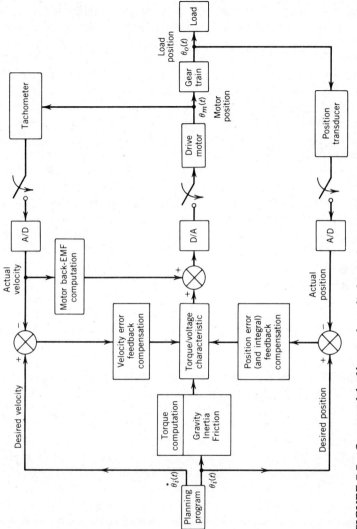

FIGURE 7.7 *Command feedforward applied to robot motion control.*

angular motion control of a single joint in a robot manipulator.[6] Since a digital computer is used for the feedforward calculations, A/D and D/A converters are used as shown to interface the analog and digital portions of the system. The trajectory that the manipulator is to follow as a function of time is computed in the planning program. Each link angle (angle of manipulator joint) is also computed. The drive motor torque required to produce the link angle/time function is then computed based on a model of the manipulator hardware. The motor torque/voltage characteristic is also modeled and thus the needed motor voltage can be computed. Since the various models are not exact, this feedforward scheme is augmented with conventional position and velocity feedback to compute correction torques to be summed with the main feedforward driving torque. Since the feedforward-derived portion of the motor torque does not require the existence of displacement or velocity *errors* for its generation (as the feedback-type torque components do), large and fast motions can be produced accurately. Duthie provides a detailed discussion of the relative advantages of such a combined feedforward/feedback system with a conventional pure feedback servo.

Some robots use conventional feedback control for each joint rotation and use feedforward only to correct for elastic deflections due to gravity loads on the robot structure and payload. Note that the majority of robots are intended to position a payload at a specific x, y, z location in space but that direct measurement of this location is difficult and *not* usually attempted. In a typical six-degrees-of-freedom robot, the angular positions of the six joint rotations are the quantities that can be conveniently measured and servo controlled. If the robot is at rest with all six angles precisely where commanded, the payload need *not* be at the desired x, y, z location due to elastic deflections of the robot members caused by gravity loads. A stiff (but lightweight) structural design may keep such elastic deflections within specifications. If not, a feedforward control mode may be useful in correcting for this effect since it is reproducible and can be calculated (or experimentally measured) as a function of the joint angles.

Figure 7.8 illustrates this concept with a simplified configuration for a two-axis (r, θ) robot whose task is to position a payload at any commanded point in a single vertical plane. Conventional servo systems position r and θ as commanded, but elastic deflection causes payload position to be in error. Translational transducers for measurement of r would normally measure the actual (elastically deflected) position and thus contribute little error. If, however, r is *inferred* from a measurement of power screw rotation angle, then axial elastic deflection causes an r error. Since the only convenient θ measurement is the joint rotation angle, elastic deflection will always cause a θ error as shown. For simplicity, assume r errors are negligible and that θ error due to bending is caused only by payload W_p acting as an end load on a simple cantilever beam

[6]B. R. Markiewicz; Analysis of the Computed Torque Drive Method and Comparison with Conventional Position Servo for a Computer-Controlled Manipulator, Jet Propulsion Lab, March 15, 1973, TM 33-601, Pasadena, California.

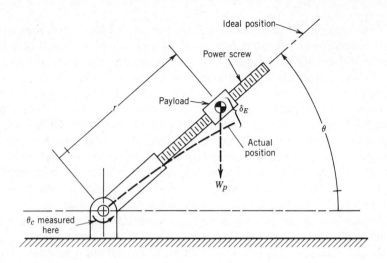

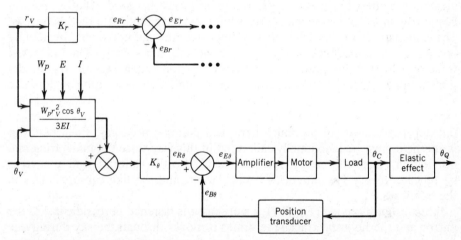

FIGURE 7.8 *Command feedforward used to correct robot elastic deflection errors.*

of length r. (The contribution of robot structure distributed gravity loads to end deflection δ_E could be calculated, but we do not wish to pursue this complexity.) Strength of materials then gives us

$$\delta_E = \frac{(W_p \cos \theta_V)r_V^3}{3EI} \tag{7.10}$$

and

$$\text{elastic correction angle} \triangleq \theta_{\text{cor}} \approx \frac{\delta_E}{r} = \frac{W_p r_V^2 \cos \theta_V}{3EI} \tag{7.11}$$

This correction is applied as command feedforward in the block diagram of Fig. 7.8.

Our final example of command-compensated control has been proposed[7] for the motion control of heliostats used in a solar power plant. Each heliostat is a large flat mirror, rotatable about two axes (azimuth and elevation) and controlled so as to accurately reflect the sun, as it "moves" through the sky daily, onto a fixed "power tower" where the heat is collected to generate steam. A large number of such heliostats are arranged around the central power tower on a flat field, so that the solar radiation impinging on a large surface area can be concentrated to a single collection point. Initial heliostat designs using feedback methods for steering the heliostats were sufficiently accurate but quite expensive. The proposed, more economical open-loop alternative uses stepping motors to position the two axes but, without feedback, has no means of verifying that the reflected solar beam is accurately pointed at the collector.

An open-loop heliostat, if constructed and installed with sufficient care could meet the pointing accuracy requirements but would also be excessively costly. In the proposed design, heliostat construction and installation accuracy requirements are modest but precision (repeatability) must be good. That is, the heliostat may have large systematic errors but must have moderate random errors. The systematic errors in each individual heliostat are "calibrated out" by initially operating it in a feedback mode with an attached "sun sensor." Data gathered while operating in this mode are computer processed to generate an error model that relates pointing error to the commanded azimuth and elevation angles. The sun sensor can now be removed and the heliostat put into tracking service. Commanded angles are now corrected by the error model before being sent to the stepping motors, giving much improved pointing accuracy. Each heliostat in the field must be individually calibrated in this way since manufacturing and installation errors will be different for each. Also the calibration may have to be repeated every few months if ageing, wear, and the like cause changes in the hardware.

In the calibrate mode (Fig. 7.9), a sun sensor is fastened perpendicular to the mirror and the heliostat tracks the sun as it moves through the sky during one day; the heliostat is *not* used for its normal function of reflecting the sun onto the power tower. As the heliostat tracking system is thus "exercised," a central computer uses astronomical data to accurately calculate the *true* azimuth and elevation angles of the sun. Because of mechanical (systematic) errors in the heliostat, the measured angles α_m and θ_m will not agree with the true angles α and θ. An error analysis (Baheti and Scott give details) relates the angles α, θ, α_m, and θ_m to the mechanical error sources. The model used takes into account three basic sources of error: mirror pedestal tilt (requires three numbers a_1, a_2, and a_3 to specify), drive-wheel radius errors (requires two numbers a_4, a_5 to specify), and elevation reference bias (requires one number a_6 to specify). The

---

[7]R. S. Baheti and P. F. Scott, Design of Self-Calibrating Controllers for Heliostats in a Solar Power Plant, Rept. 78CRD148, General Electric Corp. R&D, Schenectady, New York, July 1978.

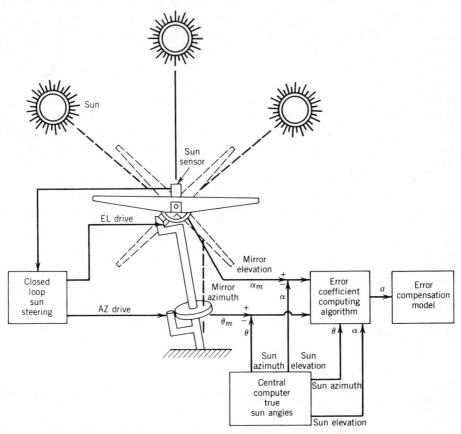

FIGURE 7.9 *Calibrate mode of heliostat system.*

error model equations turn out to be

$$(\theta_m - \theta) \sin \alpha = a_1 \cos \alpha \sin \theta - a_2 \sin \alpha + a_3 \cos \alpha \cos \theta \qquad (7.12)$$
$$+ \, a_4 \, \theta \sin \alpha$$

$$\alpha - \alpha_m = a_1 \cos \theta - a_3 \sin \theta - a_5 \alpha - a_6 \qquad (7.13)$$

As angles α, α_m, θ, θ_m are exercised over their full range during one day's operation, many sets of simultaneous values are recorded. If we add Eq. 7.12 and 7.13 we get a single equation of form

$$y = a_1 x_1 + a_2 x_2 + a_3 x_3 + a_4 x_4 + a_5 x_5 + a_6 x_6 \qquad (7.14)$$

Each simultaneous set of angle data gives a numerical value of y and corresponding numerical values for all the x's. If six sets of data were used we would have six equations in six unknowns (the a's) and could solve for the a's. Because of random errors and model approximations it is better to use *many* (not just six) sets of data and a least-squares approach to find the "best" values for the *a*'s. (Most readers have used least-squares methods to fit a straight line to scattered data. This problem is just a generalization of the line-fitting problem.) Standard multiple regression[8] programs are available on most computers to solve such problems. Having found the error model, the sun sensor is removed and the heliostat is put into its open-loop mirror steering mode (Fig. 7.10), where the error model now compensates the commanded angles, for much improved accuracy. Computer simulation of such a system showed improvements of 20 to 1 and Baheti and Scott felt that at least 10 to 1 could be realized in actual heliostats.

The method just explained is a specific example of a general design principle that is of increasing importance to machine and process designers. It is the substitution of software and computer power (today readily available) for mechanical perfection. That is, it may be more cost effective to use a less perfect mechanical design and then use computed corrections, *tailored to the individual "defects" of each produced machine,* to achieve a high level of overall performance.

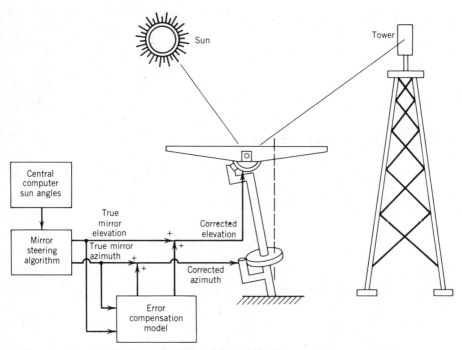

FIGURE 7.10 *Mirror steering mode of heliostat system.*

---

[8]E. O. Doebelin, *System Modeling and Response,* Wiley, New York, 1980, pp. 300–320.

PROBLEMS

7.1 The text suggests three methods (**a, b, c**) for adding feedback to the feed-forward system of Fig. 7.5. Draw and explain a block diagram showing implementation of method **a.**

7.2 Repeat problem 7.1 for method **c.**

7.3 A linearized dynamic model of the heat exchanger of Fig. 7.5 is shown in Fig. P7.1.

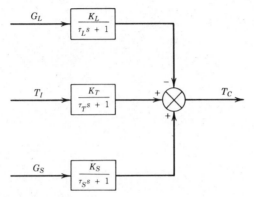

FIGURE P7.1.

   a. Devise, block diagram, and explain a disturbance compensation control if only steady-state performance is of interest.

   b. Repeat Part a if both dynamic and steady-state compensation is important. Assume all three time constants are different.

   c. Repeat Part b if $\tau_L = \tau_T$.

7.4 Modify the system of Fig. 7.6 to provide both steady-state and dynamic compensation, using the heat exchanger model of Fig. P7.1.

7.5 Modify the system of Fig. 7.8 to include correction for elastic deflections in both the r and θ directions, assuming the r measurement is inferred from power-screw rotation angle.

7.6 For a gas-fired open-loop home heating system, make a comprehensive list of all the disturbances that might cause the room temperature to deviate from its desired value. For which of these might it be practical to make measurements and compensations?

CHAPTER 8
On-Off Control and Nonlinear System Analysis

8.1 Forms and Characteristics of On–Off Control

This chapter begins our systematic study of the various basic modes of feedback control. By *mode of control* I mean the nature of the behavior of the controller G_1 in the general system diagram of Fig. 5.1. Some common control modes are on–off, proportional, integral, and derivative. The sequence of our treatment follows essentially that which a designer would use in *selecting* a control mode for a particular application. That is, good design, in general, uses the simplest (and thus usually the least expensive and most reliable) hardware that will meet system performance specifications. We should thus try the simplest mode first and go to more complex ones only as the simpler ones are proven inadequate by analysis. In the control system field, on–off controls are generally the simplest possible from a hardware viewpoint. Their analysis, due to nonlinearity, has in the past been difficult or impossible; however, today digital simulation allows us to easily get essentially exact results for any specific form of system with given numerical values. This quick and easy computer-aided analysis tool becomes also a design approach since the effect of parameter and structure changes can be so easily investigated.

On–off controls can take various forms. Figure 8.1a shows perhaps the most basic, a two-position controller without a neutral zone (dead space). Note the simplicity of controller action; manipulated variable M can take on only two possible values, depending on whether actuating signal E is positive or negative. The controller gives the same corrective effort irrespective of whether E (system "error") is small or large. Although the nonlinearity of the system prevents application of Routh or Nyquist stability criteria, it is easily seen that the system is unstable and will go into a limit-cycle oscillation. This is because M is never off; it is always on in either a positive or a negative sense. Thus controlled variable C is bound to be driven back and forth in a cyclic manner. Another viewpoint on stability is from the energy aspect. The controller can supply energy and/or material to the process at only two discrete rates. If neither of these *precisely* matches the demand of the process, the controller must continually shuttle back and forth between a supply that is too large and one that is too small, again giving a limit-cycling instability. For the reasons just enumerated, on–off controls very often limit cycle and the designer must evaluate the frequency and amplitude of the limit cycle to judge whether such behavior is acceptable. Most residential heating–cooling systems, for example use on–off control since the limit-cycling behavior is acceptable both in terms of temper-

221

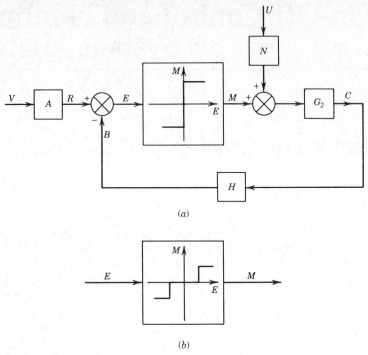

(a)

(b)

FIGURE 8.1 *Two types of on–off controller.*

ature fluctuations being small enough to be comfortable and cycling rates being slow enough to not wear out the switching hardware (solenoid gas valve, etc.) prematurely.

Figure 8.1b shows a variation called a three-position controller with a neutral zone. Here the positive and negative "on" positions are augmented by an "off" position or neutral zone where the controller supplies no energy to the process. There are now three possible discrete rates of energy supply, and limit cycling will again usually occur because none of the three exactly matches process needs. In a small number of applications, some system operating conditions *do* correspond to a zero-energy demand and the system will settle stably into the neutral zone without limit cycling since process supply and demand are balanced at this point. Such cases are, however, relatively rare.

8.2   THE DESCRIBING-FUNCTION METHOD

In Fig. 8.1a the system is assumed to be linear (with constant coefficients) except for the on–off controller. If U and V are given as specific functions of time, and if all initial conditions are known, we will know what M is at time zero. The equation relating B to M and U is then analytically solvable, and with V given, we can also get E. This solution is valid until E crosses zero, whereupon we must stop using this solution and start a new one with the new value of M caused by E crossing zero. The final conditions of the first solution become the initial conditions of the next. One can thus "piece together" a complete time history of system response since, for any stretch of time for which M is constant, the equations are linear with constant coefficients. The necessary matching of final and initial conditions, however, makes such an approach extremely tedious and complex.

To aid in the design of on–off controls and other nonlinear systems (including many that *cannot* be solved in the piecewise fashion just explained) we have available the *describing-function method*. Of all the analytical methods developed over the years for nonlinear systems, this one is generally agreed upon as being the most practically useful. It is an approximate method but experience with real systems and computer simulation results shows adequate accuracy in many cases. The method predicts whether limit-cycle oscillations will exist or not and gives numerical estimates of oscillation frequency and amplitude when limit cycles are predicted. It also usually provides design information with regard to how the system must be changed to prevent limit cycling. Basically the method is an approximate extension of frequency response methods (including Nyquist stability criterion) to nonlinear systems.

We limit ourselves to those situations (such as in Fig. 8.1) in which a single nonlinearity is embedded in a control loop of otherwise linear components. The first step in a describing-function study is always to remove the nonlinear element from the control system so it can be studied as a separate device. With the nonlinearity isolated, we apply to its input a sine wave $A_i \sin \omega t$ and, using the given characteristics of the nonlinearity, determine what its output waveform will be. The output can take many different forms, depending on the nature of the nonlinearity; however it is generally a periodic but *nonsinusoidal* function. A linear element would, of course, produce a pure sinusoidal output with frequency exactly equal to the input frequency (see Fig. 8.2).

The basic assumption made in describing-function analysis is that the nonsinusoidal waveform leaving the nonlinear element will, by the time it proceeds around the rest of the control loop and arrives at the nonlinear element input, have become nearly sinusoidal, as assumed when we determined the nonlinear element output waveform. If we think of the nonlinear element input waveform in terms of its Fourier series representation, we require that the fundamental frequency component (first harmonic) be much stronger than any of the higher harmonics. This requires that, in traversing the path through the linear parts of the system from nonlinearity output back to nonlinearity input, the higher harmonics will have been effectively low-pass filtered, relative to the first harmonic. When the linear part $G_L(i\omega)$ of the system does indeed provide a sufficiently strong filtering effect, then the predictions of describing-function analysis usually are a good approximation to actual behavior. Figure 8.3 illustrates these concepts.

In the describing-function method we consider the nonlinearity as an isolated

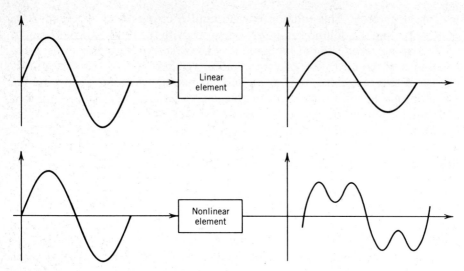

FIGURE 8.2 *Sinusoidal response of linear and nonlinear elements.*

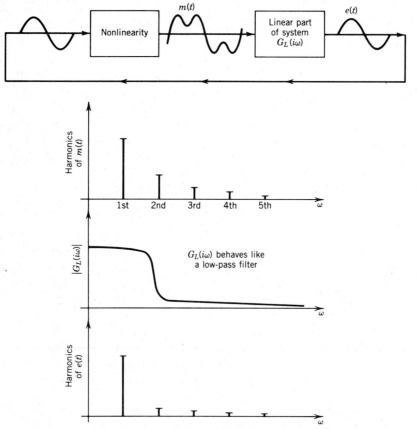

FIGURE 8.3 *Low-pass filtering assumed in describing-function analysis.*

element and approximate its output by taking only the first harmonic component $A_o \sin(\omega t + \phi)$ of its Fourier series. This will be a pure sine wave of the same frequency as the assumed sinusoidal input $A_i \sin \omega t$. We can then form the ratio of output sine wave $A_o \underline{/\phi}$ and input sine wave $A_i \underline{/0°}$ and treat this as an approximate sinusoidal transfer function for the nonlinear element. This approximate transfer function is called the describing function. It differs from a linear system transfer function in that its numerical value will vary with input amplitude A_i. Also, it may or may not depend on frequency ω. When embedded in an otherwise linear system the describing function can be combined with the "ordinary" sinusoidal transfer function of the rest of the system to obtain the complete open-loop function $(B/E)(i\omega)$; however we will get a *different* $(B/E)(i\omega)$ for every different amplitude A_i. We can check all of these different $(B/E)(i\omega)$'s for stability, using the Nyquist criterion. If *any* of them indicate closed-loop instability, we predict that the system *can* limit cycle and also can estimate the amplitude and frequency of the limit cycle.

Let us take an on–off controller with a neutral zone (dead space) as in Fig. 8.4a as an example. If A_i is less than the dead space d, then the controller produces no output; the first harmonic component of the Fourier series is of course zero, and the describing function is also zero. For $A_i > d$, the controller produces the "square wave" output $f_o(t)$. Using the Fourier series methods of Chapter 2 and details from Fig. 8.4b, we get

$$\left. \begin{array}{c} \text{average value of } f_o(t) = 0 \\[2mm] a_n = 0 \end{array} \right\} \text{ by symmetry} \tag{8.1}$$

$$b_n = \int_{-(\pi/\omega)}^{\pi/\omega} f_o(t) \sin \frac{n\pi t}{\pi/\omega} \, dt = 4 \int_{t1}^{\pi/2\omega} b \sin \frac{n\pi t}{\pi/\omega} \, dt \tag{8.2}$$

For $n = 1$, this gives the amplitude A_o of the first harmonic component as

$$A_o = \frac{4b\sqrt{(A_i/d)^2 - 1}}{\pi(A_i/d)} \qquad A_i \geq d \tag{8.3}$$

Since a_1, (the Fourier series cosine coefficient) is zero, the first harmonic component of $f_o(t)$ is exactly in phase with $A_i \sin \omega t$ and the describing function G_d is given by

$$G_d = \frac{A_o \underline{/0°}}{A_i \underline{/0°}} = \frac{4b}{\pi A_i} \sqrt{1 - \left(\frac{d}{A_i}\right)^2} \tag{8.4}$$

For a given controller, b and d are fixed and G_d is a function of input amplitude A_i, which is graphed in Fig. 8.5 together with peak location and value found by standard calculus maximization procedures. Note that this particular G_d is independent of frequency ω. Also, for a given A_i, G_d is just a pure real positive number, and thus plays the role of a steady-state gain in a block diagram. However this gain term is unusual in that it *changes* when A_i changes.

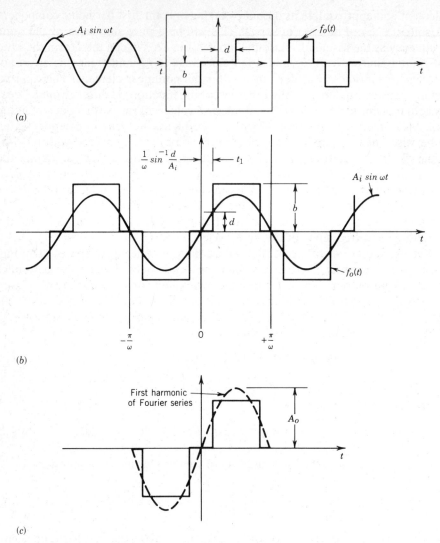

FIGURE 8.4 *Fourier series analysis of three-position controller.*

The describing function for the simpler on–off controller of Fig. 8.1a can be obtained from Eq. 8.4 by letting d approach zero.

$$G_d = \frac{4b}{\pi A_i} \qquad (8.5)$$

This describing function is also a pure gain term but now the gain approaches infinity as A_i goes to zero (see Fig. 8.6). Let us use such a controller in an otherwise linear system with three first-order components as in Fig. 8.7. For any steady command V less than 5.0, it should be clear without use of describing-function methods or other analysis that this system *must* limit cycle. (What

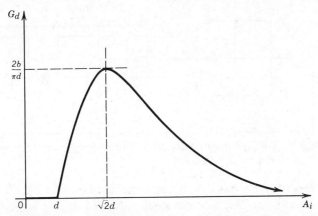

FIGURE 8.5 *Describing function of three-position controller.*

happens for steady V's greater than 5.0?) We should also note that an exact analytical solution is possible (though tedious) by "piecing together" solutions of the linear differential equation relating C to M. (Whenever E crosses zero, we switch M to the proper value and use the final conditions of the preceding solution as initial conditions for the next.) Rather than pursuing this tedious approach, we will use the approximate describing-function analysis and then check its validity by CSMP digital simulation of the exact equations.

Since the block between E and M in Fig. 8.7 is approximated as a gain term G_d

$$\frac{B}{E}(i\omega) = \frac{5G_d}{(i\omega + 1)(0.1\,i\omega + 1)(0.1\,i\omega + 1)} \tag{8.6}$$

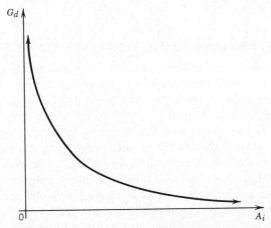

FIGURE 8.6 *Describing function of two-position controller.*

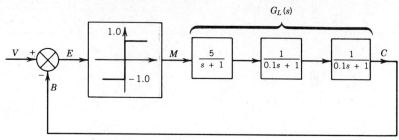

FIGURE 8.7   *Two-position controller with third-order linear elements.*

For large values of A_i (A_i has now become the amplitude of E), G_d is small and the system would be stable; whereas small values of A_i give large values of G_d and instability (see Fig. 8.8). Actually, the closed-loop system will *self-adjust* its amplitude such that the $(B/E)(i\omega)$ curve goes *exactly* through the -1 point on the Nyquist plot, a point which, for a strictly linear system, would correspond to oscillations of fixed amplitude, neither increasing nor decreasing. The reason for this behavior is as follows. Suppose A_i is at a value that puts $(B/E)(i\omega)$ exactly through -1, when a disturbance momentarily causes A_i to increase. An increase in A_i causes G_d (and thus loop gain) to decrease to a value corresponding to stability (see Fig. 8.6). Since the system is "momentarily" stable, its amplitude starts to decrease, bringing it back to the original amplitude where $(B/E)(i\omega)$ went exactly through -1. If the amplitude should momentarily *decrease* below the value putting $(B/E)(i\omega)$ through -1, this causes G_d to increase to a value corresponding to instability, which causes the amplitude to start to build up, returning the system to its original operating point. This means that sustained oscillations (limit cycling) are possible *only* at the one amplitude which gives a G_d value putting $(B/E)(i\omega)$ exactly through -1. This can be expressed as

$$G_d G_L = -1 \tag{8.7}$$

$$G_L = -\frac{1}{G_d} \tag{8.8}$$

This requirement (Eq. 8.8) for the existence of limit-cycle oscillations holds for *all* describing function applications, not just the present example.

Equation 8.8 is usually interpreted graphically as follows. We can plot the frequency response curve $G_L(i\omega)$ on the polar (Nyquist) graph. Likewise, a graph of the negative reciprocal $(-1/G_d)$ of the describing function can be overlaid on the $G_L(i\omega)$ graph. For Eq. 8.8 to be satisfied, the two graphs must intersect. If they do not, the describing function method predicts that limit-cycle oscillations can not occur. If the graphs do intersect, the point of intersection on the $G_L(i\omega)$ graph gives a numerical value for the *frequency* of the limit cycle; whereas the same point on the $(-1/G_d)$ graph gives us the predicted *amplitude* of the oscillation. For the system of Fig. 8.7 the graphs are shown in Fig. 8.9. Although, in general, Eq. 8.8 often *must* be solved graphically, in the present

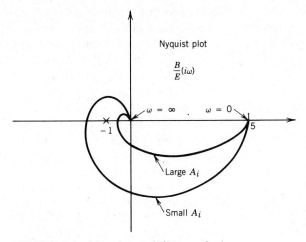

FIGURE 8.8 *Nyquist stability analysis.*

example a more accurate analytical approach is possible. Since $-1/G_d$ is a negative real number, it is clear that if an intersection is to occur it *must* happen at $-180°$ phase angle, thus we can just run a frequency response computer program for $G_L(i\omega)$ and find the ω that gives $-180°$. In our example this occurs at $\omega = 10.95$ rad/s. Furthermore, at $\omega = 10.95$, $|G_L(i\omega)| = 0.206$, thus we can use Eq. 8.5 and 8.8 to get the amplitude.

$$|G_L G_d| = 1.0 \qquad (0.206)\left(\frac{4}{\pi A_i}\right) = 1.0 \qquad A_i = 0.262 \qquad (8.9)$$

We see in this example that describing-function analysis predicts that the signal E in Fig. 8.7 can exhibit a limit-cycle oscillation of frequency 1.74 Hz

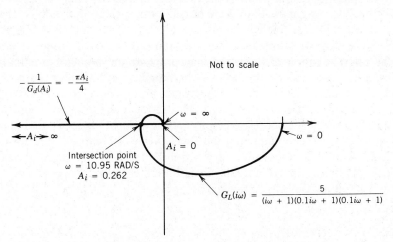

FIGURE 8.9 *Describing-function predictions for limit cycle.*

and amplitude 0.262. Describing-function analysis generally does *not* tell us what conditions (command V or disturbance U inputs, initial conditions) are required to get the limit cycle started. The basic assumption that the input signal to the nonlinearity is of form $A_i \sin \omega t$ implies that U and V are zero, since nonzero U and/or V values usually result in the nonlinearity input signal containing components *in addition* to the assumed sine wave. Thus when *checking* describing-function predictions against digital simulation, the best agreement will be obtained if we set U and V to zero and excite the system with some nonzero initial condition, or else apply some sort of U and/or V *momentarily* and then "turn it off." Actually, in practical design work, the *possibility* of limit cycling (irrespective of what it takes to start it) is usually so undesirable that one looks for ways to *prevent* it rather than taking time to investigate conditions necessary for initiation. Finally, for situations where the nonlinearity input contains signals in addition to $A_i \sin \omega t$ (such as U and/or V not being zero), the method of *dual-input describing functions*[1] may be useful.

To gain some further insight into on–off control behavior and describing-function analysis, let us modify the system of Fig. 8.7 by letting the linear portion be, successively, second order with

$$G_L(s) = \frac{5}{(s + 1)(0.1s + 1)} \tag{8.10}$$

and first order with

$$G_L(s) = \frac{5}{s + 1} \tag{8.11}$$

Figure 8.9 now looks like Fig. 8.10 where we see that intersection of $G_L(i\omega)$ and $(-1/G_d)$ is now impossible since the phase angle of neither Eq. 8.10 nor 8.11 can be more lagging than $-180°$. Describing-function analysis thus seems to predict no limit cycling, whereas the fact that M *must* be either $+1.0$ or -1.0 *dictates* that the system oscillate. One possible interpretation would be that the first- and second-order linear systems provide less of the low-pass filtering assumed in the describing-function method than did the third-order system and thus the approximation has become inaccurate to the point of predicting no limit cycle when actually one occurs. Another interpretation would note that the curves actually *do* "intersect" at the origin, predicting a limit cycle of infinite frequency and infinitesimal amplitude. This latter interpretation, even though it predicts a *physically* impossible result, agrees with the rigorous mathematical solution of the differential equations which can be carried out using the "piecewise-linear" solution method mentioned earlier. That is, if we set $V \equiv 0.0$, give C some nonzero initial value, and then solve the resulting linear equations in piecewise fashion (switching M between $+1.0$ and -1.0 whenever

---

[1]A. Gelb and W. E. Vander Velde, *Multiple-Input Describing Functions and Nonlinear System Design,* McGraw–Hill, New York, 1968.

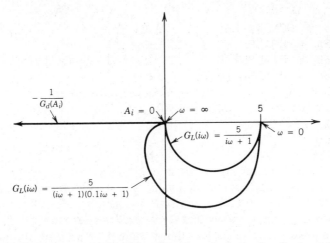

FIGURE 8.10 *Describing-function analysis for first- and second-order linear elements.*

E crosses zero), we find that C oscillates about zero with everdecreasing amplitude and everincreasing frequency.

We now want to use CSMP digital simulation to check the accuracy of the describing-function approximate results for some of the situations previously analyzed. We will write one program that will include the three cases (G_L = third order, second order, and first order) of the system of Fig. 8.7. We will let V be a unit step input that lasts from $t = 0$ until $t = 2.5$ and is then removed. This will give us information both on systems with $V = 0$ (the situation *assumed* by the describing-function method) and $V \neq 0$ (a situation *not* conforming to describing-function assumptions). Our CSMP program goes as follows.

```
TITLE   ON OFF CONTROL STUDIES
METHOD RKSFX      requests fixed-step-size integration method
PARAM   K = 5.      sets gain at 5.0
        V1  = STEP (0.0)-STEP(2.5)      defines input
        E1  = V1-C1      summing junction
        M1  = FCNSW(E1, -1., 0.0, 1.0)      on–off controller
        C11 = REALPL (0.0, 1., M1)      first-order system, i.c. = 0.0,
                                                τ = 1
        C1  = K*C11      gain of K (completes model of first system)
        V2  = V1      start model of second system
        E2  = V2 - C2
        M2  = FCNSW (E2, -1., 0.0, 1.)
        C2A = REALPL (0.0, 1., M2)  }
        C2B = REALPL (0.0, 1., C2A)  }  GL second order
        C2  = K*C2B      complete model of second system
        V3  = V1      start model of third system
```

```
           E3 = V3 - C3
           M3 = FCNSW (E3, -1., 0.0, 1.)
           C3A = REALPL (0.0, 1., M3)
           C3B = REALPL (0.0, .1, C3A)      G_L third order
           C3C = REALPL (0.0, .1, C3B)
           C3 = K*C3C     complete model of third system
  TIMER  FINTIM = 5., DELT =.001, OUTDEL  = .005
  OUTPUT C1, C2, C3
  PAGE GROUP
```

Figure 8.11 gives the results of this study. We see that C1 (system where G_L is first order) goes to the commanded value of 1.0 without overshoot and then appears to follow this *perfectly* (no visible oscillation). Actually, examination of the *numerical* printout shows that C1 is *oscillating* around 1.0, but the amplitude is so small (relative to 1.0) that it cannot be seen on the graph. The *frequency* of this limit cycle (which theoretically should be infinite) is of course incorrectly predicted by the digital simulation. Actually, *both* the amplitude (which should be infinitesimal) and the frequency are determined in this case by the approximate nature of the computing algorithms in CSMP, which cannot

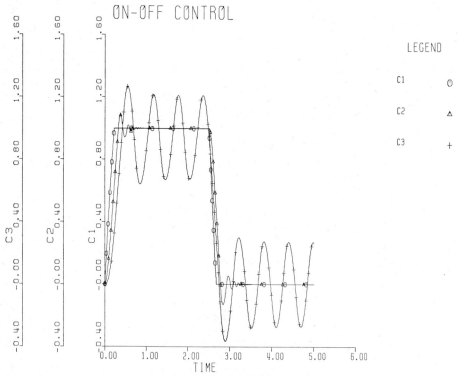

FIGURE 8.11 *Digital simulation for two-position on–off control systems.*

handle variations of infinitesimal size and infinite frequency. Note, however, that the *graphical* result of Fig. 8.11 is *not* misleading. An oscillation around 1.0 of infinitesimal amplitude and infinite frequency *does* look just like Fig. 8.11. When G_L is made second order (C2 in Fig. 8.11) we initially get visible overshooting and oscillation at noninfinite frequency, which *is* accurately computed by CSMP. As time goes by, however, we again (as predicted by theory) go to "infinitesimal" amplitude and "infinite" frequency, which is not correctly computed by CSMP but "looks OK" graphically.

When G_L is made third order we finally are dealing with mathematics that the CSMP numerical algorithms can handle with high accuracy over the entire time range of the problem. We see in Fig. 8.11 that after an initial overshoot to 1.26, a unit step input results in a limit cycle between 1.21 and 0.67, giving an amplitude of 0.27, quite close to the describing-function prediction of 0.262. The oscillation is however not symmetrical about 1.0, thus the input to the nonlinearity contains a "dc bias" in addition to the oscillatory component. This violates the assumption of the describing function analysis, but seems not to cause much error in this example. The limit-cycle frequency measured from Fig. 8.11 is 1.67 Hz and is again in good agreement with the predicted 1.74 Hz. When, at time = 2.5, V is made zero, we find the resulting limit cycle is now symmetrical around zero with amplitude of 0.27 and frequency of 1.72 Hz, even closer to the prediction, since the sinusoidal input assumption is now more closely realized. For both limit cycles observed in Fig. 8.11, the oscillatory wave form appears quite sinusoidal to the eye (though of course it is *not* a sine wave); thus this aspect of the describing-function assumptions seems to be adequately satisfied.

Let us now review and interpret the analytical and simulation results obtained thus far. When we use low (first and second) order dynamic models for the linear portion of the system, whether we employ exact analytical methods, digital simulation, or describing-function approximations, we predict physically impossible events, that is, infinite switching frequencies. Recall that whenever we choose to model a real physical system as, say, first order, we realize that this can be a good approximation only over a certain low-frequency range. If the system is subjected to higher frequency inputs, the first-order model cannot be accurate. We have just found that our on–off feedback systems actually generate such high frequency behavior when G_L is "too simple" (first or second order). We must thus conclude that to get realistic predictions of closed-loop behavior, the linear open-loop model must be of "sufficiently high" order. In our current example this appears to be third order, or perhaps more correctly, $G_L(i\omega)$ must have phase angle more lagging than $-180°$.

As our next example let us modify the system of Fig. 8.7 by giving the controller a dead space d as in Fig. 8.4. Since the maximum value of G_d is now $2b/\pi d$ (rather than ∞), the graphs of $-1/G_d$ and G_L may or may not intersect, and thus limit cycles may or may not occur (see Fig. 8.12). It is clear that if an intersection *is* to occur, it must again happen when the angle of $G_L(i\omega)$ is $-180°$, thus the describing function predicts a frequency of 1.74 Hz, just as for the

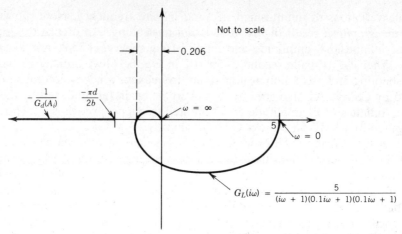

FIGURE 8.12 *Describing-function analysis of three-position on–off control system.*

controller without dead space. It is also clear that limit cycling is predicted only for deadspace d smaller than the value given by

$$\frac{\pi d}{2b} = \frac{\pi d}{(2)(1)} = 0.206 \qquad d = 0.131 \qquad (8.12)$$

If we replace the third-order G_L with the second-order or first-order models of Eq. 8.10 and 8.11, intersections are now impossible, even at $\omega = \infty$. We again use CSMP to check these describing-function predictions, using the same type of V input as in the previous program. A partial listing of the program is as follows.

```
V1 = STEP (0.0) - STEP (4.0)
V3 = V1
E3 = V3 - C3
M33 = DEADSP (-.1, .1, E3)    dead space of ±0.1 } on–off
M3 = FCNSW (M33, -1., 0.0, 1.)    b = 1.0       } controller
C3A = REALPL (0.0, 1., M3)   ]                    with dead
C3B = REALPL (0.0, .1, C3A)  } third-order G_L    space
C3C = REALPL (0.0, .1, C3B)  ]
C3 = 5.*C3C
```

The complete program includes two more models; for G_L first order (controlled variable called C1 and second order (controlled variable called C2).

This program chose $d = 0.1$ since Eq. 8.12 predicts limit cycling *should* occur for the third-order G_L for such a d value. Figure 8.13 confirms this prediction for both $V = 1.0$ and $V = 0.0$. For $V = 1.0$, the first and second order G_L *do* limit cycle (with "infinite" frequency and "infinitesimal" amplitude) even though

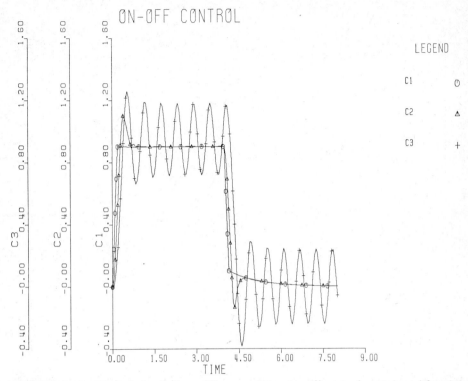

FIGURE 8.13 *Digital simulation of three-position on–off control systems with ±0.1 dead space.*

no intersection of G_L and $-1/G_d$ is possible. This occurs because C can be maintained near 0.9 *only* if M switches between $+1.0$ and 0.0. When $V = 0.0$ the first- and second-order G_L do *not* limit cycle but rather settle stably into the dead space with $M = 0.0$, since no control effort is required to maintain C in the neighborhood of zero. In Fig. 8.14 we rerun this program with $d = 0.16$, a value predicted to *not* cause limit cycling. For the third-order G_L and $V = 1.0$ we see that limit cycling *does* occur. Describing function analysis fails here since the nonlinearity input has a dc bias, violating the assumption of pure sinusoidal input. It is also clear that C can be maintained near 1.0 only if M oscillates. The first- and second-order G_L models limit cycle (infinite frequency, infinitesimal amplitude) for the same reason. When V goes to zero, the third-order G_L model oscillates a few cycles but then settles stably into the dead space, with $M = 0.0$, as predicted by describing-function analysis. The first- and second-order G_L models also do not limit cycle. In this example we see that describing-function analysis correctly predicts some aspects of system stability but is misleading with respect to others.

In the previous example, for G_L third order, $d = 0.16$, and $V = 1.0$, describing-function analysis incorrectly predicts no limit cycling, though, of course,

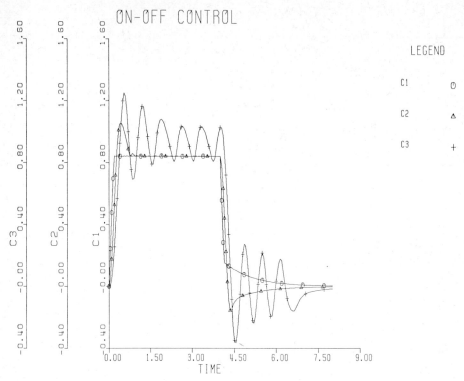

FIGURE 8.14   *Digital simulation of three-position on–off control systems with ± 0.16 dead space.*

one of the basic assumptions *was* violated. As noted earlier, limit cycling *had* to occur since *C* cannot be maintained near 1.0 without *M* periodically turning on. A system design change to overcome this problem substitutes an integrator for the first-order system $1/(s + 1)$ (see Fig. 8.15) creating a version of integral control called *single-speed floating*. Because of the integrator, signal *MI* (and thus also *C3*) can be maintained at a nonzero value even if *M* returns to zero. We now get the $-180°$ angle for G_L at $\omega = 10.00$ rad/s (1.59 Hz) and $|G_L|$ is 0.25 at this frequency. We then predict limit cycling for *d* smaller than the value given by

$$\frac{\pi d}{2b} = \frac{\pi d}{(2)(1)} = 0.25 \qquad d = 0.159 \tag{8.13}$$

A CSMP simulation was run for $d = 0.13$ (limit cycling predicted) and for $d = 0.19$ (no limit cycling predicted). Figures 8.16*a* and 8.16*b* show that describing function predictions are now accurate for both $V = 1.0$ and $V = 0.0$. (The measured frequency in Fig. 8.16*a* is 1.59 Hz, essentially perfect agreement with the prediction.)

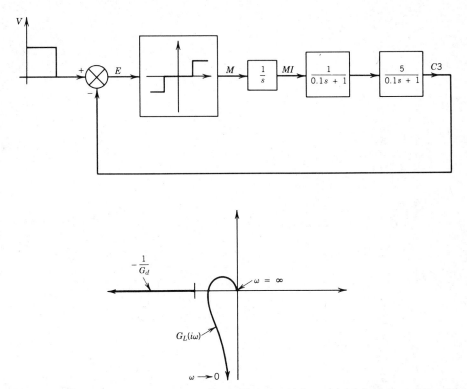

FIGURE 8.15   *Three-position controller with integral dynamics (single-speed floating control).*

For controllers with dead space, G_d has the shape shown in Fig. 8.5 and the plot of $-1/G_d$ (as in Fig. 8.12 and 8.15) is really a *double* line (see Fig. 8.17*a*). An intersection with G_L is thus really *two* intersections, at two *different* values of amplitude A_i, since both these amplitudes (A_{i1} and A_{i2} in Fig. 8.17*b*) give the *same* equivalent gain value. Only one of these amplitudes, however, corresponds to a limit cycle that can be *maintained* in a real-world system. Although an oscillation *exactly* at amplitude A_{i1} does give a Nyquist plot for G_dG_L that goes exactly through the -1 point and thus corresponds to an equivalent linear system with pure imaginary closed-loop roots, any momentary decrease in amplitude (caused by some small disturbance which will always be present in a real system) results in a lower G_d value, a *stable* closed-loop system, and a decay of the oscillation to zero. A momentary *increase* in amplitude causes an increase in G_d and thus a loop gain corresponding to an *unstable* closed-loop system (roots in right half plane). This instability causes the amplitude to increase even more, causing the operating point to move toward A_{i2}. We see thus that amplitude A_{i1} cannot be sustained since the *slightest* disturbance away from A_{i1} results in either a decay or a buildup of the oscillation. When we examine the

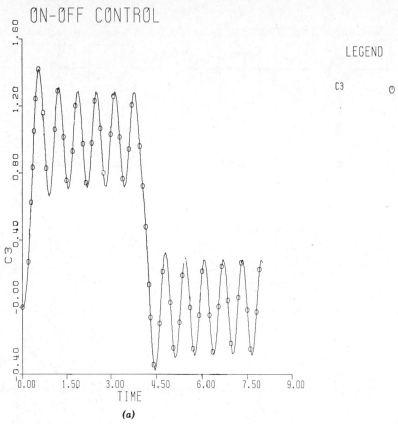

(a)

FIGURE 8.16   *Digital simulation of single-speed floating control.*

operating point at A_{i2} we see that any increase in amplitude causes a decrease in loop gain, momentary stability, and a *return* toward A_{i2}; whereas a decrease in amplitude causes an increase in loop gain, momentary instability, and a buildup of amplitude back to A_{i2}. Thus when disturbances momentarily "bump" the amplitude away from A_{i2} there is a *restoring* tendency that returns the system to the A_{i2} amplitude, making this amplitude the only one that will be observed in a real system.

8.3   UNINTENTIONAL NONLINEARITIES

Although nonlinear on–off controllers such as we have just discussed are intentionally designed into systems, we must also often deal with various (usually undesirable) nonlinear effects present in hardware that is intended to be linear. Examples of such "parasitic" nonlinearities include various forms of nonviscous friction, saturation, backlash or hysteresis, preload, and dead space. Several of

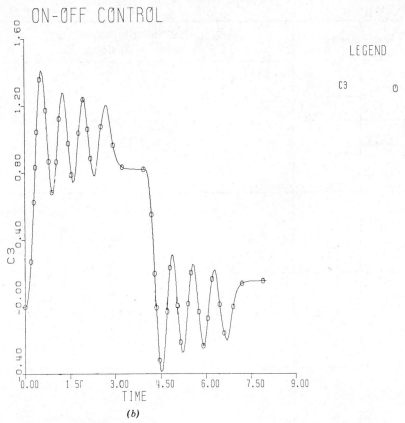

FIGURE 8.16 (Cont.)

these effects will be studied in later chapters where they arise naturally in various contexts. Here we propose to treat only one of these since a single example should be sufficient to show that the describing function method applies to this class of nonlinearities in essentially the same way as it did to our on–off controllers. The control system designer rarely needs to *derive* the describing function (as we did in Fig. 8.4 and associated text) for a particular nonlinearity since most of the common nonlinearities were analyzed and "cataloged"[2] many years ago. (Table 8.1 gives a modest sampling.) We thus mainly need to know how to *use* the describing function, together with G_L of the linear part of the system, as explained in Eq. 8.8 and Fig. 8.9.

We choose for our example the so-called *friction-controlled backlash* since it is of common occurence in mechanical systems and its behavior brings out certain features of describing-function application not encountered in our earlier ex-

---

[2]D. Graham and D. McRuer, *Analysis of Nonlinear Control Systems,* Wiley, New York, 1961, Table 4-1, pp. 128–133; A. Gelb and W. E. Vander Velde, *Describing Function* pp. 519–538.

TABLE 8.1

Nonlinearity	Describing Function (input $= A_i \sin\omega t$)
Two-position on-off	$$G_d = \frac{4b}{\pi A_i}$$
Three-position on-off	$$G_d = 0 \qquad A_i < d$$ $$G_d = \frac{4b}{\pi A_i}\sqrt{1 - \left(\frac{d}{A_i}\right)^2} \qquad A_i \geq d$$
Two-position with hysteresis	$$|G_d| = \frac{4b}{\pi A_i}$$ $$\underline{/G_d} = -\sin^{-1}\left(\frac{h}{A_i}\right) \qquad A_i \geq h$$
Three-position with hysteresis	$$G_d = 0 \qquad A_i < d + h$$ $$|G_d| = \frac{4b}{\pi A_i}\cos\left(\frac{A + B}{2}\right) \qquad \underline{/G_d} = \frac{A - B}{2} \qquad A_i \geq d + h$$ $$A = \sin^{-1}\left[\frac{2d + h}{2A_i}\left(1 - \frac{h}{2d + h}\right)\right]$$ $$B = \sin^{-1}\left[\frac{2d + h}{2A_i}\left(1 + \frac{h}{2d + h}\right)\right]$$

TABLE 8.1 (Cont.)

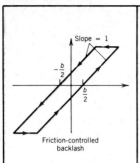

Friction-controlled backlash

$$G_d = 0 \qquad A_i < \frac{b}{2}$$

$$|G_d| = \sqrt{A^2 + B^2} \qquad \underline{/G_d} = \tan^{-1}\frac{A}{B} \qquad A_i \geq \frac{b}{2}$$

$$R = \frac{b}{A_i} \qquad A = \frac{R}{\pi}(R - 2)$$

$$B = \frac{1}{\pi}\left[\frac{\pi}{2} + \sin^{-1}(1 - R) + (1 - R)\sqrt{2R - R^2}\right]$$

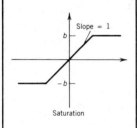

Saturation

$$G_d = \frac{2}{\pi}\left\{\sin^{-1}\left(\frac{b}{A_i}\right) + \frac{b}{A_i}\left[1 - \left(\frac{b}{A_i}\right)^2\right]^{1/2}\right\}$$

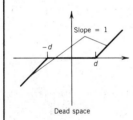

Dead space

$$G_d = 0 \qquad A_i < d$$

$$G_d = \frac{2}{\pi}\left\{\frac{\pi}{2} - \sin^{-1}\left(\frac{d}{A_i}\right) - \frac{d}{A_i}\left[1 - \left(\frac{d}{A_i}\right)^2\right]^{1/2}\right\} \qquad A_i \geq d$$

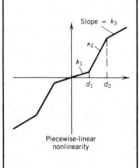

Piecewise-linear nonlinearity

$$G_d = k_3 + \frac{2}{\pi}(k_1 - k_2)\left\{\sin^{-1}\left(\frac{d_1}{A_i}\right) + \frac{d_1}{A_i}\left[1 - \left(\frac{d_1}{A_i}\right)^2\right]^{1/2}\right\}$$

$$+ \frac{2}{\pi}(k_2 - k_3)\left\{\sin^{-1}\left(\frac{d_2}{A_i}\right) + \frac{d_2}{A_i}\left[1 - \left(\frac{d_2}{A_i}\right)^2\right]^{1/2}\right\}$$

$$A_i \geq d_2$$

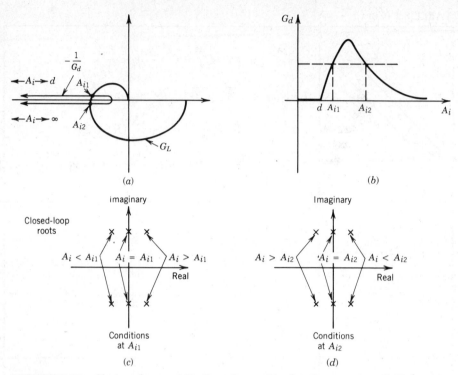

FIGURE 8.17   *Choice of correct limit-cycle amplitude when two possibilities exist.*

amples. In Fig. 8.18 the output member (displacement x_o) is subject to dry friction force sufficiently large that x_o *instantly* stops whenever x_i leaves contact, and x_o follows x_i *perfectly* when x_i makes contact. No elasticity is present and there is no "bouncing" at the instants when input and output members make or break contact. Under these assumptions the input–output relation is as shown in Fig. 8.18b. For describing-function analysis, we let $x_i = X_i \sin \omega t$, whereupon x_o will be as shown in Fig. 8.18c. Even before we carry out the Fourier series calculations, it is clear that the fundamental component of x_o will have a *phase shift* with respect to x_i, a feature not present for any of our earlier describing functions. Also, the describing function will not depend on frequency since the x_i/x_o relation of Fig. 8.18c is unchanged when ω changes.

In evaluating the Fourier series for x_o we have

average value = 0, by symmetry

$$
a_1 = 2\left[\int_0^{t_1} \left(X_i \sin \omega t - \frac{b}{2} \right)(\cos \omega t)dt + \int_{t_1}^{t_2} \left(X_i - \frac{b}{2} \right)(\cos \omega t)dt \right.
$$
$$
\left. + \int_{t_2}^{\pi/\omega} \left(X_i \sin \omega t + \frac{b}{2} \right)(\cos \omega t)dt \right]
$$

(8.14)

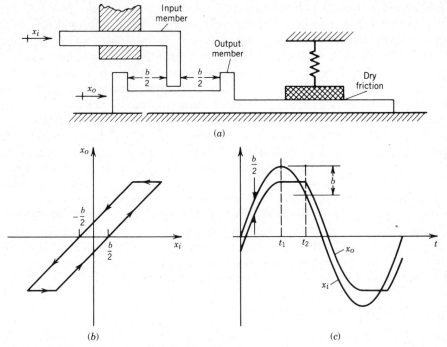

FIGURE 8.18 *Friction-controlled backlash.*

$$b_1 = 2\left[\int_0^{t_1}\left(X_i \sin \omega t - \frac{b}{2}\right)(\sin \omega t)dt + \int_{t_1}^{t_2}\left(X_i - \frac{b}{2}\right)(\sin \omega t)dt \right.$$
$$\left. + \int_{t_2}^{\pi/\omega}\left(X_i \sin \omega t + \frac{b}{2}\right)(\sin \omega t)dt\right] \tag{8.15}$$

$$t_1 = \frac{\pi}{2\omega} \qquad t_2 = \frac{\sin^{-1}\left(1 - \dfrac{b}{X_i}\right)}{\omega} \tag{8.16}$$

Carrying out these operations finally leads to the following results[3] for G_d.

$$R \triangleq \frac{b}{X_i} \qquad A \triangleq \frac{R}{\pi}(R - 2) \qquad 0 \le R \le 2 \tag{8.17}$$

$$B \triangleq \frac{1}{\pi}\left[\frac{\pi}{2} + \sin^{-1}(1 - R) + (1 - R)\sqrt{2R - R^2}\right] \tag{8.18}$$

$$|G_d| = \sqrt{A^2 + B^2} \qquad \underline{/G_d} = \tan^{-1}\frac{A}{B} \tag{8.19}$$

---

[3]H. E. Merritt, *Hydraulic Control Systems,* Wiley, New York, 1967, p. 286.

Note that G_d is given as a function of the nondimensional ratio $R = b/X_i$ and we can thus tabulate or plot a *single* table or graph of G_d that will be usable for *any* numerical value of b (see Fig. 8.19).

Since the graph of $-1/G_d$ is entirely in the third quadrant, the linear part of G_L of any feedback system with the backlash nonlinearity must extend into this quadrant if an intersection is to be possible. A simple first-order G_L thus does not permit limit cycling; however a second-order type may or may not, depending on its damping ratio and steady-state gain. Checking for an intersection must be done graphically/numerically since no analytical solution for limit-cycle amplitude or frequency is now possible. A computer program that tabulates G_L and $-1/G_d$ is useful in searching for intersections and is not difficult to write. Once the general region of an intersection is found, we can use smaller increments of b/X_i and ω to pinpoint the intersection as accurately as we wish.

In the system of Fig. 8.20 with G_{L2}, no intersections exist and digital simulation confirms the absence of limit cycling. For G_{L1}, *two* intersections exist, one at $\omega = 97$ ($b/X_i = 0.21$) and one at $\omega = 31.5$ ($b/X_i = 1.93$). Only the limit cycle at $\omega = 97$ can be sustained in a real system, however, since it exhibits restoring tendencies when the amplitude is momentarily disturbed; whereas the oscillation at $\omega = 31.5$ either decays to zero or builds up when disturbed (see Fig. 8.21). A CSMP simulation was run as follows.

```
V  =  5.*(STEP(0.0)  -  STEP(.8))
E  =  V-C
M1  =  CMPXPL(0.0,0.0,  .2,30.,  E)
M  =  M1*900.*10.
C  =  HSTRSS(0.0,-1.,1.,M)
```

The backlash effect is modeled with the single CSMP module HSTRSS (hysteresis). The first entry (0.0) inside the parenthesis is the initial value of C; whereas the next two ($-1.$, $1.$) give $-b/2$ and $+b/2$. When this simulation was run, limit cycling occurred for both $V = 5.0$ ($0 \leq t \leq 0.8$) and $V = 0.0$ ($0.8 \leq t$). For $V = 0.0$ (as the describing function assumes), ω was 95 and b/X_i was 0.24, both in good agreement with predictions.

8.4   ON–OFF CONTROL OF AN AIRCRAFT ROLL-STABILIZATION SYSTEM

Various on–off control techniques have been used to achieve simple, reliable, light-weight, and energy-efficient flight controls for aircraft and missiles.[4] In the remotely controlled vehicle of Fig. 8.22, pitch and yaw steering commands are

---

[4]M. Rogers and G. Shapiro, Closed-Loop Flip-Flop Control Systems, *Jour. Aerospace Sci.,* Nov. 1960, pp. 841–853; R. R. Lunstrom and R. I. Whitman, Analytical Investigation of Flicker-Type Roll Control for a Mach Number 6 Missile, NASA MEMO 4-23-59L, May 1959.

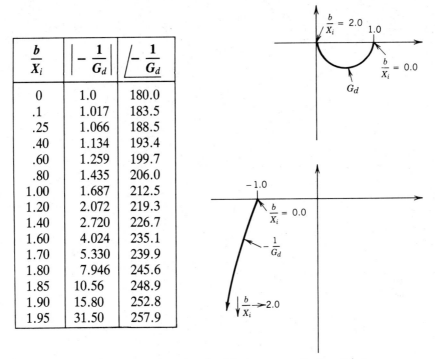

$\dfrac{b}{X_i}$	$\left\| -\dfrac{1}{G_d} \right\|$	$\left/ -\dfrac{1}{G_d} \right.$
0	1.0	180.0
.1	1.017	183.5
.25	1.066	188.5
.40	1.134	193.4
.60	1.259	199.7
.80	1.435	206.0
1.00	1.687	212.5
1.20	2.072	219.3
1.40	2.720	226.7
1.60	4.024	235.1
1.70	5.330	239.9
1.80	7.946	245.6
1.85	10.56	248.9
1.90	15.80	252.8
1.95	31.50	257.9

FIGURE 8.19 *Describing function for friction-controlled backlash.*

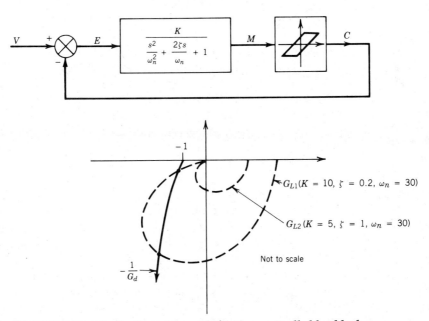

FIGURE 8.20 *Feedback system with friction controlled backlash.*

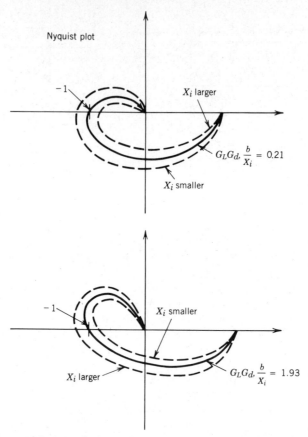

FIGURE 8.21 *Sustainable and unsustainable limit cycles.*

sent to a set of cruciform fins at the vehicle's tail. If the pitch fins are to affect only pitch and the yaw fins only yaw, the vehicle's roll angle must be kept nominally at zero. Both launching transients and wind gusts during flight tend to disturb the vehicle's roll angle ϕ_C from the desired value $\phi_V = 0$. Such disturbances can be counteracted by applying an opposing aerodynamic rolling moment T_a created by a set of differential ailerons. Since the right and left ailerons are applied in opposite senses, the vertical forces cancel (thus not disturbing the pitch motion) and a pure rolling moment is applied to the vehicle. The vehicle roll dynamics are modeled as a roll moment of inertia J and roll viscous damping B. The response of aileron angle δ_a to commands is modeled as a first-order lag with time constant τ_a.

The simplest on–off roll stabilization system possible uses a position gyroscope to sense ϕ_C, a conducting/insulating ring (fixed to the vehicle frame) to switch a relay between two positions, and a solenoid-actuated aileron drive mechanism that positions the ailerons at either a positive or negative deflection (a neutral

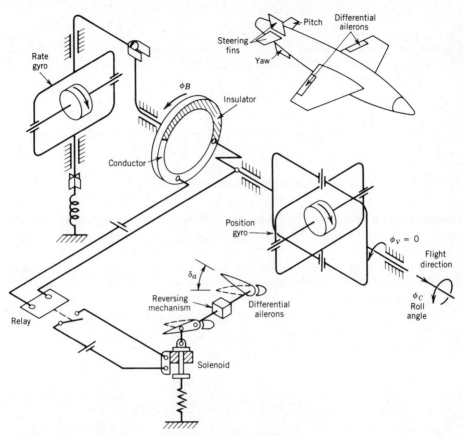

FIGURE 8.22   *On–off aircraft roll stabilization system.*

position is not used). (Figure 8.22 shows an improved system in which the switching ring is moved by a rate gyroscope, rather than being fixed to the vehicle frame.) For the simpler system (no rate gyro), Fig. 8.23 shows a block diagram. Numerical values for a proposed design are

$$\delta_{ac} = \pm 0.2 \text{ rad} \qquad J = 0.8 \text{ ft-lb}_f\text{-s}^2$$

$$\tau_a = 0.04 \text{ s} \qquad B = 20.\text{ft-lb}_f/(\text{rad/s})$$

$$K_T = 200 \text{ ft-lb}_f/\text{rad}$$

A CSMP digital simulation was run with ϕ_V a step command of $+1.0$ rad applied at $t = 0$ and removed ($\phi_V = 0$) at $t = 1.5$ s ($T_U = 0$ for this whole time period), and a disturbing torque $T_U = +20$ ft-lb$_f$ applied at $t = 3.0$ and thereafter maintained. Figure 8.24 shows the results of this simulation; when ϕ_V and T_U are both zero, a limit cycle of 3° amplitude at 3.8 Hz occurs in ϕ_C. Note that δ_{ac} switches symmetrically between positive and negative values and ϕ_C appears quite sinusoidal. When T_U is applied at $t = 3.0$, the average value of ϕ_C is

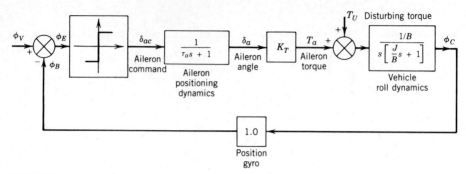

FIGURE 8.23 *Roll stabilization with position gyro only.*

forced away from $\phi_V = 0$ and δ_{ac} now spends more time negative than positive, since it must "fight out" T_U. This asymmetrical δ_{ac} also causes ϕ_C to be less sinusoidal.

In a practical application of this system the $\pm 3°$ roll oscillation was found to unduly interfere with vehicle pitch/yaw steering, requiring a design modification. In the system of Fig. 8.23, a reduction in limit cycle amplitude requires that the frequency be increased. This in turn requires reduction of τ_a and/or J/B since

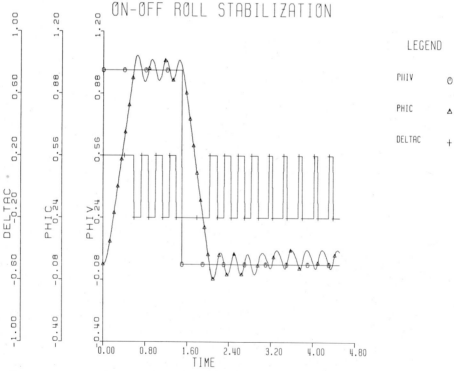

FIGURE 8.24 *System performance with position gyro only.*

we want the intersection of $-1/G_d$ and G_L (which occurs where $\underline{/G_L} = -180°$) to be at a higher frequency. In practice, reduction in τ_a and/or J/B may be impossible or undesirable, requiring another design approach. A classical technique of linear control theory (discussed in chapter 11) uses feedback of derivatives of the controlled variable to augment stability. The simplest and most-used version of this method uses the first derivative of the controlled variable, in our case this would be roll rate $\dot{\phi}_C$. This approach can be successful also with on–off control, as we now demonstrate. Since instruments called rate gyroscopes[5] are readily available for measuring $\dot{\phi}_C$, we add one of these to our system (as in Fig. 8.22) rather than attempting to differentiate the position gyro ϕ_C signal. Note that the position gyro is still necessary to provide a reference for $\phi_C = 0$; the rate gyro *augments* the ϕ_B signal as in Fig. 8.25. Although the rate gyro is a second-order instrument, to simplify describing-function design calculations we approximate it in Fig. 8.25 as a "perfect" (zero-order) means of measuring $\dot{\phi}_C$.

When we combine the position and rate gyro signals into the term $\tau_{RG}s + 1$, we can see why this design change can increase limit-cycle frequency and thereby decrease the oscillation amplitude. The frequency response of $(i\omega\tau_{RG} + 1)$ has a *leading* phase angle, opposing the lags in the rest of the system. By choosing τ_{RG} (the "amount" of derivative feedback) properly, we can force the frequency at which $\underline{/G_L(i\omega)}$ is $-180°$ to any value we wish. In fact, using describing-function methods, we can *choose* the allowable oscillation amplitude in ϕ_C and then design τ_{RG} so as to get this value. If the allowable oscillation in ϕ_C is $\pm 0.5°$, we find that $\tau_{RG} = 0.016s$ is needed. A CSMP simulation to check these pre-

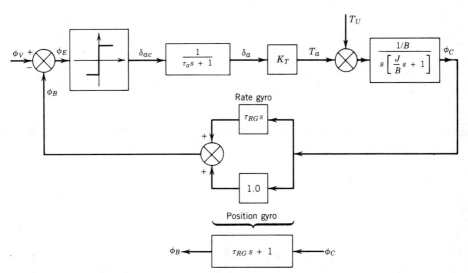

FIGURE 8.25   *System with position and rate feedback.*

[5]E. O. Doebelin, *Measurement Systems,* 3rd Ed., McGraw–Hill, New York, 1983, pp. 338–347.

dictions, using the same ϕ_V and T_U inputs as for Fig. 8.24, gave the results of Fig. 8.26. For $\phi_V = T_U = 0$, the limit-cycle frequency was 8.3 Hz with amplitude of 0.47°. To see whether the dynamics of a "real" rate gyro would change these results appreciably, the simulation was rerun using a second-order rate gyro model with $\omega_n = 400$ rad/s and $\zeta = 0.3$. These values were chosen from a catalog of available rate gyros, keeping in mind that the rate gyro dynamics should not contribute much attenuation and phase lag at the predicted cycling frequency of 8.3 Hz. Figure 8.27 shows that the real rate gyro does not seriously degrade performance; frequency is 7.7 Hz, aplitude 0.63°.

8.5 RESIDENTIAL HEATING SYSTEM USING ON–OFF CONTROL WITH SECONDARY FEEDBACK

Most residential heating and cooling systems utilize one form or another of on–off control. For example, gas-fired heating systems often use a thermostat with a bimetallic temperature sensor and a snap-action switch to control a simple solenoid-actuated on–off gas valve. Thermostats mass produced at modest cost exhibit significant hysteresis between the turn-on and turn-off temperatures,

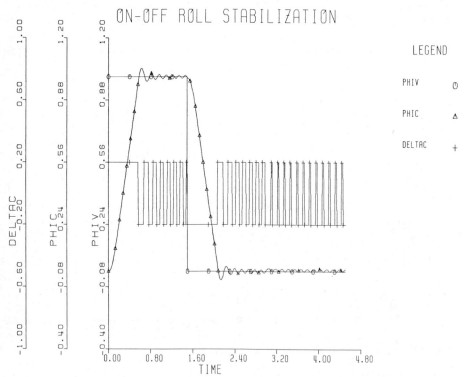

FIGURE 8.26 *System performance with position gyro and ideal rate gyro.*

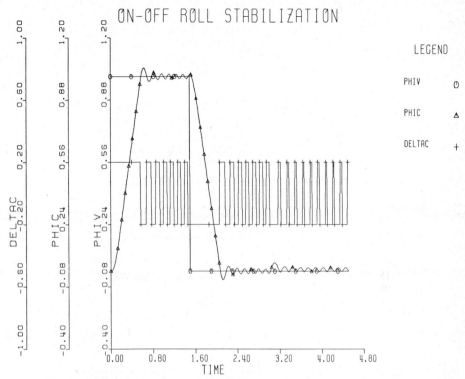

FIGURE 8.27 *System performance with position gyro and "real" rate gyro.*

leading to larger amplitude limit cycling of the controlled temperature. Rather than designing thermostats with less hysteresis, it has been found more economical to use the technique of secondary feedback[6] to reduce limit-cycle amplitude. A small electric heater is located inside the thermostat where it can effect the temperature of the bimetallic sensor used to measure room temperature (see Fig. 8.28). This heater turns on and off in synchronism with the furnace and is adjusted so that, when on, it raises the sensor's temperature by an amount equal to the hysteresis $2h$ between the turn-on and turn-off points of the thermostat. The thermostat switching action is also biased to the negative side as in Fig. 8.29. Thus if the furnace has been on and controlled temperature T_C is rising toward desired value T_V, the sensor actually feels a temperature $T_C + 2h$ and switches off when $T_E = 0$, the same behavior as a thermostat with *no* hysteresis. When the furnace is off, the auxiliary heater is also off and the sensor feels only T_C. When T_C drops to T_V ($T_E = 0$), the furnace switches on, again exactly as if there were *no* hysteresis. Thus the benefits of a hysteresis–free thermostat are achieved by addition of a simple and inexpensive heater, rather than a costly new mechanism.

---

[6]W. K. Roots, Fundamentals of Temperature Control, Academic Press, New York, 1969, Chap. 5.

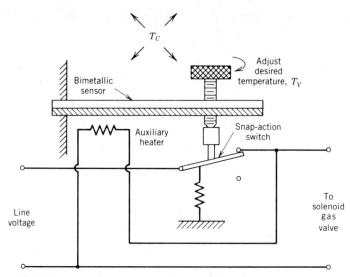

FIGURE 8.28 *Thermostat with secondary feedback.*

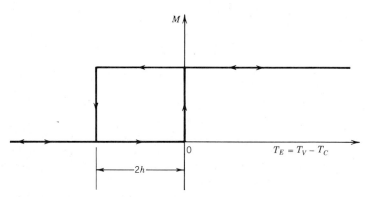

FIGURE 8.29 *Thermostat with biased hysteresis.*

To demonstrate the performance of this type of system we will run a CSMP simulation of a complete heating installation, modeled as in Fig. 8.30. The response of room temperature to furnace fuel flow is complex but experiments show that combination of a dead-time τ_{DT} and a first-order system with time constant τ_p is often adequate. Response to disturbing (outdoor) temperature T_U is modeled as first order with τ_p. Sensor and auxiliary heater dynamics are usually fast enough (relative to τ_p) to neglect, but we include first-order models to allow study of this effect if desired. We will here only study the effect of the *amount K_{sfb}* of secondary feedback since this is an adjustment made by the installer and is sometimes incorrect. All other system parameters are kept fixed at nominal values. A CSMP program goes as follows.

```
        TV  =  70.     desired temperature is 70°F
        XR  =  KA*TV  ⎫  thermostat reference input displacement
PARAM  KA  =  .02    ⎬  is scaled at 0.02 in./°F
        XE  =  XR - XB  ⎭
        XB  =  KA*TB1              ⎫  bimetallic sensor
        TB1  =  REALPL(TCO,TAUPFB,TCA)  ⎬  dynamics
INCON  TCO  =  30.     initial value of sensor temperaure
PARAM  TAUPFB  =  .05     sensor time constant
        SFB1  =  KSFB*M           ⎫  auxiliary heater
        TH  =  REALPL(ISFB,TAUSFB,  SFBI)  ⎬  dynamics
PARAM  KSFB  =  (0.0, 2.,4.,8.)     four different amounts of
                                    secondary feedback
PARAM  TAUSFB  =  .05     auxiliary heater time constant
        TCA  =  TC  +  TH     sensor input temperature
        M1  =  HSTRSS (.8, - .08,0.0,  XE)     Initial value of
                                                XE = 0.8, biased
                                                hysteresis between
                                                - 0.08 and 0.0.
        M  =  FCNSW (M1,0.0,0.0,1.)     on–off function
        MA  =  DELAY(5,.05,M)     process dead time of 0.05 hr
        M2  =  KP*MA             ⎫  Process first-order
        M3  =  REALPL(TCM,TAUP,M2)  ⎬  dynamics
INCON  TCM  =  0.0     Initial value
PARAM  KP  =  90.     Process gain
PARAM  TAUP.  =  1.5     Process time constant of 1.5 h
        TU  =  30.     outdoor temperature 30°F
        U3  =  REALPL(TUO,TAUP,TU)     process disturbance
                                        dynamics
INCON  TUO  =  30.     initial value of U3
        TC  =  M3  +  U3
TIMER FINTIM  =  3., DELT  =  .01,  OUTDEL  =  .01
OUTPUT  TV,  TC
```

In this system the optimum value of secondary feedback has K_{sfb} = 4.0 since this biases the measured T_C by 4°F, the value of the hysteresis $2h$ in Fig. 8.29. If *no* secondary feedback is used (K_{sfb} = 0.0), the response is as in Fig. 8.31*a*, a limit cycle between 67.6 and 76.8°F. The bias above desired value 70°F could be removed by making the hysteresis symmetrical (− .04, .04 rather than − .08, 0.0) but the peak-to-peak amplitude would still be 9.2°F. For K_{sfb} = 2.0 (Fig. 8.31*b*) this amplitude is reduced to 7.3°F. For the optimum value of 4.0 (Fig. 8.31*c*) it is further reduced to 5.7°F and the *average* value of T_C now equals T_V. *Excessive* secondary feedback (K_{sfb} = 8.0, Fig. 8.31*d*) further reduces limit-cycle amplitude to 3.4°F but produces an error of 1.1°F between T_V and the average value of T_C. Such a "steady-state" error could be removed by inten-

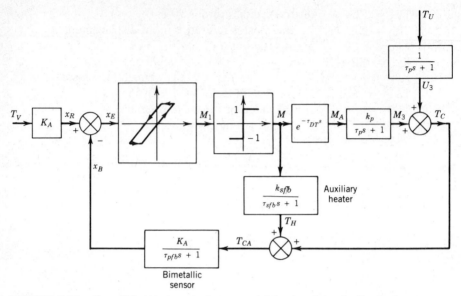

FIGURE 8.30 *On–off heating control system with secondary feedback thermostat.*

tionally "misadjusting" T_V, that is, we could "ask" for $T_V = 71.1°F$ when we really want 70.0.

BIBLIOGRAPHY

Books

1. H. E. Merritt, "Hydraulic Control Systems," Wiley, New York, 1967, Chap. 10.

2. J. E. Gibson, "Nonlinear Automatic Control," McGraw–Hill, New York, 1963.

3. D. Graham and D. McRuer, "Analysis of Nonlinear Control Systems," Wiley, New York, 1961.

4. A. Gelb and W. E. Vander Velde, "Multiple-Input Describing Functions and Nonlinear System Design," McGraw–Hill, New York, 1968.

5. G. J. Thaler and M. P. Pastel, "Analysis and Design of Nonlinear Feedback Control Systems," McGraw–Hill, New York, 1962.

6. J. G. West, "Analytical Techniques for Nonlinear Control Systems," Van Nostrand, Princeton, N.J., 1960.

7. R. L. Cosgriff, "Nonlinear Control Systems," McGraw–Hill, New York, 1958.

8. E. P. Propov, "The Dynamics of Automatic Control Systems," Addison Wesley, Reading, Mass., 1962, Part IV.

PROBLEMS

8.1 Derive the describing function for the following entries in Table 8.1:

    a. Two position with hysteresis.

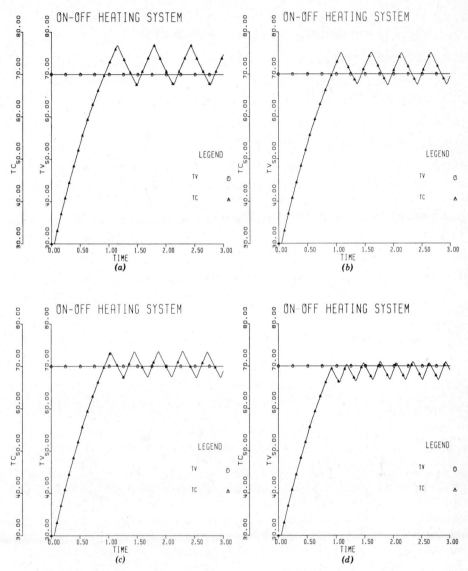

FIGURE 8.31 *Heating system performance with optimum and nonoptimum secondary feedback.*

    b. Three position with hysteresis.

    c. Saturation.

    d. Dead space.

    e. Piecewise-linear nonlinearity.

    f. Three-position on–off.

8.2 Derive the expression for the peak of the curve in Fig. 8.5.

8.3 Use the describing function to predict the limit-cycle amplitude of $C3$ in Fig. 8.13 and compare with digital simulation results.

8.4 Repeat Problem 8.3 for Fig. 8.16*a*.

8.5 Use describing-function analysis to predict limit-cycle frequency and amplitude of PHIC in Fig. 8.24 and compare with digital simulation results.

8.6 Repeat Problem 8.5 for Fig. 8.26.

8.7 Repeat Problem 8.5 for Fig. 8.27.

8.8 In the system of Fig. P8.1 what is the largest allowable value of d if the steady-state error between V and C is not to exceed 0.01 for a steady V? Using this value, what is the largest allowable value of K if we want a gain margin of 2.5? Use the describing function.

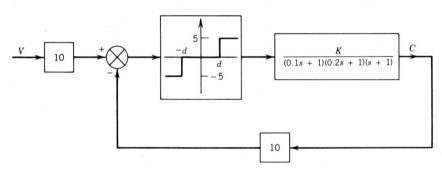

FIGURE P8.1

8.9 Check the accuracy of the describing-function predictions of Problem 8.8 using digital simulation.

8.10 Using describing-function methods, investigate the limit-cycling behavior of the system of Fig. P8.2.

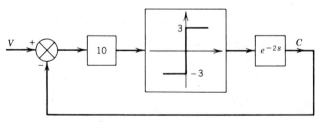

FIGURE P8.2

8.11 Check the results of Problem 8.10 using digital simulation.

8.12 In the system of Fig. 8.25, use the describing function to calculate the τ_{RG} value needed to achieve a ϕ_C limit cycle of 0.5° amplitude.

8.13 Linear, constant-coefficient models for feedback control systems always predict infinite amplitudes when absolute instability occurs. Such infinite amplitudes require system power supplies of infinite capacity and thus can, of course, not really occur. Rather, one or more system components saturates, resulting in a *nonlinear* system that limit cycles with a *finite* amplitude. The frequency and amplitude of such a limit cycle can be estimated by describing-function methods. The system of Fig. P8.3 is absolutely unstable for small-signal linear (unsaturated) operation. Use the describing function to calculate the limit-cycle behavior.

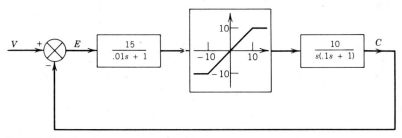

FIGURE P8.3

8.14 Check the results of Problem 8.13 using digital simulation.

CHAPTER 9
Proportional Control and Basic Feedback System-Design Procedures

9.1 GENERAL CHARACTERISTICS OF PROPORTIONAL CONTROL

By definition, the ideal proportional controller has for the transfer function G_1 in Fig. 2.1 the simple gain constant K_p; that is, manipulated variable M is directly proportional to actuating signal E. (In the process control field, *proportional band* may be used to specify the controller setting. A 10% proportional band (*PB*) means that a 10% change in the controller input causes a full-scale (100%) change in controller output. The conversion relation is thus $K_p = 100/PB$, assuming that controller input and output signals are both the same type of physical variable, i.e., both voltages or both air pressures.) A real controller will of course have in addition some dynamics, but we assume these are negligible relative to other system dynamics. Such a control mode seems quite reasonable; corrective effort is made proportional to system "error," large errors engender a stronger system response than do small ones. Unlike an on–off controller, a proportional controller can vary in a continuous fashion the energy and/or material sent to the controlled process, thus exactly matching its needs, rather than having to cycle back and forth between two (or three) fixed levels of supply. Thus, although limit cycling behavior is often tolerated in on–off control systems, it never would be in a proportional control. This does not mean that proportional control systems can never be unstable, they definitely can, however instability would not be acceptable and the system would be redesigned for stability. This stability (lack of limit cycling) is the main advantage of proportional control over on–off.

A disadvantage of proportional control relative to on–off is the generally greater complexity, higher cost, and lower reliability of the hardware. Valves for controlling fluid flow make a good example. The on–off solenoid gas valve used in a residential heating system is quite inexpensive and normally gives many years of service with no maintenance. An electrohydraulic servovalve, on the other hand, is an expensive, precisely machined component requiring meticulous cleanliness in installation and operation to avoid jamming due to microscopic dirt particles in the hydraulic fluid.

We shall find that proportional control also exhibits nonzero steady-state errors for even the least-demanding commands and disturbances. Such steady-state errors (sometimes called "droop" or "offset") are, of course, not a fatal defect since any realistic performance specification always allows *some* steady-state error. However, other modes of control (notably integral) are theoretically able to *completely* remove steady-state errors, thus this aspect of proportional control behavior is significant when choosing a control mode.

9.2 PROPORTIONAL CONTROL OF A LIQUID-LEVEL PROCESS

We will use the example system of Fig. 9.1 to develop an understanding of proportional control characteristics, first getting results specific to this example

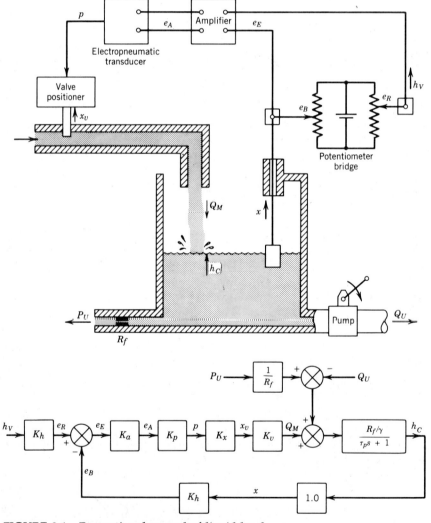

FIGURE 9.1 *Proportional control of liquid level.*

and then extending these to proportional control in general. The system shown is designed to maintain tank level h_C at desired value h_V in the face of disturbances P_U and Q_U. Disturbance flow rate $Q_U(t)(\text{ft}^3/\text{s})$ represents fluid drawn off from the tank (to be used elsewhere in the processing operation) by a variable-displacement pump. Disturbance pressure $P_U(t)$ represents a variable back pressure at another point in the process, which influences tank outflow rate through the linearized flow resistance $R_f[\text{psi}/(\text{ft}^3/\text{s})]$. We will assume an initial equilibrium operating point (all variables steady) at which inflow Q_M exactly matches the two outflows and $h_C \equiv h_V$. The necessary valve flow rate Q_M can be achieved with $h_C \equiv h_V$ ($e_E = 0$) because the electropneumatic transducer has a zero adjustment which allows pressure p to be at the middle (9 psig) of its 3-to-15-psig range when $e_A = 0$, and the valve positioner has a zero adjustment which allows the valve opening to be put anywhere desired, with the pressure p at 9 psig. With these initial "trimming" adjustments made, all the variables named in Fig. 9.1 are considered as small perturbations away from the initial steady state. Thus if the initial steady liquid level were, say, 10 ft and the instantaneous value during dynamic operation were 11 ft, h_C would be $+1.0$ ft, whereas an instantaneous level of 9 ft would make $h_C = -1.0$ ft. This way of defining system variables for analysis is desirable since it accomodates linearized models of nonlinear components and easily deals with both positive and negative changes in commands and disturbances. We shall use it almost universally in the remainder of the book.

The block diagram of Fig. 9.1 makes certain assumptions that must be explained. Tank process dynamics are found by applying conservation of volume.

$$Q_M - \frac{\gamma h_C - P_U}{R_f} - Q_U = A_T \frac{dh_C}{dt} \tag{9.1}$$

where

$$A_T \triangleq \text{tank cross-sectional area} \qquad \gamma \triangleq \text{fluid specific weight}$$

$$(\tau_p s + 1)h_C = \left(\frac{R_f}{\gamma}\right)Q_m + \left(\frac{1}{\gamma}\right)P_U - \left(\frac{R_f}{\gamma}\right)Q_U \tag{9.2}$$

where

$$\tau_p \triangleq \text{process time constant} \triangleq \frac{A_T R_f}{\gamma} \tag{9.3}$$

The float level sensor (analyzed in detail in Chap. 2, Fig. 2.27) is modeled in its simplest (zero-order) form. This is justifiable if its dynamics are negligible relative to the process time constant τ_p. We assume a tank area large enough to allow this. The potentiometer bridge and electronic amplifier obviously are fast enough to be treated as zero order in this system. Figure 9.2 shows internal details of the electropneumatic transducer, a component widely used in process control for interfacing between electrical and pneumatic portions of a system. It produces a pneumatic output signal closely proportional ($\pm 0.5\%$ nonlinearity) to an electrical input, typical ranges being $\pm 5\text{V}$ and 3 to 15 psig. We first note that this device is *itself* a feedback system, thus it must be properly designed

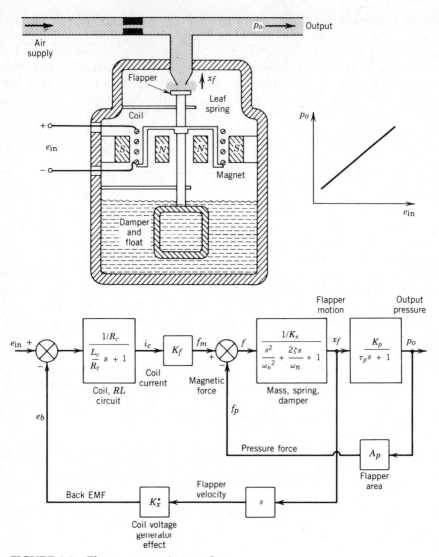

FIGURE 9.2 *Electropneumatic transducer.*

for stability and accuracy. If we are, however, *purchasers* (rather than designers) of such a device, we may assume it has been properly designed and we are concerned only with its overall dynamics from e_A to p. Using the block diagram of Fig. 9.2, one gets a fourth-order closed-loop differential equation, however experimental frequency response tests show typically a flat amplitude ratio out to about 5 Hz. This response is again very fast relative to τ_p so we model the electropneumatic transducer as zero-order also.

Figure 9.3 shows that the pneumatic valve positioner is also itself a complete feedback system, but we again assume the role of purchaser rather than designer

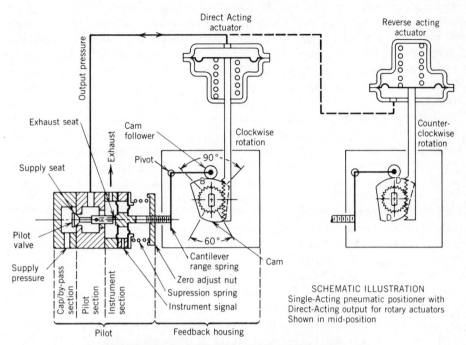

SCHEMATIC ILLUSTRATION
Single-Acting pneumatic positioner with
Direct-Acting output for rotary actuators
Shown in mid-position

OPERATING PRINCIPLE

Referring to the schematic illustration, with air pressure applied to the supply port of the Pilot, and the instrument signal at mid-range, the unit is in force balance. The instrument signal acts upon a Diaphragm Assembly creating a force toward the Pilot Valve. An opposing force from the Suppression Spring counteracts this force. The Cantilever Range Spring is in its mid-zero force position. An increase in the instrument signal causes an increased force towards the Pilot Valve and moves it open, increasing the output pressure to the Actuator. This increase causes the Actuator to rotate further and rotates the Cam clockwise. A Cam Follower rises and twists the Cantilever Range Spring in the counter clockwise direction resulting in a force at the end of the Cantilever Range Spring directed away from the Pilot Valve. The Actuator continues to rotate until the force of the Cantilever Range Spring equals the increase in the instrument signal above its mid-point. A decrease in the instrument signal from its mid-point causes the Diaphragm Assembly to move away from the Pilot Valve and open the Exhaust Seat. As the output pressure decreases, the Actuator rotates counter clockwise and rotates the Cam counter clockwise. The Cam Follower drops which twists the Cantilever Range Spring in the clockwise direction, resulting in a force directed towards the Pilot Valve. The Actuator continues to rotate until the force of the Cantilever Range Spring equals the decrease in the instrument signal below its mid-point.

FIGURE 9.3 *Pneumatic valve positioner with characterizing-cam feedback.*
Source: Conameter Corporation, Gibbsboro, New Jersey.

and inquire only into the overall dynamics relating x_v to p. These are again quite fast relative to τ_p so we model this component as zero order. We also wish to point out that this valve positioner allows one to "characterize" the static calibration curve between p and x_v and thus obtain desired linear or nonlinear relationships between p and manipulated flow rate Q_M. If a linear x_v/p relation is wanted, the cam in the positioner feedback path is given a linear rise with rotation. Various useful nonlinear relations can be obtained by giving the cam the appropriate shape. For example, suppose the flow resistance R_f is nonlinear with $h_C = C_1 + C_2 Q_M^2$ in steady state (P_U, Q_U assumed constant). This process nonlinearity makes the performance of the level control system dependent on the value of the operating point for h_C. For example, the system could be stable at one operating point and unstable at another because of changes in the equivalent linearized gain R_f/γ. Such nonuniform behavior is undesirable and can be eliminated or reduced by characterizing the valve positioner so that the Q_M/p relation is nonlinear with the *inverse* nonlinearity of the h_C/Q_M relation, making the h_C/p relation linear and thus keeping loop gain (and thereby system performance) uniform at various operating points.

The relation between Q_M and x_v in Fig. 9.1 is assumed statically linear and dynamically instantaneous. Static linearity is a good approximation for small excursions around a given operating point. Linearity for large excursions can be obtained for the path from p to h_C using a characterized positioner as earlier described. The dynamic response of Q_M to x_v is not instantaneous due to fluid inertia and compliance, however this response is again much faster than the tank-filling dynamics associated with τ_p so we model it as zero order.

Having justified the block diagram of Fig. 9.1 we now use it directly to obtain the differential equation of the closed-loop system, relating h_C to h_V, P_U, and Q_U.

$$\left[(K_h h_V - K_h h_C) K_a K_p K_x K_v + \frac{1}{R_f} P_U - Q_U \right] \frac{R_f/\gamma}{\tau_p s + 1} = h_C \qquad (9.4)$$

and

$$(\tau_s s + 1) h_C = \frac{K}{K + 1} h_V + \frac{1}{\gamma(K + 1)} P_U - \frac{R_f}{\gamma(K + 1)} Q_U \qquad (9.5)$$

where

$$\tau_s \triangleq \text{closed-loop system time constant} \triangleq \frac{\tau_p}{K + 1} \qquad (9.6)$$

$$K \triangleq \text{system loop gain} = \frac{K_h K_a K_p K_x K_v R_f}{\gamma} \qquad (9.7)$$

For the linear model of Eq. 9.5, superposition holds and we can investigate the effects of various types of commands h_V and disturbances P_U and Q_U separately. For h_V a step input of size h_{Vs} (perturbations P_U, Q_U are held at zero) we get

$$h_C = \frac{K}{K + 1} h_{Vs}(1 - e^{-t/\tau_s}) \qquad (9.8)$$

Figure 9.4 shows the characteristic first-order system response with speed determined by τ_s and a steady-state error given by

$$\frac{h_{E,ss}}{h_{Vs}} = \frac{(h_V - h_C)_{ss}}{h_{Vs}} = \frac{1}{K + 1} = \frac{100}{K + 1}\% \tag{9.9}$$

Since $\tau_s = \tau_p/(K + 1)$, we see that both speed of response and steady-state error are improved if we increase loop gain K. For example, $K = 99$ gives a 1% steady-state error and makes the closed-loop response speed of the tank *100 times* faster than the open-loop response. For step disturbance inputs P_{Us} and Q_{Us} ($h_V \equiv 0$) we get, respectively

$$h_C = \frac{1}{\gamma(K + 1)}P_{Us}(1 - e^{-t/\tau_s}) \tag{9.10}$$

$$h_C = \frac{-R_f}{\gamma(K + 1)}Q_{Us}(1 - e^{-t/\tau_s}) \tag{9.11}$$

and again see that increasing K improves both speed of response and steady-state errors.

The steady-state errors discovered in the present example are in fact characteristic of proportional control in general. The basic reason for this behavior can be explained as follows. For the initial equilibrium operating point assumed, we are able to get *zero* steady-state error ($h_C \equiv h_V$) because the zero adjustments on various components (such as the valve positioner) allow us to open the valve exactly the right amount to supply the tank outflows with $e_E = 0$. When we then ask h_C to go to a *new* value h_{Vs}, it takes a *different* valve opening to reach equilibrium at the new h_C because the outflow through R_f will be larger than that for the initial equilibrium state. When flow rate Q_M is *proportional* to e_E, a new Q_M can only be achieved if e_E is different from zero, which requires $h_C \neq h_V$; thus there *must* be a steady-state error. Similar reasoning holds for step changes in disturbances. Thus for any initial equilibrium condition we can "trim" the system for zero error but subsequent steady commands and/or disturbances must cause steady-state errors.

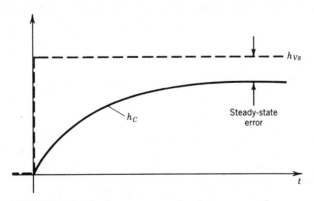

FIGURE 9.4 *System response to step command.*

The steady-state error due to step commands can theoretically be eliminated from proportional control systems by intentionally "misadjusting" h_V. That is, in Eq. 9.8 if we really want h_C to go to h_{Vs}, we instead ask for $[(K + 1)/K]h_{Vs}$. This trick does in fact work, but depends on our knowing what K is. If K is not accurately known or varies, then the trick is not 100% effective. Also, the trick does not help in reducing steady-state errors caused by disturbances. When we later study *integral control,* we will find that this mode of control removes steady-state errors due to both commands and disturbances and does not depend on knowledge of any specific numerical values for system parameters.

Equations 9.8, 9.10, and 9.11 indicate that all aspects of system behavior are improved by increasing loop gain. This is true up to a point but cannot be carried to extremes because the system will go unstable; however our present model gives no warning of instability. This defect in our model is caused by our neglect of dynamics in some of the system components. A simple Routh criterion analysis will provide a general rule as to the model complexity necessary for meaningful stability predictions. If the closed-loop system characteristic equation is first or second order, since τ(first-order system), or ζ and ω_n (second-order system) are usually positive numbers, instability is mathematically impossible. For a third-order system ($AD^3 + BD^2 + CD + E = 0$) however, even though A, B, C, and E are positive, when $AE > BC$ instability will occur. *Thus, if we want to make valid stability predictions we must include enough dynamics in our system model so that the closed-loop system differential equation is at least third order.* An exception to this rule involves systems with dead times, where instability can occur even when the dynamics (other than the dead time itself) are zero, first, or second order.

Since our present liquid-level control model is first order, it is not valid for stability predictions. However it does correctly predict system behavior as long as loop gain is not made "too large." Equation 9.6 shows that as K is increased, closed-loop system response gets faster and faster. At some point the neglected dynamics (such as in the sensor, electropneumatic transducer, and/or the valve positioner) are no longer negligible and the model of Fig. 9.1 becomes inaccurate. That is, we originally neglected these dynamics *relative to* τ_p, but in the *closed*-loop system, response speed is determined by τ_s, *not* τ_p. We can clearly illustrate this behavior by running a digital simulation on the three systems of Fig. 9.5a.

```
PARAM   K=(1.,5.,10.)
PARAM   TAU1=1., TAU2=.1, TAU3=.05
        V  = STEP(0.0)
        E1 = V-C1
        M1 = K*E1
        C1 = REALPL(0.0,TAU1,M1)
        E2 = V-C2
        M2 = K*E2
        X2 = REALPL(0.0,TAU2,M2)
```

```
         C2  =  REALPL(0.0,TAU1,X2)
         E3  =  V−B3
         M3  =  K*E3
         X3  =  REALPL(0.0,TAU2,M3)
         C3  =  REALPL(0.0,TAU1,X3)
         B3  =  REALPL(0.0,TAU3,C3)
TIMER    FINTIM = 1.8, DELT=.001, OUTDEL=.005
OUTPUT   C1,C2,C3
PAGE GROUP
```

For $K = 1$, Fig. 9.5b shows that neglecting one or two "fast" time constants does not radically change the closed-loop step response. In Fig. 9.5c and 9.5d, however, it becomes clear that for high gain the system response is significantly affected by the inclusion of the additional dynamics in the model.

In addition to neglecting some "fast" dynamics, the model of Fig. 9.1 further differs from reality because of its perfect linearity, which assumes no limit on the magnitude of corrective control action that can be applied by the system. Actually, the amplifier, electropneumatic transducer, and valve positioner all exhibit saturation, limiting their output when input becomes too large. All real systems must exhibit such power limitations and one of the consequences is that the closed-loop response speed improvement indicated by Eq. 9.6 will not be realized for large signals. To demonstrate this important general fact we simulate the systems of Fig. 9.6a.

```
PARAM    K=(1.,10.)
         V=STEP(0.0)
         E1=V−C1
         M1=K*E1
```

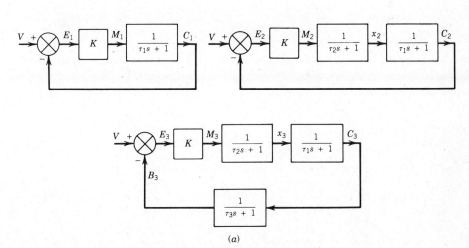

(a)

FIGURE 9.5 *Effect of loop gain on adequacy of simplified models.*

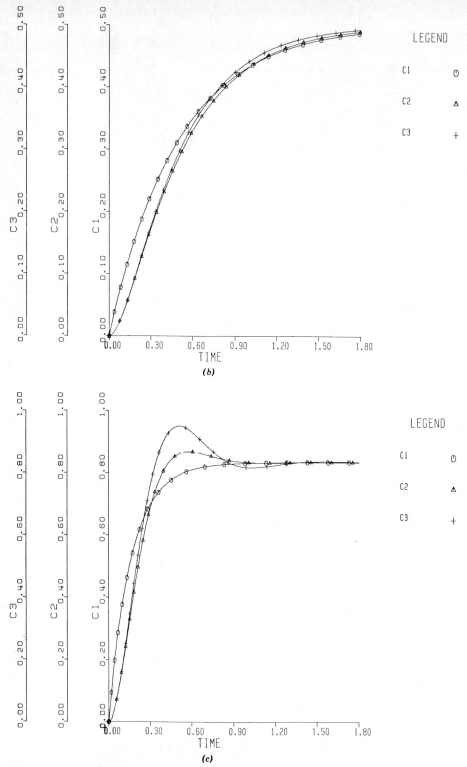

FIGURE 9.5 (Cont.)

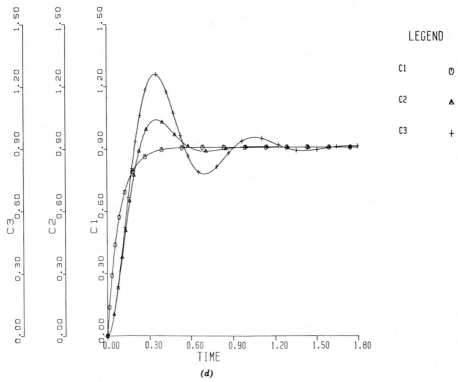

LEGEND

C1	⊙
C2	▲
C3	+

FIGURE 9.5 (Cont.)

```
          M1L =LIMIT( −1.,1.,M1)
          C1 =REALPL(0.0,1.0,M1L)
          E2 =V − C2
          M2 =K*E2
          C2 =REALPL(0.0,1.0,M2)
OUTPUT    C1(0.0,1.),C2(0.0,1.),M1L(0.0,2.)
```

All the CSMP statements used here should be familiar from earlier examples, except the OUTPUT statement, which specifies the plotting range of each variable, rather than letting the program make this choice.

In Fig. 9.6b ($K = 1.0$), loop gain and the size of command V are small enough so that M1L never saturates, thus C1 and C2 are identical. In 9.6c we have tried to improve system speed and steady-state error (according to Eq. 9.6, 9.9) by raising K to 10. With V still at 1.0, M1L immediately saturates, limiting the response speed of C1 and preventing the closed-loop system speed improvement predicted by Eq. 9.6 and demonstrated by C2 in Fig. 9.6c. When C1 finally approaches V, M1L comes out of saturation and C2 achieves the same steady-state value as C1, thus the gain increase has provided the desired improvement in steady-state error, but has not realized the predicted speed benefit. Note that

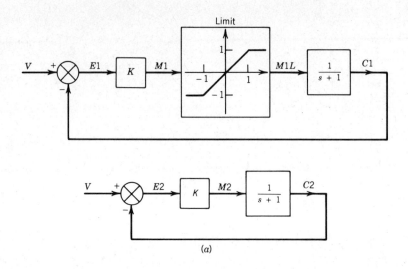

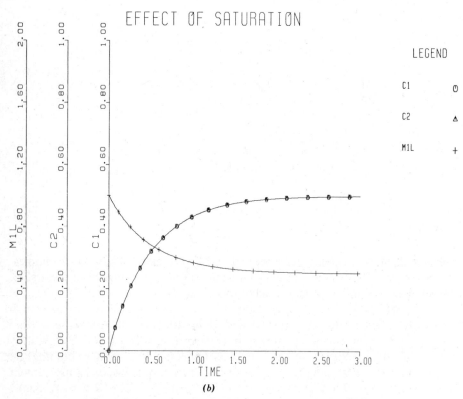

FIGURE 9.6   *Effect of saturation on closed-loop speed improvement due to high gain.*

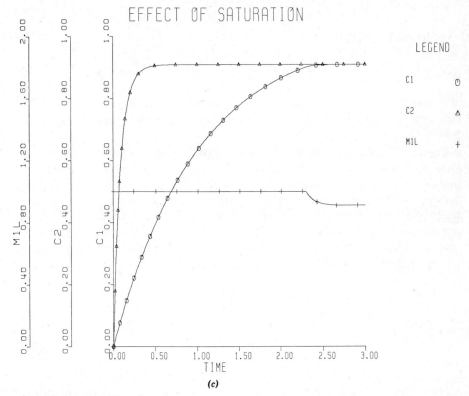

FIGURE 9.6 (Cont.)

if V were much larger than 1.0, then even the steady-state behavior suffers from saturation since C1 cannot exceed M1L in steady state.

Returning now to the strictly linear model of Fig. 9.1, let us find steady-state errors for the more difficult ramp commands and disturbances. A procedure generally useful for all types of systems and inputs is to rewrite the closed-loop system equation with system error $(V - C)$, rather than controlled variable C, as the unknown. For Eq. 9.5 we get

$$h_E \triangleq h_V - h_C$$

$$(\tau_s s + 1)(h_V - h_E) = \frac{K}{K + 1} h_V + \frac{1}{\gamma(K + 1)} P_U - \frac{R_f}{\gamma(K + 1)} Q_U \qquad (9.12)$$

$$(\tau_s s + 1)h_E = \left(\tau_s s + \frac{1}{K + 1}\right)h_V - \frac{1}{\gamma(K + 1)} P_U + \frac{R_f}{\gamma(K + 1)} Q_U \qquad (9.13)$$

For any chosen commands or disturbances the steady-state error will just be the particular solution of Eq. 9.13, step inputs giving the same results found earlier (Eq. 9.9 to 9.11). Ramp inputs may be specified as

$$h_V = \dot{h}_V t \qquad P_U = \dot{P}_U t \qquad Q_U = \dot{Q}_U t \qquad (9.14)$$

whereupon Eq. 9.13 gives

$$h_{E,ss} = \left(\frac{t + K\tau_s}{K + 1}\right)\dot{h}_V + \left(\frac{\tau_s - t}{\gamma(K + 1)}\right)\dot{P}_U + \left(\frac{R_f(t - \tau_s)}{\gamma(K + 1)}\right)\dot{Q}_U \qquad (9.15)$$

We see that ramp inputs cause steady-state errors that increase linearly with time, the rate of increase being proportional to ramp slope and inversely proportional to $K + 1$.

Figure 9.7a, the Nyquist plot for this system, shows stability for any value of loop gain. The root locus of Fig. 9.7b (obtained from Eq. 9.5) agrees with this prediction. The predicted closed-loop system behavior for high-loop gain is correct for the open-loop first-order math model used, but *not*, of course, for the real system since high gain brings forth neglected dynamics as discussed earlier. If we let the open-loop go to second order, simple analytical results are still available, so let us make the block diagram of Fig. 9.1 more realistic by taking $(x_v/p)(s) = K_x/(\tau_{vp}s + 1)$, giving for Eq. 9.5

$$\left(\frac{s^2}{\omega_n^2} + \frac{2\zeta s}{\omega_n} + 1\right)h_C = \frac{K}{K + 1}h_V + \frac{(\tau_{vp}s + 1)}{\gamma(K + 1)}P_U - \frac{R_f(\tau_{vp}s + 1)}{\gamma(K + 1)}Q_U \qquad (9.16)$$

where

$$\omega_n \triangleq \sqrt{\frac{K + 1}{\tau_p\tau_{vp}}} \qquad \zeta \triangleq \frac{\tau_p + \tau_{vp}}{2\sqrt{(\tau_p\tau_{vp})(K + 1)}} \qquad K \triangleq \text{same as Eq. 9.7} \qquad (9.17)$$

We see that now the closed-loop characteristic equation is second order and to get fast response (large ω_n) for given lags τ_p and τ_{vp} we must increase loop gain K. When the model was first order, increasing K caused no penalty. Now however we see that damping ζ(relative stability) will suffer, unless compen-

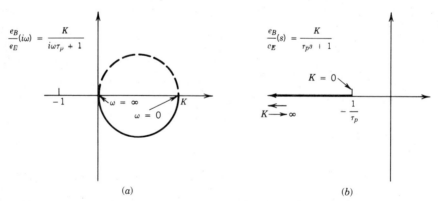

(a)   (b)

FIGURE 9.7 *Nyquist plot and root locus for first-order liquid level control model.*

sating changes in τ_p and τ_{vp} are possible. If τ_p and τ_{vp} are fixed and a certain minimum ζ is required, there is now a definite upper limit on K. For example, if $\tau_p = 60s$, $\tau_{vp} = 1.0s$, and we desire $\zeta = 0.6$, then $K = 42.1$, making $f_n = \omega_n/2\pi = 0.135$ Hz. Steady-state errors for step commands and disturbances are the same as for the first-order model, thus Eq. 9.9 predicts a 2.3% error for step commands. (Ramp-input steady-state errors require a new analysis and are left for the chapter-end problems.)

The Nyquist plot of Fig. 9.8a shows that high loop gain cannot cause absolute instability but can give poor relative stability (low phase margin). Since the closed-loop characteristic equation is only second order we can easily express the two roots $s_{1,2}$ as

$$ s_{1,2} = \frac{-(\tau_v + \tau_p) \pm \sqrt{(\tau_v - \tau_p)^2 - 4K\tau_v\tau_p}}{2\tau_v\tau_p} \tag{9.18} $$

and graph the root locus as Fig. 9.8b. Note that when K is large enough to cause complex roots, the relative stability (ζ) and response speed (ω_n) of any such root pair are directly measurable from the graph. Clearly, large K causes a decrease in ζ but *absolute* instability is impossible.

9.3 EFFECT OF GAIN DISTRIBUTION ON STEADY-STATE ERRORS

Although the closed-loop system characteristic equation (and thus stability and dynamic response) are unaffected by changes in the the "distribution" of gain

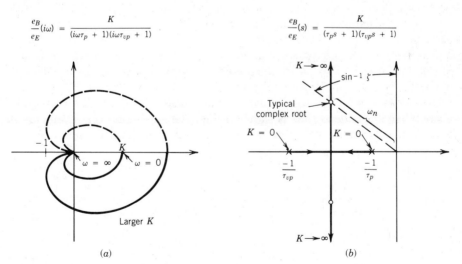

FIGURE 9.8 *Nyquist plot and root locus for second-order liquid level control model.*

from E to B (so long as *loop* gain is fixed), steady-state errors for disturbances *are* sensitive to this distribution. For the system of Fig. 9.9

$$\left(\frac{\tau}{1 + K_1 K_2 K_3}s + 1\right)E = \frac{1}{1 + K_1 K_2 K_3}(\tau s + 1)V - \frac{K_2 K_3}{1 + K_1 K_2 K_3}U_1$$
$$- \frac{K_3}{1 + K_1 K_2 K_3}U_2 \qquad (9.19)$$

For step inputs V_s, U_{1s}, and U_{2s}

$$E_{ss} = \frac{V_s}{1 + K_1 K_2 K_3} - \frac{U_{1s}K_2 K_3}{1 + K_1 K_2 K_3} - \frac{U_{2s}K_3}{1 + K_1 K_2 K_3} \qquad (9.20)$$

Note that error due to U_1 is K_2 times as large as that due to U_2. This is because U_1 enters the system "upstream" of K_2. While the loop gain $K_1 K_2 K_3$ will be set to get good relative stability, the *distribution* is sometimes open to the designer's choice. If this is the case, and if we know the relative magnitudes of the disturbances, an analysis as in eq. 9.20 allows choice of the optimum distribution of gain among K_1, K_2 and K_3, so as to minimize steady-state errors. Gain distribution also effects transient errors in a similar manner.

9.4 PROPORTIONAL CONTROL OF A FIRST-ORDER SYSTEM WITH DEAD TIME

Although we showed earlier that usually third-order closed-loop dynamics were necessary to allow stability predictions, this is not the case when dead time is present. In fact, an open-loop transfer function that is *only* dead time can cause closed-loop instability, however this case does not correspond to many practical systems so we do not study it here. Rather we investigate the combination of a dead time $e^{-\tau_{DT}s}$ and a first-order lag $1/(\tau s + 1)$. By proper choice of τ_{DT} and τ, this model can be made to adequately represent the dynamics of many industrial processes. Such processes often consist of a cascade of several or many first-order lags. The step response of such a cascade, and its approximation by a single lag and a dead time is shown in Fig. 9.10. If the process is already in existence, experimental step tests allow measurement of τ_{DT} and τ. At the

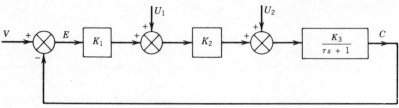

FIGURE 9.9 *Effect of gain distribution on steady-state errors.*

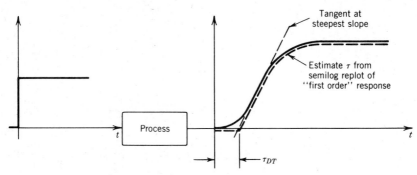

FIGURE 9.10 *Approximation of complex process response with dead-time-plus first-order lag.*

process design stage, theoretical analysis[1] allows estimation of these numbers if the process is characterized by a cascade of known first-order lags.

If we wish to analytically study closed-loop systems like that of Fig. 9.11 by solving the differential equation, we must approximate the dead-time dynamics with one of the models discussed in Chapter 2. Using the simplest of these $(e^{-\tau_{DT}^D} \approx -\tau_{DT} D + 1)$ we get

$$[(\tau - K\tau_{DT})D + 1 + K]C = K(-\tau_{DT}D + 1)V \qquad (9.21)$$

This model predicts instability when $K > \tau/\tau_{DT}$. For the more complicated dead-time approximation $(2 - \tau_{DT}s)/(2 + \tau_{DT}s)$, the closed-loop characteristic equation becomes

$$\tau_{DT}\, \tau s^2 + (2\tau + \tau_{DT} - K\tau_{DT})s + 2(K + 1) = 0 \qquad (9.22)$$

which predicts instability when $K > 2(\tau/\tau_{DT}) + 1$. The gain value for marginal stability can be found *precisely* from the Nyquist criterion since we know the frequency response of a dead time exactly. Figure 9.12 shows that for marginal stability we require that $(B/E)(i\omega)$ go precisely through $1\underline{/-180°}$. The phase angle part of this requirement can be stated as

$$-\pi = -\omega_0\tau_{DT} - \tan^{-1}\omega_0\tau \qquad (9.23)$$

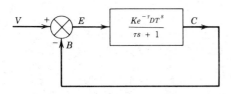

FIGURE 9.11 *Proportional control of dead-time-plus-first-order lag.*

[1]F. G. Shinsky, "Process Control Systems, Sec. Ed.," McGraw–Hill, New York, 1979, pp. 33–38.

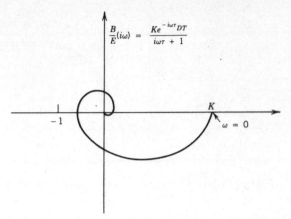

FIGURE 9.12 *Nyquist plot to determine gain for marginal stability.*

and fixes (for a given τ and τ_{DT}) the frequency ω_0 at which $(B/E)(i\omega)$ passes through $1\underline{/-180°}$. Equation 9.23 is trancendental and has no analytical solution, however we will carry out a numerical solution. Once ω_0 is found, the gain K for marginal stability is obtained by requiring $|(B/E)(i\omega_0)| = 1$.

$$1.0 = \frac{K}{\sqrt{(\omega_0\tau)^2 + 1}} \qquad K = \sqrt{(\omega_0\tau)^2 + 1} \qquad (9.24)$$

To make our numerical solution as general as possible we let

$$\tau_{DT} = n\tau \qquad 0.1 \le n \le 1.0 \qquad (9.25)$$

The range of n chosen covers the most common values encountered in modeling complex systems by the method of Fig. 9.10; however our computer program can easily use any other values. Equation 9.23 (and similar equations) can be solved numerically by writing it as

$$F = -n\omega_0\tau - \tan^{-1}\omega_0\tau + 3.14159 = 0 \qquad (9.26)$$

choosing a value for n, and then searching through a range of values of $\omega_0\tau$ until the function F crosses zero. If the increments of $\omega_0\tau$ are small enough, interpolation between the $\omega_0\tau$ values for $F = 0^+$ and $F = 0^-$ should give an accurate value for $\omega_0\tau$. Rather than writing such a program ourselves, we use one called ROOTS, available as part of the high-level engineering language SPEAKEASY, which we have used for other purposes earlier in the text. The complete program is as follows.

```
PROGRAM GAINSET     tells SPEAKEASY we want to use program mode
MARGINAL STABILITY FOR FIRST ORDER PLUS DEADTIME     title
DOMAIN COMPLEX    allows use of complex numbers
ANGLES RADIANS    angles will be given in radians
```

```
FOR N = .1,1.,.1      starts a FOR-LOOP (similar to FORTRAN DO-
                      LOOP) on n, n goes from 0.1 to 1.0 in steps of 0.1
Z1 = GRID (0.0,10.,.1)          sets up the range and increments of
Z2 = GRID (20.,200.,10.)        Z (Z ≜ ω₀τ) over which we will
Z = ARRAY (Z1,Z2)               search for a zero of F(Z)
F = -N*Z - ATAN(Z)+3.14159      defines F(Z), Eq. 9.26
R = ROOTS(F:Z)        invokes the zero-finding subroutine
K = SQRT (R*R+1)      computes gain from Eq. 9.24 once
                      R(R ≜ ω₀τ) has been found
PRINT N,R,K  prints results
ENDLOOP N    end of FOR-LOOP on N, program recycles until all N values
             have been run
END
EXECUTE GAINSET
```

Table 9.1 shows the results of this program, which can be compared with the K values for marginal stability predicted by the two dead-time approximations given earlier. The response of C to a step command V_s, using either of the two dead-time approximations, can be found analytically but is left for the chapter-end problems. For the exact dead-time model, the *steady-state* response to a step command of V_s is easily found since $C/E = K$ when all variables are constant. This gives

$$(V_s - C)_{ss}K = C_{ss} \qquad C_{ss} = \frac{K}{1 + K} V_s \qquad E_{ss} = \frac{1}{1 + K} 100\% \qquad (9.27)$$

TABLE 9.1

τ_{DT}/τ	$\omega_0\tau$	K
0.1	16.4	16.4
0.2	8.44	8.50
0.3	5.80	5.89
0.4	4.48	4.59
0.5	3.67	3.81
0.6	3.13	3.29
0.7	2.74	2.92
0.8	2.45	2.64
0.9	2.22	2.43
1.0	2.03	2.26

We see the steady-state error typical of proportional control. Design values of K must of course be *less* than those for marginal stability given in Table 9.1. A design criterion sometimes used[2] in industrial process control is *quarter-amplitude damping,* wherein each cycle of transient oscillation is reduced to one fourth the amplitude of the previous cycle. Shinskey associates this behavior with a gain margin of 2.0 for the frequency response. This is not exactly correct but is a useful approximation. If we apply it to the results of Table 9.1 for, say, $\tau_{DT}/\tau = 0.2$, we get a design gain value of 4.25, giving large steady-state errors. For this reason, processes of this type often use integral or proportional-plus-integral control, which reduces steady-state errors without requiring large K values.

To see what the exact step response looks like, we can use CSMP digital simulation. Using $\tau = 5$s, $\tau_{DT} = 1$s and a range of K values the program is as follows.

```
V  =  STEP (0.0)
E  =  V - C
M1 =  DELAY (200, 1.0,E)     dead time = 1.0 = 200 DELT
M2 =  REALPL (0.0,5.,M1)     first order system, τ = 5
PARAM K = (8.5,4.25,2.13,1.06)
      C  =  K*M2
```

Figure 9.13 shows that the damping for $K = 4.25$ is somewhat stronger than the quarter-amplitude criterion, thus the design procedure is conservative for this example. The transient frequency (measured from two successive peaks) is about 0.22 Hz. This could have been estimated from ω_0 in Table 9.1 as 8.44/(5 × 6.28) = 0.27 Hz. Once we have run CSMP and noted a peak overshoot of about 33%, we could assume an equivalent (second-order system) ζ of about 0.35 to improve future transient frequency estimates by using $\omega_0 \sqrt{1 - \zeta^2} = 0.94\omega_0 = 0.25$ in our example.

Although our SPEAKEASY program gets essentially exact results for any n values we want, a simple analytical approximation is useful for $n < 0.1$. Table 9.1 shows that in this range the first-order system contributes nearly $-90°$ to Eq. 9.26, thus the dead time provides the remaining $-90°$, giving

$$\omega_0\tau \approx \frac{1.57}{n} \approx K \qquad n < 0.1 \qquad (9.28)$$

9.5   PROPORTIONAL TEMPERATURE CONTROL OF A THERMAL SYSTEM WITH DEAD TIME

The stirred vessel of Fig. 9.14 is always full of process liquid and contains a liquid mass M_l. Liquid enters at a constant mass flow G and variable temperatures T_{U1} and leaves with temperature T_C (the controlled variable) and flow

---

[2]Ibid., pp. 8, 11.

FIGURE 9.13   *Effect of loop gain on step response.*

rate G. The electrical resistance heater is controlled by an SCR (silicon-controlled-rectifier) amplifier and produces a heating rate of Q_M watts. Basically a solid-state switch, the SCR amplifier controls heating rate by varying the point in the ac power cycle at which the heater is connected to (or disconnected from) the power line (see Fig. 9.15). Because the line frequency (60 Hz) is very high relative to thermal system frequency response, we take Q_M to mean the *average* power over the cycle, rather than the instantaneous electrical power, and assume that Q_M follows amplifier input voltage e_E instantly. The nonlinear *steady-state* relation (static calibration curve) between e_E and Q_M can be linearized at any chosen operating point, giving heater gain K_h. Heater mass, specific heat, and heat transfer area are M_h, C_h, A_h and heat transfer coefficient to liquid is U_h. Ambient temperature T_{U2} is uncontrolled and thus is a disturbance input, as is T_{U1}.

Assuming an initial equilibrium operating point and taking all variables as perturbations, conservation of energy gives for the heater

$$Q_M - U_h A_h (T_h - T_C) = M_h C_h \frac{dT_h}{dt} \qquad (9.29)$$

and for the vessel

$$U_h A_h (T_h - T_C) + G C_i T_{U1} - G C_i T_C - U_t A_t (T_C - T_{U2}) = M_l C_l \frac{dT_C}{dt} \qquad (9.30)$$

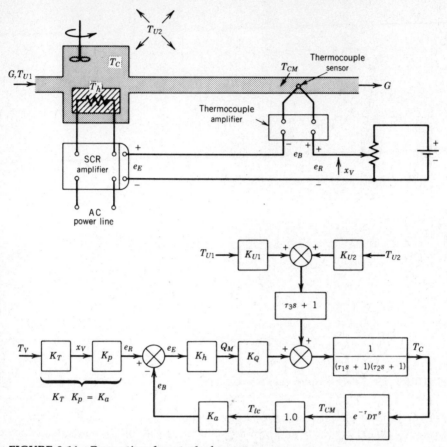

FIGURE 9.14   *Proportional control of temperature.*

where the tank wall is assumed to be pure thermal resistance $1/U_t A_t$ (no energy storage) and liquid specific heat is C_l. Elimination of T_h leads to

$$(\tau_1 D + 1)(\tau_2 D + 1)T_C = K_Q Q_M + K_{U1}(\tau_3 D + 1)T_{U1} + K_{U2}(\tau_3 D + 1)T_{U2} \quad (9.31)$$

where

$$\tau_1 \tau_2 \triangleq \frac{M_l C_l \tau_3}{G C_l + U_t A_t} \qquad \tau_3 \triangleq \frac{M_h C_h}{U_h A_h} \quad (9.32)$$

and

$$\tau_1 + \tau_2 = \frac{\tau_3(U_h A_h + G C_l + U_t A_t) + M_l C_l}{G C_l + U_t A_t} \quad (9.33)$$

where

$$K_Q \triangleq \frac{1}{G C_l + U_t A_t} \qquad K_{U1} \triangleq \frac{G C_l}{G C_l + U_t A_t} \qquad K_{U2} \triangleq \frac{U_t A_t}{G C_l + U_t A_t} \quad (9.34)$$

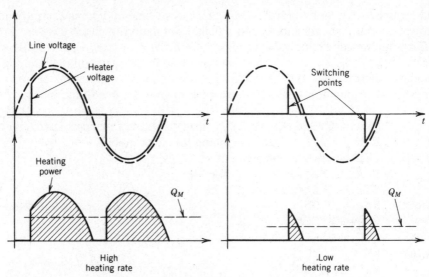

FIGURE 9.15   *SCR control of heating rate.*

Although we generally prefer to measure the controlled variable as directly as possible, location of the thermocouple temperature sensor (output a few millivolts) in the vessel encounters problems of electrical noise due to high-power SCR switching, and vibration caused by the tank stirrer, thus we locate it in the pipeline downstream of the tank, causing a dead time τ_{DT} between T_C and T_{CM}. There is also a thermal lag between fluid temperature T_{CM} and thermocouple metal bead temperature T_{tc}, however we assume this to be negligible relative to process dynamics. The thermocouple amplifier provides reference junction compensation, linearization, and voltage amplification, so that E_B (a few volts) is proportional to T_{tc}. We enter desired temperature T_V as a proportional voltage e_R from a potentiometer and subtract e_B from e_R by proper series connection.

Using the dead-time approximation $e^{-\tau_{DT}} \approx (-\tau_{DT}D + 1)$, the closed-loop system equation is

$$\left(\frac{D^2}{\omega_n^2} + \frac{2\zeta D}{\omega_n} + 1\right)T_C = \frac{K}{1 + K}\,T_V + \frac{K_{U1}}{1 + K}\,(\tau_3 D + 1)T_{U1} \tag{9.35}$$

$$+ \frac{K_{U2}}{1 + K}\,(\tau_3 D + 1)T_{U2}$$

where

$$\omega_n \triangleq \sqrt{\frac{1 + K}{\tau_1 \tau_2}} \qquad \zeta \triangleq \frac{\tau_1 + \tau_2 - K\tau_{DT}}{2\sqrt{\tau_1 \tau_2 (1 + K)}} \qquad K \triangleq K_a K_h K_Q \tag{9.36}$$

Use of the dead-time approximation simplifies the closed-loop system equation and allows easy gain setting if we use ζ (relative stability) as the design criterion and assume τ_1, τ_2, and τ_{DT} are known. Since dead-time approximations are not always accurate, we will shortly run a digital simulation to check this.

For a numerical example let us take $\tau_1 = 600s$, $\tau_2 = 60s$, $\tau_{DT} = 5s$. If we design for $\zeta = 0.6$, we get $K = 6.48$ and $\omega_n = 0.0144$ rad/s (period is $435s$/cycle). To check these results and get additional information on system behavior we run a CSMP simulation for both approximate and exact dead-time models.

```
           TVA = TVS*STEP(0.0)      step command for approximate model
           TVE = TVA      step command for exact model
PARAM      TVS = 1.0      step size is 1.0
           EEA = KA*(TVA  -  TCMA) ⎫  actuating signals for
           EEE = KA*(TVE  -  TCME) ⎬  approximate and
PARAM      KA = 0.05               ⎭  exact models
           QMA = KH*EEA
           QME = KH*EEE
PARAM      KH = 10.0
           MA = KQ*QMA
           ME = KQ*QME
PARAM      KQ = 12.96
           UA = KU1*TU1+KU2*TU2
PARAM      KU1 = 0.2,KU2 = 0.8
           TU1 = TU1S*STEP(0.0)
           TU2 = TU2S*STEP(0.0)
PARAM      TU1S = 0.0, TU2S = 0.0      no disturbance present
           LLA = LEDLAG(TAU3,TAU1,UA) ⎫  the portion of $T_C$ due to
PARAM      TAU3 = 90.,TAU1 = 600.     ⎬  $T_{U1}$  and  $T_{U2}$  passes
           TCUA = REALPL(0.0,TAU2,LLA)⎬  through transfer func-
PARAM      TAU2 = 60.                 ⎭  tions $(\tau_3 s + 1)/(\tau_1 s + 1)$
           TCQ1A = REALPL(0.0,TAU1,MA) ⎫ and $1/(\tau_2 s + 1)$
           TCQ1E = REALPL(0.0,TAU1,ME) ⎪
           TCQ2A = REALPL(0.0,TAU2,TCQ1A) ⎬ portion of $T_C$ due
           TCQ2E = REALPL(0.0,TAU2,TCQ1E) ⎪  to $Q_M$
           TCA=TCUA  +  TCQ2A    approximate $T_C$
           TCE=TCUA  +  TCQ2E    Exact $T_C$
           TCAD = DERIV(0.0,TCA)    derivative of $TCA$, initial value =
                                    0.0
           TCMA = -TAUDT*TCAD + TCA    approximate dead time
           TCME = DELAY(500,5.0,TCE)   Exact dead time
PARAM      TAUDT = 5.0
TIMER      FINTIM = 30.0, DELT = 0.01, OUTDEL = 0.2
OUTPUT     TCA,TCE,TCMA,TCME
PAGE GROUP
```

The DERIV (derivative) statement used in the approximate dead-time model is usually avoided because of potential noise problems, however T_C should be quite smooth in this example so we expect no difficulty.

Our first run explores the accuracy of the dead-time approximation by focusing on the first 30 s of response after a unit step command of T_V. In Fig. 9.16 we see that TCMA initially (first 10 s) goes off in the wrong (negative) direction; however after about 30 s it becomes a reasonable approximation to TCME. Even this initial error does not have much effect on the accuracy of TCA, which we see approximates TCE closely. Changing FINTIM to 600s, we get Fig. 9.17 showing the entire course of the transient response and making clear that the dead-time approximation is nearly perfect. Percent overshoot is about 9%, which agrees well with $\zeta = 0.6$. Damped natural period is hard to measure when so little oscillation is present, however the predicted value $2\pi/(\omega_n\sqrt{1 - \zeta^2}) = 545s$ appears reasonable on Fig. 9.17. Since an increase in dead time (other dynamics unchanged) causes a decrease in accuracy of the approximation, we rerun the problem with $\tau_{DT} = 50s$, giving Fig. 9.18. Some increase in error is apparent here, but accuracy is still acceptable for most purposes.

Returning to our basic system with $\tau_{DT} = 5s$, we now explore the response to disturbances. Equation 9.35 shows that, except for the gain factor, T_{U1} and T_{U2} cause *identical* responses in T_C, thus we need study only one, say T_{U2}. For a step disturbance this is easily done by setting TVS = 0.0 (desired temperature is zero), TU1S = 0.0, and TU2S = 1.0, giving Fig. 9.19. The peak error between T_V and T_C is about 0.14, whereas steady-state error is 0.107. The 0.107 figure is easily checked analytically from Eq. 9.35.

Since disturbances T_{U1} and T_{U2} are both measureable, the feedback system

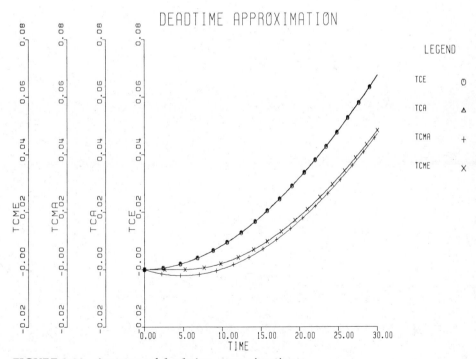

FIGURE 9.16   *Accuracy of dead-time approximation.*

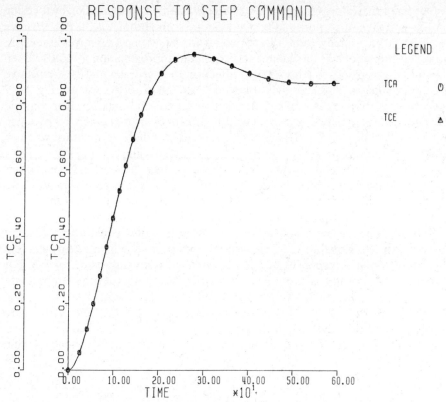

FIGURE 9.17   *Response to 1.0° F step command,* $\tau_{DT} = 5s$.

of Fig. 9.14 can be augmented with a feedforward system as in Fig. 9.20. Ideally the feedforward dynamics should be $\tau_3 s + 1$ but the derivative signal might accentuate noise, so we include the low-pass filter $1/(\tau_4 s + 1)$ with $\tau_4 = 0.1\tau_3$ even though this prevents perfect dynamic compensation. We add the following statements to our earlier program.

```
            UAFF=KU1FF*TU1+KU2FF*TU2
PARAM       KU1FF=0.2, KU2FF=0.8
            LLFF=LEDLAG(TAU3,TAU4,UAFF)
PARAM       TAU4=9.0
            ERFF=KHQ*LLFF
PARAM       KHQ=0.007716
            EEFF=KA*(TVA-TCMFF)-ERFF
            MFF=KQ*KH*EEFF
            LLAFF=LEDLAG(TAU3,TAU1,UA)
            TCUAFF=REALPL(0.0,TAU2,LLAFF)
            TCQ1FF=REALPL(0.0,TAU1,MFF)
```

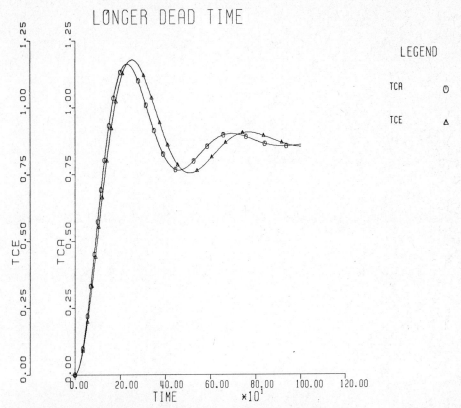

FIGURE 9.18 *Response to 1.0° F step command, τ_{DT} = 50s.*

```
              TCQ2FF=REALPL(0.0,TAU2,TCQ1FF)
              TCFF=TCUAFF + TCQ2FF
              TCMFF=DELAY(500,5.0,TCFF)
   OUTPUT     TCE,TCFF
```

(Note that we are using the exact dead-time model for our system with feed-forward.) In choosing the form of disturbance, we wish to study both momentary and sustained disturbances, so we choose T_{U1} and T_{U2} as in Fig. 9.21. These are written in CSMP as

TU1 = .08*RAMP(250.) − .08*RAMP(350.)

TU2 = .08*RAMP(0.0) − .16*RAMP(25.) + .08*RAMP(50.)

Running our CSMP simulation with these disturbances gives Fig. 9.22, which shows that transient errors are much reduced and steady-state errors are completely eliminated. (The *total* elimination of error depends on a *perfect* match of numerical parameter values in the feedforward and main system signal paths, a condition we can only approximate in practice.) The transient errors remaining

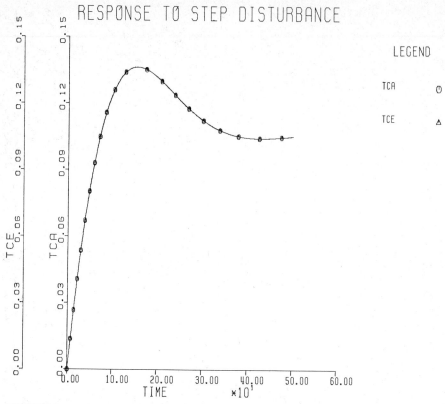

FIGURE 9.19   *Response to 1.0° F step disturbance of T_{ID}, $T_V = 0$, $\tau_{DT} = 5s$.*

in Fig. 9.22 can be reduced by making τ_4 smaller; Fig. 9.23 shows a rerun with $\tau_4 = 1.0s$, giving almost perfect results. Whether a practical system could use such a small τ_4 value depends on the noisiness of the U_{AFF} signal.

9.6   GAIN SETTING USING FREQUENCY-RESPONSE DESIGN CRITERIA

Although the design of a control system requires the choice of a numerical value for *every* system parameter, the choice of loop gain plays a central role and is often left as the final choice. We are then faced with a situation in which all parameters (τ's, ω_n's, ζ's, τ_{DT}'s. etc.) other than loop gain have been tentatively chosen and we need to pick an "optimum" value for loop gain. The most common approach is to choose gain so as to achieve a desired degree of relative stability and then to *check* other system specifications such as speed of response and steady-state errors. If these are not all satisfactory, we choose a more appropriate set of τ's, ω_n's, etc. (guided by the deficiencies of our first design) and then again set the gain for desired relative stability. If iterations of this

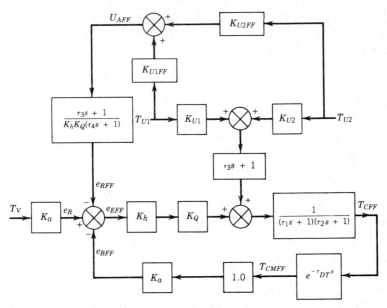

FIGURE 9.20 *Combined feedback and disturbance feedforward system.*

procedure do not produce a design that meets *all* specifications with physically realizable numerical parameters, we must then change system *configuration* (rather than just changing *numbers* in a given configuration) and repeat the gain-setting and parameter-adjustment procedures. This outline of the overall design procedure shows that gain setting for a desired level of relative stability is an often-used operation and thus should be made as systematic as possible.

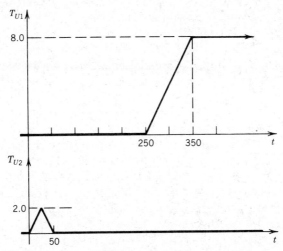

FIGURE 9.21 *Model for disturbing temperatures.*

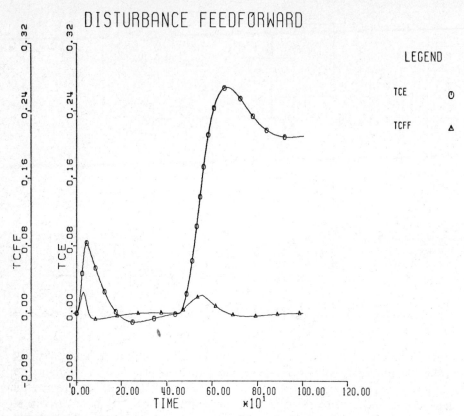

FIGURE 9.22   *Performance improvement due to feedforward, $\tau_4 = 9s$.*

Two approaches, frequency response and root locus, are in common use. This section develops the frequency-response methods.

For the simple systems studied thus far in this chapter, special gain-setting methods were unnecessary since performance characteristics were analytically available from the low-order differential equations. As we encounter higher-order systems, we lose this capability and the special gain-setting methods become useful. The frequency-response methods are based on the fact that many control systems exhibit a closed-loop amplitude ratio curve that exhibits a single peak (resonant frequency), such as in Fig. 9.24. Theory and experience have shown that if we choose loop gain such that the peak amplitude ratio M_p is about 1.3, then the transient response will usually exhibit a good compromise between speed of response and relative stability. For a simple secord-order system, $M_p = 1.3$ requires $\zeta = 0.42$, which in turn gives about a 25% overshoot for the step response. For higher-order systems no such precise relation between M_p and the transient response exists, however the correlation is good enough to make $M_p = 1.3$ a useful (but *not* infallible) design criterion.

One approach to implementing an M_p design criterion is to write an interactive computer program that accepts numerical values for system τ's, ω_n's, etc., and a tentative value for K, and then computes and displays a graph of $|(C/V)(i\omega)|$ versus ω. The designer notes the value of M_p and then iterates the process until the desired M_p is achieved. Depending on the availability and cost of computer resources, such an approach might be preferred. Before computer-aided design became as common as it is today, control system designers developed systematic graphical tools for gain setting. These methods, alone or in conjunction with computer aids, are still useful. Two separate methods are available, one for unity feedback systems and one for nonunity feedback. For unity feedback a one-step procedure without trial and error is possible; whereas nonunity feedback systems require trial and error. By *unity feedback* we mean a system in which $A(s)$ and $H(s)$ have no dynamics but are each equal to the same constant K_1 (see Fig. 9.25). When this is the case, the system block diagram can be condensed into the form shown in Fig. 9.25c, where K is the loop gain (which

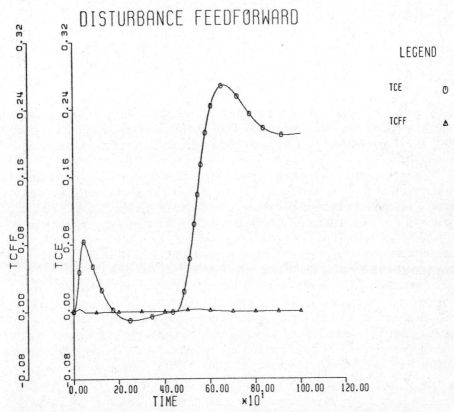

FIGURE 9.23 *Near-ideal performance with near-ideal feedforward dynamics,* $\tau_4 = 1s$.

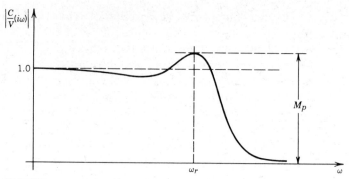

FIGURE 9.24   *Closed-loop amplitude ratio response.*

we must select to make $M_p = 1.3$). We may write

$$\frac{C}{E_1}(i\omega) = \frac{C}{V - C}(i\omega) = KG(i\omega) \tag{9.37}$$

$$\frac{V - C}{C}(i\omega) = \frac{V}{C}(i\omega) - 1 = \frac{1}{KG(i\omega)} \tag{9.38}$$

$$\frac{V}{C}(i\omega) = \frac{1}{KG(i\omega)} + 1 \tag{9.39}$$

This equation shows that $(V/C)(i\omega)$, the *inverse closed-loop transfer function*, can be obtained from the *inverse open-loop transfer function* $1/KG(i\omega)$ by adding $1\underline{/0°}$ at every frequency. Using polar plots we get the graphical interpretation of Fig. 9.26.

We see that by plotting inverse open-loop data on the set of polar coordinates with origin at (0,0) we can then read out inverse closed-loop data graphically from a second set of polar coordinates (the dashed circles and radial lines) with origin at $(-1, 0)$. Although the inverse polar plots are most convenient for establishing this concept, for actual numerical work a method based on the *direct logarithmic rectangular (Bode) plots* is preferred. It can be shown[3] that the coordinates for plotting the direct open-loop $(C/E_1)(i\omega)$ data are now a rectangular grid with phase angle on the horizontal axis and dB amplitude ratio on the vertical; whereas the direct closed-loop $(C/V)(i\omega)$ data are read from a superimposed set of amplitude ratio and phase-angle curves. These curves are *not* any familiar forms (circles, ellipses, etc.) but since the graph paper is commercially available, we need not concern ourselves with plotting these curves. Figure 9.27 shows the graph paper, which is called a *Nichols chart*.

[3]J. J. D'Azzo and C. H. Houpis, *Feedback Control System Analysis and Synthesis,* 2nd Ed., McGraw–Hill, New York, 1966, pp. 385–388.

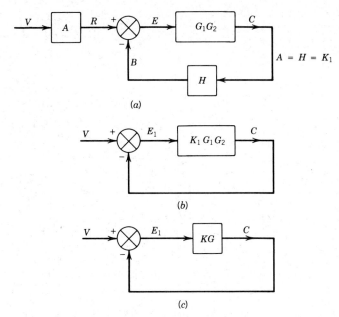

FIGURE 9.25 *Definition of unity feedback.*

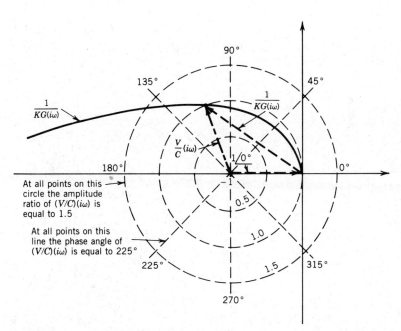

FIGURE 9.26 *Graphical relation between inverse open-loop and inverse closed-loop response.*

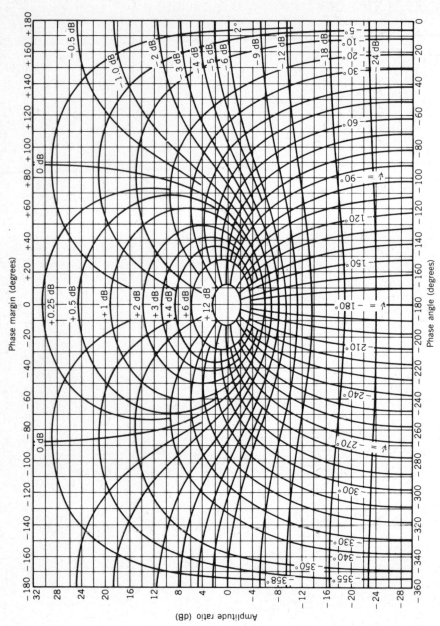

FIGURE 9.27 *Nichols chart.*

For unity-feedback systems the gain-setting procedure is as follows.

1. Choose a desired M_p value. This is usually 1.3, but if one wanted less oscillation and was willing to accept slower response and larger steady-state errors one might choose 1.1 or 1.2. If more oscillation is tolerable and we want maximum speed and minimum steady-state error, we might try 1.4 or 1.5. Convert the chosen M_p value to dB (1.3 = 2.28 dB) and locate on the Nichols chart the "egg-shaped" curve near the center that has this value of closed-loop amplitude ratio. (This may require visual interpolation). Accent this curve with a colored marker.

2. We now need to construct a standard Bode plot (amplitude ratio in dB, phase in degrees, versus frequency on a log scale) for the open-loop transfer function $KG(i\omega)$. This requires a trial value for K. Since all system parameters other than K are known, it is usually not very hard to do a Routh analysis for marginal stability. A gain margin of 2.5 is a reasonable starting point, so use K equal to 40% of the value for instability. If this procedure seems too tedious, actually *any* trial value of K that puts the data within the range of the vertical scale on the Nichols chart is acceptable.

3. Now transfer the open-loop data, one frequency at a time, from the two curves of the Bode plot onto the Nichols chart as a single curve of amplitude ratio versus phase angle, labelling each point with its frequency.

4. Since closed-loop amplitude ratio increases as we go toward the egg-shaped "center" on the curvelinear Nichols chart coordinates, the peak value of $(C/V)(i\omega)$ will be found where our plotted curve is *tangent* to the highest value of amplitude ratio. This tangency must occur *below* the 0 dB, $-180°$ point at the "center" of the Nichols chart. (If our curve crosses $-180°$ with dB > 0.0, the Nyquist criterion indicates instability.) Since the vertical axis (open-loop dB) on the Nichols chart is the same as the vertical axis on the dB Bode plot, changes in K do nothing but vertically shift our plotted curve on the Nichols chart. If our trial value of K gave a tangency different from that desired, we simply vertically shift the curve until it *is* tangent to the desired (say 2.28 dB) curve. This is most easily done with two transparent Nichols charts, one with the color-accented 2.28 dB curve and the other with the system's $KG(i\omega)$ curve. Shifting one sheet with respect to the other until tangency is obtained, one notes the displacement (in dB) of the two vertical scales. This displacement is the dB by which the assumed K must be changed in order to get the desired closed-loop peaking.

Although this gain-setting procedure is introduced in the chapter on proportional control, it is a *general* method, useful for all linear control systems of the type shown in Fig. 9.25. Systems with dead times are handled *without* approximation since we know the exact frequency response of a dead time.

As an example of this gain-setting procedure, let us take the system of Fig. 9.1 for a case when the dynamics of the electropneumatic transducer are taken as $K_p/(\tau_e s + 1)$ and the dynamics of the valve positioner are taken as $K_x/(\tau_x s + 1)$, giving an open-loop transfer function

$$\frac{e_B}{e_E}(s) = \frac{K}{(\tau_e s + 1)(\tau_x s + 1)(\tau_p s + 1)} \tag{9.40}$$

and a closed-loop transfer function

$$\frac{h_C}{h_V}(s) = \frac{K}{\tau_e \tau_x \tau_p s^3 + (\tau_e \tau_x + \tau_e \tau_p + \tau_x \tau_p)s^2 + (\tau_e + \tau_x + \tau_p)s + K + 1} \tag{9.41}$$

From Routh criterion, stability requires

$$1 + K < \left(\frac{1}{\tau_p} + \frac{1}{\tau_x} + \frac{1}{\tau_e}\right)(\tau_p + \tau_x + \tau_e) \tag{9.42}$$

If all three τ's are of similar magnitude, the allowable gain is quite small, as we can see if we set them all equal to τ, giving $K < 8$. If one or two of the τ's are much smaller than the third, then large gain is possible. For example, if $\tau_x = \tau_p$ and $\tau_e = 0.01\tau_p$, we get $K < 204$. If we take, say $\tau_p = 200s$, $\tau_x = 2s$, $\tau_e = 0.5s$, we get $K < 507$ and our trial gain value for making the Bode plot would be $203(46.1$ dB$)$. Figure 9.28 shows this Bode plot and Fig. 9.29 shows the corresponding Nichols chart. If we choose $M_p = 1.3$, the Nichols chart gain-setting procedure gives $K = 90$. Using this gain, the closed-loop frequency response (obtained from Nichols chart, or, more accurately, from a computer program) is given in Fig. 9.30a; whereas CSMP gives the unit step response of Fig. 9.30b, showing the effect of $\pm 10\%$ gain changes.

Another useful aspect of the Nichols chart is highlighted in Fig. 9.31. The "ideal" closed-loop amplitude ratio is 1.0 (0 dB), thus in "designing" the open-loop Bode plot we can strive to make our curve fall close to this contour on the

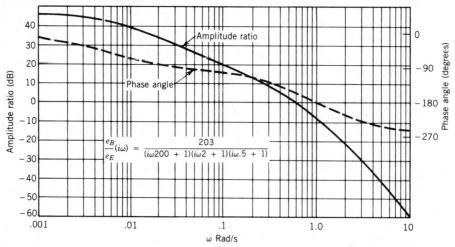

FIGURE 9.28   *Open-loop frequency response (Bode plot).*

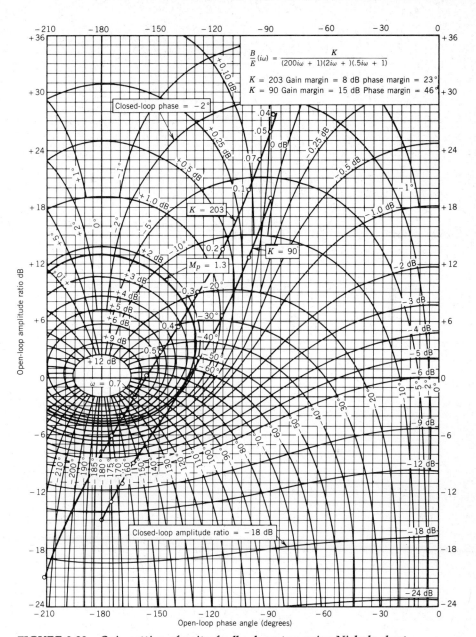

FIGURE 9.29 *Gain setting of unity-feedback system using Nichols chart.*

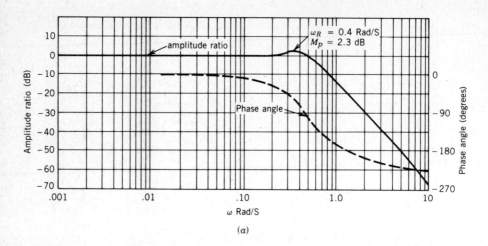

(a)

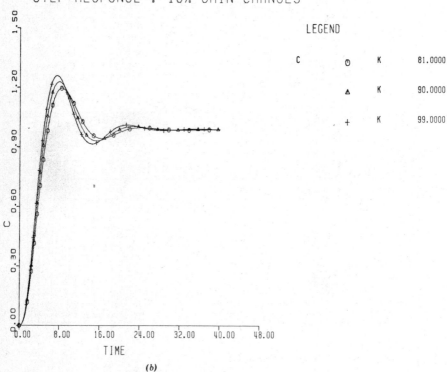

(b)

FIGURE 9.30   *Closed-loop frequency response and step response.*

Nichols chart. Since realistic specifications always allow some deviation from perfection, a practical design might strive for 0 dB ± 1 dB, say, giving an acceptable region in the form of a "funnel" as shown in Fig. 9.31. We then adjust open-loop gain and dynamics until our curve stays within the funnel over the prescribed range of frequencies.

In Fig. 9.25, when the dynamics in $A(s)$ and/or $H(s)$ are *not* negligible, the system is called *non-unity-feedback* and the one-step Bode/Nichols gain-setting procedure must be replaced by another, which involves trial and error. Of the several available procedures, we present the *inverse polar plot* scheme, a convenient method. Since trial and error is involved, interactive computer-aided approaches are appealing and I will present one in addition to the "manual" graphical method. In Fig. 9.32 we use block diagram manipulations to reduce the original configuration (9.32a) to the simplest form (9.32c), from which we write

$$KG(i\omega) \triangleq AG_1G_2(i\omega) \qquad H_1(i\omega) \triangleq \frac{H(i\omega)}{A(i\omega)} \tag{9.43}$$

$$\frac{C}{E_1}(i\omega) = \frac{C}{V - B_1}(i\omega) = KG(i\omega) \tag{9.44}$$

$$\frac{V - B_1}{C}(i\omega) = \frac{V}{C}(i\omega) - \frac{B_1}{C}(i\omega) = \frac{V}{C}(i\omega) - H_1(i\omega) = \frac{1}{KG(i\omega)} \tag{9.45}$$

$$\frac{V}{C}(i\omega) = \frac{1}{KG(i\omega)} + H_1(i\omega) \tag{9.46}$$

Our design criterion is the same as for the Bode Nichols unity-feedback situation; that is, we want the peak value of $(C/V)(i\omega)$ to be some desired value, usually 1.3. A peak value of C/V clearly means a *minimum* value of V/C, the *inverse closed-loop relation*, in Eq. 9.46. The gain-setting procedure goes as follows.

1. Decide on the M_p value desired and draw a circle with center at the origin and radius of $1/M_p$ (usually 0.77).
2. Use Routh criterion to find gain for marginal stability. Make a polar plot of $1/KG(i\omega)$ using a trial value of K equal to 40% of this value.
3. Plot $H_1(i\omega)$ on the same graph. (Note that its steady-state gain is often 1.0. Why?)
4. Add $1/KG$ and H_1 at each frequency, as required by Eq. 9.46, to get V/C. Our goal is to make the V/C curve *tangent* to the $1/M_p$ circle, since this makes the minimum value of V/C equal to $1/M_p$.
5. Change K in direction and amount suggested by the graph and repeat procedure until tangency is achieved.

Figure 9.33 illustrates the method, showing a situation in which the trial value of K was too low, making the minimum of V/C too large and thus M_p too small.

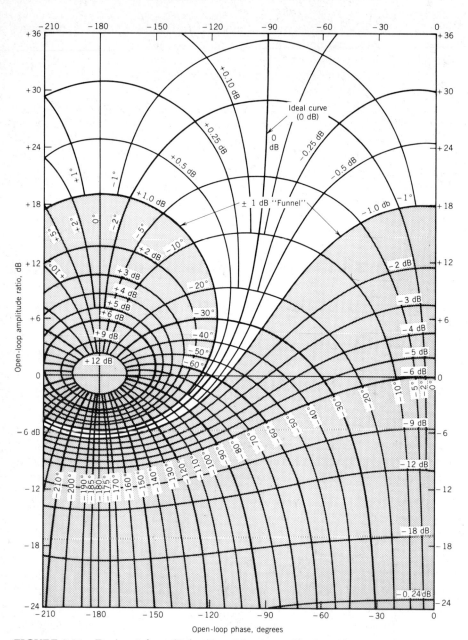

FIGURE 9.31 *Design " funnel" for closed-loop amplitude ratio.*

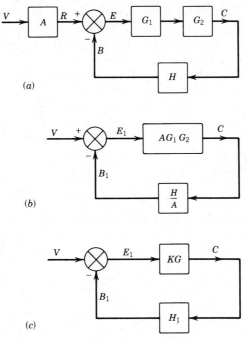

FIGURE 9.32   *Nonunity-feedback system.*

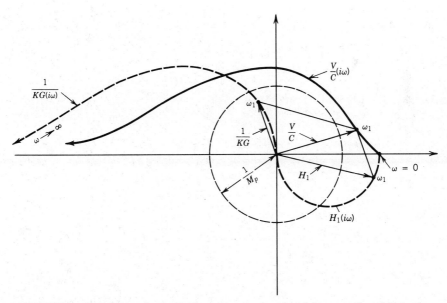

FIGURE 9.33   *Relation between inverse open-loop and inverse closed-loop for nonunity-feedback systems.*

As a practical example of a nonunity feedback system we will study the electrohydraulic rotary speed control system of Fig. 9.35, which uses the *two-stage electrohydraulic servovalve* of Fig. 9.34. The spools of servovalves rated at up to about 5 GPM can be directly positioned by solenoid-type magnetic actuators called force (or torque) motors and are called *single-stage* valves. Above this flow rating the forces required to rapidly and accurately position the main spool are beyond the capabilities of electromagnetic positioners and miniature hydromechanical servomechanisms internal to the servovalve perform this function. Large electrohydraulic servovalves will thus have an electromechanical first stage and one or more hydromechanical stages. The largest (~1000 GPM) servovalves may be three-stage devices in which the third stage uses electrical feedback (often an LVDT) of the main spool position to create a miniature electrohydraulic servo internal to the valve. The two-stage valve of Fig. 9.34 uses a magnetic force motor to position a flapper between a pair of differentially connected nozzles, creating a pressure difference across the ends of the main spool, which then moves until the flapper is again in its neutral position midway between the nozzles. The main spool thus quickly and accurately follows the flapper position produced by the force motor.

In the speed control system of Fig. 9.35, this valve controls a fixed-displacement rotary hydraulic motor and geared load. Speed is measured using a digital

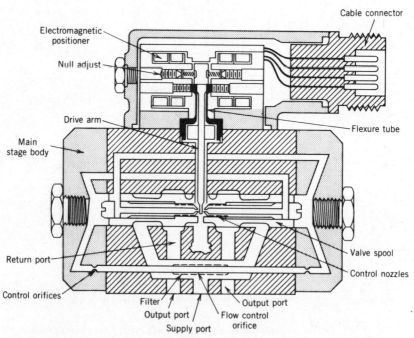

FIGURE 9.34 *Two-stage electrohydraulic servo valve.*
Source: Koehring Pegasus Division, Troy, Michigan.

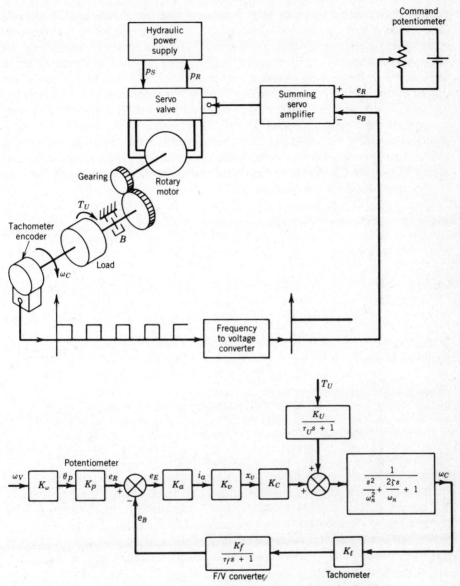

FIGURE 9.35 *Electrohydraulic speed control with nonunity feedback.*

shaft encoder of the tachometer type[4] and an electronic frequency-to-voltage converter.[5] The tachometer encoder produces a pulse train whose frequency is proportional to speed, whereas the F/V converter outputs a time-varying dc

---

[4]E. O. Doebelin, Measurement Systems, 3rd Ed., McGraw–Hill, New York, 1983, pp. 294–297.
[5]Ibid., p. 765.

voltage proportional to pulse rate. Low-pass filtering is required in the F/V output circuit and this produces first-order-lag dynamics in the speed measurement, giving a non-unity-feedback system. A 10-turn rotary potentiometer allows manual input of desired speed ω_V in terms of e_R. The summing servo amplifier is of the current output ("transconductance") type that produces an output current (servovalve force-motor current) in proportion to the differential $(e_R - e_B)$ voltage input. Such an amplifier uses a current-feedback technique to suppress the lagging inductance/resistance dynamics of the force-motor coils, giving faster valve response and contributing to the accuracy of our zero-order model assumption between e_E and main valve spool position x_v. From x_v and disturbing torque T_U to load speed ω_C we use the models of Eq. 2.170 and 3.6.

The closed-loop system equation is

$$\left[\frac{\tau_f}{\omega_n^2}s^3 + \left(\frac{2\zeta\tau_f}{\omega_n} + \frac{1}{\omega_n^2}\right)s^2 + \left(\frac{2\zeta}{\omega_n} + \tau_f\right)s + (1 + K_L)\right]\omega_C$$
$$= K_L\frac{K_\omega K_p}{K_f K_t}(\tau_f s + 1)\omega_V + K_U(\tau_U s + 1)(\tau_f s + 1)T_U \tag{9.47}$$

where

$$K_L \triangleq \text{loop gain} \triangleq K_f K_t K_a K_v K_c \tag{9.48}$$

In this example the usual step command steady-state error associated with proportional control can be removed by setting $K_\omega K_p/K_f K_t$ equal to $(1 + K_L)/K_L$. To set K_L using the method of Fig. 9.33 we first do a Routh stability study, which gives for marginal stability

$$K_L - 2\zeta\left(2\zeta + \frac{1}{\omega_n\tau_f} + \omega_n\tau_f\right) \tag{9.49}$$

The function $1/\omega_n\tau_f + \omega_n\tau_f$ has its minimum value of 2 at $\omega_n\tau_f = 1$; thus the lowest value of K_L would be $4\zeta^2 + 4$. Large values of K_L are possible for $\omega_n\tau_f$ values either much larger or much smaller than 1.0, assuming ζ is not very small. For our numerical example we will take $\omega_n = 50$ rad/s, $\zeta = 0.75$ and $\tau_f = 0.5s$, giving a trial K_L value of 6.

We will implement the gain-setting procedure with an interactive computer program based on the convenient SPEAKEASY language used several times earlier. Similar programs could be produced using other languages.

```
1.0 PROGRAM        tells SPEAKEASY we are using program mode
2.0 DOMAIN COMPLEX; ANGLES DEGREES        use complex numbers; an-
                                          gles in degrees
3.0 GRAPHICS(TEK4010,SFACT:1.3)        want to use the SPEAKEASY
                                       graphics package with a Tek-
                                       tronix 4010 terminal and a
                                       graph size 1.3 times the nor-
                                       mal
```

```
 4.0 SETITLE ('GAIN SETTING FOR NONUNITY FEEDBACK SYSTEM')
 5.0 W=GRID(0,60,1); S=W*1I    sets frequency range and increment
 6.0 OMEGA=GRID(0,360,5);  ⎫  terms involving OMEGA are
       SSS=GRID(-20,20)    ⎬  used to plot the 1/Mp circle.
 7.0 X=1/1.3*COS(OMEGA)    ⎪  The SSS statement is used later
 8.0 Y=1/1.3*SIN(OMEGA)    ⎭  to plot the coordinate axes
 9.0 YY=2,-2;XX=3,-3    used later (Step 17.0) to set range of
                          Y(2,-2) and X(3,-3) for all the polar plots
10.0 A:PAUSE,INITIALIZE K,TYPE 'K=?'    inserts a PAUSE into
                                        the program where it
                                        asks you to type in the K
                                        that you want to try
11.0 INVS=(.0004*S*S + .03*S+1)/K    computes 1/KG
12.0 REINV=REAL(INVS); IMINV=IMAG(INVS)    gets real and
                                           imaginary parts
                                           of 1/KG
13.0 H=(K-1)/(K*(.5*S+1)    computes H₁ using gain of (K-1)/K to
                            give zero steady-state error for a step
                            command
14.0 REH=REAL(H); IMH=IMAG(H)    gets real and imaginary parts of
                                 H₁
15.0 CLTF=INVS+H    Computes (V/C) as 1/KG + H₁
16.0 RE=REAL(CLTF); IM=IMAG(CLTF)    gets real and imaginary
                                     parts of V/C
17.0 LINECODE = -1; GRAPH (YY:XX)    plots two points at 3, 2
                                     and -3, -2; the real pur-
                                     pose here is to fix the scale
                                     of all the upcoming graphs;
                                     That is, the range of x will
                                     be -3 to +3 and the range
                                     of y will be -2 to +2
18.0 LINECODE=1; ADDGRAPH (0:SSS) ⎫  plots coordinate
19.0 LINECODE=1; ADDGRAPH (SSS:0) ⎭  axes through 0,0.
20.0 LINECODE=2; ADDGRAPH (IMH:REH)    plots H₁ as dashed line
21.0 LINECODE=3; ADDGRAPH (IMINV:REINV)    plots 1/KG as cen-
                                           ter-line
22.0 LINECODE=5; ADDGRAPH (Y:X)    plots 1/Mp circle as dashed line
23.0 LINECODE=1; ADDGRAPH (IM:RE)    plots V/C as solid line
24.0 BELL; PAUSE    Rings bell and pauses so you can make a hardcopy if
                    desired
25.0 ARDTF=ABS(1/CLTF); PHADTF=PHASE(1/CLTF)    computes
                                                amplitude
                                                ratio and
                                                phase of
                                                C/V
```

```
26.0 GRAPHICS(TEK4010,SFACT:1)        returns graph scale factor to
                                      "normal"
26.1 GRAPH(ARDTF:W); LINECODE=2; OVERLAY;
     GRAPH(PHADTF:W)       makes rectangular graph of amplitude ratio and
                           phase of C/V
26.2 GRAPHICS(TEK4010,SFACT:1.3)      changes graph scale back to
                                      1.3 to get ready for next pass
                                      through the program
27.0 FREE K INVS REINV IMINV H REH IMH CLTF RE IM
ARDTF PHADTF     clears memory of numerical values to make room for next
                 pass through program
28.0 BELL; BELL; BELL; PAUSE      inserts a pause to allow you to make
                                  a hardcopy if desired
29.0 GO TO A     returns to line 10.0 to try another K value
```

This program can be used for *any* nonunity-feedback gain-setting problem since only statements 5.0, 11.0, and 13.0 need to be changed to suit the particular example. This editing is easily done using DELETE and INSERT capabilities of SPEAKEASY. Note that since we made the gain of H_1 equal to $(K - 1)/K$ in line 13.0 in order to get zero steady-state error for a step command, in this example K is the gain of AG_1G_2 but is *not* the loop gain. Starting with $K = 7$, four iterations of the program were necessary to find that $K = 1.3$ would give $M_p = 1.3$. The closed-loop frequency response for this K value is shown in Fig. 9.36a, whereas 9.36b shows the corresponding step response as obtained from CSMP. If we measure speed of response by time to peak, the system appears fast; however if 5% settling time is used, the slow return to steady state after the peak (associated with the $0.5s$ feedback time constant) gives a much less favorable picture. Also, although the "trick" of making feedback gain equal to $(K - 1)/K$ is effective in making steady-state error for a step command equal to zero, the very low loop gain of 0.3 means that the system is *not* very effective in rejecting disturbances. If gain K should change (not an unlikely occurrence) even the zero steady-state error for step commands is lost. Actually, when loop gain is this low, the system is almost operating open loop, having all the drawbacks associated with open-loop systems. Thus this "bare bones" design would probably be unacceptable in most applications and would require compensation (addition of other control modes, such as integral) before performance specifications could be met. We will return to this example in later chapters to show how performance can be improved.

The computer program given for setting gain in non-unity-feedback systems (Fig. 9.32) can also be used for unity-feedback systems (Fig. 9.25) since we can easily set KG equal to $K_1G_1G_2$ and H_1 equal to 1.0. This substitutes a trial-and-error procedure for the one-step Bode-Nichols gain setting, however if our computer graphics system has fast response, the trial-and-error approach may be the most cost effective.

9.7 GAIN SETTING USING ROOT-LOCUS DESIGN CRITERIA

Historically, frequency-response techniques were developed earlier, but the root-locus method of Evans[6] quickly achieved popularity, and knowledge of both these approaches will be helpful to a control system designer. I introduced the root-locus concept briefly in Section 6.5 and now wish to develop more details. In comparison with the frequency response methods, two features should be mentioned at the outset. First, a *single* root-locus procedure is applicable to both unity and nonunity-feedback systems. Second, when a system includes dead times, root locus techniques are most convenient if the dead time is *approximated* as the ratio of two polynomials in s (see Section 2.7), whereas frequency response methods handle dead times exactly, with no difficulty.

For the general system of Fig. 5.1 the closed-loop characteristic equation is

$$1 + G_1G_2H(s) = 0 \tag{9.50}$$

If the open-loop transfer function G_1G_2H with loop gain of K is written in the

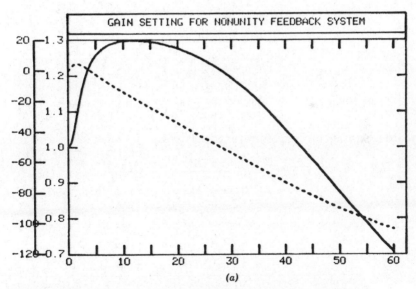

PAUSE
: >

FIGURE 9.36 *Closed-loop frequency response and step response.*

[6]W. R. Evans, Graphical Analysis of Control Systems, *Trans. AIEE,* Vol. 67, 1948. pp. 547–551; W. R. Evans, "Control System Dynamics," McGraw–Hill, New York, 1954.

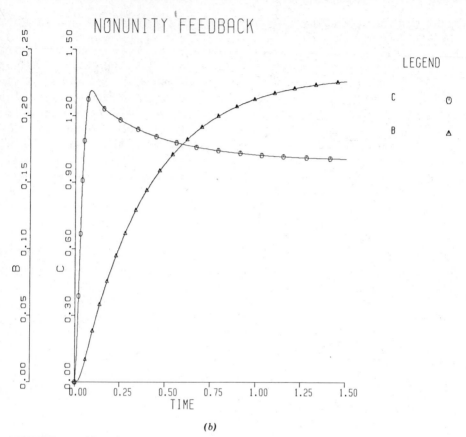

FIGURE 9.36 (Cont.)

standard form used for frequency-response studies, we might have, for example

$$G_1G_2H(s) = \frac{K(\tau_1 s + 1)}{s(\tau_2 s + 1)\left(\dfrac{s^2}{\omega_n^2} + \dfrac{2\zeta s}{\omega_n} + 1\right)} \qquad (9.51)$$

whereas the root-locus standard form would be

$$G_1G_2H(s) = \frac{k\left(s + \dfrac{1}{\tau_1}\right)}{s\left(s + \dfrac{1}{\tau_2}\right)\left(s + \zeta\omega_n + i\omega_n\sqrt{1 - \zeta^2}\right)\left(s + \zeta\omega_n - i\omega_n\sqrt{1 - \zeta^2}\right)} \qquad (9.52)$$

where

$$k \triangleq \text{static loop sensitivity} \triangleq \frac{K\tau_1\omega_n^2}{\tau_2} \qquad (9.53)$$

The characteristic equation would be

$$\frac{\tau_2}{\omega_n^2} s^4 + \left(\frac{2\zeta\tau_2}{\omega_n} + \frac{1}{\omega_n 2}\right)s^3 + \left(\frac{2\zeta}{\omega_n} + \tau_2\right)s^2 + (K\tau_1 + 1)s + K = 0 \quad (9.54)$$

The root locus is defined as the path in the complex plane of the roots of the characteristic equation, as K is varied from zero to infinity, when the numerical values of all other system parameters are assumed fixed and known. Since digital computer polynomial root-finding algorithms, such as the SPEAKEASY POLYROOT, are widely available, one can easily construct around such a general-purpose root finder a program for finding and graphing the root locus. Special-purpose root locus programs are also commercially available. A SPEAK-EASY POLYROOT program for Eq. 9.54 might go as follows.

```
:__ GRAPHICS (TEK4010)      calls graphing package
:__ PROGRAM RTLCS      want to use SPEAKEASY program mode
EDIT INPUT MODE  ⎫
 1 PROGRAM        ⎭  computer returns message
 2 DOMAIN COMPLEX      request complex number capability
 3 ADRTS=0      initialize value of ADRTS
 4 K=-0.05      initialize gain value
 5 TAU1=1.0; TAU2=.5; ZETA=.5; WN=1.0      enter parameter
                                                values
 6 A=TAU2/(WN*WN)              ⎫  compute equation
 7 B=(2*ZETA*TAU2/WN) + (1/WN*WN))  ⎬  coefficients
 8 C=(2*ZETA/WN) + TAU2        ⎭
 9 FOR M=1, 40, 1      requests computing loop to be iterated 40 times
10 K=K+.05      increment gain
11 D=(K*TAU1)+1  ⎫  compute equation coefficients
12 E = K         ⎭
13 COFS=A1D(5:E,D,C,B,A,)      defines coefficients of characteristic
                               equation as a one-dimensional array
                               (A1D)
14 ROOTS=POLYROOT(COFS)      calls root finder
15 ADRTS=(ADRTS, ROOTS)      accumulates all the root values for all the
                             K's
16 ENDLOOP M      terminates loop started in Statement 9
17 R=REAL(ADRTS)  ⎫  get real and imaginary parts of all roots
18 I=IMAG(ADRTS)  ⎭
19 LINECODE=-1      specifies plotting of points only, no connecting lines
20 GRAPH(I:R)      asks for root locus graph
21 END
PROGRAM RTLCS IS NOW DEFINED  ⎫  computer message
MANUAL MODE                   ⎭
:__ KEEP RTLCS      saves the program for future use
```

:__ EDIT   requests edit command mode
EDIT COMMAND MODE   computer message
:% RUN   requests program to be run

Statements 4, 9, and 10 give a gain range of 0 to 2.0. The upper value 2.0 is a little beyond the gain for marginal stability, as calculated from Routh criterion. Equal increments of K (as used here) may not always give the best point spacing on the graph. An alternative rule, $K = 1.15(K + 0.05)$, may give improved (more even) spacing. Starting K near zero gives a "complete" graph, but in actual design studies one focuses on selected regions of the root locus where good system performance is obtained, thus computer usage can be economized by restricting calculations to these regions.

Although computerized numerical root-finding schemes serve well for *analyzing* a given system, the graphical root-locus construction methods developed by Evans are more useful for providing the insight necessary for *designing* systems to meet specifications. I now list, and in some cases prove, a set of rules that allow rapid construction of the root locus. Note that the method works with the *open-loop* transfer function but provides results for the *closed-loop* system, just as was the case for the frequency-response methods discussed in the previous section. The great advantage of working with the open-loop transfer function is that, in most cases, it is already *factored* into such simple forms as integrators and first- and second-order systems etc. This is because the open loop is formed by simply cascading the several hardware components, and thus their individual transfer functions remain separate and visible. For the closed-loop function the individual parameters became "scrambled" and the effect of a specific hardware item on the total response is not apparent.

All the root-locus construction rules derive from the fact that, for a point s to be on the root locus, we must have

$$G_1 G_2 H(s) = -1 = 1\underline{/180°} \tag{9.55}$$

Actually, the requirement for a 180° angle is the critical one since the magnitude of 1.0 can always be achieved by choosing a proper gain value. Let us now take the general form of $G_1 G_2 H(s)$ as

$$G_1 G_2 H(s) = \frac{k(s - z_1)(s - z_2) \cdots (s - z_m)}{(s - p_1)(s - p_2) \cdots (s - p_n)} \quad n > m \tag{9.56}$$

Rule 1. Starting and Ending Points
Writing Eq. 9.50 as

$$(s - p_1)(s - p_2) \cdots (s - p_n) + k(s - z_1)(s - z_2) \cdots (s - z_m) = 0 \tag{9.57}$$

we see that if $k = 0$, Eq. 9.57 is satisfied by s equal to any of the poles of the open-loop transfer function. *Thus the root locus always starts, for $k = 0$, at the open-loop poles.* When $k \to \infty$, Eq. 9.55 can be satisfied only if s approaches one of the zeros. *Thus the root locus ends, for $k \to \infty$, at the open-loop zeros.*

Since, for a math model to give the physically required result of zero amplitude ratio at infinite frequency (sinusoidal transfer function) we have $n > m$, some of the root locus paths that start at the n poles have no zeros on which to end. These paths proceed toward $s = \infty$ along certain asymptotes given in Rule 5. Clearly $s \rightarrow \infty$ *does* satisfy Eq. 9.55, since its effect on G_1G_2H is the same as the affect of s approaching a zero.

Rule 2. Number of Branches

If $n > m$, the characteristic equation will be of nth degree, and thus there will be n roots for any particular value of gain. The branches start at the open-loop poles, which number n; thus one root will be found on each branch. Some of the branches may, however, coalesce for certain values of gain, since repeated roots are possible. These branches will always separate again when the gain is changed.

Rule 3. Loci on the Real Axis

The angle requirement of Eq. 9.55 may be written as

$$\underline{/s-z_1} + \underline{/s-z_2} + \cdots + \underline{/s-z_m} - \underline{/s-p_1} - \underline{/s-p_2} - \cdots \qquad (9.58)$$
$$- \underline{/s-p_n} = (2r + 1)\, 180° \qquad r = 0, 1, 2, \ldots$$

In testing a "search point" on the real axis to see whether it is on the root locus, any complex poles or zeros in Eq. 9.58 can be ignored since they come in symmetrical pairs and their angles cancel each other. Any real poles or zeros to the *left* of search point s each contribute $0°$ to Eq. 9.58 whereas those to the right contribute $180°$. *Thus for a real-axis search point to be on the root locus there must be an odd number of real poles plus zeros to its right.*

Special cases occasionally arise that require modification of this rule. In the vast majority of cases, k is a positive number and all open-loop poles/zeros are in the left half plane, giving the rule as just stated. Suppose, however, that the numerator of Eq. 9.56 has a term $(-3s + 1)$ and k is positive. (Note that this would occur, for example, if we used the simplest approximation for a 3-s dead time.) To make such a term consistent with the notation of Eq. 9.58 it would be written as

$$3(s - .333)\,(1\underline{/180°}) \qquad (9.59)$$

and the 3 would be absorbed into k in Eq. 9.56. Since the $1\underline{/180°}$ is added to the left side of Eq. 9.58, the right side must now be taken as $0°$, $360°$, $720°$, ... Our rule now requires an *even* number of real poles plus zeros to the right of a real-axis search point for it to be on the root locus (zero is considered an even number). Note that if the term $(-3s + 1)$ appeared in the denominator of Eq. 9.56 our new rule applies also. When there are *several* pole/zero terms of this type, the "original" rule holds if there is an even number of such terms; whereas the "new" rule applies for an odd number. Cases with complex poles/zeros in the right half plane follow the original rule.

Suppose we have a term of form $(3s - 1)$. Note that the steady-state gain (set $s = 0$) for such a term is negative and if the term were included with a set of "normal" (left half plane, positive gain) terms the complete open-loop function would have negative gain, making the feedback *positive*, which is usually unstable. Occasionally, however, such systems are stable and useful so we should treat this case. For example, if

$$A(s) = H(s) = 1.0 \qquad G_1G_2H(s) = \frac{k(s + 2)}{(s - 0.333)(s + 1)} \qquad k > 0 \quad (9.60)$$

then

$$[s^2 + (0.667 + k)s + (2k - 0.333)]C = k(s + 2)V \qquad (9.61)$$

which is stable for $k > 0.167$. Or if

$$A(s) = H(s) = 1.0 \qquad G_1G_2H(s) = \frac{k(s) + 2)}{s(s - 0.333)} \qquad k > 0 \quad (9.62)$$

then

$$[s^2 + (k - 0.333)s + 2k]C = k(s + 2)V \qquad (9.63)$$

which is stable for $k > 0.333$. Examination of Eq. 9.58 for terms of the form $(s - a)$ where $a > 0$ shows that such terms follow our original rule for real-axis loci; that is, the root locus exists where there is an odd number of real poles plus zeros to the right of the search point.

Our final case (also of rare occurence) has all the open-loop poles and zeros in the left half plane but takes k to be a negative number. This can be treated as in Eq. 9.59 (k positive but the term $1\underline{/180°}$ added), so the modified rule (even number of poles/zeros to the right) applies here.

Rule 4. Symmetry
Since any complex roots of the characteristic equation always occur in pairs of the form $a \pm ib$, the locus of roots must be symmetrical about the real axis. Thus if one gets the upper half of the locus, the lower half is obtained immediately by reflection about the real axis.

Rule 5. Asymptotes
When there are more poles than zeros, some of the branches have no zeros on which to end and must go off to infinity. It can be shown[7] that

1. The branches that go off to infinity become asymptotic to certain straight lines.

---

[7]D'Azzo and Houpis, *Feedback Control*, Sect. 7–8, 7–14.

2. The number of branches that go to infinity (and therefore the number of straight-line asymptotes) is equal to $n - m$, the excess of poles over zeros.

3. The asymptotic lines all radiate out from a single point s_{cg} on the real axis (called the center of gravity of the roots), where

$$s_{cg} = \frac{\sum \text{poles} - \sum \text{zeros}}{n - m} \qquad (9.64)$$

4. If all the open-loop poles/zeros are in the left half plane and k is positive, the angular locations θ_r of the radial asymptotic lines are given by

$$\theta_r = \frac{(2r + 1)180°}{n - m} \qquad r = 0, 1, 2, \ldots, (n - m - 1) \qquad (9.65)$$

For

a. All open-loop poles/zeros in the left-half plane, $k < 0$, or
b. An odd number of open-loop pole/zero terms of form $(-as + 1)$, $a > 0, k > 0$.

the rule changes to

$$\theta_r = \frac{(r)360°}{n - m} \qquad r = 0, 1, 2, \ldots, (n - m - 1) \qquad (9.66)$$

Rule 6. Slope at Breakaway and Break-in Points
When two real roots coalesce into a repeated real root and then break away from the real axis into two complex conjugate roots, or when two complex conjugate roots break in to the real axis to become a repeated real root, the root locus slope is (with rare exceptions) $\pm 90°$ at such breakaway or break-in points. This behavior largely precludes the existence of "cusps" (see Fig. 9.37).

Rule 7. Angle of Departure at Complex Poles
Since branches of the root locus start at each complex pole, it is helpful to know the slope of the root locus as it leaves the pole. To determine this slope (all left half plane poles/zeros, $k > 0$)

1. Measure the angle from a horizontal line through the pole in question to each other pole, taking counterclockwise as positive.

2. Repeat Step 1 for all the zeros.

3. Subtract the sum of the angles to the poles from the sum of the angles to the zeros.

4. Subtract 180° from the result of Step 3. The result will be the angle of the tangent to the root locus at the pole in question.

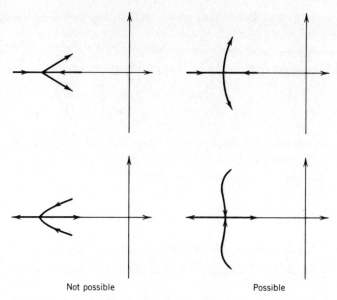

Not possible   Possible

FIGURE 9.37   *"Cuspless" behavior of root locus at breakaway and break-in points.*

For the two exceptions (a, b) noted in Rule 5, the departure angle is 180° away from the direction calculated here.

Rule 8. Angle of Arrival at Complex Zeros
Follow the four steps of Rule 7; however in evaluating the *final* angle, take *clockwise* as positive. Exceptions a, b again give a direction 180° away from that calculated "normally."

These eight rules allow the designer to sketch rapidly much of the root locus for a proposed new design. With experience, *all* of the root loci can often be sketched with rough accuracy sufficient for preliminary evaluation of design ideas. Evans, the inventor of the root-locus method, also invented a simple graphical aid called the Spirule, which allows one to rapidly "fill in the gaps" left in the root locus after using all the rules. The Spirule is available[8] for a few dollars. To use it, one guesses a search point (guided by the eight rules), where-upon the Spirule *quickly* measures the angle of Eq. 9.55. A systematic procedure is to make a "horizontal traverse" of search points in a region (such as near an asymptote) where the root locus is likely to exist. A sequence of angle readings might be 165°, 176°, 184°. Since we are looking for 180°, the fourth trial (between the 176° and 184° points) would probably be successful. Once a point on the locus is found, that branch is easily "tracked" by taking successive horizontal

---

[8]The Spirule Co., 9728 El Venado, Whittier, Calif. 90603.

traverses with the Spirule at about 1/4 in. vertical spacing. Spirule accuracy deteriorates when closer than about 1/4 in. from the real axis, however the $\pm 90°$ rule of Fig. 9.37 allows one to "fake" the curve into the real axis for breakaway or break-in points. (A rule for *calculating* breakaway or break-in points is available but requires considerable work, thus use of it is not justified since the $\pm 90°$ rule gives adequate accuracy.)

The complete root locus can be obtained by application of only the 180° angle requirement. However, if we wish to know what values of k or K are associated with any particular point on the locus, we must invoke the magnitude requirement

$$\frac{k\, |s - z_1|\, |s - z_2| \cdots |s - z_m|}{|s - p_1|\, |s - p_2| \cdots |s - p_n|} = 1 \qquad (9.67)$$

where s is the point *on* the root locus at which we wish to know k. By direct length measurement on the root locus graph we can get a numerical value for each term $|s - z|$ and compute their product P_z; similarly for each term $|s - p|$, product P_p, giving

$$k = \frac{P_p}{P_z} \qquad (9.68)$$

and then

$$K = \frac{k(-z_1)(-z_2) \cdots (-z_m)}{(-p_1)(-p_2) \cdots (-p_n)} \qquad (9.69)$$

Of course, when computer root finders (rather than graphical methods) are used to get the root locus, k and K for each point are predetermined.

Before explaining how to set loop gain with the root-locus method we want to display a number of root-locus graphs for typical systems. Figure 9.38 shows the tank-level control of Fig. 9.1 with successively more comprehensive dynamic models, starting with process dynamics τ_p only, and then including valve positioner τ_{vp}, electropneumatic transducer τ_{ep}, and finally first-order float dynamics τ_f. The dead-time approximations of eq. 9.21 and 9.22 are used to get the root loci of Fig. 9.39. Note that the simpler dead-time approximation is incapable of predicting oscillatory (but stable) responses known to exist in the exact model (Fig. 9.13); whereas the more complicated approximation produces such results for intermediate gain values. This same approximation applied to the system of Fig. 9.14 produces the root locus of Fig. 9.40a. Figures 9.40b, 9.40c, and 9.40d correspond, respectively, to Eq. 9.52, 9.60, and 9.62.

I shall now explain the root-locus gain-setting procedure. Just as in the frequency-response technique, the design criterion is a desired level of closed-loop system relative stability. However we now implement this by requiring that the dominant (closest to the origin) root pair of the closed-loop system characteristic equation exhibit a desired damping ratio ζ, rather than designing for a certain M_p in the closed-loop amplitude ratio. If we base our choice of desired ζ on a

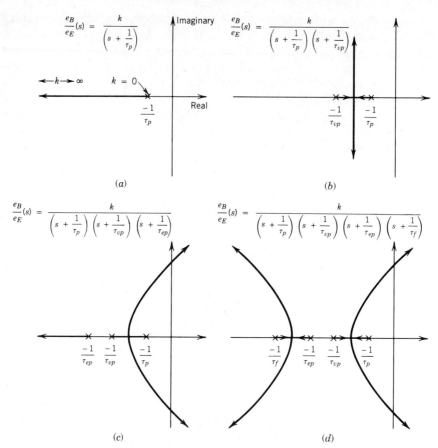

$$\frac{e_B}{e_E}(s) = \frac{k}{\left(s + \frac{1}{\tau_p}\right)}$$

$$\frac{e_B}{e_E}(s) = \frac{k}{\left(s + \frac{1}{\tau_p}\right)\left(s + \frac{1}{\tau_{vp}}\right)}$$

(a)

(b)

$$\frac{e_B}{e_E}(s) = \frac{k}{\left(s + \frac{1}{\tau_p}\right)\left(s + \frac{1}{\tau_{vp}}\right)\left(s + \frac{1}{\tau_{ep}}\right)}$$

$$\frac{e_B}{e_E}(s) = \frac{k}{\left(s + \frac{1}{\tau_p}\right)\left(s + \frac{1}{\tau_{vp}}\right)\left(s + \frac{1}{\tau_{ep}}\right)\left(s + \frac{1}{\tau_f}\right)}$$

(c)

(d)

FIGURE 9.38   *Root locus examples for one to four real poles, no zeros.*

simple second-order system and stay consistent with an M_p of 1.3, then the design value of ζ is 0.42. If the particular application cannot tolerate much overshoot, we would use a higher ζ, say 0.6 to 0.8, whereas $\zeta = 0.3$ might be considered if minimum peak time were desired and overshoot not critical. Using the system of Fig. 9.29 as an example, Fig. 9.41 shows the procedure. Application of the rules, plus a few minutes with the Spirule gets us the locus shown. Choosing $\zeta = 0.42$, we draw in the line of constant damping ratio, its intersection with the locus giving us the desired design point, whose ω_n we find to be 0.46 rad/s. Note that in general this design procedure guarantees the damping ratio of the *chosen* pair of roots but we must accept all the other roots wherever they occur on the locus. If the "designed" roots do not dominate the total response, performance may be quite different from that expected. In our present example, measurements based on Eq. 9.68 and 9.69 show that $K = 90$ is required to place the complex roots at the design point. The third root is somewhere to the left of -2 and could be found by a trial-and-error search to locate $K = 90$ on

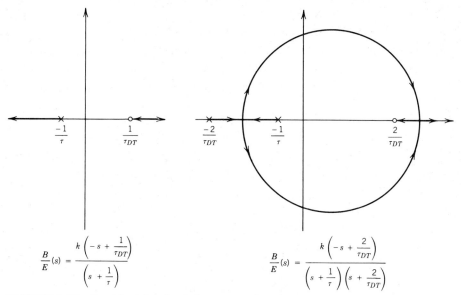

$$\frac{B}{E}(s) = \frac{k\left(-s + \dfrac{1}{\tau_{DT}}\right)}{\left(s + \dfrac{1}{\tau}\right)}$$

$$\frac{B}{E}(s) = \frac{k\left(-s + \dfrac{2}{\tau_{DT}}\right)}{\left(s + \dfrac{1}{\tau}\right)\left(s + \dfrac{2}{\tau_{DT}}\right)}$$

FIGURE 9.39   *Root loci for dead-time approximations.*

that branch. Alternatively, we could divide the quadratic factor ($\zeta = 0.42$, $\omega_n = 0.46$) associated with the "designed" roots into the cubic characteristic equation. This gives the third root as -2.2.

To get some idea of the degree of accuracy to be expected from these graphical techniques (scale: $1.0 = 10$ cm), consider that a computer program gave the exact results

$$K = 86 \qquad \zeta = 0.42 \qquad \omega_n = 0.45 \qquad \text{third root} = -2.1$$

We can also compare this root-locus design with our earlier frequency-response ($M_p = 1.3$) design. There K was set at 90, and Fig. 9.30b shows that the root-locus design ($K = 86$) would give a very similar step response. Note that for a non-unity-feedback system (such as in Fig. 9.35), the root-locus method would design the *same* gain value as for a unity-feedback-system that had the *same* open-loop dynamics (see Fig. 9.42). This can be considered a minor defect in the root-locus method of gain setting since the behavior (say the step response) of C in Fig. 9.42a will *not* be the same as in Fig. 9.42b for the same K value. That is, the root-locus method treats these two systems *identically* with regard to gain setting, whereas they are *not* the same system. One can see this in Fig. 9.36b, where B can be interpreted as C in Fig. 9.42a, whereas C can be interpreted as C in Fig. 9.42b. This is not, however, a serious defect since *all* aspects of system performance are checked and appropriate gain adjustments are made before the gain value is finalized in *any* practical design procedure. Conceptually, however, the frequency-response gain-setting methods are superior, in this respect, since they do recognize the difference between unity- and non-unity-feedback systems by designing different gains for them.

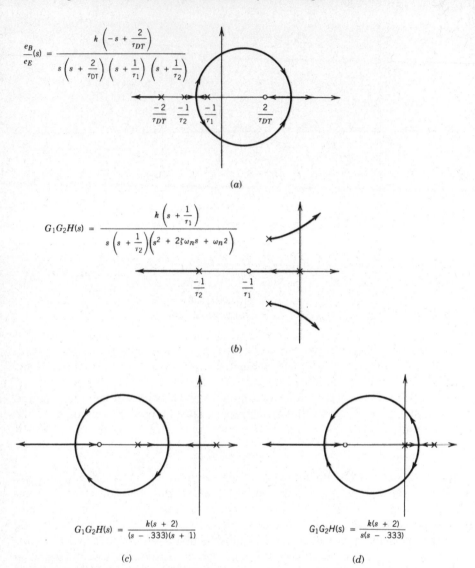

$$\frac{e_B}{e_E}(s) = \frac{k\left(-s + \frac{2}{\tau_{DT}}\right)}{s\left(s + \frac{2}{\tau_{DT}}\right)\left(s + \frac{1}{\tau_1}\right)\left(s + \frac{1}{\tau_2}\right)}$$

(a)

$$G_1G_2H(s) = \frac{k\left(s + \frac{1}{\tau_1}\right)}{s\left(s + \frac{1}{\tau_2}\right)\left(s^2 + 2\zeta\omega_n s + \omega_n^2\right)}$$

(b)

$$G_1G_2H(s) = \frac{k(s + 2)}{(s - .333)(s + 1)}$$

(c)

$$G_1G_2H(s) = \frac{k(s + 2)}{s(s - .333)}$$

(d)

FIGURE 9.40 *Some more root loci.*

9.8 DOMINANT ROOTS AND IMPERFECT POLE/ZERO CANCELLATION

This chapter serves both to explain the nature of the proportional mode of control and to introduce certain analysis and design tools of general utility. This section uses the Laplace transform to develop some particularly useful graphical interpretations of the root locus. For a system as in Fig. 9.32, with a unit step input of V

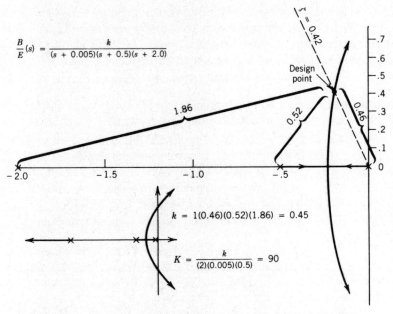

$$\frac{B}{E}(s) = \frac{k}{(s + 0.005)(s + 0.5)(s + 2.0)}$$

$$k = 1(0.46)(0.52)(1.86) = 0.45$$

$$K = \frac{k}{(2)(0.005)(0.5)} = 90$$

FIGURE 9.41   *Measurements to find gain at desired design point.*

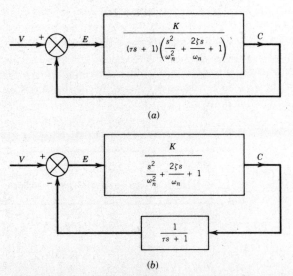

(a)

(b)

FIGURE 9.42   *Unity- and non-unity-feedback systems with same open-loop dynamics.*

$$[1 + KGH_1(s)]C(s) = [KG(s)]V(s) = \frac{1}{s} KG(s) \tag{9.70}$$

$$C(s) = \left[\frac{KG(s)}{1 + KGH_1(s)}\right]\frac{1}{s} \tag{9.71}$$

If any dead times present are approximated in our usual ways, the forward and feedback transfer functions will have the general forms

$$KG(s) = \frac{K(s - z_I)(s - z_{II}) \cdots (s - z_a)}{s^n(s - p_I)(s - p_{II}) \cdots (s - p_b)} \tag{9.72}$$

$$H_1(s) = \frac{K_H(s - z_{HI})(s - z_{HII}) \cdots (s - z_{HA})}{s^N(s - p_{HI})(s - p_{HII}) \cdots (s - p_{HB})} \tag{9.73}$$

giving

$$C(s) = \frac{1}{s} \left\{ \frac{\dfrac{K(s - z_I) \cdots (s - z_a)}{s^n(s - p_I) \cdots (s - p_b)}}{+ \dfrac{KK_H[(s - z_I) \cdots (s - z_a)][(s - z_{HI}) \cdots (s - z_{HA})]}{s^N(s - p_{HI}) \cdots (s - p_{HB})}} \right\} \tag{9.74}$$

$$C(s) = \frac{1}{s} \left\{ \frac{Ks^N[(s - z_I) \cdots (s - z_a)][(s - p_{HI}) \cdots (s - p_{HB})]}{s^{n+N}[(s - p_I) \cdots (s - p_b)][(s - p_{HI}) \cdots (s - p_{HB})]} + KK_H[(s - z_I) \cdots (s - z_a)][(s - z_{HI}) \cdots (s - z_{HA})] \right\} \tag{9.75}$$

The denominator of the expression inside the braces, set equal to zero, would be the characteristic equation of the closed-loop system. The poles of the function $C(s)$ would be the one at $s = 0$ (due to the step input), plus the roots p_1, $p_2, \ldots, p_v$ of the characteristic equation; whereas the zeros of $C(s)$ would be the zeros of G plus the poles of H_1.

$$
\begin{array}{c}
\overbrace{\text{Zeros of } G \text{ plus poles of } H_1} \\
C(s) = \dfrac{K(s - z_1)(s - z_2) \cdots (s - z_w)}{(s - 0)(s - p_1)(s - p_2) \cdots (s - p_v)} \\
\underbrace{\qquad\qquad\qquad\qquad\qquad\qquad}
\end{array} \tag{9.76}
$$

pole at origin plus roots of characteristic equation

If G has no zeros and H_1 no poles, the numerator is just K.

Using the standard partial-fraction expansion for inverse Laplace transforming, and assuming no repeated roots in the characteristic equation (feedback systems are rarely designed to have repeated roots)

$$C(s) = \frac{K_{01}}{s} + \frac{K_{11}}{s - p_1} + \cdots + \frac{K_{k1}}{s - p_k} + \cdots + \frac{K_{v1}}{s - p_v} \qquad (9.77)$$

To find a particular K_{k1}

$$(s - p_k)C(s) = \frac{K(s - z_1) \cdots (s - z_w)(s - p_k)}{s(s - p_1) \cdots (s - p_k) \cdots (s - p_v)} \qquad (9.78)$$

$$= \frac{(s - p_k)K_{01}}{s} + \cdots + K_{k1} + \cdots + \frac{(s - p_k)K_{v1}}{(s - p_v)}$$

Now let $s = p_k$.

$$K_{k1} = \frac{K(p_k - z_1) \cdots (p_k - z_w)}{(p_k - 0)(p_k - p_1) \cdots (p_k - p_v)} \qquad (9.79)$$

$$= \frac{K[\text{product of vector distances from zeros of } C(s) \text{ to } p_k]}{[\text{product of vector distance from other poles of } C(s) \text{ to } p_k]}$$

This last result has several useful graphical interpretations.

Dominant Roots

We saw in the simulation results of Fig. 9.5 that in a system with both fast and slow components, closed-loop response may be little affected by fast roots in the characteristic equation. This result can now be generalized analytically. Let us first state the concept of dominant roots and then consider an example to demonstrate its plausibility.

If some roots of the characteristic equation are relatively close to the origin and others are relatively far away, the contribution of the distant roots to the transient response is often negligible, and total system response is dominated by the roots near the origin.

To appreciate the truth of this statement, consider the root locus of Fig. 9.43. In reference to Eq. 9.76, $C(s)$ has no zeros and has poles at $s = 0$, p_1, p_2, and p_3. We wish to show that the contribution of p_1 to the total system response is negligible compared to that of p_2 and p_3. Using the values of K_{01}, K_{11}, and K_{21} measured in Fig. 9.43, we can write the step response of C as

$$c(t) \approx 0.97 + 1.2e^{-0.457t}\sin(0.759t + 234°) - 0.0082e^{-10.1t} \qquad (9.80)$$

Clearly, the rightmost term not only is very small relative to the other two but also disappears very quickly, thus the response is dominated by the root pair p_2, p_3. The *reason* behind these results is that the coefficient K_{11} of the distant root has in its denominator three large numbers (distances), whereas the coefficients of the close-in roots have only one large number. Clearly, this result is a general one for root loci with these characteristics and is not limited to this

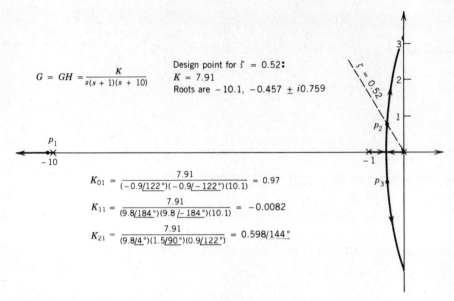

$$G = GH = \frac{K}{s(s+1)(s+10)}$$

Design point for $\zeta = 0.52$:
$K = 7.91$
Roots are $-10.1, -0.457 \pm i0.759$

$$K_{01} = \frac{7.91}{(-0.9\underline{/122°})(-0.9\underline{/-122°})(10.1)} = 0.97$$

$$K_{11} = \frac{7.91}{(9.8\underline{/184°})(9.8\underline{/-184°})(10.1)} = -0.0082$$

$$K_{21} = \frac{7.91}{(9.8\underline{/4°})(1.5\underline{/90°})(0.9\underline{/122°})} = 0.598\underline{/144°}$$

FIGURE 9.43   *Dominant roots demonstration.*

particular numerical example. Note, however, that if a "large" number of "closely grouped" distant roots exist, then the closein roots may *not* be dominant since each additional distant root contributes a *small* number to the distant-root denominator and a *large* number to the close-in root denominators.

Imperfect Pole/Zero Cancellation

A second practical use of Eq. 9.79 relates to those root loci where a system closed-loop root lies close to a zero of G or a pole of H_1. Perhaps the most common and important instance of this occurs when one employs a widely used design technique of feedback systems called *cancellation compensation* (discussed in detail in later chapters). Here one designs a controller with some zeros that (ideally) lie exactly "on top of" some basic system poles with undesirable dynamics, thus "cancelling" these bad effects from the closed-loop system response. In practice, the poles and zeros can never *exactly* cancel since they are determined by two independent pieces of hardware whose numerical values are neither precisely known nor perfectly fixed. Since imperfect cancellation is bound to occur, one wonders how bad the imperfection can get before we lose the essential benefits of cancellation. Our graphical interpretation of Eq. 9.79 is quite helpful in understanding this situation.

Let us modify the basic system of Fig. 9.43 by adding a pair of open-loop complex poles at $-0.9 \pm i\,0.917$ and then a "cancelling" pair of complex zeroes $-1.0 \pm i\,0.866$, which (as in a real-world situation) do not perfectly cancel the poles. In Fig. 9.44, note that the new pole/zero pairs have only a

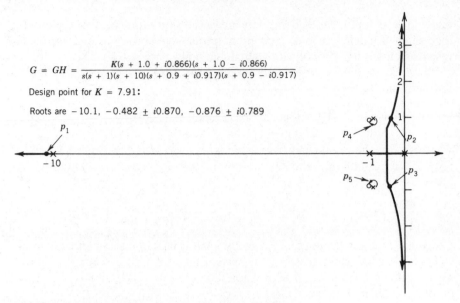

$$G = GH = \frac{K(s + 1.0 + i0.866)(s + 1.0 - i0.866)}{s(s + 1)(s + 10)(s + 0.9 + i0.917)(s + 0.9 - i0.917)}$$

Design point for $K = 7.91$:

Roots are -10.1, $-0.482 \pm i0.870$, $-0.876 \pm i0.789$

FIGURE 9.44 *Imperfect-cancellation demonstration.*

slight effect on the shape of the root locus paths that contain p_2 and p_3 and the location of the roots thereon. This is because the two *new* paths are very short segments going directly from the new poles to the adjacent new zeros. (If the "cancellation mismatch" were greater, more radical changes in the entire root locus are possible.) For *any* degree of mismatch there are, of course, two *additional* roots p_4 and p_5 contributing to the total system transient response; however if the mismatch is slight, the coefficient K_{41} for these new roots will be so small that it will make their effect negligible and the system will behave essentially as that of Fig. 9.43. This is because the numerator of Eq. 9.79 will have one very small term because of the closeness of p_4 to the zero at $-1.0 + i\,0.866$. As the mismatch is reduced, this term approaches zero and, for perfect cancellation, the two new roots do not even exist. Because of inaccuracy in measuring the small "vectors" involved here, our graphical approach serves mainly as a vehicle for understanding the situation; computer-aided numerical calculations are needed to obtain engineering accuracy.

Approximate Characteristics of Closed-Loop Step Response When Dominant Roots Exist

If, because of near/far root distribution, imperfect (but adequate) cancellation compensation, and/or other causes, a system's closed-loop response to a step command is dominated by a single pair of complex roots $\sigma \pm i\omega_d$, then this response can be quickly estimated from a few measurements on the root locus. This dominance of apparently complicated systems by a single root pair is actually fairly common, as evidenced by measured step responses (of real systems) that show a single obvious oscillation frequency, and measured frequency-response curves $|(C/V)(i\omega)|$ with a single obvious peak.

When a dominant root pair exists, Eq. 9.77 gives $c(t) \approx K_{01} + 2|K_{11}|e^{\sigma t} \sin(\omega_d t + \phi)$

$$c(t) \approx K_{01} + 2|K_{11}|e^{\sigma t}\cos(\omega_d t + \underline{/K_{11}}) \qquad (9.81)$$

where K_{11} is the coefficient associated with $\sigma + i\omega_d \triangleq p_0$ and $\phi \triangleq 90° + \underline{/K_{11}}$. The constants K_{01} and K_{11} are obtained from measurements on the root locus.

$$K_{01} = \frac{K(-z_1) \cdots (-z_w)}{(-p_1) \cdots (-p_v)} \qquad K_{11} = \frac{K(p_0 - z_1) \cdots (p_0 - z_w)}{p_0(p_0 - p_1) \cdots (p_0 - p_v)} = |K_{11}| \underline{/K_{11}} \qquad (9.82)$$

The value of K_{01} is (from theory) exactly 1.0 if $(C/E)(s)$ has any integrators and will otherwise be close to 1.0 if the system has reasonable accuracy. Thus, if we measure K_{01} from Eq. 9.82, it should be close to 1.0, giving a good check on our graphical accuracy.

A useful *speed of response* criterion T_p (time to first peak) can be found by setting dc/dt equal to zero.

$$\frac{dc}{dt} = 0 = 2|K_{11}|[-\omega_d e^{\sigma T_p}\sin(\omega_d T_p + \underline{/K_{11}}) + \sigma e^{\sigma T_p}\cos(\omega_d T_p + \underline{/K_{11}})] \qquad (9.83)$$

$$\tan^{-1}\left(\frac{\sigma}{\omega_d}\right) = \omega_d T_p + \underline{/K_{11}} = -\left(\underline{/p_0} - \frac{\pi}{2}\right)$$

$$T_p = \frac{1}{\omega_d}\left[\frac{\pi}{2} - \underline{/K_{11}} - \underline{/p_0}\right] \qquad (9.84)$$

Relative stability can be judged by the overshoot.

$$O \triangleq c(T_p) - K_{01} = 2|K_{11}|e^{\sigma T_p}\cos(\omega_d T_p + \underline{/K_{11}}) \qquad (9.85)$$

$$O = 2|K_{11}|e^{\sigma T_p}\cos\left(\frac{\pi}{2} - \underline{/K_{11}} - \underline{/p_0} + \underline{/K_{11}}\right)$$

$$O = 2|K_{11}|e^{\sigma T_p}\cos\left(\frac{\pi}{2} - \underline{/p_0}\right) = 2K_{11}\,e^{\sigma T_p}\frac{\omega_d}{\omega_n} \qquad (9.86)$$

where $\omega_n \triangleq |p_0|$.

Settling time T_s is also often useful. For a 5% settling time, if we approximate the error $v - c$ by its exponential envelope $e^{\sigma t}$ we get

$$e^{\sigma T_{s,5\%}} \approx 0.05 \qquad T_{s,5\%} \approx \frac{3}{\sigma} = \frac{3}{\zeta\omega_n} \qquad (9.87)$$

Finally, the number $N_{5\%}$ of oscillation cycles to "settle" within 5% of the final value is given by

$$N_{5\%} = \frac{T_{s,5\%}}{2\pi/\omega_d} = \frac{2\sqrt{1 - \zeta^2}}{\pi\zeta} \tag{9.88}$$

9.9   SOME USEFUL NONLINEAR "VARIATIONS" OF PROPORTIONAL CONTROL

We shall see in later chapters that when proportional control alone cannot meet all performance specifications, it can be augmented by linear control modes that utilize signals related to time derivatives and/or integrals of the actuating signal or the controlled variable. In this section I wish to show some nonlinear control modes that, since they do not employ time derivatives or integrals, are more closely related to proportional control and are thus logically discussed here rather than later.

Our first nonlinear control mode of this section has been described[9] as *adaptive gain control,* although the use of the word *adaptive* is the subject of controversy and might be challenged by some. Maselli discussed eight different classes of applications in the process control field.

1. To reduce unnecessary manipulation of variables as a result of natural process or measurement noise.
2. To produce overdamped averaging control without letting the process exceed specified limits.
3. To reduce the need for precise controller tuning to achieve adequate feedback control.
4. To provide an effective means for greatly reducing plant control loop interactions.
5. To provide external electrical gain adjustments to enable the user to have start-up, event, or time-based gain changes. To provide a capability for specific specialized algorithms.
6. To achieve tighter, more responsive, regulatory control at a specific set point.
7. To obtain comparable feedback-control-loop performance over a wide range of loads or set points.
8. To obtain reasonable feedback-control-loop performance on an extremely nonlinear process.

---

[9]S. Maselli, Adaptive Gain Control, Taylor Instrument Co., Rochester, N.Y., #98210 Issue 2; Taylor Micro-Scan 1300 Adaptive Gain Microprocessor Controller, Taylor Instrument Co., Rochester, N.Y., #98211, 1978.

I now illustrate the first of these types of applications, using a CSMP digital simulation.

Although the adaptive gain concept may take various forms, most commonly it involves a controller with a nonlinear static characteristic made up of several straight-line segments of different slopes. In the example of Fig. 9.45 a low-gain region in the neighborhood of zero error reduces unnecessary control actions for small, unimportant error signals caused mainly by process measurement noise. For a linear controller this low gain would also be in force for large disturbances, leading to excessive error. Our nonlinear control increases the gain by a factor of five when the error exceeds 1.0 in absolute value, fighting against large disturbances much more effectively. To show the improvement possible with the dual-gain controller, our CSMP program compares the system of Fig. 9.45 with a strictly linear system with gain equal to the small-signal gain of the nonlinear controller.

METHOD   RKSFX   use fixed-step size Runge–Kutta integrators (needed for random noise generator)

VL = 0   desired value for both linear and nonlinear system taken as zero

EL = VL − BL   summing junction of linear system

EPLUS = 1.0   ⎫ used on graphs to show small-signal

EMINUS = −1.0   ⎬ limits of error

BL = CL + UM   adds measurement noise UM to controlled variable CL of linear system

UMP = IMPULS (0.0, .05)   ⎫ statements that

UMG = GAUSS(1,0.0,0.8E6)   ⎪ generate the

UMH = ZHOLD(UMP,UMG)   ⎬ random measure-

UMF = CMPXPL(0.0,0.0,.67,35.,UMH)   ⎪ ment noise

UM = CMPXPL(0.0,0.0,.67,35.,UMF)   ⎭ signal UM

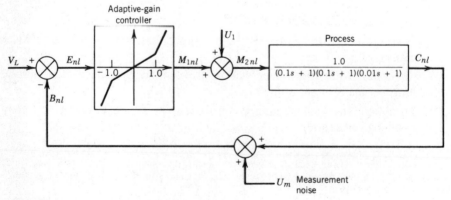

FIGURE 9.45 *"Adaptive-gain" control.*

PARAM $\left.\begin{array}{l} \texttt{M1L=KC*EL} \\ \texttt{KC=1.0} \end{array}\right\}$ linear system controller with gain KC of 1.0

PARAM $\left.\begin{array}{l} \texttt{M2L=M1L+U1} \\ \texttt{U1=KU*STEP(US)} \\ \texttt{KU=10.,US=1.0} \end{array}\right\}$ inject step disturbance $U1$ of size KU, at time US, into linear control system

$\left.\begin{array}{l} \texttt{M3L=REALPL(0.0,.1,M2L)} \\ \texttt{M4L=REALPL(0.0,.1,M3L)} \\ \texttt{CL=REALPL(0.0,.01,M4L)} \end{array}\right\}$ third-order process with controlled variable CL for linear control system; end of linear system statements

$\texttt{ENL=VL-BNL}$ start nonlinear system statements with summing junction

$\texttt{BNL=CNL+UM}$ add measurement noise to controlled variable CNL of nonlinear system

$\left.\begin{array}{l} \texttt{M1NL=AFGEN(CURVE1,ENL)} \\ \texttt{AFGEN CURVE1=-11.,-51.,} \\ \quad\texttt{-1.,-1.,1.,1.,11.,51.} \end{array}\right\}$ define nonlinear controller using arbitrary function generator

$\texttt{M2NL=M1NL+U1}$ inject step disturbance $U1$

$\left.\begin{array}{l} \texttt{M3NL=REALPL(0.0,.1,M2NL)} \\ \texttt{M4NL=REALPL(0.0,.1,M3NL)} \\ \texttt{CNL=REALPL(0.0,.01,M4NL)} \end{array}\right\}$ third-order process with controlled variable CNL for nonlinear control system

TIMER $\quad$ $\texttt{FINTIM=2.0,DELT=.001,OUTDEL=.01}$

OUTPUT $\quad$ $\left.\begin{array}{l} \texttt{EL,ENL,EPLUS,EMINUS} \\ \texttt{CL,CNL,UM,U1} \end{array}\right\}$ request graphic output

OUTPUT

We have not previously used the CSMP random signal generator so these statements require explanation.[10] The basic capability allows generation of a Gaussian ("normal") random signal with flat power spectral density (frequency content) from zero frequency to a selectable cutoff frequency ω_{co} (see Fig. 9.46a. Once ω_{co} is chosen to suit the physical problem, the second numerical parameter in IMPULS is $0.5505/\omega_{co}$; 0.05 in our case since we chose $\omega_{co} = 11.01$ rad/s. The first numerical parameter in IMPULS is always 0.0. In GAUSS, the first parameter is always integer 1, the second is $102.3\ \omega_{co}^4$ times the desired mean value of our random signal, in our example 0.0. The third parameter is the standard deviation SD of GAUSS and is given by

$$SD = 244\ \omega_{co}^4\ \sqrt{\phi_{WN}\omega_{co}} \qquad (9.89)$$

where $\phi_{WN} \triangleq$ power spectral density of UM in its flat range.

We select $\phi_{WN} = 0.0909$, giving SD $= 0.8 \times 10^6$. The RMS value of UM is given by $\sqrt{3.89\phi_{WN}\omega_{co}}$, in our case 1.97.

The arbitrary function generator AFGEN is also a new CSMP module for us so requires some explanation. We usually use it to approximate a smooth nonlinear function with a set of straight-line segments by listing the x–y coordinates

---

[10]E. O. Doebelin, *System Modeling and Response*, Wiley, New York, 1980, pp. 210–216.

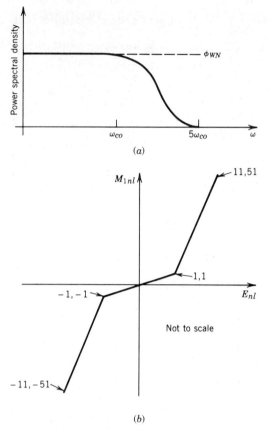

(a)

(b)

FIGURE 9.46 *Random signals and arbitrary functions in CSMP.*

of the breakpoints in the AFGEN CURVE1 statement, working from left to right on the x axis. In our present example the straight-line segments give us *exactly* what we want in our nonlinear controller (see Fig. 9.46b). Note that the slope (gain) changes by 5 to 1 in our example but any other ratio could easily be tried.

Figures 9.47a and 9.47b show the results of the simulation. In 9.47b, note that since the error stays between ± 1 limits before U1 comes on at $t = 1.0$, both linear and nonlinear system behave identically. When U1 drives the error beyond ± 1 after $t = 1.0$, the nonlinear system controls this disturbance more effectively, keeping the controlled variable closer to the zero desired value. Note, however, that the (random) effect of UM on CNL seems to be larger for $t > 1.0$ than for $t < 1.0$. This is because ENL has stayed out of the ± 1 range where low gain is used. It may be necessary to augment our nonlinear controller with a small amount of integral control (see next chapter) to bring the mean value of ENL back into the low-gain range when the constant disturbance U1 is present.

Our next nonlinear controller uses a concept employed in the disk memory head-positioning servomechanism of Fig. 1.6. There a translational dc motor ("voice coil") positions the read/write heads over the rotating memory disk at a location commanded by the computer. A velocity transducer measures head velocity so that an inner-loop velocity servo may be implemented. In the main (position) loop a "velocity curve generator" computes the difference between commanded and actual position, takes its square root, and uses this as a velocity command to the inner velocity loop. As will be shown, this nonlinear control mode attempts to enforce a *constant* acceleration (or deceleration) during the entire motion. If we set the desired acceleration close to the maximum possible from our motor, we are utilizing the motor's capabilities in an optimum fashion for both large and small changes in head position. A linear controller optimized for, say, the largest required head displacement will be needlessly slow for smaller commands. In the nonlinear controller

$$\text{command velocity} \triangleq V_c = \pm K_g \sqrt{|x_V - x_C|} \qquad (9.90)$$

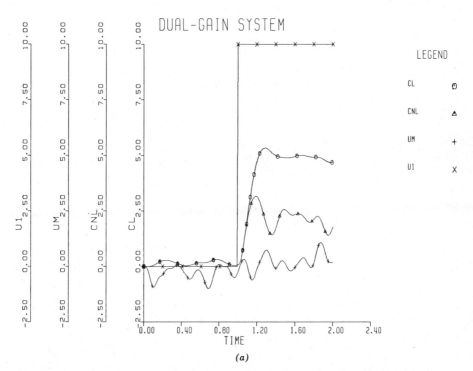

FIGURE 9.47 *Performance improvements for nonlinear dual-gain controller.*

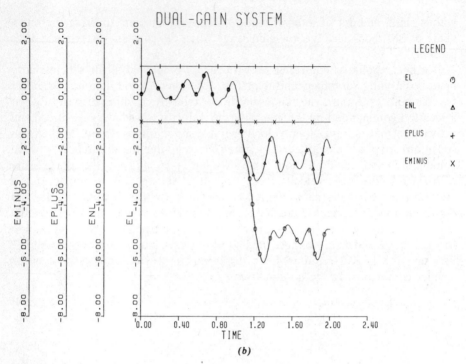

FIGURE 9.47 (Cont.)

$$K_g \triangleq \text{proportionality factor}$$

$$x_V \triangleq \text{desired position} \qquad x_C \triangleq \text{actual position}$$

and

$$\dot{V}_c = \pm K_g \frac{1}{2\sqrt{|x_V - x_C|}} (-\dot{x}_C) \qquad (9.91)$$

If the velocity servo were *perfect*, $\dot{x}_C$ would equal V_c and

$$\dot{V}_c = \pm K_g \frac{1}{2\sqrt{|x_V - x_C|}} (\pm K_g \sqrt{|x_V - x_C|}) = \pm \frac{K_g^2}{2} \qquad (9.92)$$

showing that acceleration $\dot{V}_c$ is constant. Since the velocity servo is *not* perfect, we will not achieve perfectly constant acceleration; however the nonlinear controller will provide significantly improved performance.

In Fig. 9.48 the output of the current amplifier is limited at $\pm 40\ A$, the motor force constant is 2.5 lb$_f$/A, and the moving mass is 0.005 lb$_f$-s^2/in., giving a maximum available acceleration of 20,000 in./s^2 and making $K_g = 200$ in.$^{1/2}$/s. Since the velocity servo is not perfect, this value of K_g cannot actually be used

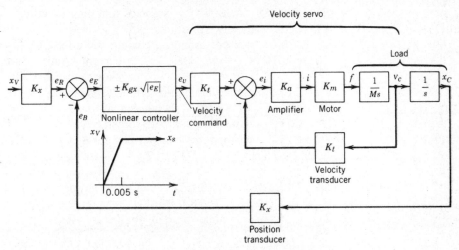

FIGURE 9.48 *Nonlinear (constant acceleration) control for disk memory.*

and one must use simulation trial and error to find the largest useable K_g value that does not cause current limiting. A similar procedure is necessary to find the maximum loop gain of the completely linear system we use for comparison. Simulation also showed that terminated ramp commands were preferable to perfect steps since they allowed use of higher gains (without current limiting) and gave faster response. Friction effects in the mechanical load are neglected, and the transconductance (current output) amplifier suppresses inductance and back-emf effects in the motor. The CSMP simulation that follows includes both the system of Fig. 9.48 and a comparison linear system that is identical except for the controller being strictly linear. We examine the response to large (2.0 in.) and small (0.5 in.) head displacement commands. The competing systems were each designed to just barely current limit (maximum acceleration 20,000 in./s^2) for the 2.0 in. command.

```
TITLE           NONLINEAR DISK-HEAD SERVO
                XV=200.*XS*(RAMP(0.0)-RAMP(.005))
                                      terminated ramp command
PARAM           XS=0.5    Small (0.5 in.) command
                EE=KX*(XV-XC)    Position summing junction
PARAM           KX=10.    position transducer sensitivity 10 V/in.
                SGN=FCNSW(EE,-1.0,0.0,1.)    change sign of
                                             command  ve-
                                             locity when EE
                                             changes sign
                EV=SGN*KGX*SQRT(ABS(EE))    nonlinear control-
                                            ler
PARAM           KGX=26.    gain (best value found by trial and error)
```

```
                    EI = KT*(EV - XCDOT)      velocity summing junction
PARAM               KT = 5.     velocity transducer gain 5 V/(in./s)
                    I = KA*EI      motor armature current (field fixed)
                    ILIM = LIMIT(-40.,40.,I)     current limit (not
                                                 shown on Fig. 9.48)
PARAM               KA = 0.2      transconductance amplifier gain 0.2A/V
                    XCDOT2 = ILIM*KM/M     Newton's law
PARAM               M = .005     load mass 0.005 lb_f-sec^2/in.
PARAM               KM = 2.5      motor force constant 2.5 lb_f/A
                    XCDOT = INTGRL(0.0,XCDOT2)     motor/load veloc-
                                                      ity
                    XC = INTGRL(0.0,XCDOT)      motor/load position
*                   START LINEAR SYSTEM MODEL      comment card
                    XVL = XV      command same as nonlinear system
                    EEL = KX*(XVL - XCL)
                    EVL = KL*EEL
PARAM               KL = 6.0      gain (best value found by trial and error)
                    EIL = KT*(EVL - XCDTL)
                    IL = KA*EIL
                    ILLIM = LIMIT(-40.,40.,IL)
                    XCDT2L = ILLIM*KM/M
                    XCDTL = INTGRL(0.0,XCDT2L)
                    XCL = INTGRL(0.0,XCDTL)
TIMER               FINTIM = .05,DELT = .00002,OUTDEL = .001
OUTPUT              XCDOT2,XCDT2L
PAGE GROUP          plot XCDOT2, XCDT2L to same scale
OUTPUT XC,XCL
PAGE GROUP          Plot XC, XCL to same scale
END
PARAM               XS = 2.0      rerun entire problem with 2.0-in. command
END
```

Figures 9.49*a* and 9.49*b* show that for the 2.0 in. command (the largest possible in this particular disk drive), the nonlinear system is somewhat (but not dramatically) better than the competing linear servo. This is because both systems were designed to just reach the limiting 20,000 in./s^2 acceleration for this size command. For any smaller commands (such as the 0.5 in. of Fig. 9.49*c* and 9.46*d*) the nonlinear system is much superior since it uses more of the available acceleration capability in trying to implement its constant-acceleration design philosophy.

Although nonlinear control laws such as the two presented in this section can be and have been implemented with analog hardware, our two specific examples use digital techniques. The adaptive gain controller is a general-purpose industrial process controller with conventional proporational, integral, and derivative control modes in addition to the adaptive gain feature, all implemented in

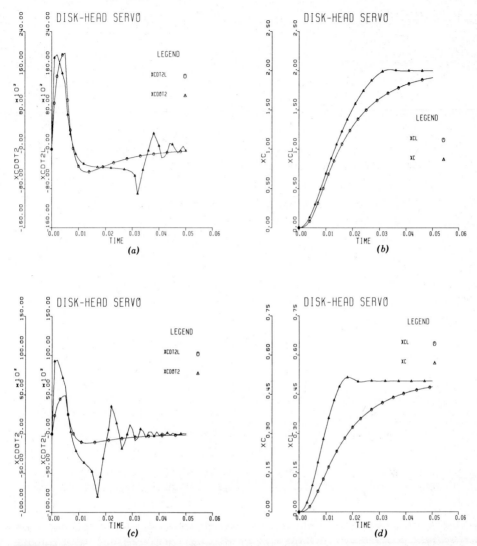

FIGURE 9.49 *Nonlinear controller improves performance over range of command sizes.*

microcomputer software. Since A/D and D/A converters at controller input and output accept and deliver analog voltage signals, the digital aspect of the instrument is really transparent to the user, so application is direct and simple. The disk memory head positioning servo is a special-purpose design for that particular application. Position feedback is from an optical encoder which provides digital position information to be compared with a digital position command. The square root is taken digitally and then D/A converted to get an analog voltage command for the velocity servo which is completely analog.

PROBLEMS

9.1 For the electropneumatic transducer of Fig. 9.2:

a. Write the closed-loop differential equation relating p_o to e_{in}. What combinations of real and complex roots are possible in the characteristic equation?

b. Relate p_o to e_{in} for static conditions. Show how use of high loop gain can give a very linear p_o/e_{in} relation even though K_p varies greatly (is nonlinear) over the 3 to 15 psig range of p_o.

c. Use Routh criterion to develop some useful design inequalities.

d. If the effect of $K_{\dot{x}}$ is negligible, redraw the block diagram and repeat the Routh analysis. Use results of Part b to explain the trade off between stability and linearity. Sketch Nyquist and root locus plots.

e. Take $K_{\dot{x}}$ as negligible but also let $K_s = 0$, making $(x_f/b)(s) = (1/B)/[s(\tau_m s + 1)]$, where $\tau_m \overset{\Delta}{=} M/B$, $M \overset{\Delta}{=}$ mass of moving parts, $B \overset{\Delta}{=}$ damping of moving parts. Repeat Parts a through e of the problem, noting that the linearity aspects of Parts b and c will require a new interpretation.

9.2 Develop a linearized dynamic feedback system model for the pneumatic valve positioner of Fig. 9.3, including enough dynamic effects to allow prediction of instability.

9.3 Analyze the system of Eq. 9.19 for steady-state errors due to ramp commands and disturbances. What can be done to minimize such errors without unduly sacrificing other aspects of system performance?

9.4 In the system of Fig. 9.9, suppose loop gain $= 100$, $10 < U_{1s} < 20$, $5 < U_{2s} < 10$, and K_1, K_2, K_3 are each adjustable in the range 1 to 10. How would you choose $K_1, K_2,$ and K_3 to minimize steady-state errors due to disturbances?

9.5 Analyze the stability of a feedback system whose open-loop transfer function is $Ke^{-\tau_{dt} s}$, using both exact (Nyquist) methods and approximate (Routh).

9.6 Solve Eq. 9.21 analytically for a step input of V and compare with both a CSMP solution of Eq. 9.21 and a CSMP solution using the exact dead-time model. Use $\tau = 5.$, $\tau_{dt} = 1$, and $K = 1$.

9.7 Repeat Problem 9.6 using the dead-time approximation of Eq. 2.177.

9.8 Repeat Problem 9.6 using the dead-time approximation of Eq. 2.178.

9.9 In the system of Fig. 9.14, let flow rate G vary with time and use linearization to develop a new process model. Draw the new feedback system block diagram.

9.10 Add a disturbance compensation control mode to the system of Problem 9.9.

9.11 Use digital simulation to study the gain mismatch problem in the system of Fig. 9.20, 9.21, and 9.23.

9.12 Repeat Problem 9.11 for the case where the gains are correct but the time constants are mismatched.

9.13 For the system of Eq. 9.42, use Nyquist criterion to show why allowable gain is low when all three τ's are similar and high when they are dissimilar.

9.14 Write a CSMP (or other digital simulation) program to produce Fig. 9.30*b*.

9.15 In the system of Fig. 9.38*d*, let the float dynamics be underdamped second order (rather than first order). Sketch the root locus.

9.16 Repeat Problem 9.15 if the float is overdamped second order.

9.17 For $K = 7.91$, using the root locations given in Fig. 9.44, compute K_{01}, K_{11}, K_{21}, K_{41} and the system step response. Run a digital simulation to compare the step response of the actual system with that of the approximate model that completely ignores roots p_4 and and p_5. Use the methods of Eq. 9.81 to 9.88 to estimate T_p, O, T_s, and $N_{5\%}$ and compare with simulation results for the exact (five-root) model.

9.18 For the following systems:

A. Set loop gain for $M_p = 1.3$ (frequency-response method).

B. Set loop gain for $\zeta = 0.42$ for dominant roots (root locus method).

C. Compute and graph $(C/V)(i\omega)$ for designs A and B.

D. Compute and graph $c(t)$ for unit step command of designs A and B.

  a. Fig. P6.10, $\tau = 1.0$s.

  b. Fig. P6.11, $\tau_{dt} = 1.0$s, $\tau = 3.0$s.

  c. Fig. P.6.12, $\tau_1 = 0.1$s, $\tau_2 = 0.05$s, $\tau_3 = 0.01$s.

  d. Fig. P6.13, $\tau_1 = 0.1$s, $\omega_n = 100$ rad/s, $\zeta = 0.8$.

  e. Fig. P6.14, $\tau = 0.8$s, $\tau_{dt} = 0.1$s.

  f. Fig. P6.15, $\tau = 0.001$s.

  g. Fig. P6.16, $\tau_1 = 0.01$s, $\tau_2 = 0.10$s.

  h. Fig. P6.18, $\tau_1 = 0.1$s, $\tau_2 = 1.0$s, $\tau_3 = 2.0$s.

9.19 Select suitable components, draw hardware schematic diagrams, draw block diagrams, and clearly explain system operation for proportional feedback control applied to the following situations.

  a. An automobile speed control ("cruise control") to maintain a set speed without driver throttle corrections. Use engine vacuum for throttle actuator power.

b. Commercially available cruise controls (as in Part a) exhibit a speed error when climbing hills. Show how a pendulum sensor can be used with disturbance feedforward to augment such a cruise control and remove this error.

c. A materials testing machine that applies dynamic forces to a specimen in response to time-varying voltage commands.

d. A pressure-vessel testing machine that applies dynamic liquid pressures to tanks in response to time-varying voltages.

e. A centrifugal pump is direct driven by a small steam turbine. We wish to control pump volume flow rate.

f. A centrifugal pump is driven from an ac induction motor through a variable-speed vee-belt drive. We wish to control pump volume flow rate. Would an ac synchronous motor be a better choice than the induction motor?

g. Rather than varying pump-drive speed as in e and f, the pump speed can be left fixed and pump flow rate varied using a discharge throttle valve. Devise such a system.

h. We want to run rotating-bending fatigue tests at elevated specimen temperatures. Devise a suitable temperature control system.

i. A small electric-powered vehicle is to steer itself to follow a white line painted on the factory floor.

j. The large rudder of a supertanker is to be remotely positioned by a helmsman's wheel.

k. The frequency of an ac generator (alternator) driven by a diesel engine is to be kept near 60 Hz.

l. A trencher for laying drainage tile in farmer's fields needs an automatic depth control.

m. Depth control for a torpedo.

n. Altitude control for an airplane.

o. Mach number control for an airplane.

p. Blade height control for a road scraper. Use a stretched wire as a reference point.

q. Lateral (steering) control of the road scraper of Part p. Can the stretched wire be used for both blade height and steering?

r. Position a pen on a plotter to record a voltage input.

9.20 Use digital simulation to compare the performance of the system of Fig. 9.45 to one that uses a nonlinear controller given by $M_{1nl} = 51\, E_{nl}\, E_{nl}^2/121\, |E_{nl}|$. Try to find a better numerical value for the controller coefficient (51/121).

9.21 Repeat Problem 9.20 for $M_{1nl} = 51\, E_{nl}^3/1331$.

9.22 Figure P9.1 shows a *unity-gain buffer amplifier,* (called "pilot relay" or "booster" in the industry), a device widely used in pneumatic control. Its function is to accept a 3 to 15 psig input pressure p_i and accurately reproduce this pressure at p_o. It takes negligible flow from the device providing p_i but can supply significant flow to the load at p_o; thus although its pressure gain is one, its power gain is more than one. It thus serves to buffer one pneumatic device from another, analogous to electronic buffer amplifiers (high input impedance, low output impedance, voltage gain ≈ 1.0) used for similar functions in electronic systems.

a. Perform a linearized dynamic perturbation analysis around a steady-state operating point of 9 psig, writing the differential equation relating $p_{o,p}$ to $p_{i,p}$. Use the following modeling:

$$G_{s,p} = K_{sf}p_{o,p} \qquad\qquad G_{n,p} = K_{np}\,p_{o,p} - K_{nx}x_p$$

$$x_p = (p_{i,p} - p_{o,p})K_d \qquad dp_{o,p}/dt = (RT/V)(dM_p/dt)$$

b. Part a should produce $(p_{o,p}/p_{i,p})(s) = K/(\tau s + 1)$. How can one choose parameters to make $K \approx 1.0$?

c. Draw a block diagram of this device and show that it is a feedback system. What must loop gain be to get $K \approx 1.0$? Does our model put any stability limit on loop gain?

d. High loop gain makes this device faster and faster, violating the pure spring model used for the moving parts. Correct this by using $x_p/(p_{i,p} - p_{o,p}) = K_d/[(s^2/\omega_n^2) + 2\zeta s/\omega_n + 1]$ and reanalyze the system. Find a stability limit on loop gain.

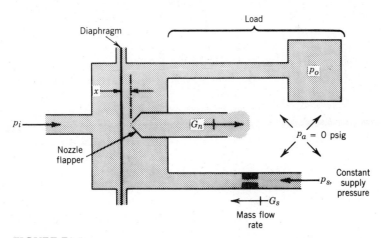

FIGURE P9.1

CHAPTER 10
Integral Control

10.1 BASIC CHARACTERISTICS OF INTEGRAL CONTROL

When a proportional controller can use large loop gain (say, 10 or more) and preserve good relative stability, system specifications, including those on steady-state error, may often be met. However, if difficult process dynamics such as significant dead times prevent use of large gains, steady-state error performance may be unacceptable. When human process operators notice the existence of steady-state errors due to changes in desired value and/or disturbance they can correct for these by changing the desired value ("set point") or the controller output bias until the error disappears. This is called *manual reset*. Integral control is a means of removing steady-state errors without the need for manual reset and, in fact, it is sometimes called *automatic reset*. We shall see that integral control can be used by itself or in combination with other control modes; indeed, *proportional plus integral control is the most common mode.*

The reason proportional control suffers from steady state errors is that a change in desired value or disturbance will require a new value of manipulated variable (flow of energy and/or material to the process) in order to achieve equilibrium at the new operating conditions. Since, in a proportional controller, manipulated variable is proportional to system error, a new value of manipulated variable is possible only if nonzero error exists. We need a control that can provide *any* needed steady output (within its design range, of course) when its input (the system error) is zero. An integrator is capable of exactly this type of behavior (see Fig. 10.1). Although integral control is very useful for removing or reducing steady-state errors, it has the undesirable side-effects of reducing response speed and degrading stability. The reduction in speed is most readily seen in the time domain, where a step input (a sudden change) to an integrator causes a ramp output, a much more gradual change. Stability degradation is most apparent in the frequency domain (Nyquist criterion) where the integrator reduces phase margin by giving an additional 90° of phase lag at every frequency, rotating the $(B/E)(i\omega)$ curve toward the unstable region near the -1 point.

Occasionally an integrating effect will naturally appear in a system element (actuator, process, etc.) *other* than the controller. These "gratuitous" integrators can also be effective in reducing steady-state errors. Although controllers with a single integrator are most common, double (and occasionally triple) integrators are useful for the more difficult steady-state error problems, although they require careful stability augmentation. Conventionally, the number of integrators between E and C in the forward path has been called the system *type number;* type 0 systems have no integrators, type 1 have one, and so forth. In

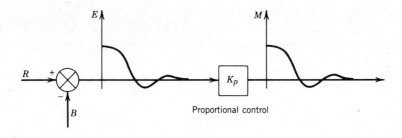

Proportional control

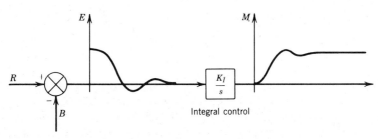

Integral control

FIGURE 10.1   *Comparison of proportional and integral control.*

addition to the *number* of integrators, their *location* (relative to disturbance injection points) determines their effectiveness in removing steady-state errors.

From a steady-state-error viewpoint, the "difficulty" of a command or disturbance is determined by the kind of manipulated-variable (M) signal required to return the error to zero in steady state. If a constant M is needed, a single integral is sufficient for zero error. If the needed M is a ramp ($\dot{M}$ constant), then two integrals are needed. A parabolic M ($\ddot{M}$ constant) requires three integrals (see Fig. 10.2). The significance of integrator *location* is brought out by Fig. 10.3. In Fig. 10.3a, with $V = V_s$ and $U = 0$ we see that C will increase until $C = V_s$, whereupon $E = 0$ and C comes to rest. With $V = 0$ and $U = U_s$, C increases until $M = -K_1C = -U_s$ making $L = 0$ and C stationary. The integrator thus gives zero steady-state error for a step command but not for a step disturbance. Mathematically

$$[(V - C)K_1 + U]\frac{K_2}{D} = C \qquad (\tau D + 1)C = V + \frac{U}{K_1} \qquad \tau \triangleq \frac{1}{K_1K_2} \qquad (10.1)$$

By relocating the integrator as in Fig. 10.3b, either or both V_s and U_s can be "cancelled" by M without requiring E to be nonzero.

$$\left[(V - C)\frac{K_1}{D} + U\right]K_2 = C$$

$$(\tau D + 1)C = V + \frac{1}{K_1}\frac{dU}{dt} = V \qquad \text{for } U = U_s \qquad (10.2)$$

We thus conclude that integrators must be located "upstream" from disturbance injection points if they are to be effective in removing steady-state errors due

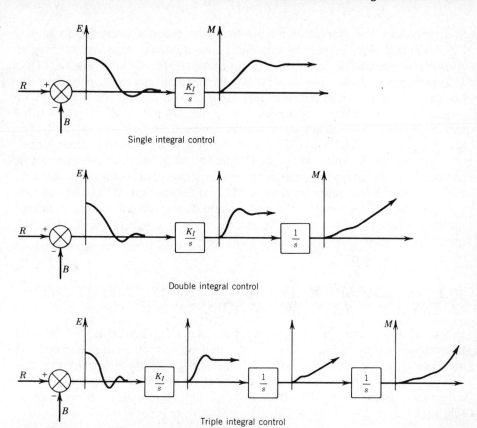

FIGURE 10.2   *Single, double, and triple integral controls.*

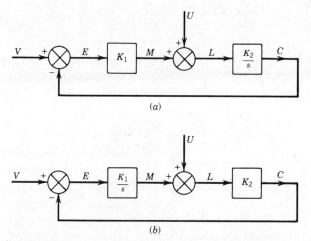

FIGURE 10.3   *Significance of integrator location.*

to disturbances; however location is not significant for steady-state errors caused by commands. With respect to dynamics, note that *any* "amount" of integral control will eventually remove the steady-state errors, but large amounts (K_1, K_2) do it more quickly (smaller τ). For commercial electronic or pneumatic process controllers, in which controller input and output are the same type of physical variable (voltage or air pressure), the "amount" of integral control may be specified numerically by giving the *integral time* T_I, where the controller transfer function is $(1/T_I)(1/s)$. Thus a small integral time means a large amount of integral control. Still another type of terminology may be encountered in combined-mode controllers such as the pneumatic proportional-plus integral of Fig. 10.4. There the controller *reset rate* in "repeats per minute" is defined (for a step input) as the number of times per minute the integral action "repeats" the proportional action; that is, reset rate $\triangleq (K_I p_s)/(K_p p_s) = K_I/K_p$ repeats/min.

10.2   HARDWARE AND SOFTWARE IMPLEMENTATION OF INTEGRAL CONTROL MODES

When controller input and output are mechanical displacements, the classical ball/disk integrator of Fig. 10.5a may be used. It was the basis of the original mechanical analog computer of the 1930s; and although no longer used in general-purpose computers, it is still manufactured and used in special-purpose computers, such as integral controllers. Assuming rigid bodies and no slippage

$$R d\theta_o = x_i \omega_d dt \qquad \theta_o = \frac{\omega_d}{R} \int x_i dt \qquad (10.3)$$

Figure 10.5b shows the familiar op-amp integrator found in electronic analog computers. For ideal conditions

$$e_o = -\frac{1}{RC} \int e_i dt \qquad (10.4)$$

Such integrators always exhibit some "drift," where output slowly changes even when $e_i \equiv 0.0$ (shorted). This will cause a steady-state error in a feed back system, analysis of which is left for Problem 10.2.

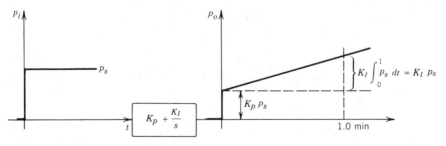

FIGURE 10.4 *Definition of reset rate.*

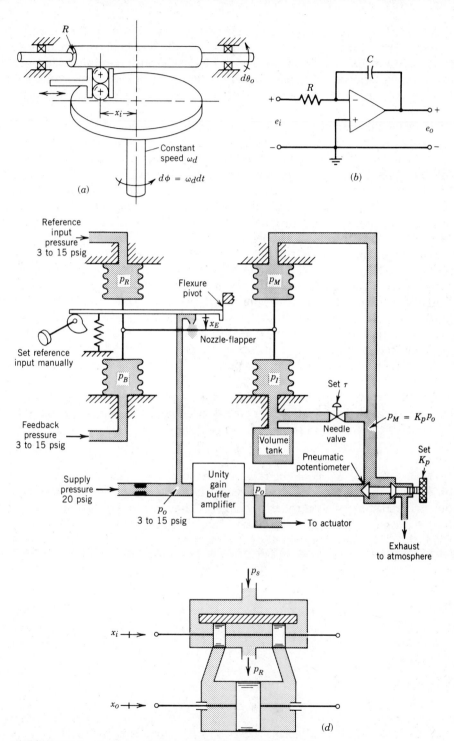

FIGURE 10.5 *Analog mechanization of integral control.*

Although the integral control mode is very widely used in pneumatic control systems, commercially available controllers provide it only in combination with proportional and/or derivative modes, not by itself. Such controllers are, however, always adjustable, so one can adjust the other modes to their minimum values, leaving the integral mode as the predominant effect if desired. Figure 10.5c shows a typical proportional-plus-integral controller, a combination capable of handling many practical control problems. Such a controller might be located 20 ft from the pneumatic sensor measuring (p_B) the process controlled variable and 300 ft from a central control room where process operators manage this loop and many others employed together for the operation of a manufacturing unit. Reference input p_R can be remotely set from a precision pressure regulator in the control room by running pneumatic tubing to the controller p_R bellows. The cam–beam arrangement also allows set-point changes to be entered manually at the controller location.

Pneumatic controllers invariably include one or more unity-gain buffer amplifiers (see Problem 9.22). That in Fig. 10.5c buffers the controller nozzle-flapper (which can supply little flow) from the controller output load, usually a diaphragm valve actuator that requires a large flow. The "pneumatic potentiometer" is a simple manually set valve that produces an output pressure p_M that is an adjustable fraction of p_o, analogous to an electrical potentiometer. In modeling the moving parts of such controllers we relate displacements to the pressure forces that cause them by pure spring elements since the slow and small motions make inertia effects negligible. The small motions also allow the use of flexure pivots rather than rolling or sliding bearings, making lost motion and friction negligible. Motion x_E of the nozzle-flapper is a superposition of the effects of the two bellows pairs $p_R - p_B$ and $p_I - p_M$.

$$x_E = (p_R - p_B)K_{b1} - (p_I - p_M)K_{b2} \qquad (10.5)$$

Constants K_{b1} and K_{b2} could be related to bellows spring constants and various linkage lengths, but we will not pursue such details since they add nothing to our understanding of controller *operation*. (They are, of course, vital in controller *design*). We neglect the nozzle-flapper time constant since it is much faster than the controller overall dynamics, giving

$$p_o = -K_a x_E \qquad (10.6)$$

Also

$$p_M = K_p p_o \qquad 0 \le K_p \le 1.0 \qquad (10.7)$$

$$\frac{p_I}{p_M}(s) = \frac{1}{\tau s + 1} \qquad (10.8)$$

where

$$\tau \triangleq R_f C_f$$

$R_f \triangleq$ fluid resistance of needle valve $C_f \triangleq$ fluid capacitance of volume

Combining all these equations we get

$$-\frac{p_o}{K_a} = (p_R - p_B)K_{b1} - \left(\frac{K_p p_o}{\tau s + 1} - K_p p_o\right)K_{b2}$$

$$= p_E K_{b1} + \left(\frac{\tau s}{\tau s + 1}\right)K_p K_{b2} p_o \qquad p_E \triangleq p_R - p_B \qquad (10.9)$$

To get the results for a mathematically ideal controller we now take K_a (a very large number) to be infinite, similar to the common infinite-gain assumption used in op-amp circuit analysis.

$$\frac{p_o}{p_E}(s) = -\left(\frac{\tau s + 1}{\tau s}\right)\frac{K_{b1}}{K_p K_{b2}} = -\frac{K_{b1}}{K_p K_{b2}}\left(1 + \frac{1}{\tau s}\right) \qquad (10.10)$$

giving perfect proportional-plus-integral dynamics.

In Eq. 10.10 the minus sign makes this a *reverse-acting controller;* that is, an increase in p_E causes a decrease in p_o. This may actually be desired, since p_o may go to a flow control diaphragm valve where increased pressure causes the valve to shut. If a *direct-acting controller* is desired, we can just interchange the p_R and p_B bellows connections. To get predominantly integral control, Eq. 10.10 shows we need to make τ small. A real controller (which cannot have infinite K_a) will approximate the behavior predicted by Eq. 10.10. The nature of the approximation is revealed by reanalyzing with K_a a finite value and comparing the frequency response of the two models. This study (left for the chapter-end problems) shows that the real device follows Eq. 10.10 except at low frequencies, where the imperfect integration causes a deviation.

The hydraulic servovalve/actuator combination of Fig. 10.5d, studied earlier in Chapter 2, is regularly used to achieve integral control. Since the actuator is capable of high-power output, the functions of controller and final control element are often combined in the same hardware. From Eq. 2.170

$$\frac{x_o}{x_i}(s) = \frac{K}{s\left(\dfrac{s^2}{\omega_n^2} + \dfrac{2\zeta s}{\omega_n} + 1\right)} \approx \frac{K}{s} \qquad (10.11)$$

If the second-order dynamics cannot be neglected as in Eq. 10.11, the integrating effect is of course still present and effective. Some two-stage servovalves (Fig. 9.34) are available without internal mechanical feedback, giving an integration internal to the valve and thus a type-2 (two integrations) controller.

When integral control is implemented digitally, the software must provide some form of numerical integration, which is always an approximation to exact analytical integration. Such numerical integration algorithms are also the heart of digital simulation languages like CSMP; however the accuracy requirements are much greater there than in digital control. CSMP gives the user a choice among seven different integrating algorithms, but integral control is usually implemented using the simplest of these, rectangular integration. Due to the need for sampling and A/D conversion, the only values of system actuating signal E available to the digital computer are the sampled and held versions of Fig. 10.6. Using rectangular numerical integration we approximate the area under the E–t curve as

$$\int_0^t E\, dt \approx \sum_{n=0}^{k} E_n T \qquad (10.12)$$

The output of a digital integral controller at any time is thus

$$K_I \sum_{n=0}^{k} E_n T = K_I T \sum_{n=0}^{k} E_n \qquad (10.13)$$

so the computer simply needs to accumulate the sum of the E_n values, call up the stored value of $K_I T$ from memory, multiply the two values and output the product, through a D/A converter, to the final control element. The sampling interval T should be short enough to not "miss" important dynamic changes in the process but not so short as to keep the computer needlessly busy. I will give more details in the chapter on digital control, but "ballpark" values for T are: flow loops 1 s, level and pressure loops 5 s, temperature loops 20 s.

Simple integrating algorithms are acceptable because the essential feature of integral control is that a persistent error should give rise to an ever-increasing corrective effort. Rectangular integration does provide this feature. Just as an analog op-amp integrator will not give the analytically predicted zero steady-state error because of drift (bias), a digital integrator suffers similarly because of the finite resolution of A/D and D/A converters. For an eight-bit A/D converter, for example, the E signal could be as large as 0.2% of full scale, one half of least significant bit (LSB), and the digital output would remain at zero. Thus an error of this magnitude would go uncorrected since the computer thinks the error is zero and provides no further correction. Even if a change in E is large enough to be detected at the controller input, no output change will be produced unless the computed product $K_I T \Sigma E_n$ changes by at least one LSB. For a given K_I, if T is too small, excessive steady-state error may result from this effect.

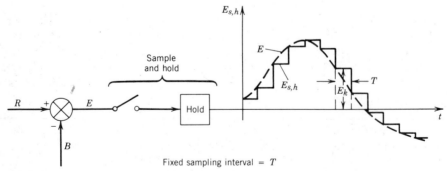

FIGURE 10.6   *Digital integral control with rectangular integration.*

When one purchases a commercial digital controller, most of the details just mentioned are taken care of by the controller manufacturer and the controller can, for the most part, be treated by the control system engineer as if it were an analog device. If a "general purpose" computer is to be programmed as a digital controller, then of course all these details must be considered.

10.3 INTEGRAL CONTROL OF DEAD-TIME-PLUS-FIRST-ORDER-LAG PROCESSES

Proportional control of this common process model was found to be difficult in Section 9.4 since loop gain was restricted by stability problems to low values, causing large steady-state error. Integral control gives zero steady-state error (for both step commands and/or disturbances) for *any* loop gain and is thus an improvement. A Nyquist stability study (as used in Section 9.4) requires no dead-time approximations and gives exact results for marginal stability gain values and cycling frequencies. Design for a gain margin of about two (quarter-amplitude damping) is then easy. The SPEAKEASY program of Section 9.4 is easily modified to get the results in Table 10.1. Note that, compared with proportional control (Table 9.1), both loop gain and speed of response (ω_0 for a given τ) are lower; however the low gain is now not as significant since we do not depend on it to reduce steady-state error.

To check the time-domain response we run a CSMP simulation on the system of Fig. 10.7a for both $K_I = 1.14$ (marginal stability) and $K_I = 0.57$ (gain margin of 2.0), giving the results of Fig. 10.7b. As predicted, $K_I = 1.14$ gives a sustained

TABLE 10.1

$\dfrac{\tau_{DT}}{\tau}$	$\omega_0\tau$	K
0.1	3.11	10.2
0.2	2.16	5.16
0.3	1.74	3.49
0.4	1.48	2.65
0.5	1.31	2.15
0.6	1.18	1.81
0.7	1.07	1.57
0.8	0.99	1.39
0.9	0.92	1.25
1.0	0.86	1.14

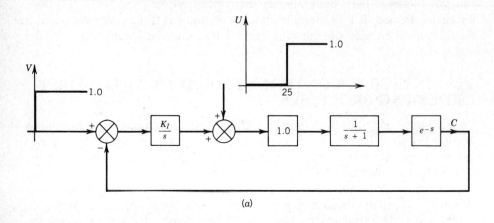

(a)

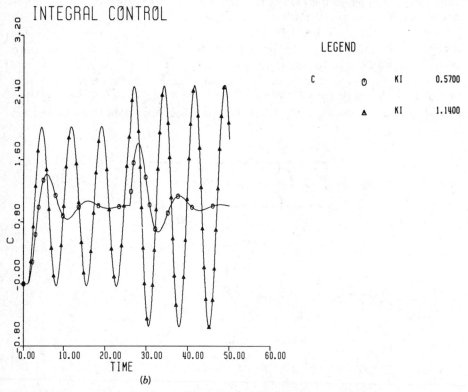

(b)

FIGURE 10.7   *Integral control of dead-time-plus- first-order lag.*

oscillation, whereas $K_I = 0.57$ gives a well-damped system with zero steady-state error for step commands and disturbances. Note that the cycling period for $K_I = 0.57$ is not much different from that for $K_I = 1.14$. This allows prediction of speed of response for all systems of this type from the $\omega_0\tau$ values of Table 10.1 since systems with quarter-amplitude damping settle down in about two cycles after a step command or disturbance. For our specific example we would predict

$$\text{settling time} \approx 2\frac{2\pi}{0.86} = 14.6\text{s} \tag{10.14}$$

Figure 10.7b shows this to be a rough but useful estimate.

10.4 INTEGRAL CONTROL OF A LIQUID-LEVEL PROCESS

Conversion of the proportional control system of Fig. 9.1 to integral control will bring out both some specific and generally useful results about the integral mode. Since both electrical and pneumatic signals are available in this system, we have a choice as to hardware implementation. Suppose we choose electronics and replace the amplifier block K_a by K_I/s. The closed-loop system differential equation is then

$$\left[(h_V - h_C)\frac{K_h K_I K_p K_x K_v}{s} + \frac{1}{R_f}P_U - Q_U \right]\frac{R_f/\gamma}{\tau_p s + 1} = h_C \tag{10.15}$$

$$(\tau_p D^2 + D + K)h_C = Kh_V + \frac{1}{\gamma}DP_U - \frac{R_f}{\gamma}DQ_U \tag{10.16}$$

$$(\tau_P D^2 + D + K)h_E = -(\tau_p D^2 + D)h_V + \frac{1}{\gamma}DP_U - \frac{R_f}{\gamma}DQ_U \tag{10.17}$$

where

$$K \triangleq \text{loop gain} \triangleq \frac{K_h K_I K_p K_x K_v R_f}{\gamma} \qquad h_E \triangleq \text{system error} \triangleq h_V - h_C$$

We note immediately from Eq. 10.17 that step changes (constant values) of h_V, P_U and/or Q_U give zero steady-state errors. For ramp inputs

$$h_V = \dot{h}_V t \qquad P_U = \dot{P}_U t \qquad Q_U = \dot{Q}_U t \tag{10.18}$$

$$h_{E,ss} = -\frac{\dot{h}_V}{K} + \frac{\dot{P}_U}{\gamma K} - \frac{\dot{Q}_U}{\gamma K/R_f} \tag{10.19}$$

We now have constant, nonzero steady-state errors whose magnitudes can be reduced by increasing K. These errors could be made zero by use of a *double* integral control; however the system would be absolutely unstable. (Chapter 11 will show how derivative control allows us to use the double integral.) Since the characteristic equation is second order we should define

$$\omega_n \triangleq \sqrt{\frac{K}{\tau_p}} \qquad \zeta \triangleq \frac{1}{2\sqrt{K\tau_p}} \tag{10.20}$$

If we take τ_p as unavailable for change, we see that increase in K to gain response speed or decrease ramp steady-state errors will be limited by loss of relative stability (low ζ). If we design for a desired ζ, the needed K is easily found and ω_n is then fixed. Because of our too-simple dynamic modeling, absolute instability is not predicted. Figure 10.8 compares proportional and integral control, showing the reduction in relative stability caused by integral control.

Let us change the controlled process in Fig. 9.1 by closing off the pipe at the left. This not only deletes P_U as a disturbance but also causes a significant change

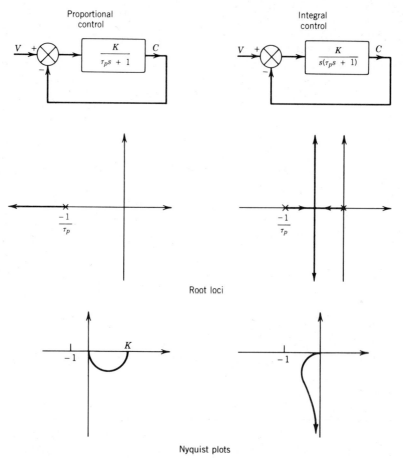

FIGURE 10.8 *Destabilizing effect of integral control.*

in process dynamics. Taking $R_f = \infty$ in Eq. 9.1 gives

$$h_C = \frac{1}{A_{TS}}(Q_M - Q_U) \tag{10.21}$$

and the block diagram of Fig. 10.9 for proportional control. The original tank process had *self regulation*. That is, considering the tank alone, if one changes Q_M and/or Q_U, the tank will itself in time find a new equilibrium level since the flow through R_f varies with level. With R_f not present, the tiniest difference between Q_M and Q_U will cause the tank to completely drain or overflow since it is now an integrator and has lost its self-regulation. From Fig. 10.9

$$(\tau s + 1)h_C = h_V - \frac{1}{KA_T}Q_U \tag{10.22}$$

$$(\tau D + 1)h_E = -\tau Dh_V + \frac{1}{KA_T}Q_U \tag{10.23}$$

We see that even with proportional control, the integrating effect in the process gives a zero steady-state error for step commands, but not for step disturbances (an example of the general case shown in Fig. 10.3). If we substitute integral control to eliminate the Q_U error, the system becomes absolutely unstable.

10.5   DOUBLE INTEGRAL CONTROL IN A SYNCHRO/DIGITAL CONVERTER

A synchro[1] is a small rotating ac machine used for precise measurement of angular displacement in many servomechanisms. Although synchros were originally developed for use with analog systems, many applications (machine tools, antennas, etc.) today use digital approaches. Even though a digital shaft-angle

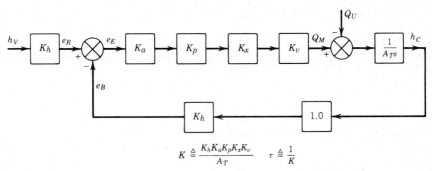

$$K \triangleq \frac{K_h K_a K_p K_x K_v}{A_T} \qquad \tau \triangleq \frac{1}{K}$$

FIGURE 10.9 *Proportional control of process with integrating dynamics.*

[1]E. O. Doebelin, *Measurement Systems,* 3rd Ed., McGraw–Hill, New York, 1983, pp. 248–252.

encoder might seem to be the logical sensor for an otherwise digital system, synchros (and the closely related resolver[2]) have a number of advantages that make them desirable. Figure 10.10 shows how a synchro/digital (S/D) converter might be used in an electromechanical position servo. (In addition to its main function of digitizing shaft position information, the S/D converter also provides a shaft *velocity* signal, a form of derivative control (discussed in Chapter 11) useful in stability augmentation.) The digital computer accepts a command angle in digital form and compares it with the actual shaft position as supplied by the S/D converter in digital form. If there is an error, the computer processes this according to its control algorithm and sends this signal through a D/A converter, to the analog power amplifier which drives the motor/load so as to reduce the error to zero.

Our main interest is in the internal details of the S/D converter itself, an all-electronic (no moving parts) feedback system that uses double integral control to achieve the necessary high accuracy (± 40 arc-sec)[3] Figure 10.11 shows its functional diagram. The ac analog voltages from the synchro are first applied to a special Scott T transformer that produces ac voltages with amplitudes proportional to the sine and cosine of synchro-shaft angle θ.

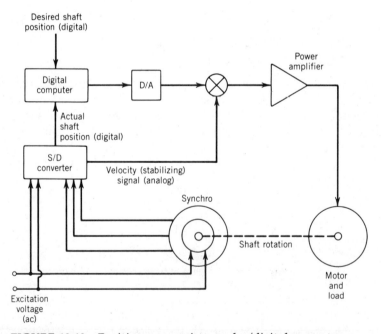

FIGURE 10.10   *Position servo using synchro/digital converter.*

[2]Ibid., p. 759.
[3]Analog Devices Model SDC/RDC 1721.

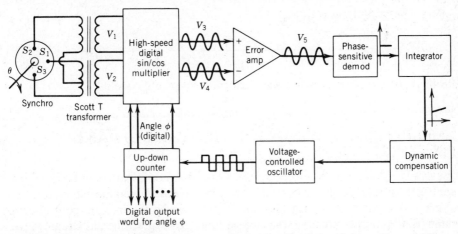

FIGURE 10.11   *Synchro/digital converter using double-integral (type-2) controller.*

$$V_1 = KE_0 \sin\omega t \sin\theta \qquad V_2 = KE_0 \sin\omega t \cos\theta \qquad (10.24)$$

Note that if a resolver (rather than a synchro) were used, the Scott T transformer would be unnecessary since the resolver produces voltages of form (10.24) directly. The voltages V_1 and V_2 are digitally multiplied by sin ϕ and cos ϕ, where ϕ is the digital output angle of the S/D converter:

$$V_3 = KE_0\sin \omega t \sin \theta \cos \phi \qquad V_4 = KE_0 \sin \omega t \cos \theta \sin \phi \quad (10.25)$$

and then subtracted to give

$$V_5 = V_3 - V_4 = KE_0 \sin \omega t(\sin \theta \cos \phi - \cos \theta \sin \phi) \qquad (10.26)$$

$$V_5 = KE_0 \sin \omega t \sin (\theta - \phi) \qquad (10.27)$$

The amplitude of V_5 is thus proportional to $\sin(\theta - \phi) \approx \theta - \phi$, the error in our feedback system. To get a dc error signal, V_5 is phase-sensitive demodulated. Our first integral control effect is achieved with an op-amp integrator whose input is the dc error signal. The next block, dynamic compensation, is necessary to stabilize the type-2 system, which would otherwise be unstable for all gains. It is an approximate proportional-plus-derivative mode $(\tau_1 D + 1)/(\tau_2 D + 1)$ which will be discussed in Chapter 12. The second integrating effect is obtained with the voltage-controlled oscillator (VCO) and digital up–down counter. A steady input voltage to a VCO produces a proportional frequency at its output, which will cause a ramp change in ϕ at the counter output. The final design[4] has

$$\frac{\phi}{\phi - \theta}(s) = \frac{55,000(0.0068s + 1)}{s^2(0.00122s + 1)} \qquad (10.28)$$

----

[4]Analog Devices Spec. Sheet C749-20-11/82, 1982, Norwood, Mass.

giving the closed-loop response

$$\frac{\phi}{\theta}(s) = \frac{0.0068s + 1}{2.22 \times 10^{-8}s^3 + 1.82 \times 10^{-5}s^2 + 6.8 \times 10^{-3}s + 1} \quad (10.29)$$

Problem 10.8 further explores the performance of this system.

10.6 INTEGRAL WINDUP (RESET WINDUP) AND ITS CORRECTION

We saw in Chapter 9 the effects of the saturation nonlinearity on proportional control performance. Integral control may be degraded even more by saturation effects. In Fig. 10.12a a large sustained error causes the integral controller to ramp its output pressure up to the 20-psig supply pressure. The diaphragm valve, sized to be wide open at 15 psig (the upper end of the normal 3 to 15 psig control range) saturates at 15 psig. The integral signal beyond $t = 7.5$ is really useless since it asks for a motion that the valve cannot produce. When the error reverses at $t = 10$, the valve cannot respond to this change until the integral signal (which has "wound up" to 20 psig) is "unwound" back to the 15-psig level at $t = 12.5$. This delayed response effect is called *reset windup* or *integral windup*. Note that this delay is in addition to the normal lagging behavior of integral control and can thus cause excessive overshooting and stability problems.

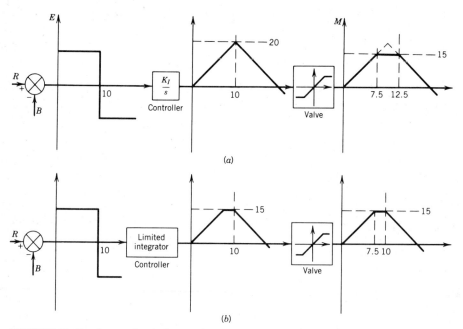

FIGURE 10.12 *Integral windup and its correction.*

Integral windup is of course not a problem in every application of integral control. If difficulty *is* anticipated, the controller can be modified in different ways to give various degrees of improvement.[5] Basically, one wants to disable the integrator whenever its output signal causes saturation in the final control element. In a digitally implemented controller such modifications are easily available, with almost unlimited versatility, through software changes. In Fig. 10.12*b* the integrator is disabled when its output pressure reaches 15 psig, preventing any windup. When the error reverses at $t = 10$ the integrator and valve immediately respond to the negative error since there is no windup that needs to first be unwound.

To provide some more details we will run a CSMP simulation on the system of Fig. 10.13, which uses the popular proportional-plus-integral control mode. (Note that windup problems are possible whether integral control is used alone or in combination.) The controlled process is modeled as $10e^{-0.1s}/(0.4s + 1)$ and final control element saturation occurs at $M3 = \pm 1.0$. We implement an anti-reset-windup feature by sensing $M3$, comparing it with the saturation level $F = 1.0$, and setting the integrator input to zero whenever saturation is present. To get a convenient comparison of the system with and without the anti-reset-windup feature, we make two runs, one with $F = 1.0$ and another with F a very large number (we use 20) so that the integrator is *never* disabled. The CSMP program goes as follows.

$$V = VS*STEP(0.0) \qquad \text{step command input}$$

PARAM $VS = 7.0 \qquad \text{step input size is 7.0}$

$$E = V - C$$

$$M1 = KP*E \qquad \text{proportional mode}$$

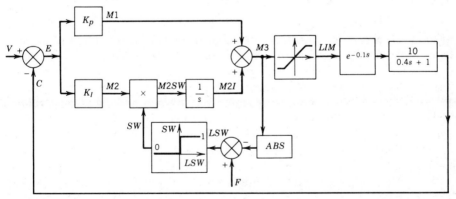

FIGURE 10.13 *Anti-reset-Windup for proportional plus integral controller.*

---

[5]F. G. Shinsky, *Process-Control Systems,* 2nd Ed., McGraw–Hill, New York, 1979, pp. 91--93, 142–144; Fisher Controls, Bulletin 34.6:4156, Marshalltown, Iowa, 1979.

```
PARAM    KP=0.1       proportional gain is 0.1
         M2=KI*E      integrator input preparation
PARAM    KI=1.0       integral gain is 1.0
         M2SW=M2*SW        multiplier used to turn integrator input M2SW on
                           and off
         LSW=F-ABS(M3)       check for saturation
PARAM    F=(1.,20.)       make two runs, with and without windup
         SW=FCNSW(LSW,0.0,0.0,1.)       SW = 0 or 1 to turn integra-
                                        tor on and off
         M2I=INTGRL(0.0, M2SW)      integral control output
         M3=M1+M2I      sum proportional and integral modes
         LIM=LIMIT(-1.,1.,M3)       final control element saturation
         CA=KPR*LIM                    ⎫
PARAM    KPR=10.                       ⎬ process model
         CB=DELAY(50.,1,CA)            ⎭
         C=REALPL(0.0, TAUP, CB)
PARAM    TAUP=0.4
TIMER    FINTIM=4.0,DELT=.002,OUTDEL=.04
```

In Fig. 10.14 we see that although both systems are absolutely stable and eventually converge on the desired value $V = 7.0$, the system with anti-reset-windup settles more quickly and with smaller transient errors.

10.7   AN ADAPTIVE FEEDFORWARD COMBUSTION TRIM CONTROLLER USING INTEGRAL CONTROL

The high cost of fuel has made the optimization of boiler combustion controls of great economic interest. Control based on the oxygen content of exhaust gases in the smoke stack has been found very effective. Figure 10.15 shows that flue gas heat losses are large both when there is insufficient air (unburned fuel goes up the stack) and when there is excess air (lots of hot air goes up the stack). By controlling air (based on the percentage of oxygen in the flue gas) within a narrow range, significant fuel savings are possible. The desired percentage O_2 for best efficiency varies significantly with boiler firing rate (load), but this complication can be handled by changing the oxygen set point in a feedforward manner, using the air damper position as an indicator of load.

A system that implements this basic concept using microprocessor technology and an unusual version of integral control is shown in Fig. 10.16. Boilers of this size (so-called package boilers) generally have a basic mechanical feedforward control system that schedules air flow in terms of firing rate by positioning the air damper with a profile cam actuated from the fuel valve. The oxygen-sensor-based system is a feedback system which augments the basic open-loop control to correct for all other disturbances and is thus called a "trim" system. The firing-rate range is broken into eight subranges, each of which has its own numerical value of oxygen set point. These values are stored in microprocessor memory. Since a single boiler may use different fuels (gas, oil) at different times, the memory is set up to hold two different eight-element arrays of set-point values, one for each fuel.

Sixteen digital integrators are provided; eight for each fuel. Each of the eight integrators is assigned to one of the eight subranges of firing rate and only one integrator is active at any time. Assuming that the boiler stays at a certain firing rate long enough, the integral controller will bring the actual percentage O_2 to the desired value for that subrange. At this time the integrator has "learned" how much to augment the action of the basic mechanical system for that particular firing subrange to attain zero error. Note that if we return to this same subrange at some later date and anything in the system has changed, in reducing the oxygen error again to zero, the integrator will "relearn" and produce a *different* final output value that takes into account the new conditions. This learning process occurs for each of the eight integrators as its subrange is exercised and gives the system its "adaptive" nature. Due to the considerable nonlinearity of combustion processes, each integrator has a fixed gain suited to its own subrange.

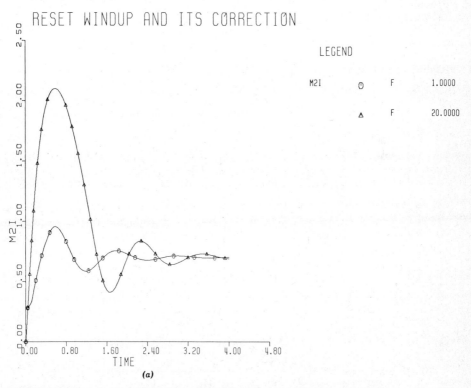

(a)

FIGURE 10.14 *System response with and without anti-reset-windup.*

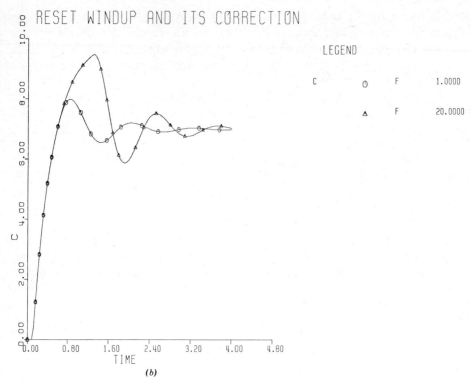

FIGURE 10.14 (Cont.)

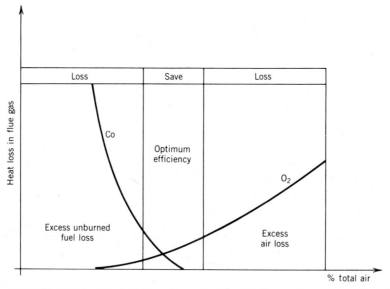

FIGURE 10.15   *Effect of air on combustion efficiency.*

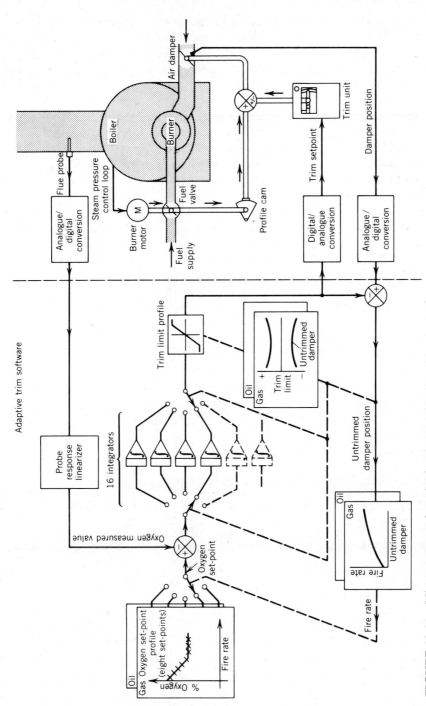

FIGURE 10.16 *Microprocessor-based adaptive combustion trim control.*
Source: Adaptive Combustion Trim Control, Energy Technology and Control Corp., Reston, VA., 1983.

Let us assume that the boiler has been running long enough to exercise all the subranges and that all the integrator outputs have found the values needed to give equilibrium at zero error. If the boiler has been running in one of the firing rate subranges and then changes to another, we stop using one integrator and switch to another. The previous integrator (now inactive) *holds* its output value until its subrange is again entered at some later time. If any conditions have changed in the interim, the integrator, seeking zero error, will change its output accordingly, updating its correction. Switching from one integrator to another with a different output value induces a transient involving all the system lags in the feedback loop. During this rather long transient, combustion efficiency is away from the optimum. Since the new integrator output value should be close to what is needed for equilibrium, feedback is not really needed during the transient, so the system is operated open loop for a timed interval. At the end of this interval the loop is closed. Even though the loop dynamics are not favorable, only a small transient is induced since the error has been reduced to near zero during the open-loop phase. This procedure is followed each time we switch integrators. The effect of the poor loop dynamics is thus minimized, giving a system that keeps the percentage of oxygen close to optimum at all times. A final feature is the trim limit profile, a safety feature that prevents the trim correction from exceeding set limits if certain failures occur.

The complex nature of this system's operation, plus the need to hold integrator output values accurately for long time periods makes a digital approach almost mandatory. This also makes any needed changes in system operation and/or numerical values quick and easy to carry out since they are accomplished by software (reprogramming).

PROBLEMS

10.1 Explain how Problem 9.1e relates to integral control.

10.2 An electronic integrator as in Fig. 10.5*b* has as input the error signal E of a feedback system and has $RC = 0.1s$. When we short the integrator input ($e_i = 0.0$) to check for drift, e_o drifts at the rate of 1 mV/s. If V and C in the feedback system are steady, what must the error E be? This drift phenomenon is present to some extent in all op-amp integrators and prevents *real* integral control systems from achieving the zero steady-state error analytically predicted for "perfect" integrators.

10.3 Analyze the controller of Fig. 10.5*c* with K_a a finite value. Compare the frequency response and step response of this (more correct) model with that given by Eq. 10.10.

10.4 Write and run a computer program to produce the results of Table 10.1.

10.5 Write and run a CSMP (or other available) simulation for the system of Fig. 10.7.

10.6 For the system of Fig. 10.7a:

  a. Set loop gain for $M_p = 1.3$. Check the step response using CSMP.

  b. Use root locus to set loop gain for $\zeta = 0.42$ for dominant roots. Plot $(C/V)(i\omega)$. Use CSMP (or other available simulation) to get the step response using an exact dead-time model.

  c. For $U \equiv 0.0$ and V a ramp input, find the exact steady-state error analytically.

  d. Show that a double-integral control gives zero steady-state error for a ramp input but is unstable for all gain values.

10.7 Convert the system of Eq. 9.19 to integral control and then carry out a complete analysis as in Eq. 10.15 through 10.20 and Fig. 10.8.

10.8 For the system of Fig. 10.11 and Eq. 10.28:

  a. Where can the velocity signal (proportional to $\dot{\theta}$) be picked off?

  b. Plot the root locus and find ζ for the dominant roots.

  c. Plot $(\phi/\theta)(i\omega)$ and find M_p. Also find the gain margin and phase margin.

  d. Plot the step and ramp responses, using CSMP or other available simulation.

10.9 Modify the simulation program for the system of Fig. 10.13 to allow study of both command and disturbance inputs. Run studies on step and ramp inputs of V and U with V and U applied separately and simultaneously.

CHAPTER 11
Derivative Control Modes

11.1 BASIC CHARACTERISTICS

Although the on–off, proportional, and integral control actions described in previous chapters can each be used as the *sole* effect in a practical controller, we will see that the various derivative control modes are always used *in combination* with some more basic control law. This is because a derivative mode produces *no* corrective effect for any *constant* error, no matter how large, and would therefore allow uncontrolled steady-state errors. Thus our discussion in this chapter cannot consider derivative modes in isolation; they will always be considered as augmenting some other mode.

Although derivative modes will be helpful in solving a variety of control design problems, one of their most important contributions is in system stability augmentation. If absolute or relative stability is the problem, a suitable derivative control mode is often the answer. (For example, every high-performance aircraft, and many other land, sea, air, and space vehicles, has its handling qualities improved by use of feedback systems employing derivative control.) This stabilization or "damping" aspect of derivative control can be easily understood qualitatively from the following discussion. Just as the invention of integral control may have been stimulated by human process operators' desire to automate their task of manual reset, derivative control hardware may first have been devised as a mimicking of human response to changing error signals. In Fig. 11.1a we assume a human process operator is given a CRT display of system error E and has the task of changing manipulated variable M (say with a control dial) so as to keep E close to zero. Ask yourself, "If I were the operator, would I produce the same value of M at time t_1 as at time t_2?" (Note that a *proportional* controller would do exactly that.) Most people agree that a stronger corrective effort seems appropriate at t_1 and a lesser one at t_2, since at t_1 the error is $E_{1,2}$ and *increasing,* whereas at t_2 it is also $E_{1,2}$ but *decreasing.* That is, the human eye and brain senses not only the ordinate of the curve but also its trend or slope. Slope is clearly dE/dt, so to mechanize this desirable human response we need a controller sensitive to error derivative. Such a control can, however, not be used alone since it does not oppose *steady* errors of any size, as at t_3, thus a *combination* of proportional plus derivative, for example, makes sense. The relation of the general concept of derivative control to the specific effect of viscous damping in mechanical systems can be appreciated from Fig. 11.1c where an applied torque T tries to control position θ of an inertia J. The damper torque on J behaves exactly like a derivative control mode in that it always opposes velocity $d\theta/dt$ with a strength proportional to $d\theta/dt$, making motion less oscillatory.

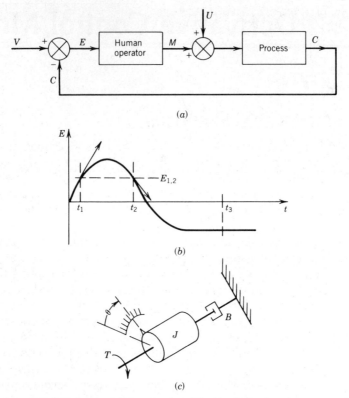

FIGURE 11.1   *Possible genesis of derivative control.*

In giving an overview of the derivative modes we can say that:

1. Although our first example worked with a derivative of the signal E, derivatives of C and, in fact, almost any available signal in the system are candidates for a useful derivative control mode.

2. First-derivative signals are most common and easiest to implement, but higher derivatives can also be useful.

3. The noise-accentuating characteristics of derivative operations may often require use of approximate (low-pass filtered) derivative signals. This prevents realization of ideal performance but significant system improvement is still often possible.

4. Derivative signals can sometimes be realized better with sensors directly responsive to the desired variable, rather than trying to differentiate an available signal. In mechanical systems, for example, an accelerometer may give a better second-derivative signal than is obtainable by double-differentiating a position transducer signal.

5. In addition to the stability augmentation already discussed, derivative modes may also offer improvements in speed of response and steady-state errors.

6. If carried to extreme, the benefits predicted for derivative control will not be realized because saturation will prevent the large corrections called for from actually being applied.

In this chapter we will explore some idealized versions of derivative control in order to make clear their basic characteristics, unobscured by confusing details. Later chapters will develop their practical implementation in a more comprehensive way.

11.2 EFFECTS OF FIRST- AND SECOND-DERIVATIVE-OF-ERROR CONTROL IN AN ELECTROMECHANICAL POSITION SERVO

Figure 11.2 shows an electromechanical servo that positions a load angle θ_C in response to a low-power mechanical command θ_V. Angle measurement and comparison is by means of a synchro pair.[1] Compared with a potentiometer bridge, these offer higher sensitivity, longer life, ruggedness, and continuous rotation capability. Their ac error signal, however, requires a phase-sensitive demodulator[2] and low-pass filter if we wish to use dc components (motor, etc.) in the rest of the system. The low-pass filter contributes a first-order lag that may or may not be negligible. The preamplifier/controller uses op-amps to obtain proportional-plus first- and second-derivative-of-error control modes. A power amplifier (dynamics negligible) boosts the preamp output power to the point where it can provide the armature current for a dc motor with PM field. Motor and load shafts are modeled with inertia, viscous friction, and disturbing torque T_U, the gear-train equivalent system method of Chapter 3 producing the transfer function shown.

Power amplifiers in servo systems are available in three major types[3]: the class AB ("smooth") transistor, the switching transistor (pulse-width modulated (PWM)), and the silicon controlled rectifier (SCR); see Fig. 11.3. Smooth transistor amplifiers use the transistors in a smooth modulating fashion, produce an output voltage that is a highpower copy of the input, have the largest bandwidth (1000 to 10,000 Hz), and require the least motor derating (oversizing) due to heating, but have poor energy efficiency (20 to 60%) and are thus used mainly below 1000 W. Switching (PWM) amplifiers use the transistors as on–off switches and are much more efficient than the modulating mode. The output voltage is a constant frequency (1 to 25 kHz), fixed amplitude waveform whose duty cycle is smoothly varied with input voltage. At zero input the output is switching back and forth between the positive and negative power supply voltages (say $\pm 30V$, 5 kHz) and this voltage does appear across the motor armature; however little armature current flows because the armature L/R time constant is too long relative to the switching frequency. The *average* current (and thus motor torque)

---

[1] E. O. Doebelin, *Measurement Systems,* 3rd Ed., McGraw-Hill, New York, 1983, pp. 248–250.
[2] Ibid., pp. 167, 244, 763.
[3] D. H. Jones, Choosing the Right Servo Amplifier, *Cont. Eng.,* Jan. 1973, pp. 40–43.

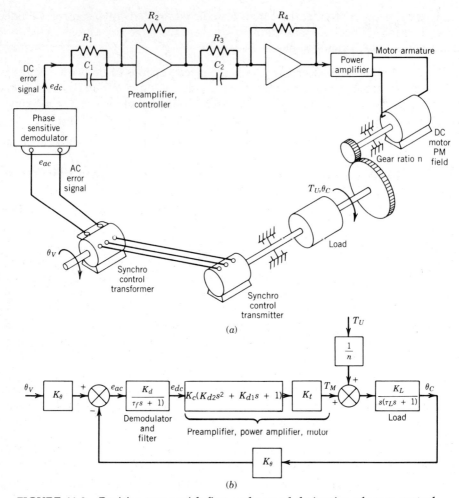

FIGURE 11.2   *Position servo with first-and-second-derivative-of-error control.*

is *exactly* zero, so the motor stands still. A positive input voltage increases the duration of the positive portion of the cycle and decreases the duration of the negative portion by an equal, proportional amount. The switching frequency is still 5 kHz but now the asymmetrical square wave has a nonzero average voltage (think of a Fourier series) that produces a proportional current and torque, causing the motor to rotate. The oscillatory terms of the Fourier series all have zero average value (causing no net rotation) and because of their high frequency cause negligible motor oscillation. A negative input voltage similarly causes rotation in the opposite direction. Switching amplifiers have good efficiency (75 to 87%), bandwidth of 500 to 2000 Hz, minor heating problems, and are economic to use in the 25 to 5000 W range. The dynamics of both smooth and PWM type amplifiers can usually be neglected in feedback systems with mechanical moving parts.

Various versions (full wave, half wave, single phase, three phase, etc.) of SCR amplifiers are available, Fig. 11.3 showing a single-phase, half-wave type. Here the input voltage smoothly modulates the point in each power-line cycle at which the SCR is made to conduct. As in the PWM amplifier, the motor responds to the average value of the waveform. Efficiency is very good (85 to 95%) but bandwidth is limited (5 to 30 Hz) and heating problems can be severe, requiring use of oversize motors. Since SCR drives are often used at high power levels (hundreds of horsepower), the modest bandwidth may not be a problem since the mechanical lags of the large moving parts may dominate.

Let us first analyze our servo system under the assumption that the synchro ac excitation frequency is high enough to allow a negligibly small τ_f.

$$\left\{(\theta_V - \theta_C)[K_\theta K_d K_c K_t(K_{d2}s^2 + K_{d1}s + 1)] + \frac{T_U}{n}\right\}\frac{K_L}{s(\tau_L s + 1)} = \theta_C \tag{11.1}$$

$$[(\tau_L + KK_{d2})s^2 + (1 + KK_{d1})s + K]\theta_C = K(K_{d2}s^2 + K_{d1}s + 1)\theta_V + \frac{K_L}{n}T_U \tag{11.2}$$

where

$$\text{loop gain} \triangleq K \triangleq K_\theta K_d K_c K_t K_L \qquad \tau_L \triangleq \frac{J}{B} \qquad K_L \triangleq \frac{n}{B} \tag{11.3}$$

For the second-order characteristic equation

$$\omega_n \triangleq \sqrt{\frac{K}{\tau_L + KK_{d2}}} = \sqrt{\frac{BK}{J + K_{d2}BK}} \tag{11.4}$$

$$\zeta \triangleq \frac{1 + KK_{d1}}{2K}\omega_n = \frac{(B + K_{d1}Kn/K_L)K_L}{2n\sqrt{K(\tau_L + KK_{d2})}} \tag{11.5}$$

From the denominator of Eq. 11.4 we see that the term $K_{d2}BK$ (which would be missing if there were no second-derivative-of-error mode) has the same dimensions as system inertia J and may thus be thought of as a pseudo-inertia effect. In determining speed of response (ω_n) it is now the "effective" inertia ($J + K_{d2}BK$) that matters, not just the physical inertia J. By manipulating K_{d2} we are thus able to change the effective inertia even though the physical inertia may be unavailable for change. At first glance it appears that K_{d2} *increases* the effective inertia, rarely a design goal; however examination of the hardware shows that K_{d2} is easily made *negative*. It thus appears that if we have minimized the physical inertia J as much as possible and ω_n is still not high enough, we can use negative K_{d2} to reduce the effective inertia and further improve response speed.

If negative K_{d2} "cancels" physical inertia, why not cancel it *completely* by making $K_{d2} = -\tau_L/K$? In Eq. 11.2 this completely removes the s^2 ("inertia") term and reduces the characteristic equation to first order. We might at this point also examine the effect of K_{d1}. In Eq. 11.5 the term $K_{d1}Kn/K_L$ has the same dimensions as system viscous friction B and may thus be interpreted as a pseudo-friction effect. If, for example, careful mechanical design reduced the

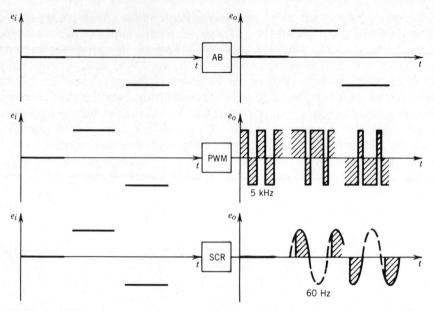

FIGURE 11.3   *Characteristics of three main types of power amplifiers.*

real friction B to nearly zero, we could still ensure proper relative stability (say $\zeta = 0.6$) by setting K_{d1} at the necessary positive value. This is the usual purpose of K_{d1}, stability augmentation by introduction of damping effects. We could however make K_{d1} *negative* and if we make $K_{d1} = -1/K$, we eliminate the s ("friction") term in Eq. 11.2, making the characteristic equation zero order if we have already "cancelled" the inertia term.

$$\theta_C = -\frac{J}{BK}\frac{d^2\theta_V}{dt^2} - \frac{1}{K}\frac{d\theta_V}{dt} + \theta_V + \frac{K_L}{nK}T_U \qquad (11.6)$$

If we apply a step command $\theta_V = \theta_{Vs}$ with $T_U \equiv 0$, the classical method of solution gives the correct result $\theta_C = \theta_V$ for $t > 0+$, but what happens between $t = 0$ and $t = 0+$ is not apparent. Similarly for a step disturbance $T_U = T_{Us}$, $\theta_C = K_L T_{Us}/nK$ for $t > 0+$. The apparent instantaneous response is of course too good to be true, but we must examine the period $0 < t < 0+$ to see why.

For the T_U step input, θ_C responds as a step also. Figure 11.4 shows that if θ_C is to be a step, then $\dot{\theta}_C$ must be an impulse and $\ddot{\theta}_C$ must be the first derivative of an impulse. In Fig. 11.2b we see that the θ_C signal is fed back through the summing junction and then through the controller where proportional, first-derivative, and second-derivative terms are formed. To produce the ideal step response of θ_C, the controller output voltage components for the derivative terms must exhibit the infinite peaks of Fig. 11.4, clearly impossible in a real system, which would saturate at a finite level. In the ideal controller, the infinite voltages also produce infinite motor torques of "just the right waveform" to take the load from rest at $\theta_C = 0$ to rest at $\theta_C = \theta_{Cs}$ in infinitesimal time. Thus

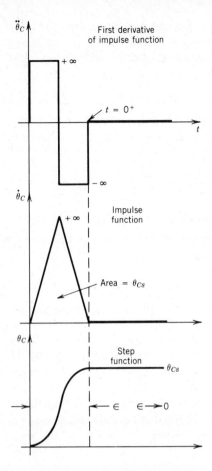

FIGURE 11.4 *Derivative controls produce infinite outputs for step inputs.*

the controller does require infinite power sources but this alone is not sufficient; it must be "smart" enough to use them in just the right way. Here, *smart* means a proportional plus first derivative plus second derivative with the coefficients adjusted to give Eq. 11.6.

The response of θ_C to a step of θ_V in Eq. 11.6 is even more bizarre since the first and second derivatives of θ_V contribute an impulse and an impluse derivative to θ_C itself. These wildly unrealistic "motions" are caused by the application of unreal infinite torques to an unreal (inertia- and friction-free) system.

We thus see that we should not try to make the coefficients of the $s^2\theta_C$ and $s\theta_C$ terms in Eq. 11.2 zero, but we *can* increase ω_n with some negative K_{d2} and *can* use K_{d1} to adjust ζ. The first-derivative term in the controller merely senses $\dot{\theta}_C$ as θ_C is fed back through the summing junction and controller, and then produces a component of motor torque proportional (positively or negatively, depending on the sign of K_{d1}) to $\dot{\theta}_C$. If K_{d1} is positive, this motor torque is felt by the load in exactly the same way as a mechanical viscous damping torque; the load cannot distinguish between them. That is, the motor is a "general purpose" torque producer and we can "make it do tricks" by clever controller design. The second-derivative component of motor torque due to K_{d2} does not have an analogous mechanical effect on the left side of the load Newton's law as does the K_{d1} term.

$$\sum \text{torques} = \text{motor torque} + \text{disturbing torque} + \text{damping torque} = J\frac{d^2\theta_C}{dt^2} \quad (11.7)$$

Rather, it combines with the right-hand side, which has the dimensions of torque but is not a torque felt by the load (as $-B\dot{\theta}_C$ is). That is, the acceleration is *determined* by the sum of the actual torques according to

$$\ddot{\theta}_C = \frac{\text{sum of torques}}{J} \quad (11.8)$$

but when one of the (motor) torque components is proportional to $\ddot{\theta}_C$, it must mathematically move to the left side as in Eq. 11.2.

$$(J + BKK_{d2})\ddot{\theta}_C = K_\theta K_d K_c K_t n(K_{d2}\ddot{\theta}_V + K_{d1}\dot{\theta}_V + \theta_V) + T_U \quad (11.9)$$

$$- (B + K_{d1}K_\theta K_d K_c K_t n)\dot{\theta}_C + K_\theta K_d K_c K_t n\theta_C$$

Thus the *effect* of K_{d2} is the same as physical inertia even though it originally enters the equation as a torque.

In Eq. 11.2, if $\theta_V \equiv 0.0$ and $T_U = T_{Us}$, this step response requires only finite acceleration and velocity if we do not set $K_{d2} = -\tau_L/K$ and $K_{d1} = -1/K$. If we set K_{d1} to get a good ζ and use negative K_{d2} to increase ω_n, only finite torques are required but they get larger as the magnitudes of K_{d2} and/or T_{Us} increase. At some point, amplifier and/or motor saturation will occur and the performance predicted by our linear model will not be obtained. (This behavior can be conveniently explored using digital simulation.) For a step command θ_{Vs} with $T_V \equiv 0.0$

$$\theta_C(s) = \frac{K(K_{d2}s^2 + K_{d1}s + 1)}{(\tau_L + KK_{d2})s^2 + (1 + KK_{d1})s + K} \cdot \frac{\theta_{Vs}}{s} \quad (11.10)$$

We could use the Laplace transform to get a complete solution for $\theta_C(t)$ but prefer just to find quickly $\theta_C(0+)$ using the initial value theorem.

$$\theta_C(0+) = \lim_{s \to \infty} \frac{K(K_{d2}s^2 + K_{d1}s + 1)\theta_{Vs}}{(\tau_L + KK_{d2})s^2 + (1 + KK_{d1})s + K} = \frac{KK_{d2}\theta_{Vs}}{\tau_L + KK_{d2}} \quad (11.11)$$

We see that θ_C undergoes a step change from $t = 0$ to $t = 0+$, thus requiring infinite motor torques as discussed earlier. Thus, although step disturbing torques sufficiently small to prevent saturation are handled by our model, step *commands* of *any* size require infinite torques and thus cannot be studied realistically.

For a ramp command $\theta_V = \dot{\theta}_{Vo}t$ we have

$$[(\tau_L + KK_{d2})s^2 + (1 + KK_{d1})s + K]\theta_E = (\tau_L s^2 + s)\theta_V - \frac{K_L}{n}T_U = \dot{\theta}_{Vo} \quad (11.12)$$

Analysis shows that we still have trouble with infinite torques at $t = 0+$, but we can compute the steady state correctly.

$$\theta_{Ess} = \frac{\dot{\theta}_{Vo}}{K} = \frac{B\dot{\theta}_{Vo}}{K_\theta K_d K_c K_t n} \tag{11.13}$$

This constant steady-state error for a ramp command does not depend on either K_{d1} or K_{d2} so it appears that this feature of performance is unaffected by these derivative control modes. Actually, we can show an indirect method for possibly gaining some improvement. If we are able, by low-friction mechanical design, to make B a small number, Eq. 11.13 shows that θ_{Ess} is reduced. In a proportional control system, a very small B would not be allowed because ζ would get too small. When first derivative control is added, Eq. 11.5 shows that B can be made as small as we wish (even zero) and system damping is still assured if K_{d1} is large enough.

We have seen so far that, at least theoretically, the second derivative mode modifies the characteristic equation such that the coefficient of the second derivative may be made to be any value without having to change the physical inertia; whereas the first derivative mode provides a similar benefit for the damping term. One can think of these control modes as providing two additional "degrees of freedom" in the design process. We have two more coefficients open to free choice and each of them provides a definite benefit. In terms of actual practical use, various approximate versions of first-derivative control are very common, whereas second derivative modes are rather rare, because of the saturation and noise-accentuating problems mentioned earlier. Coefficient K_{d1} is normally made positive to increase damping effects. Although negative K_{d2} ("inertia" reduction) would seem most desirable, positive K_{d2} has been proposed[4] as a means of getting desired low natural frequencies in hydraulic feedback-type suspension systems for high-speed trains without having to increase car mass.

If τ_f in the demodulator filter is not negligible, the effects of K_{d1} and K_{d2} change. Such a change in role can take place in any derivative control application so we wish to call it to the reader's attention. Equation 11.1 becomes

$$\left\{ (\theta_V - \theta_C) \left[\frac{K_\theta K_d K_c K_t (K_{d2}s^2 + K_{d1}s + 1)}{\tau_f s + 1} \right] + \frac{T_U}{n} \right\} \frac{K_L}{s(\tau_L s + 1)} = \theta_C \tag{11.14}$$

and

$$[\tau_f \tau_L s^3 + (\tau_f + \tau_L + KK_{d2})s^2 + (1 + KK_{d1})s + K]\,\theta_C$$
$$= K(K_{d2}s^2 + K_{d1}s + 1)\theta_V + \frac{K_L}{n}(\tau_f s + 1)T_U \tag{11.15}$$

---

[4]N. K. Cooperrider, Suspension Systems for Tracked Air Cushion Vehicles, *Gen'l. Electric R & D Center Report No. 69-C-014*, Schenectady, New York, Dec. 1968.

We see that K_{d1} influences the s term in the same way as before but now we do *not* want to make the s^2 term disappear by setting $K_{d2} = -(\tau_f + \tau_L)/K$. Since the s^3 term does *not* disappear it leaves a gap in the characteristic equation, indicative of absolute instability. A different viewpoint notes that our op-amp controller really cascades two first-order terms to get the desired second order, thus $K_{d2}s^2 + K_{d1}s + 1 = (\tau_1 s + 1)(\tau_2 s + 1)$. Using this in Eq. 11.14 and applying the design technique called *cancellation compensation,* we set $\tau_1 = \tau_f$ and $\tau_2 = \tau_L$ to get

$$\left\{(\theta_V - \theta_C)\left[\frac{K_\theta K_d K_c K_t(\tau_f s + 1)(\tau_L s + 1)}{(\tau_f s + 1)}\right] + \frac{T_U}{n}\right\}\frac{K_L}{s(\tau_L s + 1)} = \theta_C \quad (11.16)$$

$$(\theta_V - \theta_C)\frac{K}{s} + \frac{K_L}{ns(\tau_L s + 1)}T_U = \theta_C$$

$$(s + K)\theta_C = K\theta_V + \frac{K_L}{n(\tau_L s + 1)}T_U \quad (11.17)$$

The cancellation compensation design has made the system command response first order and the disturbance response overdamped second order.

The root loci of Fig. 11.5 give a particularly good interpretation of the behavior. In Fig. 11.5a, τ_1 and τ_2 are taken very small and instability occurs for intermediate gain values (conditional stability). In Fig. 11.5b we have cancelled τ_f with τ_1 but $\tau_2 \neq \tau_L$, giving a second-order system that is always stable. Fig. 11.5c uses even larger τ_1 and τ_2 values, but no cancellation, giving an always-stable third-order system. The double cancellation ($\tau_1 = \tau_f$, $\tau_2 = \tau_L$) of Eq. 11.16 gives Fig. 11.5d. For very large τ_1 and τ_2 we get Fig. 11.5e, a third-order system with always-negative real roots. All these variations show improved stability behavior compared to the basic system without any derivative control (Fig. 11.5f). Speed of response and steady-state errors (due to higher allowable loop gain) may also be better. The Nyquist plots (Fig. 11.6) show clearly how the numerator dynamics (zeros) of $(B/E)(i\omega)$ with their leading phase angles overcome the lags of the basic system to rotate the curve counterclockwise away from the critical -1 instability region.

Although the partial or full cancellation designs of Fig. 11.5b and 11.5d have an appealing simplicity and are often used, good results can be obtained without cancellation. Comparative performance studies using digital simulation allow selection of the best design in specific applications. When cancellation *is* attempted, note that it need not be perfect to be effective, as discussed in Section 9.8.

11.3 EFFECTS OF FIRST- AND SECOND-DERIVATIVE-OF-CONTROLLED VARIABLE CONTROL IN AN ELECTROMECHANICAL POSITION SERVO

When controller dynamics are located in the forward path (from E to C) we shall use the name *forward-path compensation* (also called series or cascade compensation); thus the derivative control modes discussed in Section 11.2 are

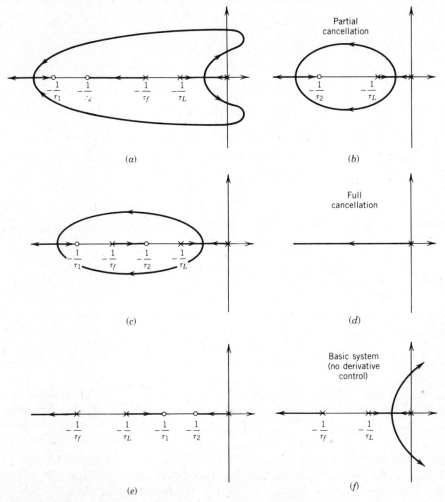

FIGURE 11.5   *Effect of derivative control on root locus.*

one form of forward-path compensation. The feedback path (*C* to *B*) often provides only a measurement of the controlled variable but sometimes useful control effects (usually some form of derivative control) may be realized here also. (This is why the *H* elements in the generalized feedback system diagram were given the more general name *feedback elements* rather than *sensing elements*.) We shall call such control modes *feedback-path compensation*.

We will change the system of Fig. 11.2 by using a proportional controller in the forward path and adding derivative modes at the controlled variable θ_C. First-derivative modes of angular position are extremely common in all kinds of servomechanisms and are usually implemented using a tachometer generator[5]

[5]Doebelin, *Measurement Systems*, p. 310.

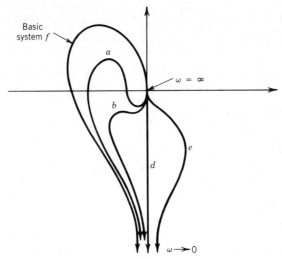

FIGURE 11.6 *Effect of derivative control on nyquist plot.*

to get a voltage proportional to shaft speed. Since tachometers tend to be noisy at low speeds, in a geared system one usually puts the tach on the motor (rather than the load) shaft; in fact the tach is sometimes geared up from the motor. Much less common, but included here to show the effect, is the use of the angular accelerometer[6] to obtain the second derivative of θ_C. As in Section 11.2, we assume the availability of perfect derivative signals so as to illustrate the main effects with a minimum of extraneous details. When derivatives of the controlled variable are used, they are usually fed back (negatively) to a summing point ahead of the final control element, so that its output will include the derivative effects. In our case, the op-amp preamplifier is easily changed to provide a multiple-summing input that accepts the tach and accelerometer output signals, giving the block diagram of Fig. 11.7 (τ_f is again neglected).

Working directly form the block diagram

$$\left\{[(\theta_V - \theta_C)K_\theta K_d - K_{d1}s\theta_C - K_{d2}s^2\theta_C]K_cK_t + \frac{T_U}{n}\right\}\frac{K_L}{s(\tau_L s + 1)} = \theta_C \quad (11.18)$$

$$[(\tau_L + K_{d2}K_cK_tK_L)s^2 + (1 + K_{d1}K_cK_tK_L)s + K_\theta K_d K_cK_tK_L]\theta_C$$
$$= (K_\theta K_d K_cK_tK_L)\theta_V + \frac{K_L}{n}T_U \quad (11.19)$$

Just as in error-derivative control, K_{d1} and K_{d2} (which can again be either positive or negative) give us complete design freedom in manipulating the characteristic equation. Comparing with Eq. 11.2, we note that the absence of derivative

---

[6]Ibid., pp. 317–328.

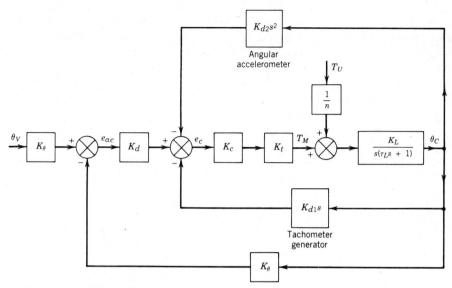

FIGURE 11.7 *Controlled-variable-derivative control.*

terms of θ_V now makes the response to steps of θ_V and T_U identical in form to that for T_U in Eq. 11.2. To define loop gain and prepare for the use of root locus or Bode/Nichols design methods, the minor loops in Fig. 11.7 must first be reduced. That is, we take $T_U \equiv 0$ and find the relation $(\theta_C/e_{ac})(s)$.

$$(K_d e_{ac} - K_{d1}s\theta_C - K_{d2}s^2\theta_C)\frac{K_c K_t K_L}{s(\tau_L s + 1)} = \theta_C \tag{11.20}$$

$$\frac{\theta_C}{\theta_E}(s) = K_\theta \frac{\theta_C}{e_{ac}}(s) = \frac{K}{s(\tau s + 1)} \tag{11.21}$$

where

$$\theta_E \triangleq \theta_V - \theta_C \qquad \tau \triangleq \frac{\tau_L + K_{d2}K_c K_t K_L}{1 + K_{d1}K_c K_t K_L} \qquad K \triangleq \text{loop gain} \triangleq \frac{K_\theta K_d K_c K_t K_L}{1 + K_{d1}K_c K_t K_L}$$

We can now redraw Fig. 11.7 as 11.8 and use standard frequency response or root locus techniques.

In comparing derivative control applied to system error with that applied to the controlled variable, we see that the effects on the characteristic equation are identical. Error-derivative control, however, has a more violent response

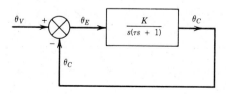

FIGURE 11.8 *Servo with minor loops reduced.*

to sudden command changes (such as steps), so it would be less favored if the application required such commands. There will also be differences in cost, weight/space, reliability and the like because the hardware used to implement the two schemes is usually quite different. Tradeoff studies of technical–economic factors must thus be carried out for each specific application in order to decide which approach in best.

Just as we saw in error-derivative control, the presence of additional dynamic terms can change the effect of controlled-variable-derivative modes. Rather than including τ_f (as we did in error-derivative control), let us keep $\tau_f = 0$ but include a time constant between e_c and T_M (example: motor armature L/R time constant τ_m). Proceeding as in Eq. 11.20

$$(K_d K_\theta \theta_E - K_{d1}s\theta_C - K_{d2}s^2\theta_C)\frac{K_c K_t K_L}{s(\tau_m s + 1)(\tau_L s + 1)} = \theta_C \tag{11.22}$$

$$\frac{\theta_C}{\theta_E}(s) = \frac{K}{s\left(\dfrac{s^2}{\omega_n^2} + \dfrac{2\zeta s}{\omega_n} + 1\right)} \tag{11.23}$$

where

$$K \triangleq \frac{K_d K_\theta K_c K_t K_L}{1 + K_{d1}K_c K_t K_L} \qquad \omega_n \triangleq \sqrt{\frac{1 + K_{d1}K_c K_t K_L}{\tau_m \tau_L}} \tag{11.24}$$

$$\zeta \triangleq \frac{\tau_m + \tau_L + K_{d2}K_c K_t K_L}{2\sqrt{(1 + K_{d1}K_c K_t K_L)(\tau_m \tau_L)}} \tag{11.25}$$

The closed-loop system is now third order with root locus as in Fig. 11.9. Although the ω_n and ζ of eq. 11.24 and 11.25 are the open-loop values, the closed-loop values are similarly influenced by K_{d1} and K_{d2}. We note that now an increase in ω_n requires a larger K_{d1} (which reduces ζ) and larger K_{d2} gives a larger ζ. The effects of the derivative control modes are thus still very useful but they are *opposite* to those in Eq. 11.19 where $\tau_m = 0.0$.

11.4 VEHICLE STABILITY AUGMENTATION

One of the most common applications of derivative control, usually implemented on the controlled variable, involves the improvement of vehicle handling dynamics by feedback means. The concept was probably originally devised for aircraft but is now used on all types of vehicles. Aircraft attitude (angular orientation) control involves three axes (pitch, roll, and yaw) of motion, however preliminary design usually considers each axis separately. A simplified Newton's law analysis[7] for the pitch axis produces the aircraft transfer function relating pitch angle θ_C to elevator angle δ_h.

---

[7]E. O. Doebelin, *Dynamic Analysis and Feedback Control*, McGraw-Hill, New York 1962, pp. 318–323.

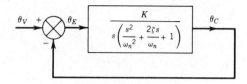

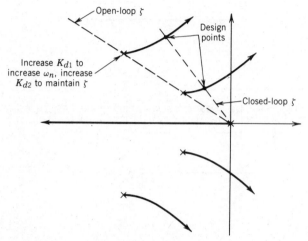

FIGURE 11.9 *Derivative control used to increase speed and maintain stability.*

$$\frac{\theta_C}{\delta_h}(s) = \frac{K_a(\tau_a s + 1)}{s\left(\dfrac{s^2}{\omega_{na}^2} + \dfrac{2\zeta_a s}{\omega_{na}} + 1\right)} \tag{11.26}$$

The numerical values of parameters K_a, τ_a, ω_{na}, and ζ_a vary from one airplane or missile to another, and for a given vehicle, with altitude and Mach number. Aerodynamic design for drag reduction often results in very poor (low) values of aircraft damping ζ_a. Use of derivative control concepts allows us to radically improve this poor stability *without changing the airplane in any way,* thus preserving the desirable low drag. Feedback control is thus a powerful tool in vehicle design since it allows the simultaneous fulfillment of conflicting performance requirements.

A basic pitch-command control system would force the aircraft pitch angle θ_C to follow the pilot's control stick rotation θ_V, using a position gyroscope[8] to measure θ_C as in Fig. 11.10. The electrohydraulic servo that powerfully positions the elevator in response to voltage command e_a is itself a complete feedback system; however, when properly designed, its closed-loop response can be adequately modeled as first order. We see from the root locus that this basic system

---

[8]Doebelin, *Measurement Systems,* pp. 338–347.

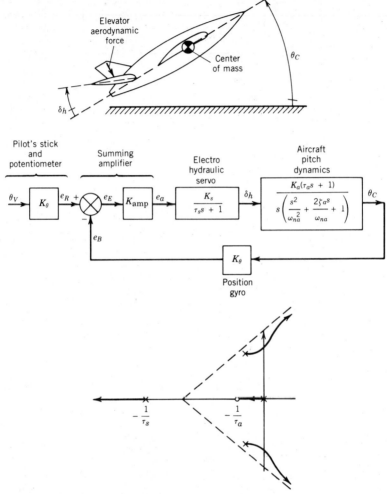

FIGURE 11.10 *Aircraft pitch angle control without stability augmentation.*

is unacceptable since the closed-loop damping is actually *worse* than that of the airplane itself, which was already bad. By adding a rate gyroscope[9] to provide measurement of $\dot{\theta}_C$ we can implement first-derivative control (see Fig. 11.11). The closed-loop system now has *better* damping than the aircraft alone, giving handling qualities acceptable to the test pilots. (This same technique is used with on–off control in the missile roll-stabilization system of Fig. 8.25).

---

[9]Ibid.

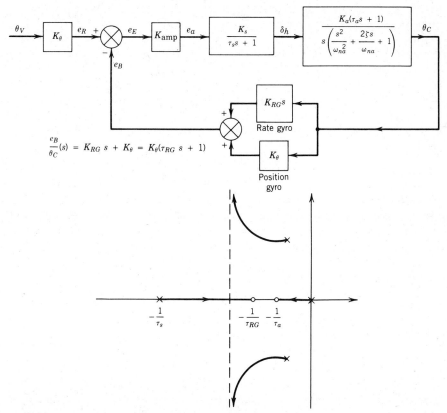

FIGURE 11.11 *Derivative control augments aircraft stability.*

Since the airplane itself has *not* been changed, the reader may find it difficult to believe that the improved damping really occurs. This "magical" aspect of derivative control can be dispelled by remembering that the elevator is a "general-purpose torque producer"; rotation of δ_h causes a proportional pitching torque on the vehicle. The derivative control simply sends a $\dot{\theta}_C$ signal to the elevator so that it produces a torque opposite and proportional to $\dot{\theta}_C$, the *same* kind of torque produced by viscous damping as the aircraft rotates through the air. Once this is clear, one can also see why this scheme will *not* work on a missile that leaves the atmosphere, since the elevator is now incapable of producing *any* torque. We can now, however, use gas jets as our torque producing device, and derivative control can again produce damped motion even though vehicle viscous damping is zero in empty space. For a vehicle (like the space shuttle) that operates both in space and in the atmosphere, a control scheme blending both aerodynamic control surfaces (elevator, rudder, etc.) and gas jets is necessary.

11.5 FEEDBACK OF INTERMEDIATE VARIABLES; STATE-VARIABLE FEEDBACK

We have now seen that useful control effects can be realized using derivatives of the E signal ("upstream" end of forward path) or the C signal ("downstream" end of forward path). Another possibility lies in feeding back "intermediate" signals in the forward path, that is, variables existing between E and C. In the system of Fig. 11.12, M_1 and M_2 are such signals. Note that since C lags M_2, then M_2 must lead C and since M_2 lags M_1, then M_1 must lead M_2, in fact

$$M_2 = \frac{sC}{K_3} \qquad M_1 = \frac{\tau_3 s + 1}{K_2}M_2 = \frac{s(\tau_3 s + 1)C}{K_2 K_3} \qquad (11.27)$$

We thus have available signals containing the first and second derivatives of C without having had to differentiate C, assuming M_1 and M_2 are measureable physical quantities. Since the forward path of many practical systems contains lagging elements, this approach to derivative control may often be possible. If we feed M_1 and M_2 back through adjustable gains K_4 and K_5 we get

$$\left[V - C - \frac{K_4}{K_2 K_3}(\tau_3 s + 1)sC - \frac{K_5}{K_3}sC \right] \frac{K_1 K_2 K_3(\tau_1 s + 1)}{s(\tau_2 s + 1)(\tau_3 s + 1)} = C \quad (11.28)$$

$$[(\tau_1\tau_2 + K_1 K_4 \tau_1\tau_3)s^3 + (\tau_2 + \tau_3 + K_1 K_4 \tau_3 + K_1 K_2 K_5 \tau_1)s^2 \qquad (11.29)$$
$$+ (1 + K_1 K_4 + K_1 K_2 K_5)s + K_1 K_2 K_3]C = K_1 K_2 K_3(\tau_1 s + 1)V$$

Typically, τ_2, τ_3, and K_2 might be fixed while τ_1, K_1, K_3, K_4, and K_5 could be chosen by the designer, with K_4 and K_5 negative if necessary. With this many degrees of design freedom, versatile adjustment of system performance is possible, subject of course to the same limitations of maximum power and saturation we encountered earlier in derivative control.

Useful intermediate variables are sometimes *hidden* when we reduce a set of simultaneous equations (used to model a piece of hardware) down to transfer functions relating only the input–output variables at the interfaces of that piece of hardware with the rest of the system. Hydraulic motion control systems are a good example. Equations 2.167 through 2.169 for a valve-controlled actuator

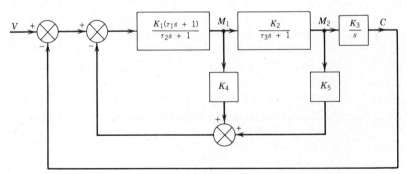

FIGURE 11.12 *Derivative control using intermediate variables.*

can be represented in full detail by Fig. 11.13a, however Fig. 11.13b would generally be employed in a motion control system design and this diagram does not show the cylinder pressures, which can be useful feedback signals, at all. Once Fig. 11.13a reminds us of the existence of p_{cl} and p_{cr}, we can use Eq. 2.169 to augment Fig. 11.13b with a feedback of the differential pressure $(p_{cl} - p_{cr})$ as in Fig. 11.14. The motivation for this is that Eq. 2.169 shows that $(p_{cl} - p_{cr})$ is directly related to the first and second derivatives of controlled variable X_C, thus we can expect some derivative control benefits from such a feedback.

$$p_{cl} - p_{cr} = \frac{M}{A_p}s^2 x_C + \frac{B}{A_p}sx_C - \frac{f_U}{A_p} \qquad (11.30)$$

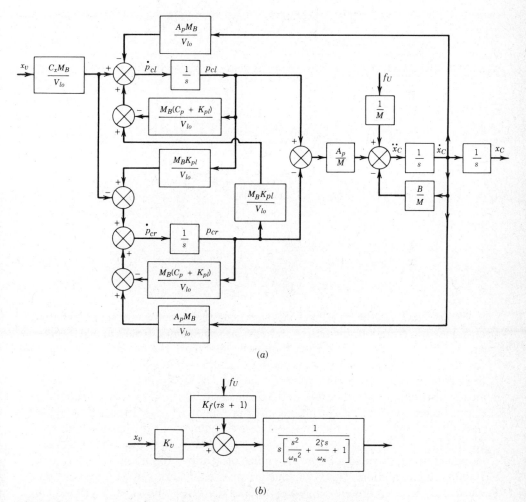

(a)

(b)

FIGURE 11.13   Hidden variables revealed by detailed block diagram.

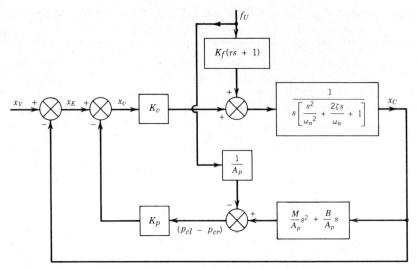

FIGURE 11.14   *Differential pressure feedback in hydraulic servo.*

Working directly from Fig. 11.14

$$\left[\left(x_V - x_C - \frac{MK_p}{A_p}s^2x_C - \frac{BK_p}{A_p}sx_C + \frac{K_p}{A_p}f_U\right)K_v \right.$$

$$\left. + K_f(\tau s + 1)f_U\right]\frac{1}{s\left[\dfrac{s^2}{\omega_n^2} + \dfrac{2\zeta s}{\omega_n} + 1\right]} = x_C \qquad (11.31)$$

and

$$\left[\frac{s^3}{\omega_n^2} + \left(\frac{2\zeta}{\omega_n} + \frac{MK_pK_v}{A_p}\right)s^2 + \left(1 + \frac{BK_pK_v}{A_p}\right)s + K_v\right]x_C$$

$$= K_vx_V + K_f(\tau s + 1)f_U + \frac{K_pK_v}{A_p}f_U \qquad (11.32)$$

We see that pressure feedback K_p augments the s^2 and s coefficients of the characteristic equation, increasing stability, as seen from a Routh analysis. Figure 11.15 shows the effect on the root locus, assuming the feedback quadratic $(M_pK_p/A_p)s^2 + (BK_p/A_p)s + 1$ has complex factors. Although pressure feedback augments stability, it has the undesirable effect of increasing the sensitivity to steady disturbing forces, because of the term $(K_pK_v/A_p)f_U$. If this is unacceptable it can be removed by processing the pressure feedback signal through a high-pass filter to "wash out" the low- and zero-frequency effects. That is, replace K_p in Fig. 11.14 by $K_p\tau_ps/(\tau_ps + 1)$. Although this changes the root locus and system closed-loop equation, proper choice of τ_p allows retention of most of the stability augmentation. This "dynamic pressure feedback" can be implemented electrically with an op-amp high-pass filter if the $(p_{cl} - p_{cr})$ signal is obtained from an electrical differential pressure sensor. A completely hydromechanical

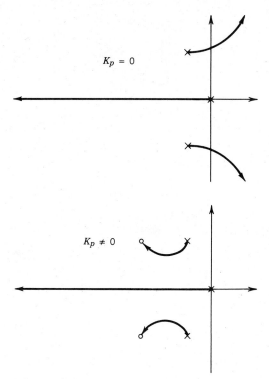

FIGURE 11.15 *Effect of pressure feedback on root locus.*

implementation can be built right into the servovalve (Fig. 11.16) and such valves are commercially available. Pressure feedback has been found particularly useful in stabilizing hydraulic servos in which actuator and/or structure compliance creates poorly damped mechanical resonances, a good example being the swiveling of huge engines used for thrust-vector control of large booster rockets.

When all the equations modeling a feedback system are diagramed in full detail (as in Fig. 11.13), those variables that appear at the outputs of all the integrators (p_{cl}, p_{cr}, $\dot{x}_c$, x_c in Fig. 11.13) are called the system *state variables*. One of the design techniques developed in "modern control theory," called state-variable feedback,[10] suggests that *all* the state variables be measured and fed back, each through its own adjustable gain. If we do this, we find that when we examine the closed-loop system characteristic equation the coefficient of each power of s includes one or more of the state-variable feedback gains in such a way that, even though unalterable system dynamics are also present, we

---

[10]S. M. Shinners, *Modern Control System Theory and Application,* 2nd Ed., Addison–Wesley, Reading, Massachusetts, 1978, p. 351.

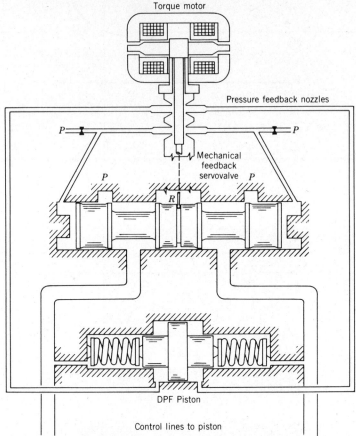

FIGURE 11.16 *Servovalve with dynamic pressure feedback.*
Source: W. J. Thayer, Transfer Functions for Moog Servovalves, *Tech. Bull. 103,* 1965; SLEW-Improves Performance of Swivel-Engine Thrust Vector Control, MR, 1962, Moog Servocontrols, East Aurora, N.Y.

have complete freedom to design the numerical value of each coefficient to be what we want. We can thus specify what type of closed-loop response we want and then adjust the various feedback gains until we get it. The reader should at this point recall that this is very reminiscent of situations encountered earlier in this chapter with various forms of derivative control. In fact, state-variable feedback is generally equivalent to controlled-variable derivative control where *all* the derivatives (up to the highest that is present in the basic system) are fed back, and is thus subject to similar power/saturation limitations if "carried to extreme."

As with most of "modern control theory," the practical use of complete state-variable feedback is rare. Since each state variable requires a sensor, this raises cost and reliability questions. One can usually meet specifications with a simpler system based on concepts from "classical" control theory. When "partial" state-variable feedback (some, not all, of the state-variables are fed back) designs are encountered in practice, we find that these are often well-known classical techniques such as derivative control, pressure feedback, and the like. Complete state variable feedback may be impossible because of inaccessibility of some of the state variables. Inaccessible process state variables may be "synthesized" (in exact or approximate form) using the *observer technique*[11] of modern control theory. Here the controller includes an adjustable math model of the process. The correct *form* of math model is *assumed* known (an important practical limitation) but numerical values of model parameters are adjusted (based on comparison of model behavior with process behavior) until the model matches the real process. This adjustment occurs on-line as the system operates, using only measurements of process input and output signals. Since the *model's* state variables *are* accessible to us, we can use these in place of the real (but inaccessible) process state variables in a state-variable-feedback control scheme.

In my opinion, the main practical utility of state-variable feedback for single input-single output systems lies in its ability to suggest candidate feedback signals that can sometimes give performance improvements similar to those of derivative control without the need to actually perform differentiation. The designer should conscientiously examine system equations and block diagrams to discover promising "intermediate" or "hidden" signals (which may or may not be state variables) and formulate feedback schemes using various combinations of these and/ or the state variables. Combining such an exploratory approach with our more definite knowledge of the basic control modes, we can come up with several alternative designs for tradeoff studies. In earlier times, analytical evaluation of such tentative and unproven design concepts would have been inhibited by considerations of time and expense, and design tended to concentrate on well-proven methods. Availability of fast, low-cost computer aided design tools such as CSMP (or other) digital simulation allows us today to be less conservative in trying out such new ideas.

11.6 IMPLEMENTATION OF DERIVATIVE CONTROL MODES

When the actuating signal is a mechanical displacement, its first derivative may be taken with the spring–damper system of Fig. 11.17a.

$$\frac{x_o}{x_i}(s) = \frac{\dfrac{B}{K_s}s}{\dfrac{B}{K_s}s + 1} = \frac{\tau s}{\tau s + 1} \tag{11.33}$$

Perfect differentiation is approached as we make τ small, however we also lose sensitivity. As mentioned earlier, the low-pass-filtered derivative $\tau s/(\tau s + 1)$

---

[11]R. Isermann, *Digital Control Systems,* Springer, New York, 1981, pp. 159–182.

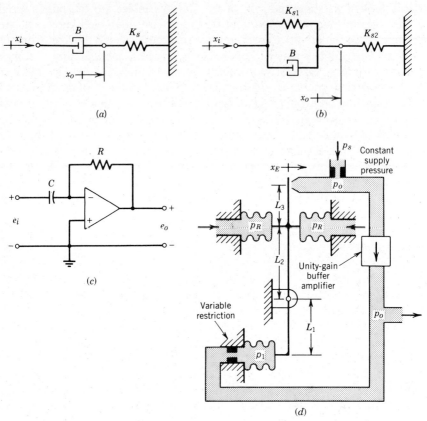

FIGURE 11.17 *Analog mechanization of derivative control.*

may actually be *preferred* in practice. A low-pass-filtered proportional-plus-derivative mode that approaches perfect proportional plus derivative as we make $K_{s2} \gg K_{s1}$ is provided by the system of Fig. 11.17b.

$$\frac{x_o}{x_i}(s) = \frac{\dfrac{K_{s1}}{K_{s1} + K_{s2}}\left(\dfrac{B}{K_{s1}}s + 1\right)}{\left(\dfrac{B}{K_{s1} + K_{s2}}s + 1\right)} = \frac{K(\tau_1 s + 1)}{\tau_2 s + 1} \qquad (11.34)$$

If a "perfect" derivative of a voltage signal is desired, the op-amp differentiation of Fig. 11.17c is available.

$$\frac{e_o}{e_i}(s) = -RCs = -\tau s \qquad (11.35)$$

Usually, low-pass-filtered differentiators are more practical; we will use them in Chapter 12. For the pneumatic proportional-plus-derivative controller of Fig.

11.17d, using assumptions similar to those used for Fig. 10.5c we can write

$$x_E = \frac{p_o}{K_a} = K_{b1} \frac{L_2 + L_3}{L_2}(p_R - p_B) - K_{b2} \frac{L_2 + L_3}{L_1} \frac{p_o}{\tau s + 1} \quad (11.36)$$

$$\frac{p_o}{p_E}(s) \approx \frac{L_1 K_{b1}}{L_2 K_{b2}}(\tau s + 1) = K(\tau s + 1) \quad (11.37)$$

Note that this controller has been set up to be direct (rather than reverse) acting. In process-control terminology τ is called the *rate time*. For a ramp input of p_E it represents the output's advance in time provided by the derivative effect as compared to pure proportional control (see Fig. 11.18). That is, a given level of output is achieved τ seconds earlier when the derivative mode is present.

In 'digital controllers the simplest version of derivative control uses the difference between the previous sampled value of actuating signal (e_{n-1}) and the current value e_n to approximate the derivative as

$$\frac{de}{dt} \approx \frac{\Delta e}{\Delta t} = \frac{e_{n-1} - e_n}{T} \quad (11.38)$$

where T is the sampling interval. Digital differentiation suffers from noisy signals just as does analog and various smoothing and filtering actions are possible. A four-point smoothing technique[12] reduces sensitivity to high-frequency noise.

$$\frac{de}{dt} \approx \frac{1}{6T}(e_n + 3e_{n-1} - 3e_{n-2} - e_{n-3}) \quad (11.39)$$

Such methods increase computer memory requirements (three previous values must be stored) and computational load, and distort the true derivative signal for rapid changes, so should be used with caution. For the ramp change of Fig. 11.19a, however, perfect results are obtained at $t = 3$.

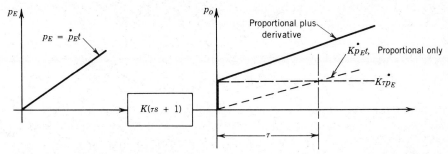

FIGURE 11.18   *Interpretation of rate time τ in proportional-plus-derivative control.*

---

[12]R. Isermann, *Digital Control Systems*, Springer, New York, 1981, p. 86.

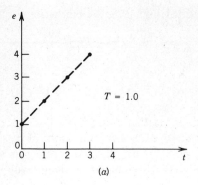

(a)

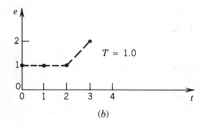

(b)

FIGURE 11.19 *Behavior of smoothed derivative algorithm.*

$$\frac{de}{dt} \approx \frac{1}{6}(4 + 9 - 6 - 1) \doteq 1.0 \qquad (11.40)$$

For a steady signal with a "noise spike" at $t = 3$ (Fig. 11.19b), Eq. 11.38 gives $de/dt \approx 1.0$, whereas the smoothed algorithm responds much less.

$$\frac{de}{dt} \approx \frac{1}{6}(2 + 3 - 3 - 1) = 0.167 \qquad (11.41)$$

Whatever algorithm is used to compute de/dt, the derivative component of control is obtained by simply multiplying de/dt by the derivative time constant (rate time) τ and the proportional gain K, as in Eq. 11.37.

11.7 PSEUDO-DERIVATIVE FEEDBACK (PDF)

Pseudoderivative feedback[13] is a control concept that takes the integral mode as the basic mode and then obtains derivative control benefits (without actually taking any derivatives) by feeding back the controlled variable downstream of the integral controller. Phelan, the inventor of this scheme, uses mainly rather simple controlled systems (K/s, K/s^2) to illustrate its behavior. In Fig. 11.20a,

[13]R. M. Phelan, *Automatic Control Systems,* Cornell Univ. Press., Ithaca, New York, 1977; F. G. Shinskey, Book Review, *Chem. Eng.*, Sept. 26, 1977.

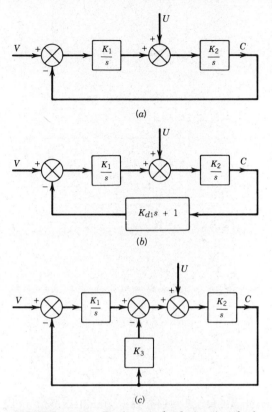

FIGURE 11.20 *Basic pseudo-derivative feed-back.*

a conventional integral control of an integral process would be absolutely unstable.

$$(D^2 + K_1K_2)C = K_1K_2V + K_2DU \qquad (11.42)$$

Addition of conventional derivative control in Fig. 11.20b provides stability.

$$(D^2 + K_1K_2K_{d1}D + K_1K_2)C = K_1K_2V + K_2DU \qquad (11.43)$$

Pseudoderivative feedback (Fig. 11.20c) gives

$$(D^2 + K_2K_3D + K_1K_2)C = K_1K_2V + K_2DU \qquad (11.44)$$

stabilizing the system without the need to take a derivative.

Application of the pseudoderivative concept to the more complicated systems of Fig. 11.21 reveals benefits here also. Since PDF is a relatively recent and little publicized idea, known practical applications are few; however the examples shown in this brief presentation show that it is worthy of consideration as a candidate control scheme, along with the more conventional methods, when one is considering alternative solutions to a control problem.

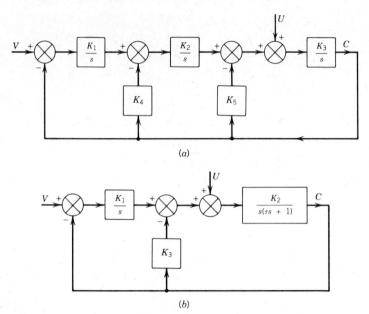

(a)

(b)

FIGURE 11.21   *PDF applied to more complex systems.*

PROBLEMS

11.1 Derive the transfer functions for the op-amp circuit and the motor/gear-train/load in the system of Fig. 11.2.

11.2 Show how the op-amp circuit of Fig. 11.2 can be modified to allow K_{d1} and K_{d2} to assume any combination of positive and negative values.

11.3 For the system of Eq. 11.2 discuss $(\theta_C/\theta_V)(i\omega)$ as:

   a. $K\rightarrow\infty$ **b.** K_{d1} and $K_{d2}\rightarrow\infty$
   Compare with a system that has $K_{d1} = K_{d2} = 0.0$.

11.4 For the system of Eq. 11.2 with $\theta_V \equiv 0.0$, T_U a step input T_{Us}, K_{d1} adjusted to keep ζ always at 0.6, $K = 10$, and $\tau_L = 0.1s$, add a torque limiter on T_M at ±10 N-m. Using digital simulation with $T_{Us} = 1$ N-m, explore system response as K_{d2} is increased from zero to beyond the point where torque saturation occurs. Choose numerical values for any unspecified parameters as needed.

11.5 For the ramp input of Eq. 11.12, show that infinite torques are required between $t = 0$ and $0+$.

11.6 For the system of Fig. 11.2, devise an alternative op-amp circuit that allows $K_{d2}s^2 + K_{d1}s + 1$ to be underdamped.

11.7 Use Routh and Nyquist criteria to study the systems of Fig. 11.5.

11.8 For the system of Eq. 11.19, can we reduce the steady-state error for a ramp command by reducing B as we did for the system of Eq. 11.2?

11.9 Choose any convenient numerical values and sketch the root locus for the system of Fig. 11.12.

11.10 Replace K_p by $K_p \tau_p s(\tau_p s + 1)$ in Fig. 11.14. Obtain the closed-loop system equation and sketch the root locus. Do a Routh stability analysis. Comment on stability and steady-state errors due to constant disturbing forces.

11.11 In Eq. 11.32 set $K_p = 0$, $\zeta = 0.05$, and choose any convenient numerical values for other system parameters. Use CSMP (or other available) digital simulation to study response to step inputs of x_V and f_U. Then explore the effect of pressure feedback by making several runs for a range of K_p values. Finally, choose a K_p value and explore the effect of τ_p when K_p in Fig. 11.14 is replaced by $K_p \tau_p s/(\tau_p s + 1)$.

11.12 Investigate the effect of inertia on the differentiator of Fig. 11.17a.

11.13 Form a double differentiator by cascading two single differentiators (Fig. 11.17a). Investigate its performance.

11.14 Reanalyze Fig. 11.17d letting K_a be finite. Compare this more correct model with that of Eq. 11.37, using frequency response methods.

11.15 Modify the system of Fig. 11.17d to give pure proportional control. Then show how such a controller can be combined with one like Fig. 11.17d to give proportional plus derivative-of-p_B control. This is sometimes used when step changes in p_R would cause unwanted large controller output in a derivative-of-p_E system.

11.16 For the signal $e = 10 \sin(2\pi t/10)$ with sampling interval $T = 1.0$, compute and tabulate the first derivative every 1.0 s, using:

  a. The exact value of the derivative (no sampling).

  b. Eq. 11.38 (assume $e \equiv 0$ for $t < 0.0$).

  c. Eq. 11.39 (assume $e \equiv 0$ for $t < 0.0$).

Repeat this study with $T = 2.0$.

11.17 Compare the root loci for the systems of Fig. 11.20a, 11.20b, and 11.20c. Compare steady-state errors for ramp inputs of V in:

  a. Fig. 11.20b.

  b. Fig. 11.20c.

  c. Fig. 11.20b with derivative control on actuating signal rather than controlled variable.

11.18 Analyze the pseudoderivative feedback system of Fig. 11.21a for stability, steady-state errors, and root locus.

11.19 Repeat problem 11.18 for Fig. 11.21b.

CHAPTER 12
Combined and Approximate Control Modes

Chapters 8 through 11 have introduced the basic control modes: on–off, proportional, integral, and derivative. Each of these has its own advantages and drawbacks, thus it should not be surprising that many practical applications are best served by some combination of basic modes. In fact, we have already encountered some examples of this. We have also, for the most part, considered the most basic or idealized versions of the modes so that their essential features could be brought out most clearly without confusing side issues. Practical versions of some controllers are not able to realize completely the ideal behavior and also may require a modified design technique. Sometimes a nonideal controller can meet specifications with simpler hardware or software. For these reasons, approximate forms of control modes should be considered.

12.1 PROPORTIONAL-PLUS-INTEGRAL CONTROL; PHASE-LAG COMPENSATION

Because it provides the steady-state-error benefits of pure integral control with faster response and improved stability behavior, proportional-plus-integral control can meet the needs of many practical processes and is the most widely used control combination. Phase-lag compensation is its approximate version realized in many practical controllers. It cannot attain the zero steady-state errors possible with perfect integral control but this is not a fatal defect because realistic error specifications *always* must allow some steady-state error.

The root loci of Fig. 12.1 make clear the stabilizing effect in proportional-plus-integral control due to the numerator term $[(K_p/K_I)s + 1]$. (Although this term, *by itself*, is a combination of proportional and first-derivative-of-error components, it would not be correct to say we are using proportional-plus-integral-plus-derivative modes since the $[(K_p/K_I)s + 1]$ term is *multiplied by*, not added to, the $1/s$ term.) Frequency response considerations would of course confirm these stability improvements since the numerator dynamics provide leading phase angle. Since the closed-loop system of Fig. 12.1b is second order, we can easily find

$$\omega_n \triangleq \sqrt{\frac{K_1 K_I}{\tau}} \qquad \zeta \triangleq \frac{1 + K_1 K_p}{2\sqrt{\tau/K_1 K_I}} \tag{12.1}$$

showing the effects of K_I and K_p on dynamic performance. For the system of Fig. 12.1d Routh criterion gives

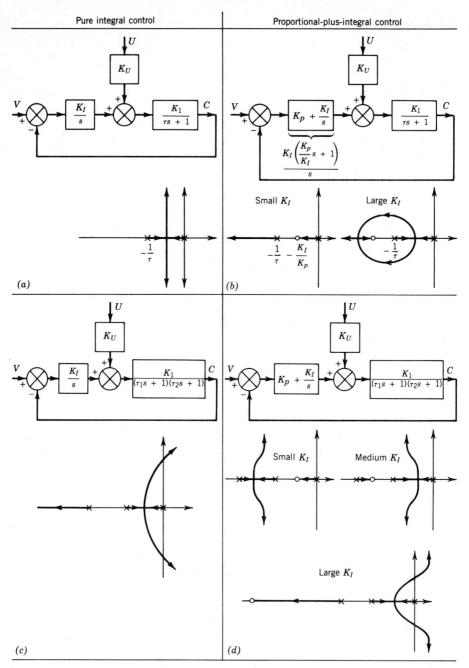

FIGURE 12.1   *Comparison of integral and proportional-plus-integral control.*

$$K_1 K_I \triangleq K_{\text{loop}} < (1 + K_1 K_p) \left(\frac{1}{\tau_1} + \frac{1}{\tau_2} \right) \text{ for stability} \qquad (12.2)$$

showing the stabilizing effect of K_p and destabilizing effect of K_I. For ramp inputs $V = \dot{V}t$ and $U = \dot{U}t$, both 12.1*b* and 12.1*d* have steady-state error.

$$E_{ss} = \frac{1 + K_1 K_p}{K_1 K_I} \dot{V} - \frac{K_U}{K_I} \dot{U} \qquad (12.3)$$

For a ramp command $\dot{V}$, using a gain margin GM in Eq. 12.2 and combining with Eq. 12.3 we get

$$E_{ss} = \frac{(\dot{V})(GM)}{\dfrac{1}{\tau_1} + \dfrac{1}{\tau_2}} \qquad (12.4)$$

If a maximum allowable E_{ss} is specified for a given $\dot{V}$, since we would rarely use $GM < 2$ and τ_1 and τ_2 would probably be given, Eq. 12.4 allows a quick check of feasibility. Whether a ramp E_{ss} is specified or not, Eq. 12.2 gives (K_1 is assumed known)

$$K_1 K_I = \frac{1 + K_1 K_p}{GM} \left(\frac{1}{\tau_1} + \frac{1}{\tau_2} \right) \qquad (12.5)$$

$$K_I = \frac{1}{K_1 GM} \left(\frac{1}{\tau_1} + \frac{1}{\tau_2} \right) + \frac{\dfrac{1}{\tau_1} + \dfrac{1}{\tau_2}}{GM} K_p \qquad (12.6)$$

$$K_I = \frac{C}{K_1} + C K_p \qquad C \triangleq \frac{\dfrac{1}{\tau_1} + \dfrac{1}{\tau_2}}{GM} \qquad (12.7)$$

To choose acceptable values of K_I and K_p we can proceed as follows. Define a range of acceptable gain margins, perhaps starting at 2, and ending at the limit given by Eq. 12.4 if an E_{ss} is specified. Starting at a low value of K_p (use zero if no other value suggests itself), pick a K_p and define, using Eq. 12.7, a corresponding range of K_I values. Using CSMP (or other) simulation, explore system response to steps of V and U for the K_I, K_p combinations defined. Repeat for several different K_p values. One hopes one or more of these combinations will satisfy system specifications. If not, assuming changes in K_1, τ_1, and/or τ_2 are possible, use Eq. 12.2 through 12.7 for guidance in selecting new values and then repeat the K_I, K_p simulation study. If all feasible combinations are explored without success, it may be time to add other control modes, such as derivative.

A CSMP simulation as described might go as follows.

```
V  = STEP (0.0)     unit step command
E  = V - C     summing junction
MP = KP*E     proportional mode
```

```
PARAM    KP  =  0.0      proportional gain, start at zero
         MI1 =  INTGRL (0.0,E) ⎫
         MI  =  KI*MI1         ⎬  integral mode
PARAM    KI =(1.0,1.25,1.67,2.5,5.0)      gain margin = 10,8,6,4,2
         M = MP + MI     proportional plus integral
         U = STEP(1.0)      delayed unit step disturbance
         U1 = KU*U
PARAM    KU = 1.0
         MC = (M+U1)*K1                    ⎫
PARAM    K1 = 5.0                          ⎬  controlled process
         C1 = REALPL(0.0,.05,MC)          ⎪
         C = REALPL(0.0,.0333,C1)         ⎭
TIMER    FINTIM = 2.0,DELT = .001,OUTDEL = .02
OUTPUT   C                   ⎫  plot five curves for five values
PAGE MERGE, GROUP           ⎬  of KI on same page, with same scale
END
PARAM    KP = 0.2                          ⎫  rerun problem with new
PARAM    KI =(2.0,2.5,3.33,5.0,10.0)       ⎬  Kp and KI set
END
PARAM    KP = 0.5                          ⎫  rerun problem with new
PARAM    KI =(3.5,4.38,5.83,8.75,17.5)     ⎬  Kp and KI set
END
/*
```

In this simulation we have taken $\tau_1 = 0.05$, $\tau_2 = 0.0333$, $K_1 = 5.0$, $K_U = 1.0$, and $E_{ss}/V \leq 0.2$, giving a maximum gain margin of 10. Figure 12.2a (pure integral control) shows that $K_I = 5.0$ is probably too lightly damped; whereas the other four values all seem to settle to the desired value at about $t = 0.8$, but with different amounts of overshoot. If the application can tolerate only small overshoot, we might choose $K_I = 1.25$, noting that response to disturbance is quite similar to that for a step command. Addition of a small amount ($K_p = 0.2$) of proportional control in Fig. 12.2b gives the expected improvement in response speed, whereas relative stability stays about the same since we have maintained the same gain margins as in 12.2a. If this response speed met specifications, $K_I = 3.33$ gives a good low-overshoot response. Going to $K_p = 0.5$ (Fig. 12.2c) does not improve the command response much but helps the transient error for the step disturbance. Note that all five command responses are now oscillatory (sometimes *without* overshooting 1.0); whereas three of the disturbance responses show negligible oscillation. A value of $K_I = 8.75$ gives good disturbance response if we can tolerate the rather large overshoot in the command response. To emphasize the cost effectiveness of this digital simulation approach to design, we note that the CSMP program was written in less than five minutes, and ran in about four seconds at a cost (1985 prices) of less than $5. The CSMP software leases for about $100/month for educational institutions.

By definition, a *phase-lag compensator* has a transfer function of form

$$\frac{q_o}{q_i}(s) = \frac{K(\tau s + 1)}{\alpha \tau s + 1} \qquad \alpha > 1 \qquad (12.8)$$

and is an approximate version of proportional-plus-integral control, as seen from the frequency response graphs of Fig. 12.3. At intermediate and high frequencies the two transfer functions are indistinguishable, but at low frequencies the lack of an integrator limits the amplitude ratio of the phase lag compensator to the finite value K. If we have conceived a basic control system, set the gain to achieve a desired relative stability, checked response speed and found it acceptable, but find steady-state errors too large, phase-lag compen-

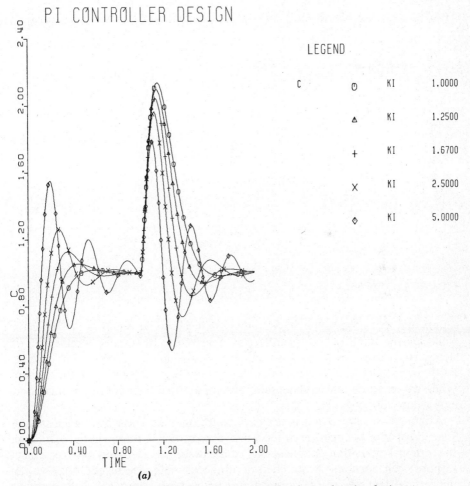

FIGURE 12.2 *Proportional-plus-integral controller design by simulation.*

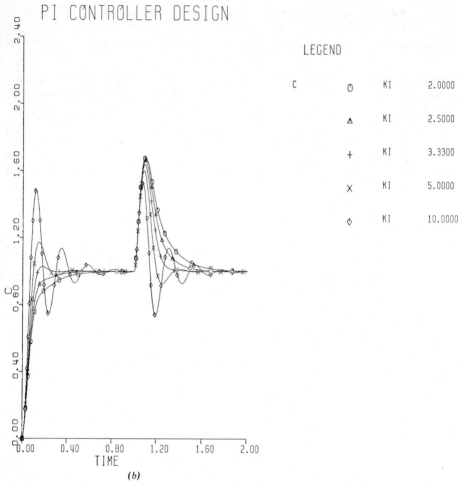

FIGURE 12.2 (Cont.)

sation may be appropriate. Generally, it will allow an increase in loop gain while leaving stability and speed largely unaffected. If we need a reduction of steady state errors by some known factor, this usually requires a gain increase by the same factor, and we should choose α in Eq. 12.8 equal to this factor. When we set gain in the compensated system we usually find it can be increased by a value close to α.

Although the choice of α is straightforward once we know how much steady-state error improvement is needed, the choice of τ is more subtle. Although the original invention of phase-lag compensation was based on frequency-response concepts and occured before root locus methods existed, today we can understand and design the compensator with either approach. Both methods will be presented since each illuminates different aspects of behavior and con-

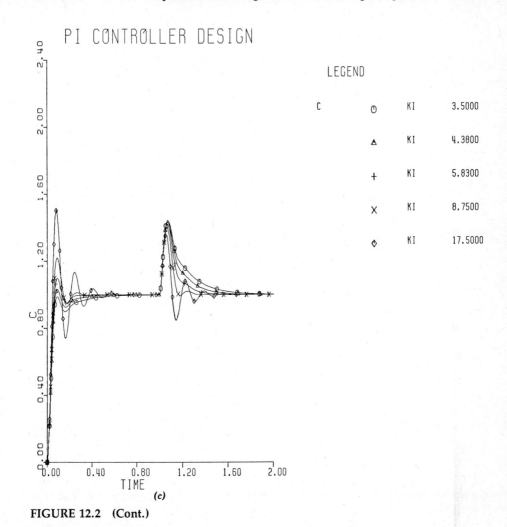

FIGURE 12.2   (Cont.)

tributes to a better overall understanding. For the frequency-response approach the direct polar plot version of the Nichols chart gives the clearest view. Recall that the Nichols chart allows us to plot logarithmic open-loop data on a rectangular grid and read out logarithmic closed-loop data on a superimposed curvelinear grid. A similar display is possible[1] using polar (nonlogarithmic) graphs.

D'Azzo and Houpis show that the contours of constant closed-loop amplitude ratio M (superimposed on a conventional polar open-loop plot of $(C/E)(i\omega)$ for

[1]J. J. D'Azzo and C. H. Houpis, "Feedback Control System Analysis and Synthesis," 2nd Ed., McGraw–Hill, New York, 1966, pp. 361–374.

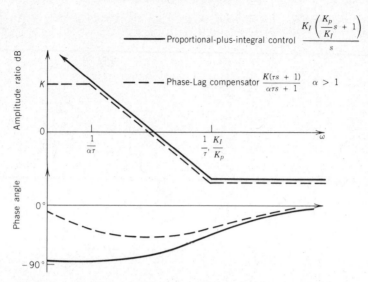

FIGURE 12.3 *Phase-lag compensator as approximate proportional-plus-integral.*

a unity-feedback system) are circles with

$$\text{center on } x \text{ axis at } x = -\frac{M^2}{M^2 - 1} \qquad (12.9)$$

$$\text{radius} = \left| \frac{M}{M^2 - 1} \right| \qquad (12.10)$$

as in Fig. 12.4, where the gain has been set to get $M_p = 1.3$. To appreciate the idea behind the phase-lag compensator, Fig. 12.5 shows a basic system whose gain has been set to get $M_p = 1.3$; but with this gain, suppose we find that steady-state errors are three times too large, suggesting phase-lag compensation with $\alpha = 3$. The polar plot for the compensator is a perfect semicircle as shown, fixed as soon as α is chosen; however the distribution of ω along this semicircle depends on τ, which we must now choose. It is chosen so that the high-frequency region (basic system closed-loop resonant frequency ω_{RB} to $\omega = \infty$) of the compensator contributes mainly an amplitude ratio $1/\alpha$ with very little phase shift. Though there is no "magic value," a phase shift of $-5°$ at ω_{RB} is a good design value, and since ω_{RB} is a known number, this fixes τ since it can be shown[2] that for a lag compensator

$$\tau^2 + \frac{\alpha - 1}{\omega \alpha \tan \phi} \tau + \frac{1}{\omega^2 \alpha} = 0 \qquad (12.11)$$

---

[2]Ibid., p. 436.

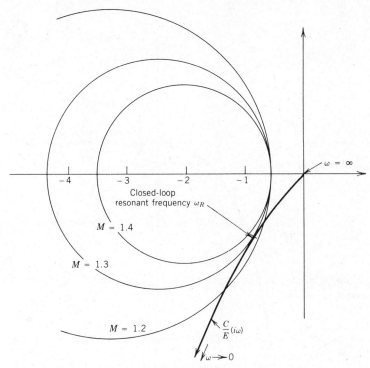

FIGURE 12.4 *Frequency-response gain setting by direct polar plot.*

With $\omega = \omega_{RB}$, $\phi = -5°$, and α chosen, this quadratic gives two values of τ of which we want the one that puts ω_{RB} at the high (rather than low) frequency end of the compensator. If ω_{RB} is, say, 0.4 and $\phi = -5°$, then $\alpha = 3$ gives $\tau = 18.9$.

To see how the compensator works, note that we multiply the amplitude ratios and add the phase angles when we combine the basic system with the compensator in Fig. 12.5. Thus, in the frequency range $\omega_{RB} \to \infty$, the compensator "shrinks" the basic system curve by the factor α but causes little shape distortion since phase is little changed (less than 5° at any ω). At lower frequencies the "compensated system, no gain change" curve approaches the basic system since the compensator approaches $1\underline{/0°}$. When we now set gain on the compensated system we can re-expand the curve by a factor of about α since this will essentially duplicate the original curve in the critical region near ω_{RB}. The new resonant frequency (and thus system speed) should be quite close to the original ω_{RB} and we again set gain for $M_P = 1.3$; so both speed and stability are close to the original system, but we have increased loop gain by about α. For frequencies below ω_{RB} the compensated system now deviates from the original curve since both amplitude ratio and phase angle of the compensator have an effect; however these changes cause no problem because they occur *away* from the critical region near ω_{RB}. That is, the new curve does *not* cut

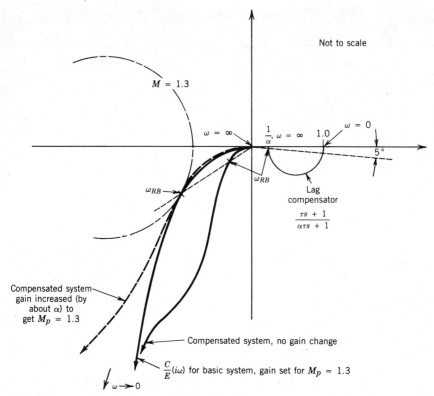

FIGURE 12.5   *Explanation of phase-lag compensator principle.*

through any closed-loop M circles with $M > 1.3$. We might also note at this point that "high-frequency attenuation compensator" might be a more descriptive name than "phase-lag" since the lagging phase, although present, is *not* the essence of the compensator's beneficial effect.

There usually is no real need to plot the curves of Fig. 12.5 in an actual design problem; that figure was for purposes of understanding what was going on. Rather, we select α from needed steady-state error improvement and τ from ϕ (take $-5°$) and ω_{RB}, which would be known from the basic system design. Final setting of α, τ, and loop gain are then efficiently done with digital simulation of the step response, which one would always check anyway, even if the compensation design had been done with frequency-response graphing. As an example, take the type-0 system designed in Fig. 9.28 and 9.29. Suppose that for loop gain of 90 ($M_p = 1.3$) we find the steady-state error for step commands to be three times too large. Since this system has $\omega_{RB} = 0.4$ rad/s and we want $\alpha = 3$ we can use $\tau = 18.9$s calculated earlier. A CSMP simulation to finalize numerical values and compare responses of the original and compensated systems follows.

```
           V  =  STEP(0.0)        start basic system
           E  =  V－C
           M1=KC*E
PARAM      KC=30.
           M2=REALPL(0.0,2.0,M1)
           M3=REALPL(0.0,.5,M2)
           U=KU*STEP(40.)
PARAM      KU=30.
           M4=(M3+U)*KP
PARAM      KP=3.0
           C=REALPL(0.0,200.,M4)     end basic system
           EC=V－CC     start compensated system
           M1C=LEDLAG(TAU1,TAU2,EC)       compensator
PARAM      TAU1=18.9,TAU2=56.7
           M2C=KCC*M1C
PARAM      KCC=90.
           M3C=REALPL(0.0,2.0,M2C)
           M4C=REALPL(0.0,.5,M3C)
           M5C=(M4C+U)*KP
           CC=REALPL(0.0,200.,M5C)       end compensated system
TIMER      FINTIM=115., DELT=.01, OUTDEL=.4
OUTPUT   C,CC
PAGE GROUP
END
```

In Fig. 12.6a this simulation shows the step command dynamics to be little affected by the compensation (as desired); however the step disturbance response appears to have a slow exponential component not clearly predicted by our design procedure, thus the need for final evaluation of preliminary designs by comprehensive simulation is again demonstrated. The predicted three-fold improvement in steady-state errors is of course achieved but is more graphically apparent in the disturbance response (why?). Variations in compensator time constants (but with α always 3) are explored in Fig. 12.6b (TAU1 = 10, TAU2 = 30) and 12.6c (TAU1 = 25, TAU2 = 75). Small changes in loop gain are also studied.

Although a special phase-lag compensation design procedure can be devised for nonunity-feedback systems, doing so is probably not worthwhile when we do the final "tuning" with CSMP. Rather we choose an α and τ in the same way as we do for unity feedback (ω_{RB} *is* known when basic system gain is set for nonunity-feedback systems) and then choose final values from step-response simulations.

For the root locus approach to phase-lag compensator design we assume the basic uncompensated system (with gain set for a desired ζ) has an open-loop transfer function:

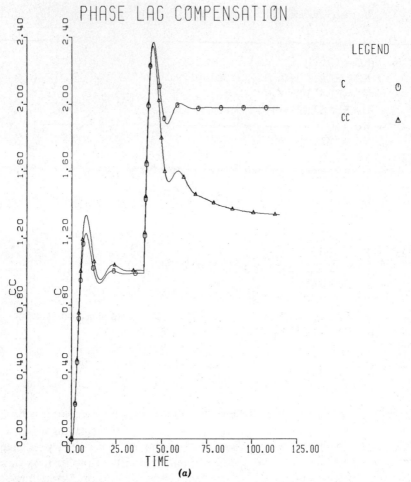

FIGURE 12.6   *Phase-lag compensation: compared with basic system, parameters varied.*

$$G_x = \frac{k(s - z_1) \cdots (s - z_w)}{(s - p_1) \cdots (s - p_v)} \tag{12.12}$$

$$\text{gain of } G_x \triangleq K_x = \frac{k(-z_1) \cdots (-z_w)}{(-p_1) \cdots (-p_v)} \tag{12.13}$$

Adding the phase-lag compensator G_c (with additional gain A), the compensated open loop has

$$G = G_c G_x = \frac{Ak}{\alpha} \left(\frac{s + \dfrac{1}{\tau}}{s + \dfrac{1}{\alpha\tau}} \right) \frac{(s - z_1) \cdots (s - z_w)}{(s - p_1) \cdots (s - p_v)} \tag{12.14}$$

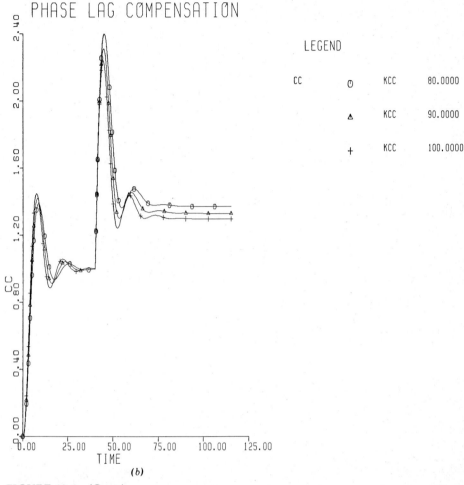

FIGURE 12.6   (Cont.)

with

$$\text{gain} = AK_x \qquad \text{static loop sensitivity} = \frac{Ak}{\alpha} \qquad (12.15)$$

The design procedure strives to:

a. In the neighborhood of the dominant roots, make the compensated root locus nearly identical to the uncompensated root locus.

b. Adjust static loop sensitivity in the compensated system to be about the same as in the uncompensated system.

c. Make negligible the effect of the new root introduced by the compensator.

If **a.** and **b.** are successful, the dominant roots of the compensated system will be nearly the same as for the uncompensated, keeping speed and stability as

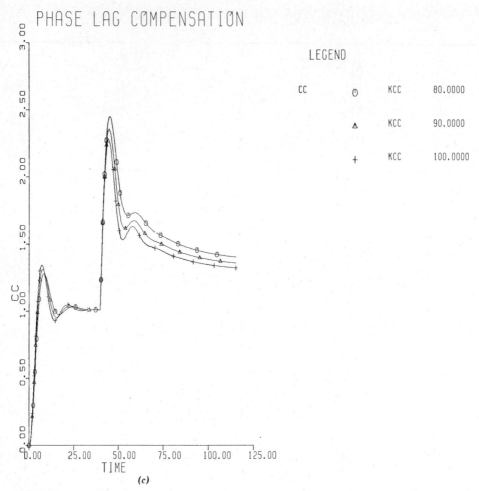

FIGURE 12.6 (Cont.)

before, whereas the added gain A will be about equal to α (Eq. 12.15). If **c.** is successful, the added root will not be dominant.

These statements are best understood in terms of an example: Let us use the same basic system as in our frequency-response design. Figure 12.7a shows the uncompensated system. Assume we need a gain increase of three times, so choose $\alpha = 3$, the same as in the frequency-response approach. Choice of τ is now, however, based on a different criterion. To attain goal **b.**, the pole/zero pair of the compensator should be "graphically close to each other." That is, when plotted to the scale of the basic system, and thinking in terms of Spirule measurements to locate the dominant portion of the root locus, the pole and zero should be close enough that their two measurements nearly cancel each other, allowing the *other* poles and zeros to locate the locus. Since α is often

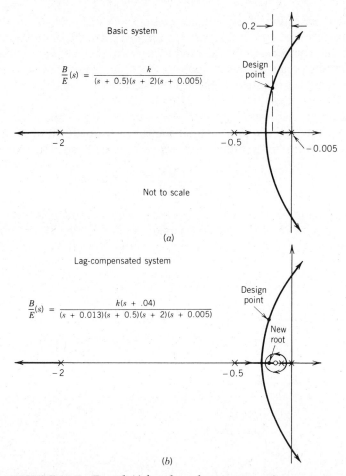

FIGURE 12.7 *Root loci for phase-lag compensation.*

as large as 10, the compensator pole and zero might be in a ratio as large as 10 to 1, however they *can* be kept "graphically close" if we choose $1/\tau$ to be "small" relative to the real part of the basic system dominant roots. Just as in the frequency-response design where we wanted a "small" phase shift at ω_{RB} and suggested $-5°$ as a reasonable value, we now need to be specific about the "smallness" of $1/\tau$. Although a range of values can give acceptable results, we will suggest using $1/\tau \approx 1/5$ of the dominant root real part, giving $\tau = 25$ for our example. (Note that our frequency-response design ($\tau = 18.9$) would also "look OK" on the root locus.) The new root is crowded in close to the zero, ensuring that its contribution will be negligible.

Just as in the frequency-response design, there is no real need to plot the graph (Fig. 12.7*b*). If the dominant roots of the basic system are known from the preliminary design, we can select τ and then finalize choice of α, τ, and gain using digital simulation of the step response. From this point of view the frequency response and root locus methods are just two different ways to understand how the compensator works, and to select trial values for α, τ, and gain as good starting points for the computer simulation. Note that the third CSMP run of our frequency response design used $\tau = 25$ and $\alpha = 3$ and thus corresponds to our root-locus design, so see Fig. 12.6c for these results.

Turning now to hardware implementation of phase-lag compensation, we first note that the proportional-plus-integral controller of Eq. 10.10 when more correctly modeled with finite (rather than infinite) nozzle-flapper gain gives

$$\frac{p_o}{p_E}(s) = -\frac{K_{b1}K_a(\tau s + 1)}{((K_aK_pK_{b2} + 1)\tau s + 1)} \qquad (12.16)$$

which is a phase-lag compensator. A passive pneumatic version is shown in Fig. 12.8a, whereas 12.8b is its electrical analog. Active electronic compensators using op-amps (Fig. 12.8c) would be largely preferred to the passive system of

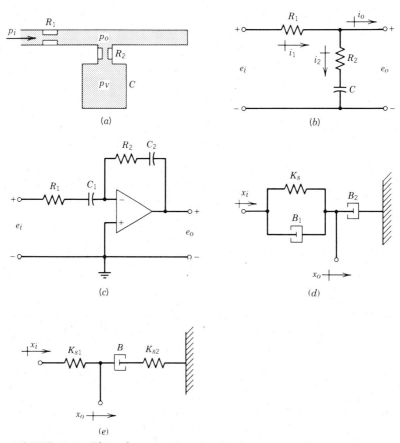

FIGURE 12.8   Phase-lag compensation hardware.

Fig. 12.8*b* today. Passive mechanical phase-lag compensators are shown in 12.8*d* and 12.8*e*. Practical pneumatic, electrical, or mechanical compensators will generally have upper limits on the values of α and τ that can be achieved. For example, the passive electrical compensator of Fig. 12.8*b* has

$$\alpha = 1 + \frac{R_1}{R_2} \qquad \tau = R_2 C \qquad (12.17)$$

Large τ requires large R_2 and/or C. Large C is physically large and heavy, which might be limiting in airborne applications. Also, large capacitors may deviate from ideal behavior because of effects such as leakage resistance, changing the transfer function. Large R_2 requires even larger R_1 since we usually want large α. To get the phase-lag transfer function we assume the current i_o is negligible relative to i_1, making $i_1 \equiv i_2$. As R_1 and R_2 get larger, they approach the input resistance of the device (often an amplifier) connected at e_o, and i_o is no longer negligible, changing the compensator transfer function. Amplifier input impedance can be increased (though not without limit) but this may raise the cost and accentuate noise problems. Pneumatic, mechanical, and active electronic compensators will suffer from similar limitations, though of course the actual numerical limits are different in each case. Although phase-lag compensation could be implemented in digital software with very large α and τ values, there is usually no real incentive to do this since the "ideal" version (proportional-plus-integral) is so easily programmed.

12.2   USE OF PHASE-LAG COMPENSATION IN DIESEL ENGINE GOVERNING

Speed control ("governing"[3] of prime movers such as steam, gas, and water turbines, steam engines, and internal combustion engines has been a major application area for feedback control from the earliest times to the present day. I here present a recent development[4] illustrating how all-mechanical lag compensation was used to improve performance of a proportional-type governor for diesel-electric generating sets. The basic proportional governor uses a fly-

---

[3]D. G. Garvey, Precision Speed Control Systems with Throttle Disturbance Forces, *ASME Paper 81-DGP-3,* 1981: M. A. Eggenberger, Basic Elements of Control Systems for Large Steam Turbine-Generators, *GET-3096E,* General Electric Co., Schenectady, N.Y., 1977; R. L. Witt, Pneumatic Load Equalizing System for Multi-Engine Applications, *ASME Paper 75-DGP-4,* 1975; L. O. Long, Governing Systems, *PMCC 73-2,* Woodward Governor Co., 1973; P. C. Callan and M. A. Eggenberger, Speed-Load Control for Reheat Turbines, *Cont. Eng.,* Jun. 1967, pp. 85–91; J. H. Bickford, How To Control Speed with Mechanical Governors, *Mach. Design,* Apr. 13, 1967, pp. 168–174; R. Oldenburger, Hydraulic Speed Governor with Major Governor Problems Solved, *ASME Paper 63-WA-15,* 1963; D. B. Welbourn, D. K. Roberts, and R. A. Fuller, Governing of Compression-Ignition Oil Engines, *Proc. Instn. of Mech. Engrs.,* Vol. 173, No. 22, 1959.
[4]S. A. Jaliwala, E. Day, and W. H. Stahl, Diesel Engine Governing with Load Feedback, *ASME Paper 81-DGP-5,* 1981 (Cummins Engine Co.).

weight speed sensor to produce a force proportional to engine speed. This force is compared with a spring force related to desired speed, the net force producing a spring deflection that positions a fuel valve to increase or decrease the engine fuel flow so as to keep speed near the desired value. When loop gain is set for adequate relative stability, this proportional control system typically exhibits 10% or more "droop." That is, the steady-state speed error for a load change from no-load to full load is about 10% of the desired speed. Sophisticated hydraulic or electromechanical governors with better performance have been available, but only at high cost. These are called isochronous governors and use integral control. The system to be described achieves about a 400% (4 to 1) reduction in steady-state error using simple and inexpensive mechanical phase-lag hardware.

The concept for governor improvement arose from recognition that the air pressure of the engine turbocharger is proportional to engine load. When load increases, the conventional governor allows a considerable drop in speed because its low loop gain does not open the fuel valve sufficiently. The new governor adds a turbocharger pressure sensor that causes an additional fuel valve opening proportional to pressure (load) and thus brings the speed closer to the set point. This, in effect, amounts to a higher loop gain, but because the turbocharger pressure has a significant dynamic lag, this portion of the feedback is delayed and stability is maintained. Although not apparent in the original form, this scheme turns out upon mathematical analysis to be a lag compensator.

In Fig. 12.9 the basic proportional governor would lack the p_c pressure sensor and the L_1/L_2 linkage. The right end (x_c) of the governor spring K_g would be positioned manually by a desired-speed screw adjustment. In the modified governor, x_c is automatically positioned by the p_c signal and desired speed is again manually set with a screw arrangement as shown. By making the pressure sensor spring K_p about 20 times stiffer than K_g, we force x_c to be determined almost entirely by p_c.

$$x_c \approx \frac{A_p}{K_p R_L} p_c \qquad (12.18)$$

To find the valve motion x_v we neglect inertia and friction, thus the flyweight force can be set equal to the governor spring force.

$$K_g(x_c - x_v) = K_x x_v + K_\omega \omega_C \qquad (12.19)$$

Actually the flyweight force is a nonlinear function of x_v and ω_C; Eq. 12.19 uses a linearized version with coefficients K_x and K_ω found by experiment.

$$x_v = \frac{K_g}{K_g + K_x} x_c - \frac{K_\omega}{K_g + K_x} \omega_C \qquad (12.20)$$

The relation between fuel valve opening x_v and pressure p_r (called "rail" pressure), for assumed constant supply pressure, has been found to be first order.

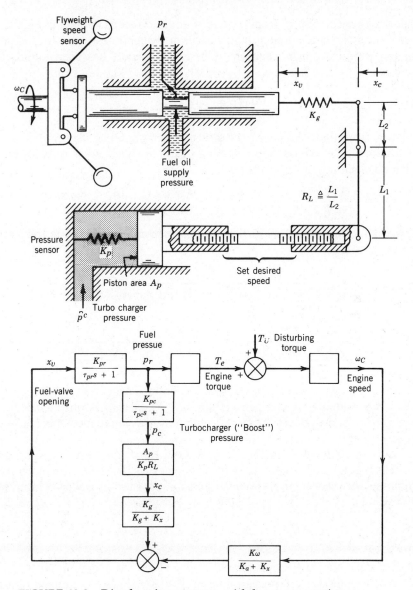

FIGURE 12.9   *Diesel engine governor with lag compensation.*

$$\frac{p_r}{x_v}(s) = \frac{K_{pr}}{\tau_{pr}s + 1} \qquad (12.21)$$

Rail pressure p_r is the pressure supplied to the engine fuel injectors to vary the fuel flow rate and thus engine power. Changing p_r also causes a change in turbocharger air pressure p_c because of engine torque and speed changes. Although the physical "path" from p_r to p_c is not as direct as implied in Fig. 12.9, one can certainly instrument p_r and p_c, run a dynamic test, and thus experimentally determine a p_c/p_r relation. A first-order model has been found adequate.

$$\frac{p_c}{p_r}(s) = \frac{K_{pc}}{\tau_{pc}s + 1} \qquad (12.22)$$

Working from these equations (or directly from the block diagram), we now want to remove the minor loop involving the p_c feedback so as to get a direct relation between ω_C and p_r. This leads to

$$\{(K_g + K_x)K_pR_L\tau_{pc}\tau_{pr}s^2 + (K_g + K_x)K_pR_L\tau_{pc}s + [(K_g + K_x)K_pR_L \qquad (12.23)$$
$$- K_gK_{pc}K_{pr}A_p]\}p_r = - K_{pr}K_\omega K_pR_L(\tau_{pc}s + 1)\omega_C$$

Since τ_{pc} is typically about 40 τ_{pr} we can approximate $(\tau_{pc} + \tau_{pr})$ in the s term below as τ_{pc} to get

$$\frac{p_r}{\omega_C}(s) = -\frac{K_{cr}(\tau_{pc}s + 1)}{(\tau_{pr}s + 1)(\alpha\tau_{pc}s + 1)} \qquad (12.24)$$

where

$$K_{cr} \triangleq \frac{K_\omega}{\dfrac{K_g + K_x}{K_{pr}} - \dfrac{K_gK_{pc}A_p}{K_pR_L}} \qquad \alpha \triangleq \frac{1}{1 - \dfrac{K_{pr}K_{pc}K_gA_p}{K_pR_L(K_g + K_x)}} \qquad (12.25)$$

Since without the p_c feedback, p_r/ω_C would be of form $K/(\tau_{pr}s + 1)$, the modified governor has added a lag compensator ($\alpha > 1.0$, always). Adjustment of α to a proper value in the governor mechanism involves a simple mechanical change in L_1 to vary R_L.

To study the complete system, we now need to obtain the engine and load dynamics. Frequency-response testing of engines of this type has shown that the relation between engine torque T_e and rail pressure p_r is well represented by a dead time

$$\frac{T_e}{p_r}(s) = K_e e^{-\tau_{dt}s} \qquad (12.26)$$

where the dead time can be estimated as 2.4 times the engine firing interval. An electrical generator can present unusual load dynamics. Assuming the presence of a fast-acting voltage regulator and a fixed electrical load (say a fixed resistor), an increase in engine speed results in a *decrease* in generator torque T_g. This occurs because the voltage regulator keeps voltage constant, and with a fixed load (say a fixed resistor), current will also not change, thus electrical power is constant. With electrical power fixed, mechanical power input $T_g\omega_C$ is also fixed, thus if ω_C increases, T_g must decrease. This speed/torque relation, $T_g\omega_C = $ constant, is nonlinear (hyperbolic) so we linearize it at the operating point to get

$$T_g = - K_{gt}\omega_C \qquad K_{gt} > 0.0 \qquad (12.27)$$

Actually we will let K_{gt} represent *all* the speed-dependent load torques (including windage and friction). The coefficient K_{gt} remains positive since the generator torque effect outweighs the frictional effects. Newton's law then gives

$$T_U + T_e + K_{gt}\omega_C = J\dot{\omega}_C \qquad (12.28)$$

$$(-\tau_L s + 1)\omega_C = -\frac{1}{K_{gt}}T_e - \frac{1}{K_{gt}}T_U \qquad \tau_L \triangleq \frac{J}{K_{gt}} \qquad (12.29)$$

where T_U represents any external disturbing torque. We can now show the complete system block diagram as in Fig. 12.10.

For a typical 300 kW generating set running at a design speed of 1800 rpm Jaliwala et al., gives numerical values.

$$\tau_{pc} = 1.0s \qquad \tau_{pr} = 0.025s \qquad \alpha = 3.33$$

$$\tau_L = 3.33s \qquad K_{gt} = 0.3 \text{ ft-lb}_f/\text{rpm} \qquad \tau_{dt} = 0.0267s$$

When the lag compensation is not used, $(\tau_{pc}s + 1)/(\alpha\tau_{pc}s + 1)$ is not present and a loop gain of -17 is found to give acceptable relative stability. Addition of the lag compensator allows loop gain to be raised to -70. Note that the negative loop gain causes ω_C to *exceed* ω_V in steady state for a constant ω_V.

$$(\omega_V - \omega_C)K_{\text{loop}} = \omega_C$$

$$\omega_C = \frac{K_{\text{loop}}}{1 + K_{\text{loop}}}\omega_V = 1.06\omega_V \qquad K_{\text{loop}} = -17 \quad (12.30)$$

$$= 1.015\omega_V \qquad K_{\text{loop}} = -70$$

For steady disturbing torques T_U with compensator present

$$(21\omega_E + T_U)(-3.33) = \omega_C = -\omega_E \qquad \omega_E = -0.048\ T_U \frac{\text{rpm}}{\text{ft-lb}_f} \qquad (12.31)$$

Negative loop gain and an *unstable* open loop ($N_p = 1$) give the unusual Nyquist diagram of Fig. 12.11, with compensated and uncompensated systems having similar (but not identical) shapes. Conditional stability is present for both cases, thus loop gain either too high or too low causes absolute instability. A CSMP simulation for a step change in T_U of 1 ft-lb$_f$ gives the results of Fig. 12.12 where *WC* is engine speed change in rpm for the basic governor (loop gain of -70) and *WC2* is the basic governor with loop gain of -17. With low loop gain this

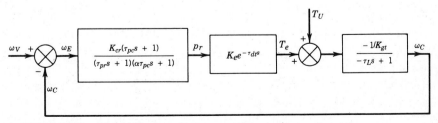

FIGURE 12.10 *Block diagram for governed engine/generator set.*

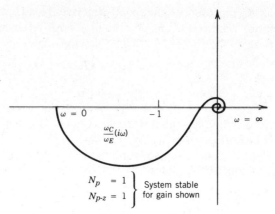

FIGURE 12.11 *Nyquist plot for governed engine/generator.*

system has adequate stability margins but unacceptable steady-state error. Higher gain corrects the steady-state error but puts the system too close to instability to be acceptable in practice. The lag compensated system (WC3) has both good stability and steady-state error. The rather large transient error and slow return to steady state do not cause any problems in this type of application.

12.3 PROPORTIONAL-PLUS-DERIVATIVE CONTROL; PHASE-LEAD COMPENSATION

Because derivative control is never used alone, we studied the proportional-plus-derivative mode at some length in Chapter 11 and thus here concentrate on the approximate version, called *phase-lead compensation* and given by

$$\frac{q_o}{q_i}(s) = \frac{K(\tau s + 1)}{\alpha \tau s + 1} \qquad \alpha < 1 \qquad (12.32)$$

If a basic system has had its gain set for desired relative stability and we then find that its response speed is too slow, lead compensation may be helpful. Also, if a basic system is structurally unstable (gain setting does *not* provide stability), lead compensation may stabilize the system. Usually, lead compensation also provides a modest gain increase, so steady-state errors are reduced whether this was a problem or not.

Although the principle of lag compensation was perhaps most clearly seen in the frequency response, why lead compensation works is most apparent from the root locus, especially when we use cancellation as our design principle, so we examine it first. In Fig. 12.13a our basic system is the same type-0 system used for the lag compensation example, but we now assume that when gain is set for desired stability we find response speed three times too slow. In choosing

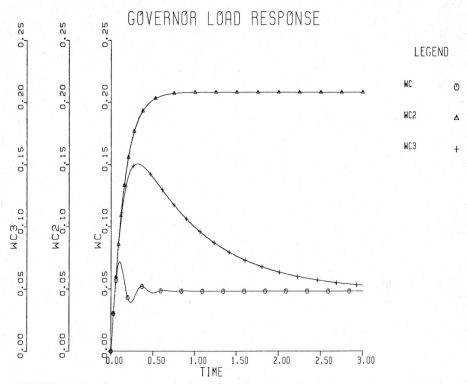

FIGURE 12.12   *Governor response with and without lag compensator.*

τ and α for the lead compensator, we find that a cancellation principle wherein we cancel the longest basic system time constant (in the case of type 1 or higher systems) or the second-longest (in type-0 systems) works about as well as more complicated approaches. Choosing τ is thus quick and easy, however α usually requires some trial and error in order to realize the needed speedup (three to one in our example). In contrast with the lag compensation, where *any* needed reduction in steady-state error can be *theoretically* achieved simply by using a large α (the limitation on α in real systems is a hardware problem), in the lead compensator the speedup achieved is *not* determined only by α but depends on the remaining (uncancelled) basic system dynamics.

In our example, choice of α can be based on the graphical approach of Fig. 12.13*a*, which shows roughly where the new dominant roots need to be to achieve the desired speedup and maintain the same closed-loop ζ. To place the new pole $-1/\alpha\tau$ we could use graphical (Spirule) trial-and-error until we got the new locus to go through the new design point; however we would still want to check the step response, so why not use simulation to do all this at once? If, in Fig. 12.13*a* we assume the new breakaway point is about half way between the origin and $-1/\alpha\tau$, choosing $-1/\alpha\tau \approx -1.6$ looks like it will put the new locus about where we want it, giving a trial value of $\alpha = 0.31$ and Fig. 12.13*b*. We

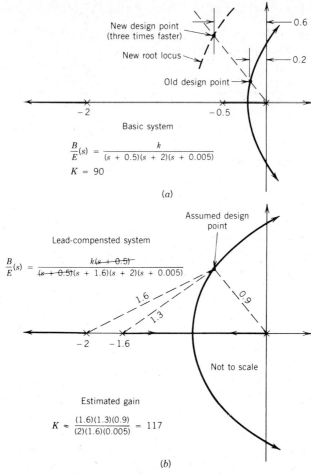

FIGURE 12.13   *Root loci for lead compensator.*

can estimate the new gain quickly from root locus measurements as shown, using the *assumed* new design point. Our simulation might then go as follows.

```
          V  = STEP (0.0)      start basic system (step command)
          E  = V-C
          M1 = KC*E
PARAM     KC = 30.
          M2 = REALPL (0.0,2.0,M1)
          M3 = REALPL (0.0,.5,M2)
          M3L = LIMIT (-500., 500,M3)     valve saturation
          U  = KU*STEP (20.)     delayed step disturbance
PARAM     KU = 30.
          M4 = (M3L+U)*KP
```

```
PARAM    KP  =  3.0
         C   =  REALPL (0.0,200.,M4)     end basic system
         EC  =  V - CC     start compensated system
         M1C =  KCC*EC
PARAM    KCC =  39.
         M2C =  REALPL (0.0,.5, M1C)
         M3C =  REALPL (0.0,TAUC, M2C)
         M3CL = LIMIT (-500., 500., M3C)     valve saturation
PARAM    TAUC = 0.63
         M4C =  (M3CL+U)*KP
         CC  =  REALPL (0.0, 200., M4C)     end compensated system
TIMER FINTIM = 40., DELT = .01, OUTDEL = .4
OUTPUT C,CC
PAGE GROUP
```

This simulation includes valve saturation (needed for a later study) but the limits are set so high (± 500) that no saturation occurs in our present runs. Figure 12.14a shows that we have achieved significant speedup; whether it is three to one depends on the criterion. Using a $\pm 5\%$ settling time, we get almost three to one. The disturbance response shows the predicted small improvement in steady-state error.

Suppose our specifications had required a 10-to-1 speedup. We still want to cancel the pole at -0.5 so $\tau = 2.0$, and we want to decrease α, but as $-1/\alpha\tau$ moves to the left of the basic system pole at -2, *this* pole (rather than $-1/\alpha\tau$) determines the dominant part of the root locus, so making α very small (even zero, which gives perfect proportional plus derivative) does not give the desired speedup. We see this limitation in Fig. 12.14b (KCC = 100) where moving $-1/\alpha\tau$ from -10 (TAUC = 0.1) to -50 (TAUC = 0.02) has almost no effect on response. We have reached the limits of this design approach. Clearly the pole at -2 is now the problem, so we might add a *second* lead compensator to cancel it (I have seen as many as *three* used successfully in practice). Cancelling the poles at -0.5 and -2.0 with two lead compensators, if we make their α's small enough, we can theoretically move the dominant locus as far to the left as we please. The limiting case of this procedure is of course a perfect proportional-plus-first-and-second-derivative mode (α's $\equiv 0.0$), which makes the closed-loop system a first-order one for which high loop gain gives as small a time constant as we wish. How closely we can approach this ideal (how small we can make the α's) in a real system is limited by noise levels and saturation. Noise problems are difficult to predict until operating hardware is available, however saturation limitations can usually be studied at the design stage, using simulation.

Our earlier CSMP simulation included valve limiting at the wide-open and shut positions so we can use this (modified for *two* lead compensators) to study how small we can make the α's (how fast we can make the system) before saturation limits response speed. We take $\tau_1 = 0.5$ and $\tau_2 = 2.0$ to achieve the

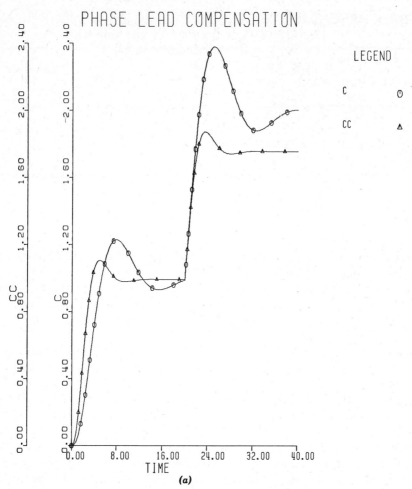

FIGURE 12.14 *Step command and disturbance responses for basic and lead-compensated systems; parameter variations.*

double cancellation, take the two α's equal (not necessary, but a reasonable simplification), and then explore a range of α's. Saturation of course depends also on the type and size of command and/or disturbance, so if worst cases for these cannot be precisely defined, the simulation cannot be totally realistic; however some useful information can usually be obtained. In Fig. 12.15a we have $\alpha\tau_1 = .1$ KTAU and $\alpha\tau_2 = .4$ KTAU, so the three cases shown there use $\alpha = 0.1, 0.04,$ and 0.02, there is no valve saturation and KCC = 400. The fastest of these three systems settles in about 0.5s with no overshoot, better than a 40-to-1 speed improvement over the basic system of Fig. 12.14a. This speed will not be realized, however, if saturation occurs in the signal M3C, whose peak value is given by the simulation as 300 for the unit step command. In Figure

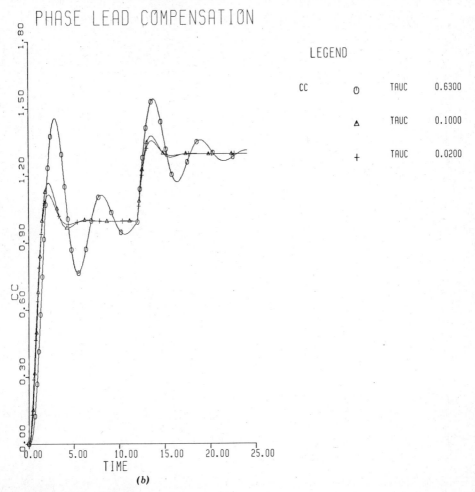

FIGURE 12.14 (Cont.)

12.15*b* (valve saturates at ±40 units of M3C) we see how saturation reduces response speed.

To show the frequency-response interpretation of phase-lead compensation we again use the direct polar plots, as we did for phase lag in Fig. 12.5. This time the name (phase lead) *does* correspond to the operating principle, as we shall see in Fig. 12.16. The basic system has had its gain set for $M_p = 1.3$ and its resonant frequency ω_{RB} is too small (system too slow) to meet our speed specification, say by a factor of three, so our design goal is to raise ω_{RB} by a factor of three without compromising stability or steady-state errors. Frequency-response considerations do *not*, for most people, suggest the cancellation type of design which was more or less obvious on the root locus, so we will not use this design principle. Rather, we will choose the compensator's τ and α such

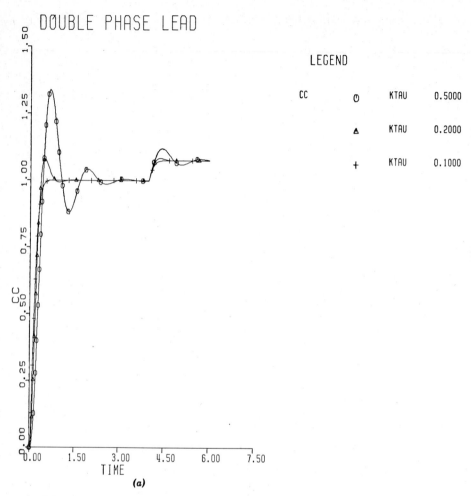

FIGURE 12.15 *Double lead compensation, with and without saturation.*

that we get sufficient phase lead that when we reset the gain for $M_p = 1.3$, the tangency will occur at a new resonant frequency ω_{RC} that will be three times larger than ω_{RB}. Just as in the root-locus approach, this will require some trial and error, which we accomplish with CSMP simulation once starting values are estimated from available theory.

From Fig. 12.16 it is clear that a phase-lead compensator has a maximum phase angle that depends on α, smaller α's giving larger maximum angles and $\alpha = 0$, the limiting case, giving 90°. It can be shown[5] that

$$\phi_{max} = \sin^{-1} \frac{1 - \alpha}{1 + \alpha} \qquad (12.33)$$

---

[5] D'Azzo and Houpis, "Feedback Control," p. 437.

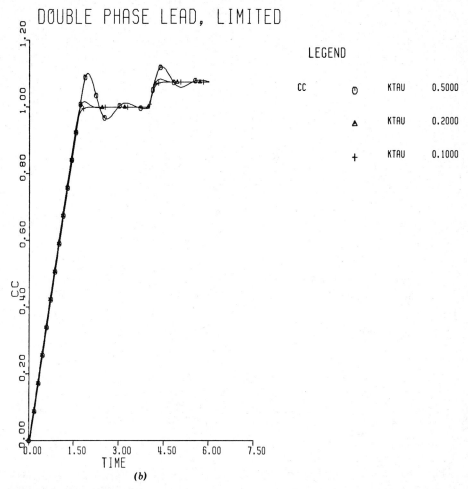

FIGURE 12.15   (Cont.)

and the frequency ω_{max} at which ϕ_{max} occurs is

$$\omega_{max} = \frac{1}{\tau\sqrt{\alpha}} \qquad (12.34)$$

One way to estimate starting values is to locate the desired ω_{RC} on the basic system open-loop curve using the known amount of speedup required. Using the same example system as we used for the phase-lag compensator we assume that a three to one speedup is wanted; since $\omega_{RB} = 0.4$, $\omega_{RC} = 1.2$. At $\omega = 1.2$ on the basic system curve, $\phi \approx -190°$, whereas $\phi \approx -140°$ at ω_{RB}. Our design concept is to use the compensator's leading phase angle to rotate the basic system curve counterclockwise until ω_{RC} is found on the same radial line as was ω_{RB}. In our present example this requires 50° phase lead. Note that

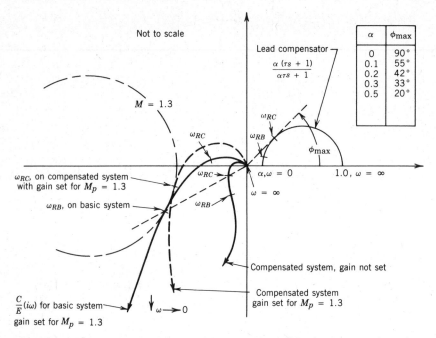

Not to scale

Lead compensator

$$\frac{\alpha\,(\tau s\,+\,1)}{\alpha\tau s\,+\,1}$$

α	ϕ_{max}
0	90°
0.1	55°
0.2	42°
0.3	33°
0.5	20°

$M\,=\,1.3$

ω_{RC}

ω_{RC} ω_{RB}

ω_{RC}

ϕ_{max}

ω_{RC}, on compensated system
with gain set for $M_p\,=\,1.3$

ω_{RC}

$\alpha, \omega\,=\,0$ $1.0, \omega\,=\,\infty$

ω_{RB}, on basic system

ω_{RB}

$\omega\,=\,\infty$

Compensated system, gain not set

Compensated system
gain set for $M_p\,=\,1.3$

$\dfrac{C}{E}(i\omega)$ for basic system

$\omega \longrightarrow 0$

gain set for $M_p\,=\,1.3$

FIGURE 12.16   *Frequency-response interpretation of lead compensation.*

the compensated system curve (with the original gain) also "shrinks" since the compensator amplitude ratio is less than 1.0. When we "reexpand" the curve (set the gain) of the compensated system until a tangency with $M_p\,=\,1.3$ is again achieved, the point of tangency must be at a higher frequency than ω_{RB}, giving a faster system. Since our criteria for choosing α and τ are rough, some adjustment will be needed to achieve the new desired resonant frequency.

For our numerical example, we need 50° of phase lead at $\omega\,=\,1.2$, which requires $\alpha\,=\,0.13$ and $\tau\,=\,2.3$, somewhat different from the starting values produced by the root-locus method using cancellation. The estimation of a trial gain value here is more difficult than for the phase-lag compensator since the compensator curve for $\omega_{RB} < \omega < \infty$ is not simply "shrunk" but undergoes both magnitude and phase changes. Since CSMP simulation studies a range of gains so easily, and since experience shows that lead compensation usually allows a modest gain increase, we will not even try to use frequency-response graphics to estimate a gain but rather we will just use CSMP step responses to study a gain range near the uncompensated value (KCC = 40, 60, 80). Figure 12.17 shows that the frequency-response design with KCC = 60 is perhaps a little faster than the first root-locus design of Fig. 12.14a, but is almost identical with the two low-α designs of 12.14b.

Turning to the hardware implementation of phase-lead compensation, a more correct modeling (finite nozzle-flapper gain) of the proportional-plus-derivative pneumatic controller of Fig. 11.17d shows it to be a phase-lead device. Although

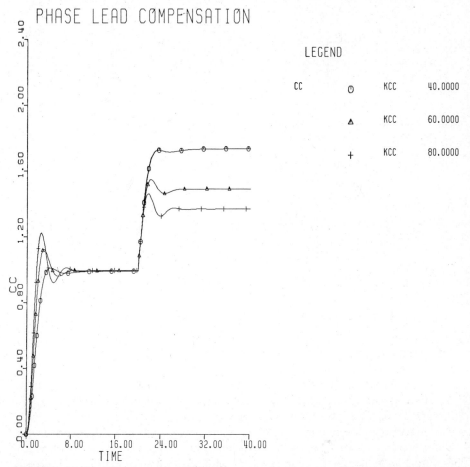

FIGURE 12.17 *Step command/disturbance response of phase-lead compensator designed by frequency-response method.*

the passive phase-lead circuit of Fig. 12.18*a* is different from the phase-lag of Fig. 12.8*b*, the active circuits (Fig. 12.8*c*, Fig. 12.18*b*) have the same form, numerical values determining whether a specific example is a lag or a lead. Spring–damper networks can be used two ways to produce phase-lead behavior, as seen in Fig. 12.18*c* and 12.18*d*. The simplest digital software, capable of producing either lag or lead behavior, approximates the derivatives with differences, giving

$$\tau_1 \frac{(m_{n-1} - m_n)}{T} + m_n = \tau_2 \frac{(e_{n-1} - e_n)}{T} + e_n \qquad (12.35)$$

where *e* is the controller input, *m* is its output, and *T* is the sampling interval between the previous sampled value ($n - 1$) and the present one (n).

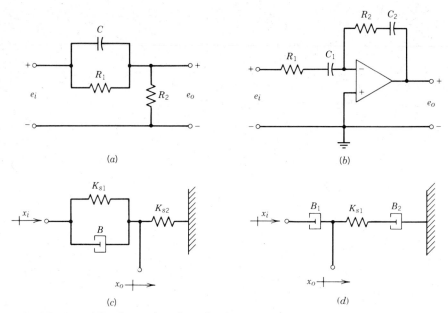

FIGURE 12.18 *Hardware for phase-lead compensator.*

In addition to the speed-improvement examples emphasized so far, phase-lead compensation is also effective in stability augmentation. A good example is the synchro/digital converter of Section 10.5. To achieve small (theoretically zero) steady-state error for constant-velocity rotation this device uses double integral (type 2) control in its basic design. An open-loop transfer function of K/s^2 makes the closed loop marginally stable for all gains since the second-order characteristic equation lacks the first-derivative term, giving the root locus of Fig. 12.19a. A phase-lead compensator with *any* values of τ and α will provide absolute stability in this situation, however the synchro/digital converter requires good relative stability and a certain speed of response. These requirements were met with $\tau = 0.0068$ s, $\alpha = 0.18$, and loop gain of 55,000, giving the root locus of Fig. 12.19b.

12.4 PROPORTIONAL-PLUS-INTEGRAL-PLUS-DERIVATIVE (PID) CONTROL; LAG/LEAD COMPENSATION

This combination of basic control modes can improve "all" aspects (stability, speed, steady-state errors) of system performance and is the most complex method available as an off-the-shelf general-purpose controller. This has been true for about 60 years. The "staying power" of PID control over the years and around the world is testimony to the basic soundness of these fundamental control actions. If we look at analog pneumatic and electronic controllers, their

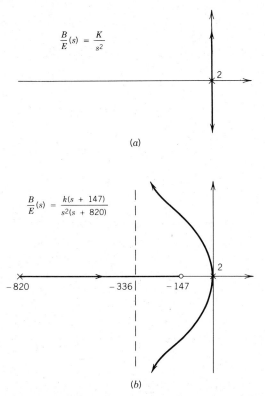

$$\frac{B}{E}(s) = \frac{K}{s^2}$$

(a)

$$\frac{B}{E}(s) = \frac{k(s + 147)}{s^2(s + 820)}$$

$-820 \qquad -336 \qquad -147$

(b)

FIGURE 12.19 *Phase-lead compensation used to stabilize type-2 system.*

microprocessor-based digital versions, or the individual control loops implemented in a large general-purpose digital process computer, over and over again we see successful applications of P, PI, PD, and PID controls. When we introduced each of the three modes we explained its utility in relatively simple qualitative fashion. This simplicity is perhaps the basis of the strength of the PID modes; they "make sense." It is also possible to justify them on more theoretical grounds.[6] In any case, although new and useful control laws will arise from time to time (as they have in the past), the PID modes will undoubtedly continue to play a central role in the future.

The ideal form of the PID controller (lag-lead compensation is its approximation) is given by

$$\frac{q_o}{q_i}(s) = K_p + \frac{K_I}{s} + K_d s \tag{12.36}$$

---

[6]T. J. Pemberton, PID: The Logical Control Algorithm, *Cont. Eng.,* May 1972, pp. 64–67, Jul., 1972, pp. 61–63.

$$= K_p + \frac{1}{\tau_I s} + \tau_d s \qquad \tau_I \triangleq \frac{1}{K_I} \qquad \tau_d \triangleq K_d \qquad (12.37)$$

Although this "summation" form of the equation displays each separate mode most clearly, to use our standard root-locus or frequency-response methods we require the form

$$\frac{q_o}{q_i}(s) = \frac{K_I\left(\dfrac{K_d}{K_I}s^2 + \dfrac{K_p}{K_I}s + 1\right)}{s} \qquad (12.38)$$

which shows that the controller introduces two zeros (which could be real or a complex pair) and a pole at the origin. If all desired values of $K_p, K_I,$ and K_d can be realized, we can place the zeros as we please for cancellation and/or noncancellation types of compensation, Fig. 12.20 showing some examples.

In Chapter 9 we studied an electrohydraulic speed-control system (Fig. 9.35) and found that with proportional control only, good performance could not be attained. We now apply PID control to a similar system to show the improvement. Our present example system, Fig. 12.21, uses pump control (rather than the valve control of Fig. 9.35) but since the transfer functions for pump and valve control have the same form, our analysis actually applies to both types of systems. Our choice of pump control is based on a desire to show as many different examples as we can and on the availability of a commercial PID controller[7] specifically designed for this application in heavy-duty construction machinery. In addition to conventional PID control, a feedforward mode useful for reducing the effects of pump speed changes is provided. The pump is driven by power takeoff from the diesel engine used to drive the construction machine, and thus engine speed is not constant and pump speed varies, causing transient disturbances to our hydraulic servo system. This can be seen from the pump-controlled actuator transfer functions (Eq. 2.154)

$$\omega_m(s) = \frac{K_\phi \phi_i(s) + K_T(\tau s + 1)T_l(s)}{\dfrac{s^2}{\omega_n^2} + \dfrac{2\zeta s}{\omega_n} + 1} \qquad (12.39)$$

where the effect of pump speed ω_p (usually taken as fixed) shows up in K_ϕ.

$$K_\phi = \frac{\omega_p d_p d_m}{d_m^2 + K_{lpm}B} = K_{\omega p}\omega_p \qquad (12.40)$$

If we consider the engine (pump) speed changes as *known* functions of time, this gives our differential equation a time-varying coefficient K_ϕ, making it analytically unsolvable; however CSMP has no trouble with this.

Since the feedback system will itself correct for disturbances such as pump speed changes, we should think of the feedforward mode as a means of improving performance beyond what feedback alone can provide or as a means of relieving

---

[7]PID Controller MCE 100A, B, *Publ. 95-8968,* Sundstrand Mobile Controls, Minneapolis, Minn., 1982.

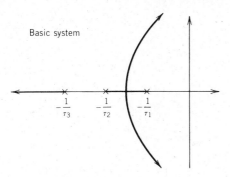

Basic system

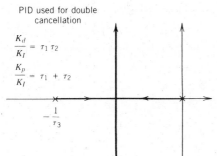

PID used for double cancellation

$$\frac{K_d}{K_I} = \tau_1\,\tau_2$$

$$\frac{K_p}{K_I} = \tau_1 + \tau_2$$

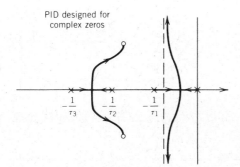

PID designed for complex zeros

FIGURE 12.20 *Effect of PID control on a three-pole, type-0 system.*

the demands on the feedback system. To design the feedforward compensation for pump speed changes, we first assume steady operation at a design point where $\phi_i = \phi_{i0}$, $T_U = T_{U0}$, $\omega_V = \omega_C = \omega_{V0}$, and $\omega_p = \omega_{p0}$. If ω_p now changes from ω_{p0} we want the feedforward system to maintain ω_C at ω_{V0} without help from the feedback action. In the initial steady state at the design point

$$M = \phi_{i0}K_{\omega p}\omega_{p0} \tag{12.41}$$

When ω_p changes to $\omega_{p0} + \omega_{pp}$ we want M to *not* change, since then ω_C would stay fixed. Under the new conditions, M is given by

$$M = (\phi_{i0} - K_{ff}K_s\omega_{pp})K_{\omega p}(\omega_{p0} + \omega_{pp}) \tag{12.42}$$

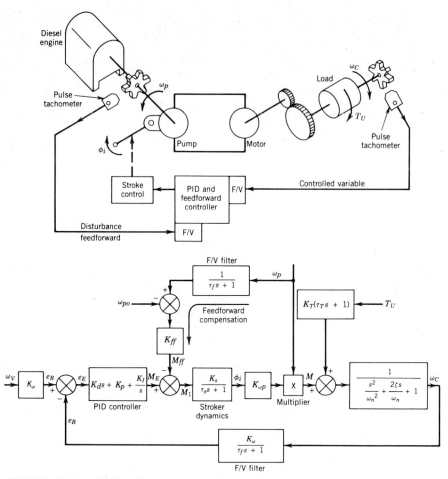

FIGURE 12.21   *PID/feedforward control applied to hydraulic speed-control system.*

Setting Eq. 12.41 and 12.42 equal leads to

$$K_{ff} = \frac{\phi_{i0}}{K_s \omega_p} \qquad (12.43)$$

We see that although ϕ_{i0} and K_s are constants, ω_p is not, and thus K_{ff} is not a constant. This (nonlinear) relation might have been anticipated since the effect of disturbance ω_p is multiplicative, not additive like T_U. It is technically possible to implement Eq. 12.43 to get "exact" compensation or we might opt for a simpler, less expensive linear approximation where K_{ff} *is* made constant at $\phi_{i0}/K_s \omega_{p0}$. This latter approach may be quite practical since the *feedback* action will trim out any remaining errors due to ω_p changes anyway. In our simulation study it is easy to compare these two (or other conceivable) alternatives.

System hardware includes pulse tachometers for shaft speed measurement, the pulse rates being converted to proportional voltages in frequency-to-voltage converters that are part of the controller. The F/V converters introduce first-order lags in their low-pass filters. PID control is realized using analog (op-amp) electronics with each mode separately adjustable. Controller output is sent to the pump stroke control. This may itself be a complete valve-controlled electrohydraulic servo. We model its closed-loop dynamics as first order.

Some analytical work will be helpful in preparing for the simulation. To get some useful results quickly we neglect the feedforward compensation, disturbing torque T_U, and F/V filter dynamics, and try a PI rather than PID controller. This gives

$$\frac{e_B}{e_E}(s) = \frac{K_I K_s K_\phi K_\omega \left(\dfrac{K_p}{K_I} s + 1\right)}{s(\tau_s s + 1)\left(\dfrac{s^2}{\omega_n^2} + \dfrac{2\zeta s}{\omega_n} + 1\right)} \tag{12.44}$$

the root locus of Fig. 12.22, and the characteristic equation

$$\frac{\tau_s}{\omega_n^2} s^4 + \left(\frac{1}{\omega_n^2} + \frac{2\zeta\tau_s}{\omega_n}\right)s^3 + \left(\frac{2\zeta}{\omega_n} + \tau_s\right)s^2 + \left(\frac{KK_p}{K_I} + 1\right)s + K = 0 \tag{12.45}$$

Application of Routh criterion produces two inequalities. The first is, for stability,

$$\frac{KK_p}{K_I} < \frac{2\zeta}{\tau_s\omega_n} + 4\zeta^2 + 2\zeta\omega_n\tau_s \tag{12.46}$$

and the second is too complicated to be of much use. Suppose we have $\omega_n = 30$ rad/s, $\zeta = 0.3$, and $\tau_s = 0.1$ s. Then $KK_p/K_I < 2.36$, and if we try $K_p/K_I = 0.05$ and a gain margin of two, we get a trial value of loop gain $K = 23.6$. Although the second Routh inequality is too complicated to evaluate in letter

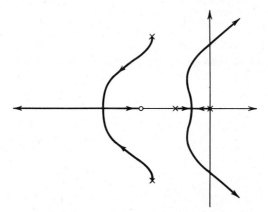

FIGURE 12.22 *Root locus for PI control.*

form, it is easy to check numerically now that we have numbers. We find no sign changes, so our trial values should be stable; however we are not sure what the gain margin really is (why?). We can now program our simulation, providing for any desired features in Fig. 12.21, but activating only those presently of interest.

```
              WV  = WVO+WVP*STEP(0.1) ⎫  Initial steady state
PARAM         WVO = 100., WVP = 1.0   ⎬  plus step command
              EE  = KW*(WV-WC)        ⎭
PARAM         KW  = .02
              MP  = KP*EE             ⎫  proportional mode
PARAM         KP  = .05               ⎭
              M1I = INTGRL (.0833, EE) ⎫
              MI  = KI*M1I             ⎬  integral mode
PARAM         KI  = 1.0                ⎭
              M1D = TRANSF(1,B,1,A,EE)        ⎫
STORAGE       B(2),A(2)                       ⎪  approximate
TABLE         B(1-2)=.001, 1.0,A(1-2)=1.0,0.0 ⎬  derivative
              MD  = KD*M1D                     ⎪  mode
PARAM         KD  = (0.0,.002,.02)            ⎭
              ME=MP+MI+MD      combine the three modes
              WP=WPO+KWR*(RAMP(TW1)-RAMP(TW2)) ⎫  pump speed
PARAM         WPO=100., TW1=.1,TW2=.2,KWR=0.0  ⎭  changes
              MFF=KFF*(WP-WPO)  ⎫  feedforward compensation, disabled
PARAM         KFF=0.0           ⎭  (KFF = 0)
              M1=ME-MFF
              MPU-KS*M1
PARAM         KS=6.0                            ⎫
              PHI=REALPL(PHIO,TAUS,MPU)         ⎬  pump stroker dynamics
PARAM         PHIO=0.5,TAUS=0.1                 ⎭
              M=PHI*KWP*WP
PARAM         KWP=2.0
              WC1=CMPXPL(.1111, 0.0, .3,30.,M)
              WC=900.*WC1
TIMER         FINTIM=0.7, DELT = .001, OUTDEL = .01
OUTPUT        WC,ME,PHI,WP,M
```

The approximate derivative mode in this simulation uses the CSMP module TRANSF, which we have not encountered before. It allows simulation of a general transfer function

$$\frac{q_o}{q_i}(s) = \frac{A_M s^M + A_{M-1} s^{M-1} + \cdots + A_1 s + A_{M+1}}{B_N s^N + B_{N-1} s^{N-1} + \cdots + B_1 s + B_{N+1}} \qquad M \le N \qquad (12.47)$$

using the STORAGE and TABLE statements to give numerical values of the A's and B's. We use it here for

$$\frac{M1D}{EE}(s) = \frac{s}{0.001s + 1} \qquad (12.48)$$

The time constant 0.001 in this filtered differentiator may need to be adjusted once we get some results. In Fig. 12.23 we see that our initial choice of loop gain, with KD = 0, gives a system that is just barely stable. Rather than reduce the gain, which *would* improve stability, we choose to retain this gain value and obtain better stability by adding the derivative mode (KD = 0.002). This value of KD was found by trial and error to give a good response that settled rapidly without excessive overshoot and oscillation. Note that *more* derivative control (KD = 0.02) gives an initially faster rise but has too much overshoot and oscillation, thus it is undesirable.

Note that we have achieved good response to commands by using all three of the PID modes. We now study the effect of the feedforward mode on pump-

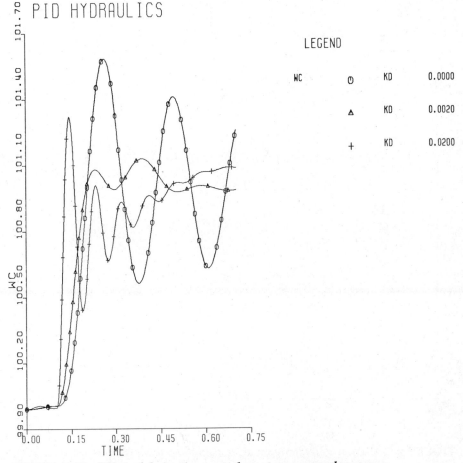

FIGURE 12.23 *Effect of derivative control on step-command response.*

speed disturbances. The statement for pump speed WP had KWR = 0.0; we now set this equal to 100 to cause pump speed to ramp from 100 at t = 0.1 to 110 at t = 0.2. Instead of KFF = 0.0, we now program

$$KFF = (.083333/WP)*Z$$
$$PARAM\ Z = (0.0,\ 1.0,\ 2.0,\ 5.0)$$

For Z = 0.0, the feedforward mode is disabled, but recall that the pump-speed disturbance *will* be corrected by the feedback action (Fig. 12.24a, Z = 0.0). For Z = 1.0, Eq. 12.43 is implemented and we see that the peak speed error caused by the disturbance is significantly reduced. It is not, however, made zero because Eq. 12.43 gives *perfect* compensation only if the pump stroke change

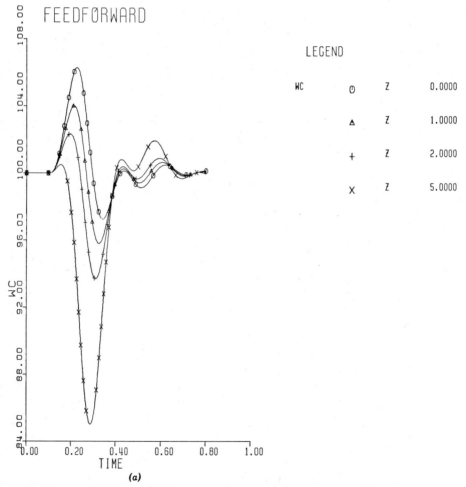

FIGURE 12.24 *Effect of feedforward on response to pump speed disturbance.*

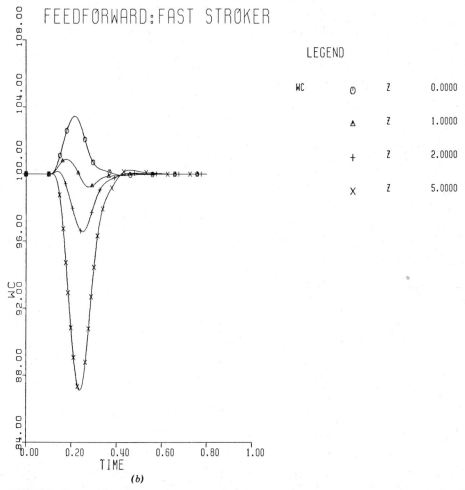

FIGURE 12.24   (Cont.)

commanded by the feedforward is instantly realized. Our pump stroker has a 0.1-s time constant so its correction will be late; whereas the disturbing effect of ω_p on pump flow is instantaneous. Making K_{FF} larger (Z = 2.00, 5.00) does not help; it reduces the positive peak error but greatly increases the negative peak error. What *would* be effective is a decrease in τ_s and/or a lead compensator in the feedforward path, so that pump stroke response to ω_p is speeded up. Figure 12.24b shows the beneficial effect of a faster pump stroker ($\tau_s = 0.02$), Z = 1.0 now giving excellent rejection of the disturbance. Finally, the division by a time-varying ω_p in Eq. 12.43 needlessly complicates the system hardware, since taking ω_p *constant* at ω_{po} gives almost identical response. That is, replacing a quantity that varies from 100 to 110, by a constant 100 will be hardly noticed, especially since we do *not* depend solely on feedforward. The feedback action will trim out any errors left by imperfect feedforward. The commercial controller (Sundstrand) cited earlier uses a constant (adjustable) K_{FF}.

The approximate version of PID control implemented in many practical controllers is called *lag/lead compensation*. Mathematically it is exactly a cascading of the phase-lag and phase-lead controllers discussed earlier.

$$\frac{q_o}{q_i}(s) = \left(\frac{\tau_g s + 1}{\alpha_g \tau_g s + 1}\right)\left(\frac{\tau_d s + 1}{\alpha_d \tau_d s + 1}\right) \qquad \alpha_g > 1 \qquad \alpha_d < 1 \qquad (12.49)$$

The effects on system performance are also a superposition of the two separate effects, thus a lag/lead controller can improve "all" aspects of performance (as can a PID): stability, speed, and steady-state errors. Selection of τ_g, τ_d, α_g, and α_d require no new discussion; we essentially design the two compensators separately, using techniques explained earlier.

Hardware for lag/lead compensation can be a cascading of lag hardware and lead hardware but one can often find equipment specifically designed for the combined modes that may be preferable in terms of cost, simplicity, and the like. The passive electrical network of Fig. 12.25a has[8]

$$\frac{e_o}{e_i}(s) \approx \frac{(R_2 C_2 s + 1)(R_1 C_1 s + 1)}{\left(\dfrac{R_2 C_2}{\alpha} s + 1\right)(\alpha R_1 C_1 s + 1)} \qquad (12.50)$$

where

$$\alpha \triangleq \frac{R_2}{R_1 + R_2} \qquad \frac{(R_1 + R_2)C_1}{R_2 C_2} \gg 1.0 \qquad (12.51)$$

which lacks versatility since different α's cannot be used for the lag and the lead. The cascaded active networks of Fig. 12.25b allow complete freedom in the choice of τ's and α's and can also provide gain if needed. Mechanical compensation is not as common as pneumatic or electronic compensation but the spring and damper system of Fig. 12.25c is the analog of 12.25a and has

$$\frac{X_o}{X_i}(s) \approx \frac{\left(\dfrac{B_1}{K_1} s + 1\right)\left(\dfrac{B_2}{K_2} s + 1\right)}{\left(\dfrac{B_1}{K_1 \alpha} s + 1\right)\left(\dfrac{B_2}{K_2} \alpha s + 1\right)} \qquad (12.52)$$

where

$$\alpha \triangleq \frac{B_1}{B_1 + B_2} \qquad \frac{(B_1 + B_2)K_2}{B_2 K_1} \gg 1.0 \qquad (12.53)$$

---

[8]J. E. Gibson and F. B. Tuteur, *Control System Components*, McGraw-Hill, New York, 1958, p. 17.

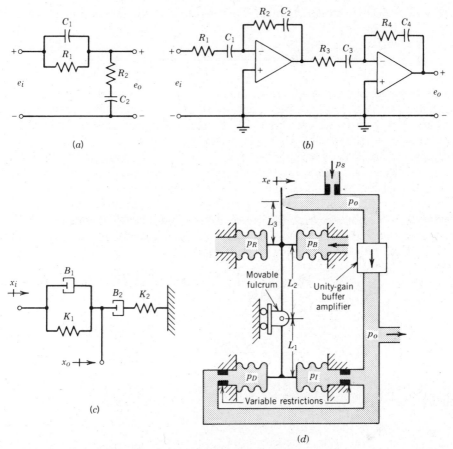

FIGURE 12.25 *Hardware for lag–lead compensation.*

Pneumatic PID controllers always use approximate integral and derivative modes and are thus really lag/lead compensators. Although design details vary with the manufacturer, Fig. 12.25d illustrates the general operation principles of most units. Since pneumatic controllers are quite common we will analyze this device in some detail.

Using the same assumptions as in earlier pneumatic controller studies we can write

$$X_e = K_{b1}\frac{L_2 + L_3}{L_1}(p_R - p_B) + K_{b2}\frac{L_2 + L_3}{L_1}(p_I - p_D) \quad (12.54)$$

$$\frac{p_D}{p_o}(s) = \frac{1}{\tau_D s + 1} \qquad \frac{p_I}{p_o}(s) = \frac{1}{\tau_I s + 1} \qquad X_e = \frac{p_o}{K_a} \quad (12.55)$$

$$\frac{p_o}{K_a} = K_{b1}\frac{L_2 + L_3}{L_1}p_E + K_{b2}\frac{L_2 + L_3}{L_1}\left(\frac{p_o}{\tau_I s + 1} - \frac{p_o}{\tau_D s + 1}\right) \quad (12.56)$$

A perfect PID controller requires $K_a = \infty$, making the left side of Eq. 12.56 zero and leading to

$$\frac{p_o}{p_E}(s) = \frac{K_{b1}L_1}{K_{b2}L_2}\left(\frac{\tau_D\tau_I}{\tau_I - \tau_D}s + \frac{\tau_D + \tau_I}{\tau_I - \tau_D} + \frac{1}{\tau_I - \tau_D}\frac{1}{s}\right) \tag{12.57}$$

The movable fulcrum allows adjustment of L_1/L_2, and needle-valve flow restrictions set τ_I and τ_D. A wide-open valve makes a $\tau \approx 0$; whereas a shut valve makes $\tau = \infty$; thus we can "turn off" modes we do not want as follows. For either $\tau_I = 0$ or $\tau_D = 0$ we get PI control, for $\tau_I = \infty$ or $\tau_D = \infty$ we get PD control, whereas $\tau_D = 0$, $\tau_I = \infty$, or $\tau_D = \infty$, $\tau_I = 0$ gives pure proportional control. Actually, there are not really two ways to accomplish the various modes since one of the two choices makes the controller itself unstable.

Such controllers are potentially unstable because they are *themselves* feedback systems, as we now see from Fig. 12.26. To see that this PID controller is really a lag/lead approximation and to discover the stability problem, we model the nozzle-flapper more correctly with finite gain.

$$K_2 p_E + K_1\left(\frac{p_o}{\tau_I s + 1} - \frac{p_o}{\tau_D s + 1}\right) = \frac{p_o}{K_a} \tag{12.58}$$

and

$$\frac{p_o}{p_E}(s) = \frac{K_2 K_a(\tau_D s + 1)(\tau_I s + 1)}{\tau_D \tau_I s^2 + [\tau_D + \tau_I + K_1 K_a(\tau_I - \tau_D)]s + 1} \tag{12.59}$$

Second-order models rarely predict instability but this one does since, for stability,

$$\tau_D + \tau_I + K_1 K_a(\tau_I - \tau_D) > 0 \tag{12.60}$$

thus

$$\frac{\tau_I}{\tau_D} > \frac{K_1 K_a - 1}{K_1 K_a + 1} \approx 1 \qquad \text{since } K_1 K_a \gg 1 \tag{12.61}$$

We now see that, for stability, $\tau_I > \tau_D$, and since half of the "choices" listed earlier for control mode selection violate this there really is no choice. A useful physical interpretation of $\tau_I > \tau_D$ can be seen in Fig. 12.26 where we note a *positive* ("unstable") feedback path from p_o to X_{fb} and a negative ("stable") path from p_o to X_{fb}. As long as $\tau_I > \tau_D$, the positive feedback is more delayed and attenuated than the negative, which thus "overpowers" it to maintain stability.

To show in what sense the lag/lead compensator of Eq. 12.59 is an approximation to the exact PID of Eq. 12.57 we compare their frequency responses by writing the PID as

$$\frac{p_o}{p_E}(i\omega) = \frac{K_{b1}L_1}{K_{b2}L_2(\tau_I - \tau_D)}\left[\frac{(i\omega\tau_D + 1)(i\omega\tau_I + 1)}{i\omega}\right] \tag{12.62}$$

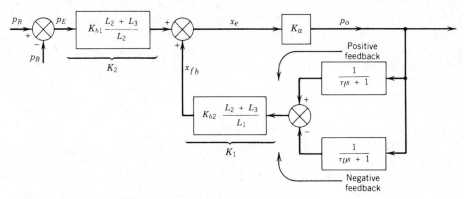

FIGURE 12.26   *Feedback nature of pneumatic controller.*

and the lag/lead as

$$\frac{p_o}{p_E}(i\omega) = K_2 K_a \frac{(i\omega\tau_D + 1)(i\omega\tau_I + 1)}{(i\omega\tau_3 + 1)(i\omega\tau_4 + 1)} \tag{12.63}$$

where $\tau_3\tau_4 \triangleq \tau_D\tau_I$ and $\tau_3 + \tau_4 \triangleq \tau_D + \tau_I + K_1 K_a(\tau_I - \tau_D)$. Since $(\tau_3 + \tau_4) >> (\tau_D + \tau_I)$, if we make, say, τ_3 very large to meet this requirement, then τ_4 must be less than τ_D to meet the requirement $\tau_3\tau_4 = \tau_D\tau_I$. Thus, $\tau_3 > \tau_I$ and $\tau_4 < \tau_D$ and we can draw Bode plots with the breakpoints in a proper sequence, as in Fig. 12.27. We see that, at intermediate frequencies, one cannot distinguish

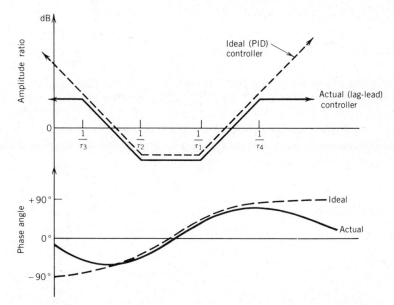

FIGURE 12.27   *Comparison of ideal and actual PID controllers.*

between the two devices; in fact if experimental frequency-response tests were run and limited to $1/\tau_3 < \omega < 1/\tau_4$, one could *not* prove that the real controller was "imperfect." Only at low frequencies does the lack of a perfect integrator cause discrepancies, as does the lack of a perfect differentiator at high frequencies. We should also note that at frequencies higher than $1/\tau_4$ a real controller's amplitude ratio must attenuate toward zero. This would require additional dynamics in our lag/lead model, perhaps considering inertia effects or pneumatic effects higher than first order.

Time-proportioning control, long popular for temperature systems, combines the simplicity and reliability of on-off control elements with the versatility and high performance of PID controllers. Here a repetitive control cycle of fixed time duration (say 10 seconds) is divided into on-time and off-time. The controller computes the sum of P, I and D control modes and adjusts the duty cycle (percent on-time) in proportion to this sum. Thus when small control effort is wanted, the duty cycle might be 1 second on, 9 seconds off, while large control effort would give 9 seconds on, 1 second off. Full scale (saturated) effort would leave the final control element on continuously. For electrical heaters, simple relays turning the AC power line on and off may be employed, the slow response of most thermal systems serving to give a smooth temperature response from the abrupt heating-rate changes. Processes without such smoothing action may not tolerate the effects produced by the on-off actions.

12.5 COMPENSATION FOR BASIC SYSTEMS WITH DOMINANT RESONANCES

We have presented lag, lead, and lag/lead compensators as being capable of providing any needed improvements in performance, as summarized in Table 12.1. The success of these designs, however, depends on the basic system dynamics being dominated by real poles. When this is not the case, the methods may fail and others are needed. Figure 12.28a shows a basic system with a pair of complex poles that are *not* dominant. Suppose we have set gain for a desired ζ and find response too slow, suggesting lead compensation. The cancellation-

TABLE 12.1 Compensator Capabilities

Compensator Type	Needed Improvement		
	Stability	Speed	Steady-State Errors
Lag			XX
Lead	XX	XX	X
Lag/lead	XX	XX	XX

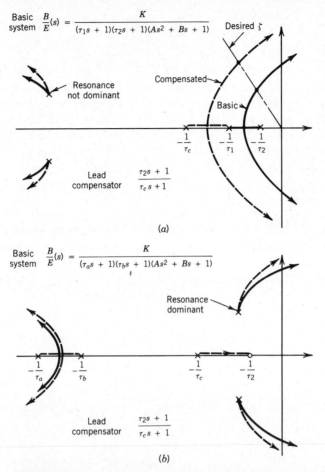

FIGURE 12.28 *Lead compensation ineffective for dominant basic-system resonance.*

type lead compensator shown appears to work and detailed design/analysis confirms this. In Fig. 12.28*b* however, where the complex poles *are* dominant, we find first that our design rules give no guidance in choosing τ and α and if we locate the compensator pole/zero arbitrarily as shown, analysis shows little or no performance improvement. Similar difficulties may be encountered with lag compensators.

Since dominant complex poles (resonances) do arise in practice, for example, in servomechanisms where the mechanical drive train has vibration problems, or in aircraft and missiles where structure bending vibration is coupled into the control system, we need compensation to deal with this. Controlled-variable derivative control may be helpful, as we showed in Fig. 11.10, 11.11, 11.14, and 11.15. Another approach uses forward-path cancellation compensation with a compensator

$$G_c(s) = \frac{As^2 + Bs + 1}{Cs^2 + Ds + 1} \tag{12.65}$$

where A and B are chosen to cancel the basic system resonance, whereas C and D are picked to give a "more desirable" (overdamped or underdamped) pole pair than did A and B (see Fig. 12.29). We of course strive for perfect cancellation, however real systems will always have some "mismatch" as in Fig. 12.29. As I have shown earlier, as long as mismatch is not excessive, the extra underdamped pole pair will contribute little to the system response.

The type of compensator just discussed is usually realized with op-amp electronics in analog systems or with appropriate software in digital systems. We now show an analog servomechanism example from the missile guidance field.[9] A gyrostabilized platform using air-bearing gyros and low-friction gimbals presents an open-loop transfer function containing a quadratic with essentially zero damping, as in Fig. 12.30a. A simple proportional control would be unstable for any gain, as the root locus shows. Although the "undesirable" quadratic fits the pattern of our general discussion, a cancellation approach is *not* advisable

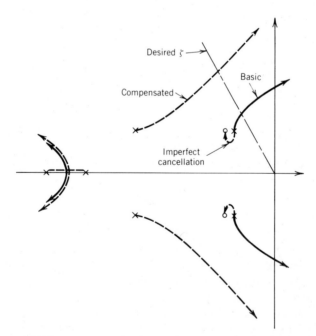

FIGURE 12.29 *Cancellation compensation effective for basic-system resonance.*

[9]J. L. Dooley, Active Compensation Networks Stabilize Guidance Platform, *Electromechanical Design*, Apr. 1965, pp. 36–42.

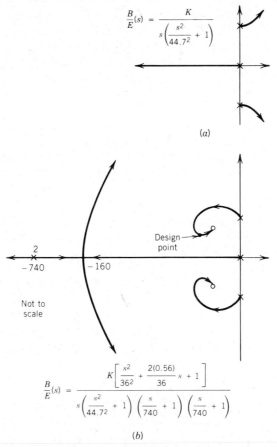

$$\frac{B}{E}(s) = \frac{K}{s\left(\dfrac{s^2}{44.7^2} + 1\right)}$$

(a)

2

−740

−160

Design point

Not to scale

$$\frac{B}{E}(s) = \frac{K\left[\dfrac{s^2}{36^2} + \dfrac{2(0.56)}{36}s + 1\right]}{s\left(\dfrac{s^2}{44.7^2} + 1\right)\left(\dfrac{s}{740} + 1\right)\left(\dfrac{s}{740} + 1\right)}$$

(b)

FIGURE 12.30 *Compensation for undamped inertial guidance platform.*

here because of pole/zero mismatch. Usually such mismatch is tolerable since the "extra" transient terms have small coefficients and are thus negligible. In our present example the coefficients would also be small but the dynamic term might be *unstable* since the "extra" locus is near the imaginary axis and might cross into the right half plane. The only safe thing to do is to place the compensator zeros well into the left half plane and "suck" the root loci emanating from the pure imaginary poles over to them. This is accomplished with the compensator of Fig. 12.30*b*; its poles were made fast and critically damped but other numerical values would also work. Testing of the actual servosystem, once it was constructed, gave the following performance.

closed-loop damping ratio	0.66	gain margin 18dB	
settling time	0.25 s	phase margin 142°	
bandwidth	38 rad/s		

Analog electronic compensators of this type can be constructed from op-amps in several ways; Fig. 12.31 shows the method used by Dooley. Various combinations of R and C values could be used. In Dooley, the constraints $R <$ 100,000 Ω, $C < 0.03$ μF were inforced, the R constraint to minimize problems of noise and amplifier input impedance, the C because of weight and space requirements related to a desire to mount these electronics directly on the platform gimbal.

Sometimes the offending resonances "wander around," requiring compensation that adapts to these changes. A good example is found in the flight control of large liquid-fueled booster rockets. These long, slender vehicles exhibit several modes of lightly damped bending vibrations whose frequencies change during flight. The frequencies depend on vehicle elasticity and mass, and the mass changes significantly as fuel is consumed since the fuel may represent 90% of the total mass. In this example, the use of fuel is accurately known, as a function of flight time, before the flight occurs, thus the variation of vibration frequency is also predictable. The control system designer thus implements compensators of the form discussed in this section but with parameter values that change with time according to a fixed schedule, "tracking" the changes in the controlled system as they occur. When structural natural frequencies cannot be accurately predicted and/or they change unpredictably, the control system has the added task of measuring these frequencies as they occur. The use of phase-locked-loop techniques has been proposed[10] as the solution to a problem of this type in the Galileo spacecraft designed to observe Jupiter.

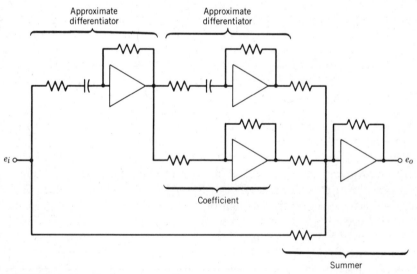

FIGURE 12.31 *Op-amp controller.*

[10]E. K. Kopf, T. K. Brown, and E. L. Marsh, Flexible Stator Control on the Galileo Spacecraft, Tech. Support Package on *NASA Tech. Brief Vol. 7, No. 2, Item #10,* Jet Propulsion Lab, Pasadena, Calif., Jul. 1983.

12.6 CONTROLLER TUNING, MANUAL AND AUTOMATIC

In servomechanism applications, system math models are often close enough to reality that controller parameters for optimum response can be fairly well predicted at the design stage and only a minor "tweaking" of adjustments is needed during experimental development. In the process-control field, however, it is not uncommon for a controller to be installed on a process with little or no analytical study being done beforehand. Such a practice may not be as irrational as it first seems. Industrial processes can be quite expensive and time-consuming to analyze, if meaningful analysis is possible at all. Process characteristics can change radically in unpredictable fashion. Also, many years of experience have shown that a PID controller is versatile enough to control a wide variety of processes. Thus an approach that selects a certain class (P, PI, PID) of controller based on past experience with certain classes of processes (pressure, flow, liquid level, temperature, etc.), and then sets controller parameters by experiment once the controller is installed, is not unreasonable. This experimental "design" of controller settings has come to be known as *controller tuning*.

Although there will always be exceptions, Shinskey[11] makes the following general recommendations with respect to the appropriate type of controller:

Process	Proportional Band (%)	Integral	Derivative
Flow	100–500	Essential	No
Pressure, liquid	50–200	Essential	No
Pressure, gas	0–5	No	No
Level	5–50	Seldom	No
Temperature and vapor pressure	10–100	Yes	Essential
Chemical composition	100–1000	Essential	If possible

Methods for adjusting the numerical values of the control modes are classified as open loop or closed loop. Closed-loop methods[12] characterize the system in terms of its *ultimate period* P_u under closed-loop pure proportional control. To find P_u one sets the amounts of integral and derivative control to their smallest values (ideally zero) and then gradually increases the proportional gain (reduces

[11]F. G. Shinskey, *Process Control Systems*, McGraw-Hill, N.Y., 1979, p. 80.

[12]J. G. Ziegler and N. B. Nichols, Optimum Settings for Controllers, *ISA Jour.*, 1964, June pp. 73–74, July pp. 75–76, Aug. pp. 63–64; D. W. Pessen, Optimum Three-Mode Controller Settings, *ASME Paper 52-A-58*, 1952; Shinskey, *Process Control Systems*, 1979, 96–99.

proportional band) until marginal stability is attained. That is, when the controller set point is slightly disturbed, an oscillation occurs that neither grows nor decays. The period of this oscillation is P_u, which we must measure. A *noninteracting* PID controller has

$$\frac{q_o}{q_i}(s) = \frac{100}{B_p}\left(1 + \frac{1}{\tau_I s} + \tau_D s\right) \tag{12.66}$$

where

$$B_p \triangleq \text{proportional band} \qquad 1/\tau_I \triangleq \text{reset rate} \qquad \tau_D \triangleq \text{rate time} \tag{12.67}$$

(The meaning of proportional band, reset rate, and rate time for *interacting*[13] controllers varies from one design to another, so one must consult the manufacturer's literature to relate settings to the values given here for noninteracting controllers.) Various authors disagree somewhat on the relationship between P_u and desirable values of B_p, τ_I, and τ_D; however, since such values are properly considered as tentative starting points for the tuning process rather than as fixed, final values, such disagreement should not cause great concern unless the disagreement is vast. Shinskey[14] suggests for a noninteracting PI controller that we set $\tau_I = 0.43P_u$ and then adjust B_p for the desired level of damping, by experiment. For quarter-amplitude damping we can expect B_p to be about twice the value for marginal stability and the new closed-loop period to be about $1.45P_u$. For a noninteracting PID controller the recommendations are $\tau_I = 0.34P_u$, $\tau_D = 0.08P_u$. If then B_p is doubled we can expect a new period of about $0.9P_u$. If the new periods deviate greatly from the values predicted here, one can decrease either or both τ_I and τ_D to increase the period or increase τ_I and/or τ_D to decrease the period. Shinskey's recommendations are based on the disturbance response of the first-order-plus-dead-time process model we have discussed earlier.

Open-loop tuning methods are less direct than closed-loop methods since they require an open-loop test of the process to establish numerical values for critical parameters in an assumed form of process model. Having these numerical values in hand, a set of rules then suggests appropriate numbers for controller parameters. The Ziegler–Nichols method,[15] one of the earliest, is still in use and requires an open-loop process, unit-step response test as in Fig. 12.32, where we measure an apparent dead time τ_{DT} and a maximum slope (reaction rate) R. Suggested trial settings are then:

P control $B_p = 100\,\tau_{DT}R$ $\qquad\qquad\qquad\qquad\qquad$ (12.68)

PI control $B_p = 110\,\tau_{DT}R \qquad \tau_I = 3.3\,\tau_{DT}$ $\qquad\qquad$ (12.69)

---

[13]Shinskey, *Process Control Systems*, 94–96.
[14]Ibid. p. 99.
[15]J. G. Ziegler and N. B. Nichols, Optimum Settings for Automatic Controllers, *Trans. ASME*, Vol. 64, 1942, pp. 759–768.

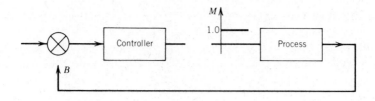

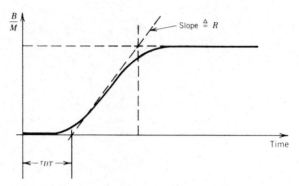

FIGURE 12.32 *Process model for Ziegler–Nichols tuning.*

PID control $B_p = 80\ \tau_{DT}R$ $\tau_I = 2\tau_{DT}$ $\tau_D = 0.5\ \tau_{DT}$ (12.70)

Another, more recent, study[16] models the process as either first-order-lag-plus-dead-time or overdamped-second-order-lag-plus-dead time. Numerical values for the dead time and time constants are obtained from appropriate open-loop response tests on the process and are assumed known. An interacting PID controller of the type below is assumed.

$$\frac{q_o}{q_i}(s) = \frac{100}{B_p}\left(\frac{\tau_I s + 1}{\tau_I s}\right)(\tau_D s + 1) = \frac{100}{B_p}\left(\frac{\tau_I \tau_D s^2 + (\tau_I + \tau_D)s + 1}{\tau_I s}\right)\quad(12.71)$$

Note that the "amounts" of the various modes of control are different from the controller of Eq. 12.66.

	Eq. 12.66	Eq. 12.71
Proportional	$\dfrac{100}{B_p}$	$\dfrac{100(\tau_I + \tau_D)}{B_p\tau_I}$
Integral	$\dfrac{100}{B_p\tau_I}$	$\dfrac{100}{B_p\tau_I}$
Derivative	$\dfrac{100\tau_D}{B_p}$	$\dfrac{100\tau_D}{B_p}$

[16]C. L. Smith, A. B. Corripio, and J. Martin Jr., Controller Tuning from Simple Process Models, *Instr. Tech.*, Dec. 1975, pp. 39–44.

Tuning recommendations are as follows. Values of τ_I and τ_D are totally unaffected by process dead time. If the process is first-order-plus-dead time, PI (not PID) control is recommended with τ_I equal to the process time constant. Note from Eq. 12.71 that this is cancellation compensation. Gain is then adjusted by trial to get the desired damping. This last step is the only place where the presence of dead time would have an effect. For a second-order-plus-dead-time process, τ_I is set to cancel the longest of the two time constants and τ_D cancels the shorter. Again, dead time has no effect except during the final gain-setting step for desired damping.

This tuning method makes no distinction between setpoint and disturbance response and may produce τ_I values too long to give optimum disturbance response when process dead time is much less than 30% of the process time constant. This is partly because the cancellation used removes the process time constant from the command response but *not* from the disturbance response. This problem was the subject of an exchange of "letters to the editor"[17] in which a first-order-plus-dead-time process with PI control, $\tau_{DT} = 1$ min, and $\tau = 80$ min was discussed. Figure 12.33 shows the two alternative tuning concepts; small

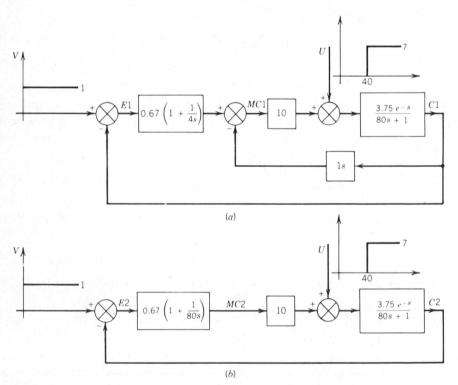

FIGURE 12.33 *Comparison of alternative tuning recommendations.*

---

[17]F. G. Shinskey, A. B. Corripio, Letters, *Instr. Tech.*, Apr. 1976, p. 6.

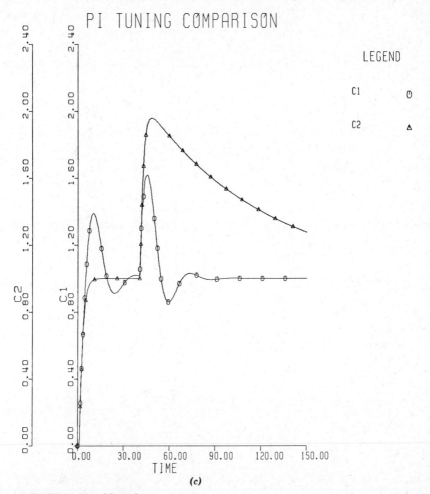

FIGURE 12.33   (Cont.)

τ_I plus some derivative control, versus τ_I set to cancel $\tau = 80$. In Fig. 12.33c, C2 has a good response to V (cancellation *is* effective) but response to U (often more important) is very sluggish since cancellation does *not* occur. For C1 both responses are more oscillatory but the response to U settles out much earlier than for C2, greatly reducing the time-integrated error. The large amount of integral control for C1 raises concern about integral windup but Fig. 12.33d shows controller output is not excessive. The C2 system's disturbance response can be improved somewhat by doubling the gain, however instability prevents further gain increase.

Figures 12.34 and 12.35 show the results of a comprehensive tuning study by Fertik. The process is characterized by a dead time T_D and a "storage time" T_{ps} (similar to a time constant) defined for a process step change as in Fig.

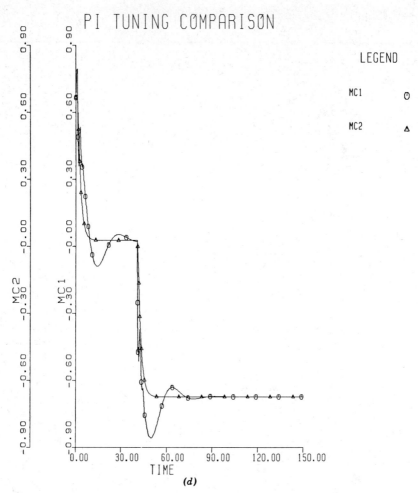

FIGURE 12.33   (Cont.)

12.34. "Process controllability," defined as T_D/T_{ps} is a measure of control difficulty; as it approaches 1.0, the process approaches pure dead time, the most difficult to control. For PI control (Fig. 12.34) note that (for $T_D/T_{ps} < 0.3$) controller integral time T_i is set *differently* to optimize set-point changes than for process disturbances, and the cancellation principle ($T_i = T_{ps}$) used by Smith et al. would hold *only* for set-point changes at $T_D/T_{ps} = 0$. Fertik states also that pure first-order-lag processes do not need derivative mode, whereas pure dead time does not benefit from it. For intermediate degrees of process controllability ($0.1 < T_D/T_{ps} < 0.5$) however, derivative mode helps, thus the PID charts (Fig. 12.35) cover only this range. Note that the PID controller is an interacting type with a filtered derivative mode. Fertik also provides useful information on digital implementation and noise problems.

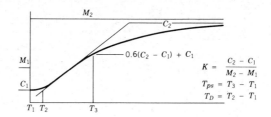

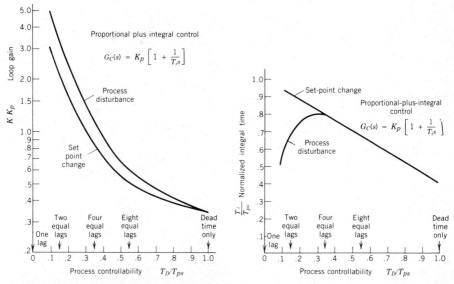

FIGURE 12.34   *PI tuning recommendations.*
Source: H. A. Fertik, Tuning Controllers for Noisy Processes, *ISA Trans.*, Vol. 14, Nov. 4, 1975, pp. 292–304.

Although I have presented controller tuning[18] as a "manual" operation performed by a human process operator, since the selection of the trial parameter values for the controller *is* mathematically defined, the potential for automatic tuning exists and is consistent with the historic trend to minimize human involvement in system control. With the computing power needed to automate the tuning process now readily available in both large-scale process computers for multi-loop systems and even microprocessor-based single-loop controllers, interest in *self-tuning* schemes has recently increased. However practical implementation must, at present, still be considered somewhat experimental until more operating experience is obtained. The main difficulty is that, although the measurements and computations needed to calculate the trial values of controller settings are mathematically defined, these trial settings usually require some

---

[18]G. K. McMillan, *Tuning and Control Loop Performance,* ISA, Research Triangle Park, NC, 1983.

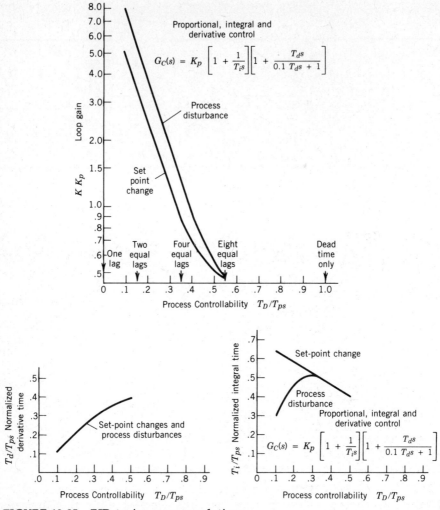

FIGURE 12.35 *PID tuning recommendations.*
Source: H. A. Fertik, Tuning Controllers for Noisy Processes, *ISA Trans.*, Vol. 14, Nov. 4, 1975, pp. 292–304.

(sometimes considerable) operator revision before acceptable response is obtained. The reduction to a mathematical algorithm of the human operator's experience and judgement used in this final tuning step can be a major problem.

   The concepts behind schemes for various degrees of self-tuning using large-scale process computers have been published[19] for some time; however the

---

[19]R. M. Bakke, Computer Helps Operator Tune Process Controllers, *Instr. Tech.*, Sep. 1968, pp. 52–56; P. W. Gollier and R. E. Otto, Self-Tuning Computer Adapts DDC Algorithms, *Instr. Tech.*, Feb. 1968, 65–70; S. Bennett and D. A. Linkens, Computer Control of Industrial Processes, Peter Peregrinus, New York, 1982, Chap. 4.

number of actual practical applications is difficult to judge from the literature since one is never sure if successful applications remain unpublished for proprietary reasons. In the case of dedicated single-loop controllers (which are ready-made, off-the-shelf items) one can be more specific since the manufacturers conspicuously advertise these for sale. I am aware of several self-tuning controllers of this type that are now on the market. Since these were all introduced very recently, their industrial acceptance remains to be documented. I will describe these controllers, all microprocessor based, giving as much detail as the manufacturers provide or space allows.

The first controller,[20] a general-purpose process controller, requires that a human operator enter nominal values for proportional, reset, and rate modes into controller memory and bring the controlled process to a stable, typical operating point with zero error. The largest acceptable values of set-point upset, and the desired 90% rise time for the controlled process (closed-loop) must also be entered. The controller will "test" the closed-loop system's response by applying small positive and negative step changes in desired value. Large changes would allow system response to be found more quickly, but would cause unacceptable perturbations in the controlled variable. Remember, the prime function of the controller is to keep the controlled variable steady; the self-tuning procedure, which is superimposed on the normal control operation, cannot be allowed to excessively degrade this operation. Once these preparations are complete, the operator can initiate self-tuning with a push button, whereupon the controller applies a positive step change, waits for system response, and then applies a negative step of equal size, exercising the closed-loop system so that its response can be measured and analyzed. From these measurements the controller develops a process model and chooses PID constants appropriate for this model and the desired 90% rise time. If a single pair of positive and negative inputs gives incomplete or unreliable data, up to five such test cycles may be invoked. Unfortunately, all the *details* of the tuning process have not been made public. When self-tuning is complete, the controller signals the operator, who can then engage the controller with the new PID constants or override them. Thereafter, the self-tuning process can be repeated, either manually at the operator's convenience or automatically with a chosen frequency, thus allowing the controller to update its behavior to suit any changes in process behavior.

Our second self-tuning controller[21] is marketed as a temperature controller; however it seems that its principle could be applied to other process variables also. It uses a variation of the closed-loop ultimate period approach we discussed earlier under manual tuning. When the self-tuning mode is engaged, the controller temporarily becomes an on–off type, guaranteeing that a limit cycle of definite period will occur. (This is not exactly what is done in manual ultimate-period tuning since there one adjusts proportional gain carefully to get sustained cycling at marginal stability.) The first limit cycle after engaging self-tuning is

---

[20]*C1.1511-IN, Self-Tuning Option,* Leeds and Northrup, North Wales, Penn., 1982; N. Andreieve, A Self-Tuning Controller that Continually Optimizes PID Constants, *Cont. Eng.,* Aug. 1981.
[21]REX-C100, Syscon International, S. Bend, Id. 1982.

used to allow transients to settle; whereas the next two measure amplitude A, period T, and fractional on-time T_o, using average values for the two cycles (see Fig. 12.36). Controller output O takes the form

$$O = \frac{1}{P}\left(e + \frac{1}{I}\int e\,dt + D\frac{db}{dt}\right) \tag{12.72}$$

Note that the derivative mode is on the measured value b of the controlled variable, not the actuating signal e. Since the controller is digital (microprocessor based), the derivative is formed from sampled values of b as we have seen earlier, except that a digital low-pass filter with a 10-s time constant is included and an absolute upper limit of 5.4°F/s is imposed. The numerical value of the proportional effect P is tuned according to

$$P = \frac{10A}{3.424\,T_o^2 - 3.896\,T_o + 3.572} \tag{12.73}$$

whereas integral effect I is given by

$$I = \frac{T}{3.951\,T_o^2 - 4.605\,T_o + 2.449} \tag{12.74}$$

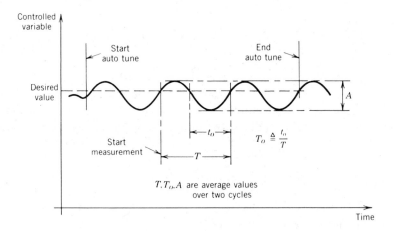

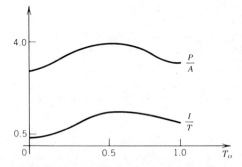

FIGURE 12.36 *Self-tuning scheme based on ultimate period.*

and $D = I/4$. The integral effect is realized digitally in the usual way except that a reset–windup feature is included. Of the three tuning formulas, only $D = I/4$ can be easily related to other tuning approaches, since this recommendation has appeared several times in earlier literature. The basis for the P and I tuning has not been made public by the manufacturer.

Another[22] approach to self-tuning retains the classic PID control modes and implements an automated version of classical manual tuning methods (Ziegler–Nichols, etc.). Self-tuning is initiated only when closed-loop system error exceeds a threshold value (outside the process noise level) indicative of significant changes in desired value and/or disturbances. The system then watches for a peak in the error signal and times the interval between successive peaks and also measures the ratio of peak amplitudes. This data is processed to obtain measures of closed-loop damping and speed that are then compared with desired levels set into the controller by the operator and adjustments of proportional, integral and derivative modes are made. Subsequent disturbance inputs initiate further cycles of self-tuning which gradually drive the controller toward the desired performance.

Our final self-tuning controller[23] abandons classical PID control for a "modern control" technique called a stochastic estimator, minimum-variance, self-tuning controller. Here a process model is continuously generated and updated from measurements of process input and output available during normal operation of the process (artificial input signals are not used to exercise the process). The digital controller has up to 10 adjustable parameters (PID has only 3) and should be adaptable to a wide range of processes and operating conditions. From the measured process model data the system synthesizes the control algorithm structure and numerical values such that the variance (mean-squared error) is minimized over time. The ASEA brochure claims "It is no exaggeration to say that PID control will be soon obsolete". Control engineers certainly need all the help they can get in solving tough problems. Time will tell whether this latest development will indeed totally displace earlier methods.

PROBLEMS

12.1 For the system of Fig. 12.1b, derive the closed-loop differential equation and calculate steady-state errors for steps and ramps of V and U.

12.2 Repeat Problem 12.1 for Fig. 12.1d and check stability using Routh criterion.

[22]T. W. Kraus and T. J. Myron, Self-Tuning PID Controller Uses Pattern Recognition Approach, *Cont. Eng.*, June 1984; Introducing Exact® Control, PUB 627, Foxboro Corp., Foxboro, Mass., 1984.

[23]Novatune Adaptive Control, CK23-2018E,ASEA Ind. Syst., Milwaukee, WI., 1985; M. J. Piovoso and J. M. Williams, Self-Tuning Control of pH, ISA/84 Intl. Conf., Houston, Oct. 1984; K. J. Astrom, Theory and Applications of Adaptive Control—A Survey, *Automatica*, Vol. 19, No. 5, pp. 471–486, 1983.

12.3 Study the use of cancellation compensation in the systems of Fig. 12.1*b* and *d*, using the proportional-plus-integral controller to form the cancelling zero.

12.4 Derive transfer functions for:

  a. Fig. 12.8*a*. Assume flow through R_1 and R_2 equal.   **d.** Fig. 12.8*d*.
  b. Fig. 12.8*b*.   **e.** Fig. 12.8*e*.
  c. Fig. 12.8*c*.

12.5 Investigate the effects of inertia on the compensators of Fig. 12.8*d* and 12.8*e*.

12.6 Derive transfer functions for the systems of Fig. P12.1 and comment on their utility.

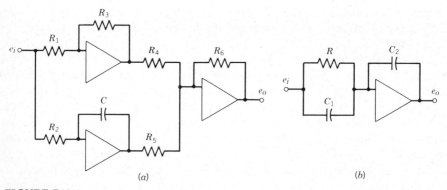

(a)   (b)

FIGURE P12.1.

12.7 For the engine governor of Fig. 12.9, using text numerical values:

  a. Compare open-loop frequency responses of the basic and lag-compensated systems.

  b. Compare the closed-loop $(\omega_C/\omega_V)(i\omega)$ frequency responses of basic and lag-compensated systems.

  c. Compare $(\omega_C/T_U)(i\omega)$ for basic and lag-compensated systems.

12.8 Analyze the device of Fig. 11.17*d* with a finite value of K_a to show it to be a phase-lead compensator.

12.9 Try to find a passive pneumatic analog for the passive electrical phase-lead network of Fig. 12.18*a*.

12.10 Analyze the mechanical phase-lead devices of Fig. 12.18*c* and 12.18*d* for $(x_o/x_i)(s)$.

12.11 For the numerical values given in Fig. 12.19*b*, what is the largest closed-loop ζ possible for any gain? If we needed a larger value of ζ than this, how would you change the numerical values of the lead compensator?

12.12 Discuss in detail the consequences of making the pump stroker in Fig. 12.21 an integrating device.

12.13 Derive Eq. 12.16.

12.14 Write a CSMP (or other available digital simulation) program to study response to step commands and disturbances for the system of Fig. 12.10.

12.15 For each basic system given in Fig. P12.2, use root locus to set gain for $\zeta = 0.4$ and use frequency response to set gain for $M_p = 1.3$. For the system designed by root locus, add (using root-locus methods) a lag compensator that allows a gain increase of four-to-one while preserving speed and stability. For the system designed by frequency response, add (using frequency-response methods) a lag compensator that achieves the same gain as achieved by the root locus, lag-compensated system. Finally, use CSMP (or other available simulation) to compare command and disturbance step responses for all four systems.

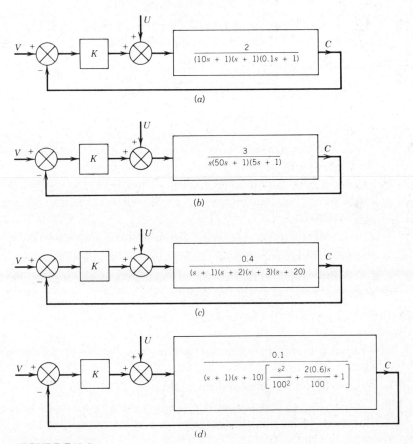

FIGURE P12.2.

12.16 Repeat problem 12.14 except now design lead compensators that double the speed of response of the basic systems while preserving relative stability and not degrading steady-state error behavior.

12.17 Design lag–lead compensation for each system of Fig. P12.2, using the specifications of Problems 12.15 and 12.16, and employing:

a. Root locus methods.

b. Frequency-response methods.

Find response to command and disturbance step inputs using CSMP (or other available) digital simulation for both the basic and compensated systems.

12.18 a. Use root-locus methods to set the gain of the system of Fig. 12.30b for a closed-loop ζ of 0.66. Then use a root finder to find all the closed-loop roots.

b. Use CSMP (or other available) digital simulation to find the response to step command V of the closed-loop system.

12.19 Derive the transfer function $(e_o/e_i)(s)$ for the compensator of Fig. 12.31. Then choose R and C values to obtain the numbers used in Fig. 12.30b.

12.20 For a basic system with

$$\frac{B}{E}(s) = \frac{C}{E}(s) = \frac{K}{(0.01s + 1)(0.015s + 1)\left[\dfrac{s^2}{10^2} + \dfrac{2(0.6)s}{10} + 1\right]}$$

a. Use root-locus methods to set gain so that dominant closed-loop roots have $\zeta = 0.4$.

b. Design a cancellation compensator (as in Fig. 12.29) to double the speed of response and maintain $\zeta = 0.4$. What is the effect on steady-state errors?

c. Use CSMP (or other available) digital simulation to study the effect of imperfect cancellation on the step command response.

12.21 Use CSMP (or other available) digital simulation to explore the effect of gain increases for the system of Fig. 12.33b.

12.22 In Fig. 12.34 an open-loop process test gives $K = 0.9$, $T_{ps} = 20s$, $T_D = 4s$.

a. Choose PI and PID control parameters optimized for set-point changes.

b. Repeat Part a for process disturbances.

c. Treating T_{ps} as a first-order lag and T_D as dead time and letting disturbance U enter process at same point as controller output, use CSMP (or other available) digital simulation to study response to step commands and disturbances for the systems of Parts a and b. Comment on the results.

CHAPTER 13
Case Studies and Special Topics

13.1 AN AIRBORNE HELIOSTAT, AN ELECTROMECHANICAL TYPE-3 SERVO SYSTEM WITH TRIPLE LEAD COMPENSATION

The case studies of this chapter illustrate applications and extensions of general principles (presented earlier in the text) to practical systems which have been constructed and successfully operated. Our first example, an airborne heliostat,[1] was designed for use with a solar spectrograph experiment that was part of the Airborne Solar Eclipse Expedition sponsored by NASA in 1965. The spectrograph was carried aloft aboard a Convair CV 990 aircraft and was to make approximately-2-min-long photographic exposures of an image of the solar corona. By flying the aircraft along the proper course, we move the spectrograph so as to always be at the point of "total eclipse." But since the aircraft was subject to rolling, pitching, and yawing motions also, some means of stabilizing the image presented to the spectrograph was necessary. The root-mean-square (RMS) angular error due to aircraft motion could not exceed 1.0 min of arc (1/60 of a degree!) over a 2-min period or else the image would be too fuzzy.

The design problem was to mount a telescope rigidly to the aircraft structure and then provide an optical-electromechanical system that would ensure that the telescope's line of sight or "look direction" was always pointed at the sun. To accomplish this a heliostat consisting of a flat mirror, servo-positioned about two perpendicular axes, was designed to reflect the solar image onto the aperture of the spectrograph. If this mirror's attitude in space could be suitably controlled, the image would be stabilized. This was accomplished by mounting a rate-integrating gyroscope[2] (an absolute angular-displacement sensor) on each axis and using its output signal as the error signal for actuating the motor that drove each axis in a servo loop. (Optical sensing means were ruled out because of the vast change in light intensity associated with total eclipse.)

Figure 13.1 shows the arrangement of the two axes of the heliostat, and since each axis presents essentially the same control problem we consider only a single axis, the yaw axis, from now on. The relation between aircraft yaw motion $\dot{\psi}_A$ and mirror yaw motion $\dot{\psi}_M$ is seen in Fig. 13.2a. As the aircraft rotates, it tends to carry the mirror along with it because of friction in the bearings and

---

[1] J. L. Whittaker, C. Burdin, and J. D. Clarke, Heliostatic Image Stabilization System, Douglas Missile and Space Sys. Div., *Report SM-48335,* Feb. 1966, Santa Monica, Calif.
[2] E. O. Doebelin, *Measurement Systems,* 3rd Ed., McGraw-Hill, 1983, pp. 344–346.

457

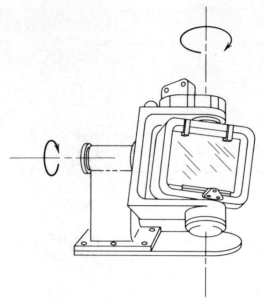

FIGURE 13.1 *Heliostat with two-axis servo drives.*

brushes of the drive motor (Fig. 13.2*b*). Another coupling effect occurs whenever the motor rotor and stator have a relative velocity and generate a back emf. This generator effect causes an armature current and thus an electromagnetic torque on the rotor. One can minimize the undesired torque by driving the motor armature with an amplifier whose output resistance (which appears in the armature current) is large enough to make this velocity-induced current negligible.

Since the angular motions of the aircraft are the cause of the undesired motions of the mirror, these motions constitute disturbance inputs to the control system, and estimates of their nature and magnitude are useful in system design. To gather such information, the aircraft was instrumented with roll, pitch, and yaw gyros, and a recorder, and taken aloft for several test flights. The recorded motions were of course somewhat random, however three frequencies seemed to contribute most of the total motion. Based on these measurements, the disturbance inputs were modeled as

$$\text{roll angle} \quad = 0.2 \sin 10\pi t + 0.4 \sin 0.4\pi t + 1.4 \sin 0.02\pi t \text{ degrees} \quad (13.1)$$

$$\text{pitch angle} = 0.08 \sin 10\pi t + 0.1 \sin 0.4\pi t + 0.2 \sin 0.02\pi t \text{ degrees} \quad (13.2)$$

$$\text{yaw angle} \quad = 0.04 \sin 10\pi t + 0.2 \sin 0.4\pi t + 0.8 \sin 0.02\pi t \text{ degrees} \quad (13.3)$$

corresponding to frequencies of 5, 0.2 and 0.01 Hz.

Initially a type-1 servo system was attempted but paper studies plus analog simulation showed that all specifications could not be met, so type-2 came under

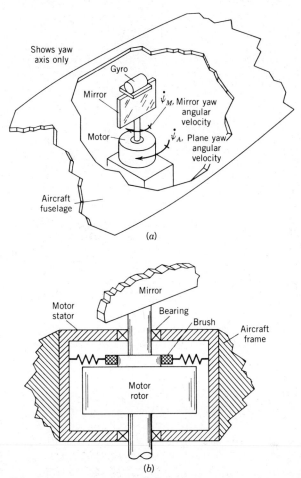

Shows yaw
axis only

Gyro

Mirror

$\dot{\psi}_M$, Mirror yaw
angular
velocity

Motor

$\dot{\psi}_A$, Plane yaw
angular
velocity

Aircraft
fuselage

(a)

Mirror

Motor
stator

Bearing

Brush

Aircraft
frame

Motor
rotor

(b)

FIGURE 13.2   *Details of a single axis.*

consideration. At the same point someone suggested the elimination of the
disturbing torque due to motor back emf by use of a high-output-impedance
power amplifier, which approximates a current source. When an armature-
controlled, PM field dc motor is driven from a voltage source e_i we have

$$e_i - K_e\dot{\theta} - i_aR_a = 0 \qquad K_t i_a = J\ddot{\theta} \text{ friction neglected} \qquad (13.4)$$

where $-K_e\dot{\theta}$ is back emf, and

$$\frac{\theta}{e_i}(s) = \frac{1/K_e}{s\left(\dfrac{JR_a}{K_tK_e}s + 1\right)} = \frac{K_1}{s(\tau s + 1)} \qquad (13.5)$$

When, on the other hand, we use an ideal current source i_a, the armature circuit
equation is not needed since i_a is not an unknown.

$$K_t i_a = J\ddot{\theta} \qquad \frac{\theta}{i_a}(s) = \frac{K_t/J}{s^2} = \frac{K_2}{s^2} \qquad (13.6)$$

Use of a current source also makes the armature resistance R_a and inductance L_a totally irrelevant. A high-output-impedance amplifier is not a *perfect* current source but can come close enough to be modeled as such. For the motor actually used, $R_a = 7.5\Omega$ and $L_a = 0.016H$, giving an L/R time constant of about $0.002s$. The power amplifier used has $R_{\text{out}} = 8000\Omega$, reducing the time constant to $2\mu s$, which can be neglected relative to other dynamics in this system. When an integral controller is combined with these motor/load dynamics, a type-3 system results.

The only significant nonlinearity in the system is the dry friction on the mirror shaft. This friction is actually the main source of error in mirror position; if it were zero, aircraft motions would not induce mirror motions. Obviously, then, this friction cannot be neglected when evaluating system accuracy. However, in designing the servo loop (setting gain, choosing compensation, etc.) the friction causes little deviation from linearity since it is so small (0.016 ft-lb$_f$) compared to the motor's full-scale magnetic torque of 1.2 ft-lb$_f$. The design approach used was to neglect the friction for the dynamic analysis, and design on a linear basis so as to get numerical values easily. Then, analog computer simulation, *including* the dry friction, was employed to check system accuracy.

The "bare bones" system (amplifier/integrator, motor/load, and gyro) of Fig. 13.3 is absolutely unstable no matter what gain is used. Three compensation schemes were investigated: dc tachometer feedback, rate gyro feedback, and stabilizing networks in the forward path. The tachometer feedback introduced additional friction and other undesirable coupling since it measures *relative* velocity of its rotor and stator, rather than absolute velocity. This, plus potential noise problems at the low speeds involved in this system, led to its rejection. These low angular rates also precluded use of rate gyros since they would be working near their threshold sensitivity much of the time, giving unreliable signals. Also, analysis showed that instability occurred at gain levels that were too low to meet accuracy requirements. Examination of lead compensator networks in the forward path showed that these could give sufficient leading phase angle to fight out the lag due to the three integrations and allow high enough gain (with good stability) to meet accuracy goals. Three such lead networks

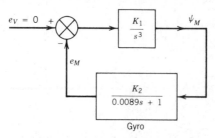

FIGURE 13.3 *Basic system (type 3).*

(cascaded) were found to be necessary since each network provides less than 90° of phase lead. This amount of lead compensation is unusual since the need for it arises from the type-3 servo system, which is quite rare. It also requires a low-noise actuating signal, which the rate-integrating gyro fortunately provides. The three lead compensators and one electronic integration were all accomplished with one op-amp and suitable resistors and capacitors.

Although many servo systems use gearing between motor and load, the high precision required in this application makes direct drive[3] attractive as a means of avoiding gear problems such as backlash. Direct-drive torque motors are PM field, armature-controlled dc motors specially designed to be directly coupled to their mechanical load. They feature high torque-to-inertia ratio, high mechanical resonance frequency, low inductance for fast response, high accuracy at low speed, and good linearity of torque with current, independent of speed or angular position. Available "frameless" as three separate parts (rotor, brush ring, PM field) in a "pancake" form, they can be "designed into" equipment to conserve space and weight. The Inland T-2950A used in this application has the following characteristics.

Torque constant	= 0.37 ft-lb$_f$/A	Resistance	= 7.5Ω
Back-emf constant	= 0.5 V/(rad/s)	Inductance	= 16 mH
Rotor inertia	= 2.9 × 10^{-4} ft-lb$_f$-s^2	Peak Torque	= 1.2 ft-lb$_f$
Dry friction	= 0.017 ft-lb$_f$	Brush life	> 10^7 revolutions
Weight	= 2.2 lb$_f$	Size 3.7 in. diameter, 1.4 in. long	

The choice of numerical values for gain and compensator time constants was made (neglecting dry friction) by use of conventional frequency-response design methods (Bode plot, Nichols chart) to achieve $M_p \approx 1.3$ at $\omega_R = 30$ rad/s. Once the system was constructed, gain was set to give $\zeta \approx 0.7$ for step commands. Figure 13.4 shows a block diagram of the complete system. Numerical values are:

K_a = 90.2 V/V	τ_5 = 0.033 s
K_c = 0.0901 V/V	τ_6 = 0.0009 s
τ_1 = 0.213 s	τ_G = 0.0089 s
τ_2 = 0.011 s	K_p = 1.0 A/V
τ_3 = 0.045 s	K_t = 0.37 ft-lb$_f$/A
τ_4 = 0.0042 s	T_F = 0.016 ft-lb$_f$
J = 0.11 ft-lb$_f$-s^2	

The power amplifier (Fig. 13.5) was designed and constructed by Douglas aircraft, whereas the op-amp used for the controller was a Zeltex 146E, a high-

---

[3]Direct Drive Servo Design Handbook, Inland Motor Corp., Radford, Virginia, 1964.

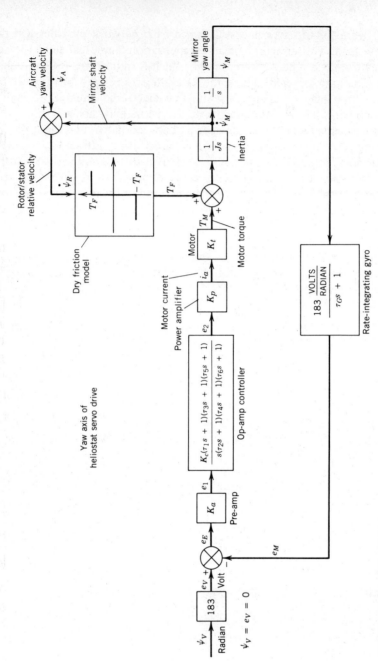

FIGURE 13.4 *Final design.*

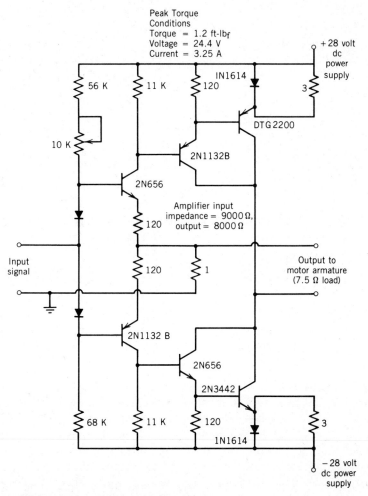

FIGURE 13.5 *Power amplifier design.*

performance, chopper-stabilized unit with 10^7 open-loop gain, and low dc offset and drift. Figure 13.6 shows the controller circuit. Analysis gives

$$Z_f = \frac{K_2(\tau_1 s + 1)}{s(\tau_4 s + 1)} \qquad Z_i = \frac{K_1(\tau_2 s + 1)(\tau_6 s + 1)}{(\tau_3 s + 1)(\tau_5 s + 1)} \tag{13.7}$$

$$\frac{e_o}{e_i}(s) = -\frac{0.0901(0.213s + 1)(0.045s + 1)(0.033s + 1)}{s(0.011s + 1)(0.0042s + 1)(0.000985s + 1)} \tag{13.8}$$

which is given by

$C_1 = 0.068 \mu F \qquad C_2 = 0.01 \mu F \qquad C_3 = 0.10 \mu F \qquad C_4 = 0.47 \mu F$

$R_1 = 487 \ k\Omega \qquad R_2 = 147 \ k\Omega \qquad R_3 = R_4 = 453 \ k\Omega \qquad R_5 = 1910 \Omega$

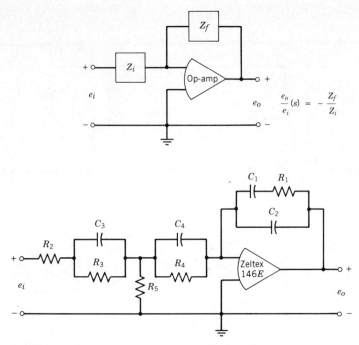

FIGURE 13.6 *Op-amp servo controller.*

Using the numerical values given, the loop gain in Fig. 13.4 is 5000, a very large number, but necessary to meet the stringent accuracy requirements, and made possible by the sophisticated triple-lead compensation. Upon completion of pencil-and-paper design and computer simulation studies, the heliostat was constructed and lab tested on an oscillating table to simulate aircraft motions, allowing final adjustment of system numerical values. Ultimate verification of performance was made by actual flight testing, which showed that heliostat gimbal motion errors were on the order of 0.03 to 0.07 arc minutes, whereas the final optical image was stabilized to 0.3 to 0.8 arc minutes, within the 1.0 arc minute design specification.

13.2 AN ANTIAIRCRAFT GUN DIRECTOR, A HYDROMECHANICAL TYPE-2 SERVO WITH MECHANICAL COMPENSATION

Figure 13.7 shows an antiaircraft gun positioning servo system in which the position command is provided by a human tracker. (Another version of such systems uses a radar tracker to replace the human, however the gun-positioning servo remains essentially the same.) We should first note that the complete gun director must provide *two* axes (azimuth and evaluation) of control to point the

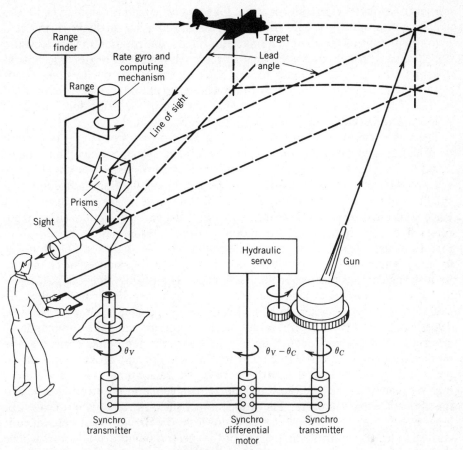

FIGURE 13.7 *40 mm antiaircraft gun director.*

gun properly. It is usually acceptable for initial design purposes to consider each axis separately, so from here on our discussion is of the azimuth axis only. (*Final* design should check for any possible problems that might arise from simultaneous motion of both axes, such as the azimuth moment of inertia becoming *variable* with elevation angle.)

In a man–machine system like that of Fig. 13.7 the designer strives to relieve the human operator of any tasks that are better performed by machinery so that concentration can be focused on those functions best suited to human performance. Computation of target lead angle *could* be assigned to the gunner's mind (duck hunters do it all the time) but this system provides a machine aid. Lead angle depends on target range, angular velocity ω_{LOS} of the line of sight, and gun ballistics. If the gunner's instructions are simply to keep the cross hairs *on* the target (not to lead it), this task can be accurately performed and we may assume that the angular velocity of the θ_V shaft is ω_{LOS}, which we measure as shown with a rate gyroscope. We also assume that an optical range finder (details not given) provides range information. Fixed ballistic data is combined with the varying range and ω_{LOS} data in a computing mechanism auxiliary to the rate gyro, the output of the computations being an angular rotation that deflects the upper prism through an angle equal to the computed lead angle. The optical system thus allows the gunner to *look* right at the target but to *point* (shaft angle θ_V) at its predicted future position where target and artillery shell should collide. The data system just described is not within the servo loop but is of interest since it shows where the servo command originates.

The servo system itself provides further aid to the operator in that the power to drive the gun turret is supplied by a hydraulic system rather than gunner muscle. Command angle θ_V is measured by a synchro, as is gun angle θ_C. A synchro differential motor receives the θ_V and θ_C information and produces a shaft rotation equal to their difference. This low-power error signal actuates a valve-controlled servo that strokes the variable displacement pump supplying a rotary hydraulic motor geared to the gun turret. Internal details are shown in Fig. 13.8 and 13.9. The synchro differential motor's shaft rotation is $\theta_V - \theta_C + \theta_B$, where $\theta_V - \theta_C$ is produced electrically and θ_B is produced mechanically by a spring/damper positive feedback linkage. The positive feedback recovers an integration lost by closing a loop around the valve/cylinder and makes the overall system type 2 when combined with the integration in the pump/motor. Type 2 is desired since typical target motions produce ramplike commands θ_V and zero tracking error for such commands means the gun points directly at the predicted target position, not behind it. To keep friction broken loose in all parts of the drive system, the valve sleeve is given a continuous low-amplitude, high frequency (33 Hz) "dither" motion x_d by a cam and push rod attached to the sleeve feedback linkage. The frequency of such dithering is not critical as long as it is high relative to system bandwidth; 33 Hz was used here since it corresponded to the rotational speed of the electric pump drive motor, whose shaft could thus be used for the dither camshaft.

Analysis of the system block diagram (Fig. 13.9) reveals some interesting features. First note that the second-order dynamics usually included in valve/cylinder and pump/motor hydraulics have been neglected for simplicity and that we ignore the dither. We can write

$$\left\{ \left[(\theta_V - \theta_C) + \frac{sK_{fb}}{K_{pm}(\tau_{fb}s + 1)}\theta_C \right] r - \frac{sK_l}{K_{pm}}\theta_C \right\} \frac{K_{vc}}{s} = \frac{s}{K_{pm}}\theta_C \qquad (13.9)$$

$$\left[\frac{\tau_{fb}}{K_{vc}K_{pm}r}s^3 + \left(\frac{1}{K_{vc}K_{pm}r} + \frac{K_l\tau_{fb}}{K_{pm}r} \right)s^2 \right.$$

$$\left. + \left[\tau_{fb} + \frac{1}{K_{pm}r}(K_l - K_{fb}r) \right]s + 1 \right]\theta_C = (\tau_{fb}s + 1)\theta_V \qquad (13.10)$$

If we set $K_l - K_{fb}r = 0$, the steady-state error for a ramp θ_V becomes zero, as desired. However if τ_{fb} were zero, we would lose the s term completely in Eq. 13.10, giving absolute instability; thus the delay in the positive feedback is essential. To see that setting $K_l = K_{fb}r$ makes the system type 2 we need to reduce the minor loops.

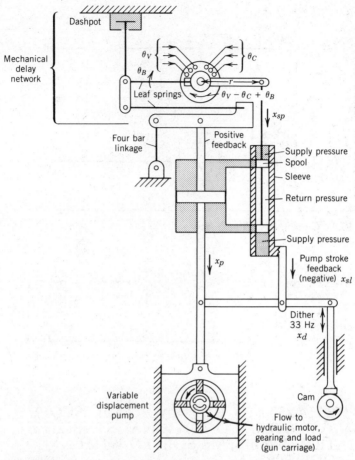

FIGURE 13.8   *Pump stroke control with delayed positive feedback and dither.*

$$\left[\left(\theta_V - \theta_C + \frac{K_{fb}}{\tau_{fb}s + 1}x_p\right)r - K_l x_p\right]\frac{K_{vc}}{s} = x_p \tag{13.11}$$

$$\frac{x_p}{\theta_V - \theta_C}(s) = \frac{K_1(\tau_{fb}s + 1)}{s(\tau_1 s + 1)}$$

where

$$\tau_1 \triangleq \frac{\tau_{fb}}{1 + \tau_{fb}K_{vc}K_l} \qquad K_1 \triangleq \frac{K_{vc}r}{1 + \tau_{fb}K_{vc}K_l} \tag{13.12}$$

We can now redraw the block diagram as in Fig. 13.10, which makes clear that we have type 2. Also note that since $\tau_1 < \tau_{fb}$ there is also lead compensation, augmenting stability. In fact, our present model is absolutely stable for *any* positive loop gain. Inclusion of the neglected dynamics in the valve/cylinder and/or pump/motor hydraulics will of course make absolute instability possible.

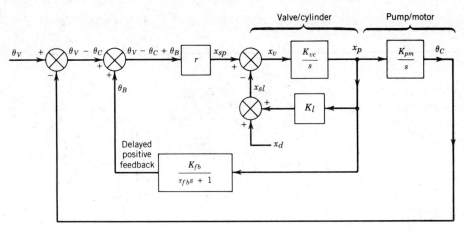

FIGURE 13.9   *Gun director block diagram.*

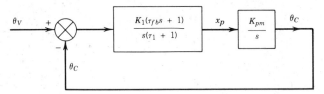

FIGURE 13.10   *Block diagram with minor loop reduced.*

13.3   THE HAYSTACK HILL RADIO TELESCOPE POINTING SYSTEM; CONDITIONAL STABILITY/ SATURATION PROBLEMS SOLVED WITH A DUAL-MODE SERVO

In 1960 the U.S. Air Force, through its project manager MIT Lincoln Laboratory, solicited proposals for design and construction of a large and precise radio telescope to be installed on Haystack Hill outside Boston, Massachusetts. Proposals from various bidders were evaluated and the contract was awarded to North American Aviation (Columbus, Ohio Division). One part of the proposal was devoted to the electrohydraulic control system for steering and pointing the antenna dish, and we base our discussion on the design presented in this proposal.[4] Design and construction proceeded over several years, with the antenna being erected and put into service in 1964, with operation continuing today (1985). Our discussion will also include problems that arose during final development and required modifications of the original design. At the time of

---

[4]G. Carroll, Haystack Antenna System Control Systems Technical Proposal, North American Aviation Columbus Division, *Rept. NA60H-220,* Apr. 20, 1960.

its erection, this antenna was the most precise instrument of its size in the world.

A so-called elevation-over-azimuth configuration was chosen to position the 120-ft-diameter dish (see Fig. 13.11). Figures 13.12 and 13.13 show mechanical details of the drive arrangement, two hydraulic motors (fed in parallel by a single servovalve) being used to drive each axis. Digital shaft angle encoders with a resolution of 0.001° are used to sense the position of each axis. Bearings for the elevation axis are grade zero Timken tapered-roller bearings, whereas the elevation axis thrust and radial loads are carried by a four-point angular contact ball bearing (141-in. diameter) with two rows of balls. (After the contract was won and design proceeded, it was decided that a hydrostatic oil bearing would instead be used for the elevation axis.) The required pointing accuracy is ±0.005° at the RF (radio frequency) axis for any combination of 0 to 1°/s tracking velocity and 0 to 0.012°/s² tracking acceleration. Control system shaft-positioning errors must be *less* than ±0.005° since dish geometry errors (caused by gravity and temperature induced distortions, etc.) can cause a misdirection of the electromagnetic line of sight (RF axis) even when the shafts are precisely positioned.

Selection of motors and gear boxes requires estimation of the torques that must be provided for friction, windage, and acceleration. Mass moments of inertia for the elevation and azimuth axis were estimated at 3×10^6 and 4×10^6 lb$_f$-ft-s², respectively, whereas dead weights were 120,000 lb$_f$ and 240,000 lb$_f$. Since the azimuth axis thus has the most severe friction and inertial loads, the proposal carried out a design for it only and suggested that if the same components were used for the elevation axis its performance should also be adequate. This design philosophy also minimizes the size of the spare parts inventory that must be kept on hand. Note also that, at the proposal stage,

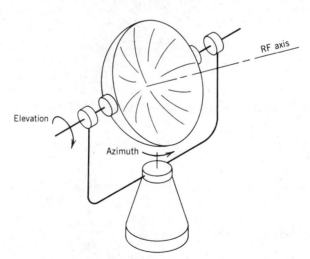

FIGURE 13.11 *Elevation-over-azimuth antenna configuration.*

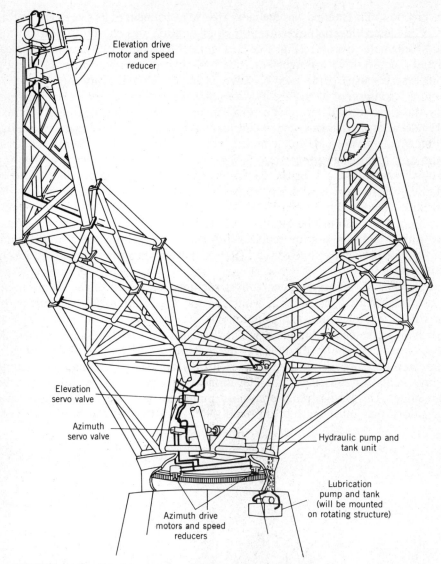

Elevation drive
motor and speed
reducer

Elevation
servo valve

Azimuth
servo valve

Hydraulic pump and
tank unit

Lubrication
pump and tank
(will be mounted
on rotating structure)

Azimuth drive
motors and speed
reducers

FIGURE 13.12 *Mechanical arrangement of drive systems.*

engineering design effort is funded by the proposing company and cannot be recovered unless the design competition is won, thus the companies must risk some of their capital in the attempt to get business.

Inquiries at the bearing manufacturer gave

$$T_f \stackrel{\Delta}{=} \text{friction torque}$$
$$= (0.002 \text{ to } 0.003)(\text{thrust load})(\text{bearing radius})$$

(13.13)

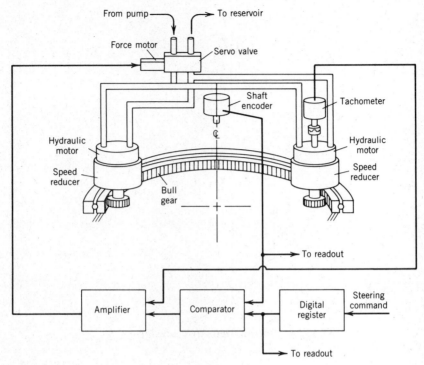

FIGURE 13.13   *Details of servosystem hardware.*

A conservative design value was taken as

$$T_f = (0.02)(240,000)(80 \text{ in.}) \left(\frac{1}{12}\right) = 32,000 \text{ ft-lb}_f \qquad (13.14)$$

Antenna specifications require a maximum acceleration of $1°/\text{s}^2$, thus the torque T_a for acceleration is

$$T_a = J\alpha = (4 \times 10^6 \text{ lb}_f\text{-ft-s}^2) \left(\frac{1°/\text{s}^2}{57.3°/\text{rad}}\right) = 70,000 \text{ ft-lb}_f \qquad (13.15)$$

Since the antenna will be inside a protective fiberglass radome, no wind gust torques are possible; however rotation of the dish in still air does cause a drag torque. One intuitively believes this is small, but since the dish is very large, you can not be sure without a calculation. The worst case would be azimuth rotation at maximum velocity with the antenna pointing horizontal, as in Fig. 13.14. Specifications call for a maximum velocity of $3.3°/\text{s}$. The drag torque is assumed to be caused by stagnation pressure of the air against the dish surface, which is treated as a flat plate (with a correction factor of 1.28 for the actual shape).

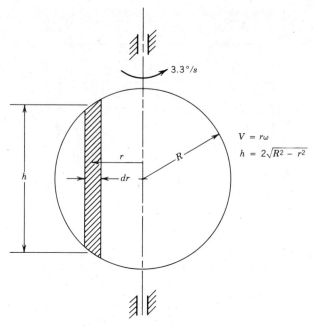

FIGURE 13.14   *Antenna windage calculation.*

$$T_w \triangleq \text{windage torque} = 2 \times 1.28 \int_0^R \left(\frac{\rho V^2}{2}\right) \underbrace{(h \ dr)}_{} \ (r) \qquad (13.16)$$

$$\underset{\text{two sides}}{\downarrow} \qquad \underset{\text{pressure}}{\downarrow} \ \underset{\text{area}}{\downarrow} \quad \underset{\text{lever arm}}{}$$

$$T_w = 2.68 \times 10^{-6} \ R^5 = 2080 \ \text{ft-lb}_f \qquad \text{for } R = 60 \ \text{ft} \qquad (13.17)$$

The azimuth drive will thus have to be capable of producing a maximum torque of

$$T_{\max} = T_f + T_a + T_w = 104{,}080 \ \text{ft-lb}_f \qquad (13.18)$$

We can now select a motor and gearbox to meet these needs. Gearbox complexity, cost, and backlash usually increase with gear ratio, thus we prefer a high-torque motor. In 1960 the selection of high-torque hydraulic motors suitable for precision servo control was not as great as it is today, thus there were not many candidates to consider and the choice was narrowed to an eight-cylinder radial piston design with a proven record of servo performance in machine tool drives and a displacement d_m of 22 in.^{3}/rev. This motor is controllable over a range of 1/60,000 to 500 rpm. When rotating at 100 rpm it can be reversed in 0.0015 s with only 0.0006 of a revolution movement after the signal to reverse. System supply pressure was chosen as 1000 psi, a conservative low value conducive to long life and one for which many standard hydraulic components are designed. Using the maximum power transfer result of Section 3.3, we want the motor pressure drop to be 667 psi at the maximum torque condition. This gives for both motors on the azimuth drive

$$\text{total motor torque} = \frac{2\Delta p d_m}{2\pi} = \frac{2 \times 667 \times 22}{12 \times 2\pi} = 354 \text{ ft-lb}_f \quad (13.19)$$

A tentative gear ratio is then

$$N = \frac{104,080}{354} = 294 \text{ to } 1 \quad (13.20)$$

Consultation with gear box manufacturers indicated that four meshes of precision helical gears with 98% efficiency per mesh will be needed. A corrected gear ratio is then

$$N = \frac{294}{0.98^4} = 320 \text{ to } 1 \quad (13.21)$$

Since actual gear ratios convenient to produce depend on tooling and other manufacturing considerations, the final choice had to accommodate this and was 325 to 1.

With motors and gear boxes chosen, we now turn to servovalve selection, which generally is determined by maximum flow rate requirements at actuator maximum velocity. The maximum angular velocity of 3.3°/s used in the windage torque calculation was not directly given in the specifications but was derived from the following requirement. The azimuth drive must be capable of making a 180° position change in one minute with maximum acceleration held to $\pm 1°/s^2$. This motion can be accomplished in various ways, so one must choose some reasonable duty cycle, such as that in Fig. 13.15, for calculation purposes. From that figure we may write

$$180° = \frac{1°}{s^2} \frac{t_1^2}{2} + t_1(60 - 2t_1) + \frac{1°}{s^2} \frac{t_1^2}{2} \quad (13.22)$$

$$t_1 = 3.3 \text{ s} \qquad \text{maximum velocity} = 3.3°/s$$

Maximum motor velocity is then 179 rpm and the servovalve flow rate (for two motors) is 34.1 gal/min. When the load is running at constant velocity the motor supplies only friction and windage torques (no acceleration) and, for both motors

$$\text{motor torque} = \frac{34080}{(325)(0.98)^4} = 116 \text{ ft-lb}_f \quad (13.23)$$

$$\text{motor pressure drop} = 198 \text{ psi}$$

With a 198 psi drop across each motor, the drop across both ports of the servovalve is 802 psi. One can now consult servovalve catalogs and choose a valve that will provide at least 34.1 gpm with an 802 psi pressure drop when wide open. Though not necessary for valve selection we can estimate the needed valve port flow area A_v from standard fluid mechanics orifice-flow equations, using an oil density of 0.000084 $(lb_f\text{-}s^2)/in.^4$ and discharge coefficient of 0.63.

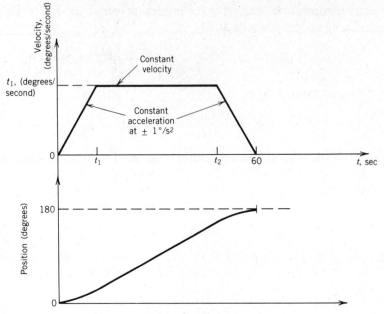

FIGURE 13.15   *Duty cycle for 180° position change.*

$$\text{volume flow rate} = 0.63\, A_v \sqrt{\frac{2\Delta p}{0.000084}} \qquad A_v = 0.097\ \text{in.}^2 \qquad (13.24)$$

At this point our "bare bones" system comprises a type-1 electrohydraulic servo as in Fig. 13.16. Our next step is to estimate the loop gain needed in this system to meet a steady-state error goal given in the specifications. This specification requires a tracking error at the RF axis of no more than $\pm 0.005°$ when the antenna has a motion given by

$$\theta_C = \omega t + \frac{\alpha t^2}{2} = (1°/\text{s})t + (0.012°/\text{s}^2)\frac{t^2}{2} \qquad (13.25)$$

Since this motion is quite slow, we simplify the system model to the pure integrator of Fig. 13.16*b*, which then gives error θ_E as

$$\theta_E = \frac{1}{K}\frac{d\theta_C}{dt} = \frac{1}{K}(\omega + \alpha t) \qquad (13.26)$$

The error component αt (caused by the constant-acceleration component of θ_C) would increase without bound, however the $0.012°/\text{s}^2$ maximum acceleration continues only until it contributes a velocity equal to the given ω of $1°/\text{s}$. Thus the maximum error is

$$\theta_{E\max} = \frac{1}{K}(\omega + \omega) = \frac{2°/\text{s}}{K} \qquad (13.27)$$

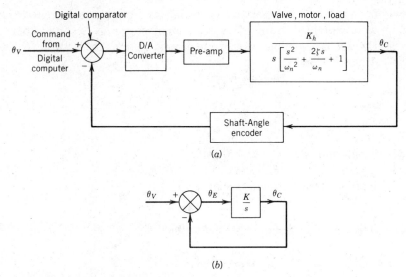

FIGURE 13.16 *Basic system block diagram with simplified version.*

This pointing error is the "vector sum" for both azimuth and elevation and is worst when the elevation axis points horizontally, giving

$$\text{total error} = \sqrt{\theta_{E,el}^2 + \theta_{E,az}^2} = \sqrt{2}\,\theta_{E,az} \qquad (13.28)$$

where we have assumed the azimuth and elevation errors to be equal. We have then

$$K = \frac{(2°/s)\sqrt{2}}{0.005} = 567\,\frac{1}{s} \qquad (13.29)$$

If we use $K = 567$, *all* the allowable error is used up by the control system, leaving none for the dish structural errors mentioned earlier. Consultation between the structural and control-system engineers on the relative severity of these two error sources resulted in a decision to divide the error budget about equally and a design goal of $K = 1200$ was set. The structural engineers also provided an estimate of the antenna structure's first natural frequency of vibration as 5 Hz. To reduce the possibility of control-system motions exciting this frequency, a closed-loop peak of 1.3 at 0.5 Hz, with rapid attenuation beyond 0.5 Hz, was set as another design goal, contingent upon this not violating any other specifications.

At this point the basic hardware items had all been selected and it was possible to check dynamics for possible simplification. The shaft encoder, D/A converter, and preamplifier produced a 1000 Hz amplitude-modulated ac signal that had to be demodulated and low-pass filtered to produce a suitable dc control signal. Because of the high ac frequency, the low-pass filter could use a small τ (0.01 s) and thus these dynamics could be neglected relative to control system bandwidth of a few hertz. Servovalve dynamics from valve current to valve opening were obtained from the valve manufacturer as second order with $\omega_n \approx 50$ Hz, again negligible. For the hydraulic system, the natural frequency ω_n was calculated from

$$\omega_n = \sqrt{\frac{2M_B d_m^2}{JV_c}} = 14 \text{ rad/s} \qquad (13.30)$$

where

$M_B \overset{\Delta}{=}$ bulk modulus = 100,000 psi

$J \overset{\Delta}{=}$ inertia referred to motor shaft = $J_{\text{motor}} + \dfrac{J_{\text{load}}}{n^2}$

$J = 0.044 + \dfrac{4 \times 10^6 \times 12}{(325)^2} = 500 \text{ in-lb}_f\text{-s}^2$

$V_c \overset{\Delta}{=}$ volume of oil under compression

$V_c = 55 \text{ in.}^3 \text{ (two motors)} + 45 \text{ in.}^3 \text{ (estimated tubing)}$

From past experience, hydraulic system damping was estimated at $\zeta = 0.3$.

The basic system thus appears as in Fig. 13.17a, for which Routh criterion gives $K < 2\zeta\omega_n = 8.4$; whereas we require $K = 1200$ to meet the steady-state error specification! This presents a challenging compensation problem. Using frequency-response design methods (Bode/Nichols plots), various combinations of lag and lead compensators, plus feedback of derivatives of the controlled variable were tried, finally leading to the system of Fig. 13.17b. The Nyquist plot (note the distorted scale) shows the compensated system to be conditionally stable but with ample (10 to 1) gain margin for both increasing or decreasing gain; whereas the closed-loop frequency response meets the design goals given earlier. Compensation in the feedback path uses *very* strong (approximate fourth) derivative signals but the tachometer generator vendor promised a very clean (0.04% ripple) tachometer signal. Analog computer simulation (digital simulation was uncommon in 1960) of this linear system model was performed at this point to check transient performance specifications, since the pencil-and-paper design was based only on steady-state and sinusoidal behavior. All specifications were met, so the linear design was accepted.

A final stage of analysis included in the proposal was a study of two nonlinearities that were felt to be potentially significant, servovalve null pressure sensitivity and gearbox backlash. With the motor and load stationary, dry friction prevents motion until the servovalve opens sufficiently to develop the needed "breakout" pressure. In an *ideal* valve, any pressure from zero to the supply pressure can be obtained, under stalled-load conditions, with an *infinitesimal* valve opening. Because of internal leakage in a real valve, however, near the null position there is a *proportionality* between valve opening and the pressure developed across a stalled load. This is called the null pressure sensitivity and

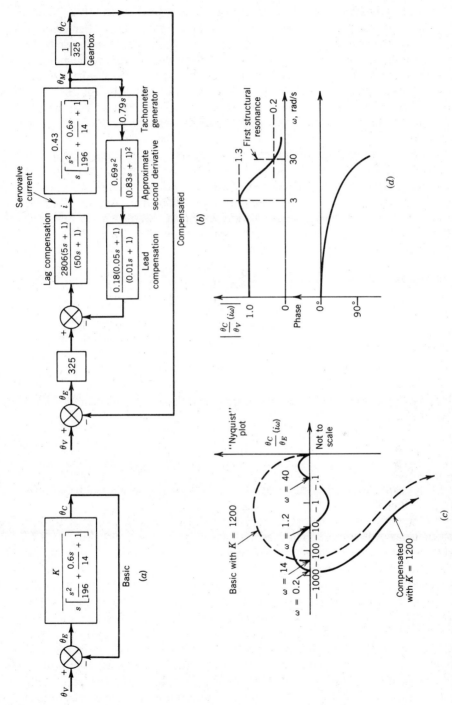

FIGURE 13.17 *Performance of basic and compensated systems.*

results in a steady-state position error since a finite error signal to the servovalve is required to develop enough pressure to break the friction loose. Errors below this critical value will *not* be corrected, thus the type-1 steady-state error behavior assumed in the linear model will *not* be achieved in reality. Experience with servovalves of the type used in this system has shown that it takes 5 to 8% of the total valve stroke to achieve full supply pressure across the load. Full stroke in the present valve required 40 mA of current, so, taking a conservative 10% value for the null range, our null pressure sensitivity is 250 psi/mA (see Fig. 13.18*a*).

To estimate the static pointing error caused by this combination of load static friction and valve null behavior, we need a number for system total static friction. We earlier computed bearing friction (and ignored motor friction) and since the motor has been selected now, we can get friction data from the manufacturer and include it in our present error estimate. Figure 13.18*b* shows measured friction torque for our motor, the breakout friction of 160 in.-lb_f being the number pertinent to our present calculation.

$$\text{load friction (referred to motor shaft)} = \frac{32{,}000}{325(0.98)^4} = 116 \text{ ft-}lb_f$$

$$\text{total friction at motor shaft} = \frac{116}{2} + 15 = 73 \text{ ft-}lb_f$$

$$\text{motor } \Delta p \text{ required to break loose} = \frac{24\pi}{22}73 = 250 \text{ psi}$$

In Fig. 3.17*b* we now need to compute how large θ_E must be to cause a valve current i of 1 mA.

$$(\theta_E)(325)(2806) = 1 \text{ mA}$$

$$\theta_E = 1.1 \times 10^{-6} \text{ rad} = 0.00063°$$

Since the resolution of the shaft encoder is 0.001° and the allowable pointing error is ± 0.005°, this error was judged acceptable.

For the gear-backlash study, the four-mesh gearbox with backlash in each mesh of gears was approximated as a single mesh of gears (325 to 1) with a single backlash. Due to the large dry friction value and the slow motions involved, the friction-controlled backlash model used in Section 8.3 of this text was judged applicable for a describing-function analysis. With G_1 and G_2 defined as in Fig. 13.19*a*, the describing function graph of Fig. 13.19*b* predicts a limit cycle of amplitude 0.44° and 3.2 rad/s frequency at the motor shaft. Although this is only 0.0005° at the antenna shaft, limit cycling of any kind was judged unacceptable and a "fix" had to be devised. Fortunately, the use of high-frequency dither to quench such limit cycles is a well-known technique. A nonlinear analog simulation was used both to check the describing-function predictions and to design and evaluate the dither effect. Limit-cycle amplitude and frequency closely checked the analytical prediction and a 60 Hz dither at the motor shaft

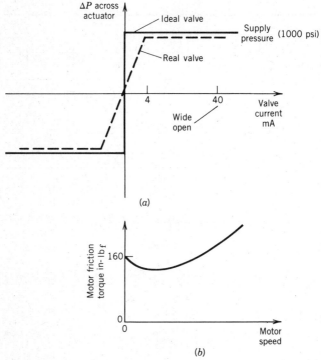

FIGURE 13.18 *Servovalve null pressure sensitivity and motor friction behavior.*

was found to quench an established limit cycle in 2 s. In the actual system, dither was introduced electrically at the servovalve, using conveniently available 60-Hz line power as the dither source.

I have now presented in abbreviated form the essential features of the design proposal submitted, in competition with others, to the project sponsor. It was selected as the most promising, a contract was awarded, and detailed design and construction went forward over a period of about four years. It is quite common in projects of this magnitude that unanticipated problems will arise, requiring design modifications. The tachometer generator, on which all the feedback compensation depends, turned out to be noisier than expected, requiring drastic reduction in the "amount" of derivative control possible. This, together with a reduction in hydraulic natural frequency from 14 to 6 rad/s (due to a larger volume of oil under compression than originally estimated) greatly reduced the allowable loop gain. To maintain the required accuracy with reduced loop gain it was necessary to go to type 2, rather than the original type 1. A type-2 system that met all the specifications under the new conditions was designed. This type-2 system, like the original type-1 system, was conditionally stable, and it was discovered that under some operating conditions, sufficient saturation occured to reduce loop gain beyond the gain margin, causing absolute

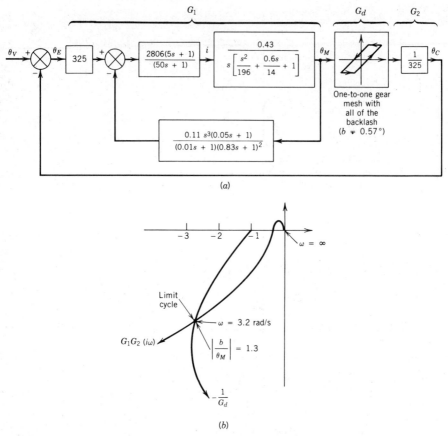

FIGURE 13.19   *Gear backlash describing-function study.*

instability. (This problem would have been present in the original type-1 system also.)

This stability problem (caused by the saturation nonlinearity) was fixed by use of another (intentional) nonlinear feature, a dual-mode controller that switches the system from type 2 to type 1 whenever system error exceeds a critical value where saturation begins to reduce loop gain dangerously. Figure 13.20 shows this final configuration. When large disturbances and/or commands cause momentarily large error and subsequent saturation, the switching logic that monitors the size of the error switches the op-amp controller to the lower, "straight through" signal path that defines a type-1 closed-loop system with a Nyquist plot that does *not* show instability for reduced gain. This stable system brings the error back toward zero, and when it is small enough to no longer cause saturation-induced gain reduction, the controller switches to the upper signal path, which adds an electronic integrator plus some lead compensation, defining a type-2 system of high accuracy and bringing the error stably toward

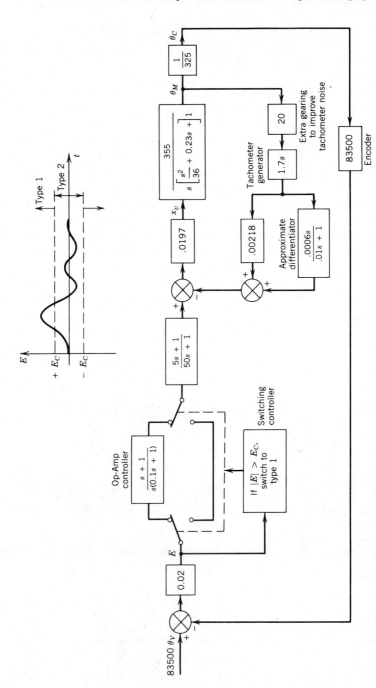

FIGURE 13.20 *Dual-mode switching servo to solve conditional stability problems.*

zero. Final design of such systems with multiple nonlinearities (saturation plus a switching controller) is best accomplished by computer simulation, once linear system analytical studies have "roughed out" the two individual (type 1 and type 2) designs.

The problem of conditional stability with saturation is not limited to the present specific example and the switching controller approach used in the Haystack antenna is a candidate solution for similar problems. Other possible approaches have been investigated,[5] Fig. 13.21 showing the basic system used as an example by Truxal.[6] The type-1 system shown there has the conditionally stable Nyquist plot of Fig. 13.22a when loop gain is properly set. If we could cause this curve to change as in Fig. 13.22b when saturation occurs, the lower effective gain would *not* cause instability. The reduction in phase lag needed to shift the curve as shown is obtained in a switching controller by removal of an integrator, giving a 90° reduction at every frequency. I now wish to show Truxal's alternative method for achieving the desired lag reduction.

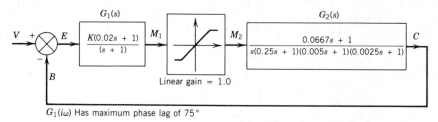

FIGURE 13.21   *Conditionally stable type-1 system with saturation.*

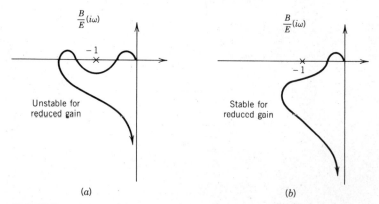

FIGURE 13.22   *Reduction in phase lag removes conditional stability.*

---

[5]R. C. Rapson, Jr., Evaluation of Several Methods for Large-Signal Stabilization of Conditionally Stable Systems with Saturation, Ph.D. Diss., Ohio State University, Mechanical Engineering, Sep., 1969.
[6]J. G. Truxal, *Control Systems Synthesis*, McGraw-Hill, 1955, pp. 595–598.

To implement this method we must be able to reconfigure the system with a minor loop as in Fig. 13.23. In designing this minor loop we make use of the well-known behavior of *any* closed-loop system with desired value V, controlled variable C, forward path G, and feedback path H.

If amplitude ratio of $GH(i\omega)$ is small relative to 1.0

$$\frac{C}{V} = \frac{G}{1 + GH} \approx \frac{G}{1 + 0} \approx G$$

If amplitude ratio of $GH(i\omega)$ is large relative to 1.0

$$\frac{C}{V} = \frac{G}{1 + GH} \approx \frac{G}{GH} \approx \frac{1}{H}$$

In Fig. 13.23, the minor loop contains the saturation, which we treat approximately as a variable gain $0 < K_{df} \leq 1.0$, using the describing function. For nonsaturating operation $K_{df} = 1.0$, whereas increasing saturation causes $K_{df} \to 0$. For nonsaturating operation we want system behavior to be identical to that of Fig. 13.21, that is

$$\frac{M_2}{E}(i\omega) = \frac{K(0.02i\omega + 1)}{(i\omega + 1)} \tag{13.31}$$

since this system *was* satisfactory when unsaturated. When saturation causes $K_{df} \to 0$, we get

$$\frac{M_2}{E}(i\omega) \approx \frac{K_1 K_2 K_{df}}{1 + K_2 K_{df} H_1} \approx K_1 K_2 K_{df} \tag{13.32}$$

which we see has *zero* phase lag, 75° *less* than the unsaturated system; thus the desired curve shift from Fig. 13.22a to 13.22b may be obtained. Since $K_{df} = 0.0$ is realized only for *infinitely* large signals, we will not actually achieve a 75° improvement; however sufficient improvement for stabilization may be possible. Since the describing function is an approximation, final design should

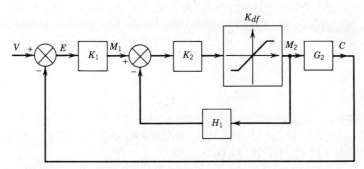

FIGURE 13.23 *Minor-loop feedback to reduce phase-lag during saturation.*

be checked and refined with simulation studies, however a preliminary design can proceed as follows.

To preserve the original behavior for unsaturated operation we require

$$\frac{K_1 K_2 K_{df}}{1 + K_2 K_{df} H_1(i\omega)} = \frac{K(0.02i\omega + 1)}{i\omega + 1} \quad \text{when } K_{df} = 1 \quad (13.33)$$

Solving for $H_1(i\omega)$ gives

$$H_1(i\omega) = \frac{\left(\dfrac{K_1 K_2}{K} - 0.02\right)i\omega + \left(\dfrac{K_1 K_2}{K} - 1\right)}{K_2(0.02i\omega + 1)} \quad (13.34)$$

and since $K_1 K_2$ must equal K to match the original gain

$$H_1(i\omega) = \frac{\dfrac{0.98}{K_2}(i\omega)}{0.02i\omega + 1} \quad (13.35)$$

This gives

$$\frac{M_2}{E}(i\omega) = \frac{KK_{df}}{1 + 0.98\, K_{df}\left(\dfrac{i\omega}{0.02i\omega + 1}\right)} = \frac{KK_{df}(0.02i\omega + 1)}{(0.98\, K_{df} + 0.02)i\omega + 1} \quad (13.36)$$

For unsaturated operation ($K_{df} = 1$) this becomes

$$\frac{M_2}{E}(i\omega) = \frac{K(0.02i\omega + 1)}{i\omega + 1} \quad \text{max lag} = 75° \quad (13.37)$$

whereas maximum saturation ($K_{df} \to 0$) gives

$$\frac{M_2}{E}(i\omega) = KK_{df} \quad \text{max lag} = 0° \quad (13.38)$$

Using this preliminary design as a starting point, simulation studies would explore the apportionment of gain K into K_1 and K_2 and system behavior for various types and sizes of command inputs, to see whether acceptable performance is attained. Just as for the switching controller approach, this scheme is potentially useful for conditional stability/saturation problems in general, not just this specific example.

13.4 SPECIALIZED CONTROLLERS FOR SYSTEMS WITH LARGE DEAD TIMES: SAMPLING CONTROLLERS AND SMITH PREDICTORS

In the process-control field the common occurrence of systems with significant dead times creates difficult control problems, as we have already seen in Sections 9.4, 9.5, 10.3, and Fig. 12.34 and 12.35. A conventional proportional-plus-integral controller, properly tuned (Fig. 12.34), may often be the best practical solution, however a number of controllers designed specifically for dead-time problems have been invented and successfully applied to real processes. In choosing among the available alternative control schemes, readers will usually have to perform a rather careful and comprehensive evaluation of *their specific* application, since the literature reveals some difference of opinion even among expert practitioners. [In evaluating opinions expressed in the literature, the reader is further cautioned to make a distinction between authors with long experience in daily practical engineering ("expert practitioners") and theoretical/academic types whose practical experience may be limited.]

An example case where even expert practitioners disagree is the *sampling* (or "wait and see") *controller.* Shinskey[7] promotes this scheme, saying, "A sampling integral controller is so effective on dead-time dominated processes that it is recommended over complementary feedback, to be used even on continuous measurements." Ross,[8] on the other hand, rates this approach the worst of four alternatives considered, using a rating scheme that considers cost, ability to handle noisy signals, step-disturbance performance, ability to handle changing dead times and process gain, and difficulty of tuning. I thus present the sampling controller as a method that *is* being successfully applied but that should be carefully evaluated against other candidate methods for each new application. Figure 13.24 shows a sampling integral controller applied to first-order-lag-plus-dead-time process. The sampling switch closes for a time interval P every TS seconds, in a periodic fashion. The interval P should be small relative to TS, and TS should be equal to or greater than the process dead time, optimum values of both P and TS being found most easily by simulation studies such as the CSMP that follows.

```
              V  =  KV*STEP(0.0)      } step command of size 1.0 at t = 0.0
PARAM         KV =  1.0              
              X  =  IMPULS (0.0,TS)   } train of "clock" pulses at TS
PARAM         TS =  2.0                 intervals, starting at time = 0
              TRIG = PULSE (P,X)      } periodic switch action as in Fig. 13.24
PARAM         P  =  0.1                 (TRIG = 1.0 when CLOSED)
              E  =  V - C    continuous actuating signal
              ES =  TRIG*E      sampled actuating signal
              M1 =  INTGRL (0.0, ES)  }
              M2 =  KI*M1             }  integral controller
PARAM         KI =  1.0              
              U  =  KU*STEP(TU)       }  step disturbance of size KU at time TU
PARAM         KU =  0.0, TU = 50.0   
              MP =  U + M2
```

[7]F. G. Shinskey, *Process-Control Systems*, 2nd Ed., McGraw-Hill, New York, 1979, pp. 103–107.
[8]C. W. Ross, Evaluation of Controllers for Deadtime Processes, ISA Trans., Vol. 16, No. 3, 1977, pp. 25–34.

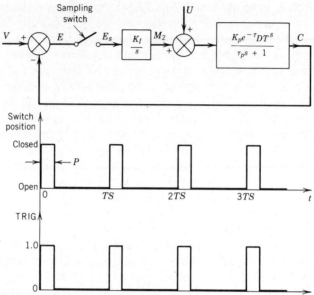

FIGURE 13.24 *Sampling controller for dead-time problems.*

```
              C1  = KP*MP                              ⎫  controlled process
PARAM         KP  = (1.0,2.0,5.0,10.0,20.0)           ⎪  with τ_p = 1.0,
              C2  = REALPL (0.0, 1.0, C1)             ⎬  τ_DT = 1.0 and a
              C   = DELAY (100, 1.0, C2)              ⎭  range of gain values
TIMER         FINTIM = 50., DELT = .01, OUTDEL = .4
OUTPUT        C
```

Figure 13.25 shows that with $P = 0.1$ and sampling interval TS equal to twice the dead time, a loop gain $(K_I)(K_p)$ of 5.0 seems to give a fast, well-damped response. Note that with an integral controller, the control interval P also contributes to the "gain" since longer P results in a larger controller output for the same value of E. The concept behind the sampling controller is that the control action will be "better informed" if it *waits* a period $TS > \tau_{DT}$ to see what the result of its previous action actually was, before instituting a new correction. A rerun of this program with $TS = 0.5$ (half the dead time) shows (Fig. 13.26) that a gain of 10 is now absolutely unstable. Returning to $TS = 2.0$ and with $K_p = 1.0$, Figure 13.27 uses an expanded time scale to show some details of the sampling controller response, particularly how it "waits" for the system to respond.

Another candidate for control of dead-time processes is the Smith predictor[9]

[9] O. J. M. Smith, A Controller to Overcome Dead Time, *ISA Jour.*, Feb. 1959, Vol. 6, No. 2, pp. 28–33; D. E. Lupfer and M. W. Oglesby, Applying Dead-Time Compensation for Linear Predictor Process Control, *ISA Jour.*, Nov. 1961, Vol. 8, No. 11, pp. 53–57; J. R. Scleck and D. Hanesian, An Evaluation of the Smith Linear Predictor Technique for Controlling Deadtime Dominated Processes, *ISA Trans.* Vol. 17, No. 4, pp. 39–46, 1978; C. W. Ross, *Deadtime Processes*.

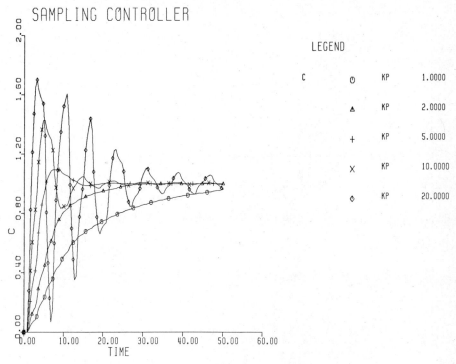

FIGURE 13.25 *Effect of loop gain on sampling controller performance.*

(also called complementary feedback), one version of which is shown in Fig. 13.28. Here the controller includes a model of the controlled process, but with the dead-time effect separated from the rest of the process dynamics. The need for a dead-time model in the controller inhibited the application of this concept before digital control became common, because good analog dead-time implementations were not available. Accurate digital realizations of dead time are easily achieved today, encouraging consideration of Smith predictors. In Fig. 13.28a, ideally $K_{pm} \equiv K_p$, $\tau_{pm} \equiv \tau_p$, and $\tau_{DTm} \equiv \tau_{DT}$; that is, the controller model is a perfect match with the process. (Mismatch of the two sets of dynamics results in degradation of control performance and is a real practical problem.) For $U \equiv 0$ and perfectly matched dynamics, note that the signal $(C - CM)$ would be exactly zero at all times since the paths from $M2$ to C and $M2$ to CM are identical. This means that the closed-loop system is effectively as in Fig. 13.28b with the dead time *outside* the loop. Thus the loop is easier to design analytically and can have a better response. The "external" dead time in the path from $M2$ to C is a fundamental limitation which cannot be removed.

In Fig. 13.28b if we take $U \equiv 0$ and assume a perfect match between model and process we get

$$(\tau_p s^2 + s + K_I K_p)C = K_I K_p e^{-\tau_{DT}s}V \qquad (13.39)$$

Clearly the characteristic equation of this closed-loop system is totally unaffected

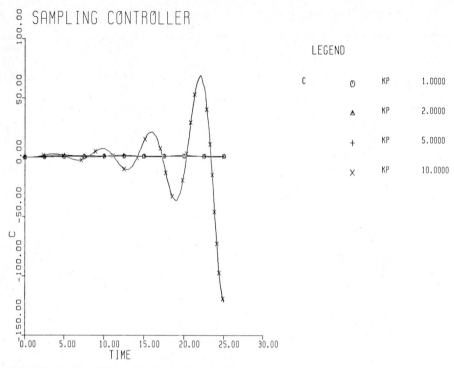

FIGURE 13.26   *Effect of gain with sampling interval equal to half the dead time.*

by the dead time τ_{DT}, allowing us to set gain for desired performance without the limitations associated with dead times. Of course the right side of Eq. 13.39 shows that the effect of any command *V will* be delayed by τ_{DT}. This aspect of the dead time cannot be avoided. We will now use CSMP simulation to compare the performance of the Smith predictor system of Fig. 13.28*a* to that of a conventional integral controller for the same process, for both step commands and disturbances.

```
                V  = KV*STEP(0.0) ⎫
PARAM           KV = 1.0          ⎬  unit step command at time = 0
                E  = V−CC    summing junction
                M1 = INTGRL(0.0,E) ⎫
                M2 = KI*M1         ⎬  integral controller
PARAM           KI = 1.0           ⎭
                MP = U+M2          ⎫
                U  = KU*STEP(12.)  ⎬  step disturbance at time = 12
PARAM           KU = 1.0           ⎭
                C1 = KP*MP                      ⎫ first-order plus
PARAM           KP = 1.0                        ⎥ dead-time process
                C2 = REALPL (0.0, 1.0, C1)      ⎬ Kp = 1.0, τp = 1.0,
                C  = DELAY (100, 1.0, C2)       ⎭ τDT = 1.0
```

SAMPLING CONTROLLER

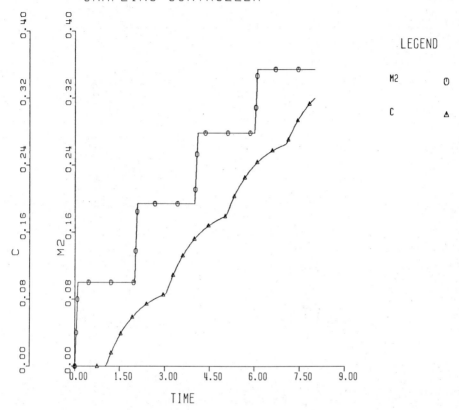

FIGURE 13.27 *"Wait-and-see" nature of sampling controller.*

```
            C1M  =  KPM*M2                          ⎫
PARAM       KPM  =  1.0                             ⎬  process model
            C2M  =  REALPL (0.0, 1.0, C1M)          ⎮  (perfect match)
            CM   =  DELAY (100, 1.0, C2M)           ⎭
            CC   =  C2M + C - CM     complete Smith predictor
            VA   =  V      start conventional system
            EA   =  VA - CA      summing junction
            M1A  =  INTGRL (0.0, EA)    ⎫ integral controller with same
            M2A  =  KI*M1A              ⎭ gain as in Smith predictor
            MPA  =  UA + M2A
            UA   =  U      same disturbance
            C1A  =  KP*MPA                           ⎫
            C2A  =  REALPL (0.0, 1.0, C1A)           ⎬  same process as for
            CA   =  DELAY (100, 1.0, C2A)            ⎭  Smith predictor
TIMER       FINTIM = 30., DELT = .01, OUTDEL = .2
OUTPUT      C,CA
```

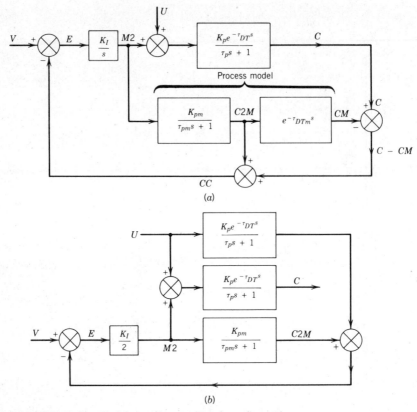

FIGURE 13.28 *Smith predictor control configuration.*

Figure 13.29 dramatically illustrates the performance advantage of the Smith predictor (controlled variable C) over a conventional integral controller (CA) with the same gain, for both step commands and disturbances. Equation 13.39 predicts a second-order transient with $\omega_n = 1$ rad/s and $\zeta = 0.5$, which is visually verified in Fig. 13.29, as is the 1-s dead time in both C and CA's response. Since one of the main difficulties encountered in applying the Smith predictor is a mismatch between the real process dynamics and their model in the controller, we run another simulation in which dead-time mismatches of $\pm 10\%$ and $\pm 50\%$ are studied. Since CSMP's designers did not provide the DELAY statement with the capability of accepting a dead time in letter form and then exercising multiple values through a multivalued PARAM statement (as we have done with many multiple-run CSMP problems before), our CSMP program requires use of a NOSORT section with conventional FORTRAN programming as follows. Right after the C2M = statement in the previous program we write:

```
PARAM      K = (1., 2., 3., 4., 5.)      want five values of τDTm
NOSORT     enter a program section that will not be sorted
```

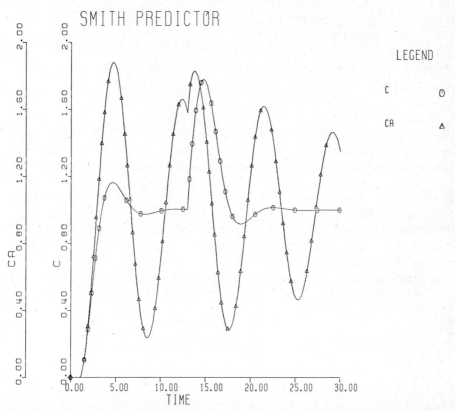

FIGURE 13.29 *Comparison of Smith predictor with conventional integral control.*

```
        IF (K−1.1) 1, 1, 2        ⎫  selects τ_DTm = 1.0 when K = 1.
1       CM=DELAY (100, 1.0, C2M)  ⎬
        GO TO 20                  ⎭
2       IF (K−2.1) 3,3,4          ⎫  selects τ_DTm = 1.1 when K = 2.
3       CM = DELAY (110, 1.1, C2M)⎬
        GO TO 20                  ⎭
4       IF (K−3.1) 5,5,6          ⎫  selects τ_DTm = 0.9 when K = 3.
5       CM = DELAY (90, .9, C2M)  ⎬
        GO TO 20                  ⎭
6       IF (K−4.1) 7,7,8          ⎫  selects τ_DTm = 0.5 when K = 4.
7       CM = DELAY (50, .5, C2M)  ⎬
        GO TO 20                  ⎭
8       IF (K−5.1) 9,9,10         ⎫  selects τ_DTm = 1.5 when K = 5.
9       CM=DELAY (150, 1.5, C2M)  ⎬
        GO TO 20                  ⎭
10      CONTINUE
20      CONTINUE
```

```
SORT        end of NOSORT section
            CC = C2M + C - CM  ⎫
            etc.               ⎬  rest of program is same as before
                               ⎭
OUTPUT    C   ⎫  plots five curves of C versus time
PAGE MERGE    ⎬  on one graph, for the five different τ_DTm values
              ⎭
```

Figure 13.30 shows the results of this study. We see that when the model dead time is too small (K = 3,4, τ_{DTm} = 0.9, 0.5) the command response deteriorates significantly; whereas τ_{DTm} too large (K = 2, 5, τ_{DTm} = 1.1, 1.5) seems to cause little trouble. This can be partially explained by noting that in Fig. 13.28a, if τ_{DTm} were *zero*, signals *C2M* and *CM* would *cancel* each other in *CC*, and the system would revert to an "ordinary" one (no Smith predictor effect *at all*). We see also that the size and shape of the peak error due to the disturbance at t = 12 is largely unaffected by mismatch errors, although system stability and settling time are adversely affected, especially by τ_{DTm} being too small. The poorer performance for disturbance inputs can again be partially explained from Fig. 13.28a where it is clear that a *U* input goes "straight through" to *C* when it first occurs and that the effect of the model path only makes itself felt *later*.

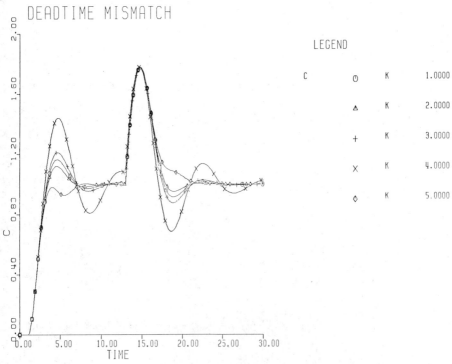

FIGURE 13.30 *Effect of process/model mismatch in dead time.*

A recent application of the Smith predictor to a British sinter plant has been described in the literature.[10] Plants of this type use several conveyor belts to transport materials between the various processing operations. In the example cited, a 7-min transport lag between an ore dispenser and a surge hopper caused a PID controller to oscillate unacceptably even when proportional band and reset time were set to the largest available values. By augmenting the PID controller with a Smith predictor, much-improved control was obtained.

Many dead-time control problems can be handled by conventional PI controllers, sampling controllers, or Smith predictors. Further information on these and other schemes not covered here may be found in the literature.[11]

13.5 ULTRAPRECISION SPEED CONTROL; PHASE-LOCK-LOOP SERVOS

Many speed-control systems using electrical or hydraulic motors employ a dc tachometer generator as the speed sensor and a dc voltage as the reference input, achieving accuracies the order of 0.1 to 1.0%. When higher accuracies are required, the tachometer generator becomes a limiting factor since changes in air gap, temperature, and brush wear cause its sensitivity (volts/rpm) to be uncertain by more than 0.1%. By substituting a digital speed sensor (tachometer encoder[12]) and a crystal-controlled reference frequency source, much higher accuracy can be achieved. One means of utilizing these components is called a phaselock-loop servo. These have been treated in detail by Geiger[13] and we now present a brief description based on this study.

Tachometer encoders use magnetic or optical principles to produce a train of electrical pulses whose frequency is accurately proportional to shaft angular velocity. In an optical tachometer a rotating glass disk has alternate transparent and opaque segments around its outer edge, with a stationary grating with the same "line" spacing adjacent. A light source projects a beam through the disk and grating and an electrooptic sensor (phototransistor) one the other side produces an electrical pulse train in response to the alternating illumination. The roughly sinusoidal waveform is converted to a more useable square wave

---

[10]M. Hague, R. A. Senior, and R. W. Smith, The Excess Weight Controller, *Meas. and Cont.*, Vol. 14, Jan. 1981, pp. 13–17.

[11]J. F. Donogue, Review of Control Design Approaches for Transport Delay Processes, *ISA Trans.*, Vol. 16, No. 2, 1977, pp. 27–34; R. F. Giles and T. M. Bartly, Gain-Adaptive Deadtime Compensation, *ISA Trans.*, Vol. 16, No. 1, 1977, pp. 59–64; W. M. Wheater, How Modeling Accuracy Affects Control Response, *Cont. Eng.*, Oct. 1966, pp. 85–87; M. Mori, Discrete Compensator Controls Dead Time Process *Cont. Eng.*, Jan. 1962, pp. 57–60; I. McFarlane, Dead-time controller (Smith Predictor) applied to interacting feedback loops on a baking oven, Trans. Inst. MC, Vol. 6, No. 1, Jan.–Mar. 1984.

[12]Doebelin, *Measurement Systems*. pp. 294–299.

[13]D. F. Geiger, *Phaselock Loops for DC Motor Speed Control*, Wiley, New York, 1981.

of the same frequency with simple electronics. Typical tachometers have 100 to 5000 lines and since this number is absolutely fixed for a given disk, a certain shaft speed *always* produces a precisely proportional frequency, irrespective of temperature, age, wear, and the like. Since the line spacing on the disk cannot be manufactured precisely uniformly, inaccuracy of *instantaneous* sensed speed within one revolution is still possible; however, average sensed speed over an integer number of revolutions is absolutely accurate. To take advantage of the tachometer encoder's speed measurement accuracy, the servo-system reference input must be similarly precise, and we use a frequency synthesizer built from a piezoelectric crystal and readily available IC chips. Such frequency sources are accurate to a few parts per million, compared with a few parts per thousand for dc voltage references used with dc tachometer generator speed sensors.

A linear system model has been found to be useful for analysis and design purposes and is presented in Fig. 13.31, using an armature-controlled dc motor as the actuator. The phaselock-loop principle could be implemented in various forms; Fig. 13.31 shows a scheme that has proven successful in practice. The controller includes three separate signal-processing paths whose outputs are combined to produce a single control voltage e_c that is filtered and put into a "transconductance" (voltage input, current output) power amplifier supplying armature current i_a to the fixed-field (usually PM) motor. The current amplifier suppresses the L/R time constant of the armature circuit, lowering the order of the open-loop transfer function denominator and thus easing compensation. We now examine each of the three controller functions separately.

To understand the action of the path that uses frequency-to-voltage (F/V) converters we first show the basic F/V converter of Fig. 13.32a. The square wave e_ω of frequency f Hz triggers the multivibrator every $1/f$ s, each time producing a pulse of fixed height E_p and duration T_p. Thus the pulse train e_{fi}

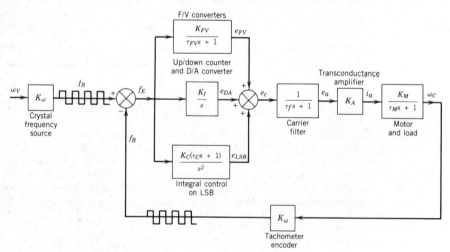

FIGURE 13.31 *Phaselock-loop speed control system.*

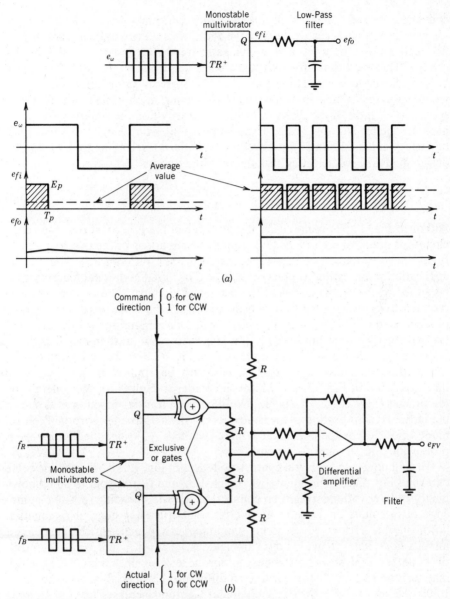

FIGURE 13.32 *Frequency-to-voltage conversion.*

at the filter input has a small average value when f is small and a large average value when f is large. The low-pass filter extracts this average value as a "dc" voltage proportional to frequency f at its output e_{fo}. Choice of the filter time constant τ_{FV} is a tradeoff between low ripple in the e_{fo} signal (requires large τ_{FV}) and fast response to changing frequency (requires small τ_{FV}). An additional dynamic lag is a dead-time effect due to the "sampled" nature of the F/V conversion process. That is, useful frequency information is contained only in the positive-going edges of the e_ω wave form, thus there can be a dead time of as much as $1/f$ s between a step change in frequency and the sensing of this change. This dead-time effect is usually made negligible by putting a lower limit on the allowable controlled speed relative to servo-system closed-loop bandwidth ω_N rad/s. When the tachometer is running at the minimum allowable speed ω_{CMIN} rad/s, if the disk has N lines per revolution the square-wave signal has a frequency $N\omega_{\text{CMIN}}/2\pi$ Hz. Experience has shown that this frequency must be kept at at least 10 times the servo-system bandwidth, leading to the following design guideline for selecting tachometer "line density" N.

$$N \geqq \frac{300\ \omega_N}{\pi(\text{rpm})_{\text{MIN}}} \tag{13.40}$$

Although Fig. 13.32a makes clear the principle of the F/V converter, the actual circuit for practical use in Fig. 13.31 requires comparison of the two frequencies f_R and f_B to develop a voltage proportional to their difference f_E. Also, if the application requires that ω_C *reverse* (*CW, CCW*), then additional features must be added to the tachometer and the command circuitry. The tachometer now requires a *second* grating and optical sensor on the same disk, arranged to produce a square wave 90° out of phase with our original square wave. These two square waves are put into a D flip-flop, whose output then provides the 1, 0 logic signals shown as "actual direction" in Fig. 13.32b. To command CW or CCW rotation, similar 1, 0 logic signals must be applied at the "command direction" line in Fig. 13.32b. I refer the reader to Geiger for more details on the practical F/V circuit of Fig. 13.32b; its overall transfer function is as shown in Fig. 13.31. This portion of the controller provides a basic proportional control mode (modified by the unavoidable filter dynamics τ_{FV}) since the voltage e_{FV} is proportional to f_E when f_E is steady.

Turning now to the integral control mode K_I/s in Fig. 13.31, we examine some details in Fig. 13.33. The square-wave signals f_R and f_B are input to an up/down binary counter (often an eight-bit counter) as shown. A leading (positive-going) edge of f_R counts the counter up one bit while a leading edge of f_B counts it down one bit. (If two edges occur "simultaneously," a special clocking circuit inhibits *both* edges to avoid count ambiguities.) Thus if f_R and f_B are identical, the counter output stays constant except for the least significant bit (LSB), which may turn on and off. If there is a fixed difference between f_R and f_B, say f_R = 10,000 Hz and f_B = 9990 Hz, the counter's output increases linearly (ramps) with time, in our numerical example at the rate 10 counts/s, thus we accomplish a digital integrating effect. For a 5000-line disk running at 3000 rpm, we get a frequency of 250 kHz, and a four parts per million (ppm) speed error (250,000 versus 250,001) causes the counter to ramp at 1 count/s, demonstrating the potential accuracy of such systems. Since our motor is driven by an analog voltage, we connect a D/A converter[14] (often 8-bit) to the counter output. The

---

[14]Doebelin, *Measurement Systems*, pp. 767–775.

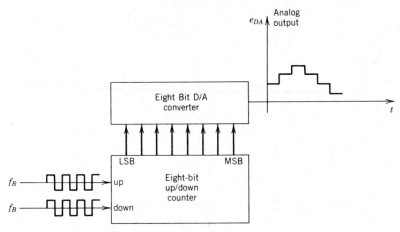

FIGURE 13.33 *Digital integration scheme.*

D/A converter output is biased so that when the counter is empty (all bits = 0), the analog output voltage is, say, -5 V and when the counter is full (all bits = 1) we have $+5$ V. For an eight-bit counter (maximum count = 256), we would change the output voltage roughly 40 mV for each count. As in any integral control mode; the D/A output will change until it is large enough to drive the motor at the desired speed; however because of the quantized nature of the D/A output, the LSB of the counter must switch on and off in a duty cycle which produces an average motor torque just sufficient to meet the needs of the load. This produces a frequency-locked condition but not a true phase lock.

The integral control mode just described seems to provide the desired correspondence between the reference speed and motor speed; however when this correspondence is approached, the counter operation becomes erratic because of the circuit arrangement which rejects coincident edges of tachometer and reference signals. Such coincident edges occur too frequently when the two frequencies are nearly equal and the LSB duty cycle is allowed to vary randomly. The third controller function in Fig. 13.31 is designed to resolve this difficulty by forcing the duty cycle of the LSB to be exactly 50%, giving the f_R and f_B signals a 180° phase difference. This ensures that the leading edges of the up-count and down-count signals will be well separated in time, giving reliable counter operation as the frequency-locked condition is approached. Figure 13.34 shows the op-amp circuit used to implement this function, using the pulse train provided by the turning on and off of the LSB as input voltage. (When the LSB is on, its voltage is constant at a value that is often the same as E_p in Fig. 13.32a since both devices (multivibrator, counter) use the same family of basic digital building blocks.) Our op-amp circuit is intended to compare the average value of the LSB pulse train with $E_p/2$. Any difference is integrated and sent to the motor drive circuit. Only when the dc component of the LSB output is exactly

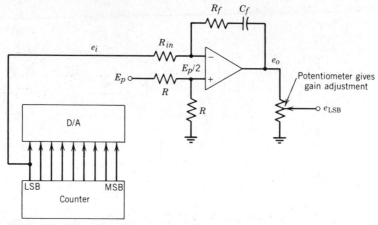

FIGURE 13.34 *Control of LSB duty cycle.*

$E_p/2$ (50% duty cycle) will the op-amp output cease to change, thus this component of the controller will enforce the desired 180° phase between f_R and f_B at the frequency-locked condition. With the op-amp positive input at $E_p/2$ volts we get

$$\frac{e_i - \dfrac{E_p}{2}}{R_{in}} = \frac{\dfrac{E_p}{2} - e_o}{R_f + \dfrac{1}{C_f D}} = -\frac{C_f D e_o}{R_f C_f D + 1} \qquad (13.41)$$

$$e_o = -\left(\frac{R_f}{R_{in}} + \frac{1}{R_{in}C_f D}\right)\left(e_i - \frac{E_p}{2}\right) \qquad (13.42)$$

There is an *additional* integrating effect between the frequency error and the e_i signal, caused by a steady "widening" of the LSB pulse when there is a frequency difference. We see in Fig. 13.35a that a fixed frequency difference results in a linear increase with time in the LSB pulse area, thus the average value of e_i grows with time, giving an integrating effect. This works properly, however, only for small changes in frequency near the locked condition since the LSB pulse can only widen so far before the leading edges of f_B and f_R become coincident (or nearly so), the width drops to zero, and the widening process repeats itself. However this control mode is *intended* to function only for small perturbations in the neighborhood of the frequency-locked and LSB-50%-duty-cycle condition (see Fig. 13.35b). Combining the integrating effect of Fig. 13.35 with the op-amp behavior of Eq. 13.42, and treating the system of Fig. 13.31 as a small perturbation model (thus $E_p/2$ does not appear, since it has no perturbation), the single integral plus double integral transfer function $K_C(\tau_C s + 1)/s^2$ is explained. Since all three of the controller modes produce signals that display some "stepiness," a final overall filtering (the "carrier" filter $1/(\tau_f s + 1)$) is provided.

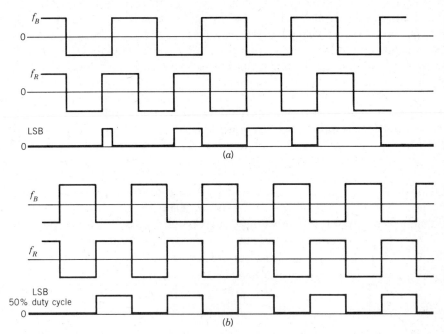

FIGURE 13.35 *(a) Integrating effect of frequency error. (b) 180° phase relation for 50% LSB duty cycle.*

For the system model of Fig. 13.31 we get an open-loop transfer function of type-2 form.

$$\frac{f_B}{f_E}(s) = \frac{K\left(\dfrac{s^2}{\omega_n^2} + \dfrac{2\zeta s}{\omega_n} + 1\right)}{s^2(\tau_{FV}s + 1)(\tau_f s + 1)(\tau_M s + 1)} \tag{13.43}$$

Geiger presents a very detailed design procedure, including many practical aspects of component selection, based on standard frequency response methods such as Bode plots. A few highlights may be of interest here. Usually ζ is set at 1.0 to give well-behaved lead compensation near ω_n, which might typically be 150 rad/s. Also, the two filter time constants τ_{FV} and τ_f are often taken equal, typical values being 0.001 s. The motor/load time constant τ_M varies significantly with load inertia, but a typical value might by 0.1 s. Laboratory measurement techniques needed for the practical troubleshooting and development of phase-lock servos are also covered in Geiger.

13.6 NONLINEAR CONTROLLERS TO STABILIZE VARIABLE-INERTIA ROBOT ARMS

Since most industrial robots are feedback systems (servomechanisms), stability considerations are an important part of their design. When a robot arm has several joints, motion about one joint pivot point can cause a change in the moment of inertia about another joint, giving a mechanical system with variable inertia. Since inertia is one of several parameters affecting stability, it is possible that a robot arm servo system could be stable for one position of the arm and unstable for another. If we reduce loop gain to recover stability in the unstable position, this lower gain may reduce accuracy in other positions. (Another "variable inertia" problem in robots occurs when the "payload" being moved by the arm is changed. A materials-handling robot, for example, has one inertia when it reaches its empty hand to grasp a machine part and a different inertia when the part is "in hand" and being positioned.) Various methods of dealing with the problem of variable inertia are in use or have been suggested. Use of high-inertia motors or high gear ratios can make the total effective inertia depend mostly on motor inertia (which is fixed) and relatively little on arm inertia, which changes. Another approach, which we explore below in simplified fashion, uses a nonlinear controller to compensate for the nonlinear (variable-inertia) load.

Consider the servosystem of Fig. 13.36a for positioning the θ_C axis of a two-axis robot shown schematically in Fig. 13.36b. (Figure 13.36a assumes angle ϕ is fixed, giving a fixed inertia J about the ϕ axis.) Routh criterion gives

$$K_{\text{loop}} \triangleq \frac{K_1}{B} < \left(\frac{1}{\tau_M} + \frac{B}{J}\right) \tag{13.44}$$

Since J varies with ϕ, gain margin will vary with ϕ, giving nonuniform performance as ϕ changes, or perhaps even instability. To study this problem, and

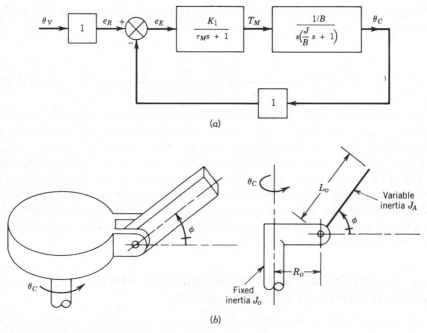

(a)

(b)

FIGURE 13.36   *Variable-inertia robot servo system.*

possible solutions, we must first be careful to note that when ϕ changes with time while θ is also varying, it is incorrect to simply let J become a time-varying number in the block diagram of Fig. 13.36a. Rather we must return to the basic Newton's law, properly formulated for a *variable* inertia.

$$\sum \text{torques about } \theta_C \text{ axis} = \text{time rate of change of angular momentum} \quad (13.45)$$
$$\text{about } \theta_C$$

$$= \frac{d}{dt}(J\omega_C) = J\dot{\omega}_C + \omega_C\dot{J} \quad (13.46)$$

If we assume that the ϕ motion is *given* as a function of time, then Eq. 13.46 is linear with time-varying coefficients since J and $\dot{J}$ are known time functions. If we took a more general view and let ϕ also be an unknown in its own equation of motion, then the equations are truly nonlinear. To get a formula for J we now approximate the arm as a geometric "line" of total mass M and length L_0 as in Fig. 13.36b. Using calculus and the definition of moment of inertia we find the variable inertia J_A to be

$$J_A = M\left(R_0^2 + R_0L_0 \cos \phi + \frac{L_0^2}{3}\cos^2\phi\right) \quad (13.47)$$

We now pose a specific problem to get some experience with this type of system. To exercise the robot in a way that brings forth the variable-inertia effects, let command θ_V and arm rotation ϕ be given as in Fig. 13.37, a representative two-axis motion. Also take numerical values as

$$J_0 = 0.005 \qquad M = 0.006 \qquad L_0 = 5.0 \qquad R_0 = 1.0$$

$$B = 1.0 \qquad K_1 = 44.29 \qquad \tau_M = 0.05$$

Since the differential equations of such systems are analytically unsolvable, we use CSMP simulation to model Eq. 13.46, 13.47, and the linear parts of Fig. 13.36a. (Details are left for problem 13.17). Simulation results given in Fig. 13.38 show the system to be stable in the low-inertia position ($\phi = 90°$) but

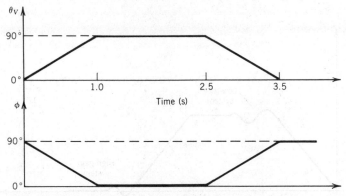

FIGURE 13.37 *Schedule for θ_V and ϕ to exercise robot.*

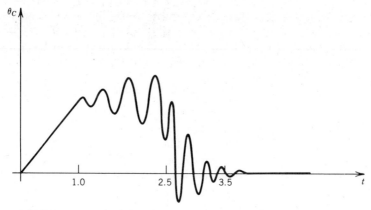

FIGURE 13.38 *Stability behavior of uncompensated variable-inertia system.*

unstable as $\phi \to 0°$. One approach to stabilization of this system is to have the servo system adjust its own loop gain, as a function of inertia, so that inequality 13.44 is always satisfied. Note that this is *not* analytically rigorous since we are using Routh criterion (strictly limited to constant-coefficient linear systems) on a variable-coefficient system. However the idea is intuitively reasonable and we will check it carefully using simulation. Since we know J as a function of ϕ and the robot already has means for measuring ϕ (the ϕ-axis servo system requires it), this approach is practically feasible, particularly if the robot controllers are microprocessor based, making available the computing power to calculate J from the measured ϕ.

One approach would be to continuously adjust K_{loop} so that $K_{loop} = 0.4$ $(1/\tau_M + B/J)$, that is, maintain a gain margin of 2.5 at all times. Using CSMP to try out this idea we get the favorable results of Fig. 13.39 (Problem 13.7c). Although our system is now stabilized, the performance is not uniform, the lowered gain at $\phi = 0°$ giving more overshoot and oscillation (and lower accuracy) than the higher-gain system at $\phi = 90°$. To maintain more uniform

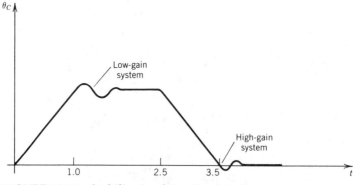

FIGURE 13.39 *Stabilization by gain scheduling.*

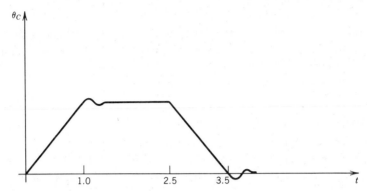

FIGURE 13.40 *Stabilization by scheduled cancellation compensation.*

performance, we might try a controller that keeps loop gain fixed but uses an "adjustable cancellation compensator" to "cancel" the variable inertia dynamics and replace them with a favorable (and *constant*) first-order lag. Such a controller would have the differential equation

$$\tau_1 \frac{dq_o}{dt} + q_o = \tau_2 \frac{dq_i}{dt} + q_i \tag{13.48}$$

where

$q_o \triangleq$ controller output $q_i \triangleq$ controller input

$\tau_1 \triangleq$ "variable time constant" $= J/B$

$\tau_2 \triangleq$ fixed time constant $= (J_o + MR_o^2)/B$

The numerical value of τ_1 is continuously changing as J changes. Implementation of such a controller again requires some computing power, as might be available with a microprocessor-based controller. Note also that the *concept* of this control mode is again a nonrigorous extension of standard linear system design methods to a situation where they do not strictly apply. However the scheme is not unreasonable and the availability of CSMP (or other) digital simulation allows us to try out our "bright idea" with a minimum of effort. Figure 13.40 shows the results of such a simulation and our hopes are confirmed; we have achieved both stability and uniform performance.

PROBLEMS

13.1 For the system of Fig. 13.4, neglect the dry friction nonlinearity and use numerical values given in the text.

a. Plot the Nyquist diagram, check stability, and find the gain margin and phase margin.

b. Plot the closed-loop frequency response.

c. Use a root-locus computer program to compute and plot the root locus for loop gain values from 0 to 20,000. For a loop gain of 5000 show all the roots and write out the *form* of the transient terms in the step-function response. (Do *not* calculate the coefficients of the terms, just leave them as letters.)

13.2 For the system of Fig. 13.4 with numerical values as given in the text:

a. Use CSMP (or other available) digital simulation to find the response to a step command, with and without the dry friction.

b. With dry friction present and $\psi_V \equiv 0.0$, take aircraft yaw motion ψ_A as:

$$\dot\psi_A = 0.4\pi \cos 10\pi t, \text{ %/s}$$

Use digital simulation to solve for ψ_M and compare with the allowable 1 min of arc.

13.3 Ignoring the dry-friction nonlinearity, for the system of Fig. 13.4:

a. Investigate the performance possible if we use only one phase lead compensator, $(\tau_5 s + 1)/(\tau_6 s + 1)$.

b. Repeat Part a if we use two compensators, $(\tau_1 s + 1)(\tau_5 s + 1)/(\tau_2 s + 1)(\tau_6 s + 1)$.

13.4 For the system of Fig. 13.9:

a. What is the effect on system performance if we model θ_C/x_p more correctly as $K_{pm}/s\left(\dfrac{s^2}{\omega_{np}^2} + \dfrac{2\zeta_p s}{\omega_{np}} + 1\right)$? Take $\zeta_p < 1.0$.

b. Repeat Part a if we *also* model x_p/x_v more correctly as $K_{vc}/s\left(\dfrac{s^2}{\omega_{nv}^2} + \dfrac{2\zeta_v s}{\omega_{nv}} + 1\right)$, with $\zeta_v < 1.0$.

13.5 Design the system of Fig. 13.10 to have a closed-loop overshoot of 25% and a 10% settling time of 2 s for a step command. You may use a combination of "pencil and paper" and simulation design tools. Then use simulation to study the following problem. At time $= 0$ let the gun be at rest with $\theta_C \neq \theta_V = 0$ when θ_V becomes a ramp $\dot\theta_{V0} t$. How long does it take for the gun to "lock on and track" the target?

13.6 In Fig. 13.7 the guns are sometimes "dual mounts," where the gun barrel axis is not on the center of rotation of θ_C. Gun recoil forces then tend to twist θ_C away from its commanded position. Explain clearly in detail how you would study this problem.

13.7 Use digital simulation to study the step and ramp response of the system of Fig. 13.17b.

13.8 Use digital simulation to study the limit cycling behavior of the system of Fig. 13.19. Then add a 60 Hz dither signal at location i and find its minimum amplitude needed to quench the limit cycle. Then check the response of θ_C to step commands of θ_V.

13.9 Use digital simulation for the system of Fig. 13.20 to:

    a. Study the step command response of each of the dual modes *separately*.

    b. Add a valve saturation at $x_v = \pm 0.03$ and repeat Part a.

    c. Now implement the switching controller and find a suitable value for the switching threshold E_c.

13.10 Use digital simulation to compare the step command responses of the systems of Fig. 13.21 and 13.23.

13.11 Use digital simulation to study the response of the system of Fig. 13.24 to step disturbances.

13.12 Modify the system of Fig. 13.24 to use proportional-plus-integral control. You may want to add a "hold" device since a sampling switch alone will allow the proportional part of the controller output to drop to zero whenever the switch is open. Study the response of your system to step commands and disturbances using digital simulation.

13.13 Derive the transfer function $(C/U)(s)$ for the system of Fig. 13.28b, assuming a perfect match between model and process.

13.14 Use digital simulation to study the behavior of Smith predictors as in Fig. 13.28a but with:

    a. Proportional-plus-integral controllers.

    b. Proportional controllers.

    c. Proportional-plus-integral-plus-derivative controllers.

13.15 Use digital simulation to study the behavior of the system of Fig. 13.28a for:

    a. Mismatches of process/model gain.

    b. Mismatches of process/model time constant.

13.16 For the phaselock servo of Fig. 13.31:

    a. Derive Eq. 13.43.

    b. Using numerical values given at the end of section 13.5, set loop gain using frequency response methods. Then find the closed-loop frequency response and use digital simulation to get the step response.

    c. Repeat Part b using root locus methods to set the gain.

13.17 For the variable-inertia robot of Fig. 13.36:

    a. Derive Eq. 13.47.

    b. Derive the needed equations and then use digital simulation with the given numerical values and inputs of Fig. 13.37 to verify the results of Fig. 13.38.

    c. Implement on CSMP the constant-gain-margin stabilization technique suggested in the text and verify the results of Fig. 13.39.

    d. Implement on CSMP the controller of Eq. 13.48 and verify the results of Fig. 13.40.

CHAPTER 14
Digital and Computer Control

14.1 INTRODUCTION

The behavior of systems using sampled data, the essential feature of digital and computer control, was investigated many years ago[1] and is part of classical control theory. Many texts devoted to this have appeared since,[2] including very recent ones with the most up-to-date treatment.[3] A thorough mastery of all the details requires extended study, as would be presented in a quarter or semester course devoted entirely to the subject, based on a suitable specialized text. However, it is possible to present considerable useful information relative to this area in a modest space and time for those readers who are unable to pursue the more comprehensive treatment but who need a practical level of familiarity with the subject. This chapter is intended to provide such a presentation.

Although I will show some digital control systems that do not employ a full-fledged digital computer, there is no question that the use of a digital computer (mainframe, mini, or micro) as the system controller provides many potential and actual benefits which make such control schemes increasingly attractive. We should note at the outset, however, that "digital control," with a few notable exceptions, does *not* provide basic new control ideas but, for the most part, is just a technically better way of *implementing* basic concepts already known from earlier times. Or, as in the case of many microprocessor-based controllers, it may make *economically* feasible, combinations of control functions long known to be advantageous, but not implemented for practical reasons.

When digital computers were first applied to industrial process control in the late 1950s they were used in a so-called *supervisory control*[4] mode and the machines were generally large-scale "mainframes" (minis and micros had not yet been developed). In supervisory control the individual temperature, pressure, flow, and the like feedback loops are locally controlled by the same electronic or pneumatic analog controllers used before the installation of the computer. The main function of the computer is to gather information on how

---

[1]J. R. Ragazzini and G. F. Franklin, *Sampled-Data Control Systems*, McGraw-Hill, New York, 1958.

[2]J. A. Cadzow and H. R. Martens, *Discrete-Time and Computer Control Systems*, Prentice-Hall, Englewood Cliffs, New Jersey. 1970.

[3]G. F. Franklin and J. D. Powell, *Digital Control of Dynamic Systems,* Addison-Wesley, 1981; R. Iserman, *Digital Control Systems*, Springer, New York, 1981; K. J. Aström and B. Wittenmark, *Computer Controlled Systems*, Prentice-Hall, Englewood Cliffs, N.J., 1984.

[4]C. L. Smith, *Digital Computer Process Control,* Intext, Scranton, 1972, Chap. 3; T. J. Williams, Computer Control Technology—Past, Present and Probable Future, *Trans. Inst. MC* **5**, (1), Jan.–Mar. 1983.

the overall process is operating, feed this into a technical–economic model of the process (programmed into computer memory), and then periodically send signals to the set points (desired values) of all the analog controllers so that each individual loop is operated in such a way as to optimalize the overall operation.

Once a certain amount of computer experience had been obtained with supervisory systems, a logical extension was *direct digital control* (DDC). Here the analog controllers were no longer used, the central computer serving as a single time-shared controller for all the individual feedback loops. Conventional control modes such as PI or PID were still used for each loop, but the digital versions of the control laws for each loop resided in software in the central computer. The computer input and output were multiplexed sequentially through the list of control loops, updating each loop's control action in turn and holding this value constant until the next cycle. One expected advantage of DDC (which unfortunately did not often materialize) was an economic savings for systems with many (50 or more) loops, where the cost of the multiplicity of analog controllers would exceed that of the central computer that replaced them. Early efforts in this direction resulted in a number of economic disasters which gave computer control a bad name for several years. The main reasons (in addition to the usual growing pains of a new technology) were that people underestimated computer programming costs and overestimated computer reliability. Because a failure in the central computer of a DDC system shuts down the *entire* system, it was necessary to provide a fail-safe backup system, which usually turned out to be a *complete* system of individual loop analog controllers, negating the expected hardware savings.

Although experience and technical developments in the intervening years have greatly eased software and reliability problems, current concepts of *distributed control* still include backup schemes[5] of one kind or another, though the backup may consist of a multiplicity of digital computers with shared responsibilities such that a single computer failure is not disastrous. Combinations of analog and digital methods based on individual (hybrid analog–digital) loop controllers are also in use and claim[6] the following benefits:

1. Individual controllers form an excellent base on which to build a hierarchical control system.
2. Such a system enjoys good reliability.
3. Such a system allows a fast plant turnaround and a smooth plant start-up.
4. Troubleshooting and maintenance of an individual-controller-based system are fast and logical.
5. In many cases, individual controllers offer the best kind of operator interface.

---

[5]T. J. Myron, Jr., Digital Technology in Process Control, *Comp. Design,* Nov. 1981, pp. 117–128.
[6]J. J. Murray, The Great Analog/Digital Debate, *ISA Trans.,* Vol. 16, No. 1, 1977, pp. 87–92.

6. For most feedback control loops, the PID algorithm implemented with analog circuitry is superior to the PID algorithm implemented in digital software. Derivative mode for instance, requires extreme resolution (14 bits in the computer input A/D converter) to approach the quality of analog performance.

As minicomputers and then microcomputers appeared on the general computer scene, they were of course assimilated into control technology. Concurrent with these changes in control equipment were changes in the type of personnel used to design, develop, and operate computer control systems. Many of the early growing pains of computer control can be charged to the need to use on control projects computer specialists recruited from the data processing field, since the control and process engineers of that era had little computer expertise. These computer experts had little or no practical process knowledge and often did not appreciate the problems involved in making a computer work reliably in the hostile environment of a factory or chemical plant. Although it might seem that a "team" made up of control engineers (with little computer savvy) and computer specialists (with little process–control background) might do a good job of computer-control system design, history shows that this was often not the case.

Control engineers have historically been, of necessity, engineering generalists since they must deal with a wide variety of controlled systems (jet engines, steam power plants, machine tools, etc.) using different forms of control equipment (pneumatic, hydraulic, mechanical, electronic, etc.); thus they are accustomed to the need to continually familarize themselves with new hardware. The digital computer, however, particularly in its early forms, was a control "component" whose application required a depth of knowledge not quickly attainable. One aspect of this problem involved programming languages. Even today, engineering students (other than those training as computer specialists) usually learn only "high level" languages such as FORTRAN, whereas the most machine-efficient programming for control applications may require use of assembly- or machine-language programming. Although knowledge of elementary FORTRAN is an easily acquired, durable, and transportable skill, low-level languages are machine specific, time variant, and difficult to learn. Also, to realize the high *machine* efficiency possible with low-level languages requires a much greater investment in programming time, thus the *overall* efficiency (including programming cost) may suffer. For example, to implement numerical integration using CSMP (a higher-level language than FORTRAN) requires only a *single line* of code (Y = INTGRL (IC,X)), while the FORTRAN coding of the Runge-Kutta integrating algorithm might take a *page* of code, and the machine language program for this operation might take *many pages* of code.

When computer speed and memory are very expensive and/or when we are developing a system that will be used in large quantities (like an automobile engine microprocessor controller) then machine language programming is justified since its large cost can be traded off against hardware costs and/or amortized over a large production run. However, current capabilities and future trends in minicomputer and microcomputer cost and performance seem to indicate that rather high-level languages are becoming more and more justifiable in all applications. This trend is welcomed by most control engineers since it allows them to reassume more complete responsibility for the overall control system design process, rather than sharing this with computer specialists. Concurrent with increasing availability of convenient high-level languages, development of modular, ready-made *interface devices*[7] for connecting the computer to the "outside" world of analog sensors and actuators makes this aspect of control system design less dependent on "electronics experts."

Returning to our historical account, the other main early application areas of digital methods was *machine tool numerical control,*[8] which developed at about the same time (R & D started about 1949, production machines first appeared around 1957) as computer control in the process industries. In a numerically controlled machine tool (lathe, milling machine, etc.), the information relative to the desired dimensions of the part being machined is entered in digital form (often punched paper tape) rather than the analog template used in tracer-controlled machines (see Fig. 1.2). Measurement of machine motions is accomplished with digital transducers (encoders) or analog transducers (synchros, resolvers, etc.) with electronic synchro/digital conversion, thus the feedback signal is also digital. The digital command and feedback signals are compared to obtain a digital error signal. Digital computing is also used to coordinate the motions of the several axes in a multiaxis machine and to perform interpolation between the specified end points of a given motion; however analog methods are still (1985) mainly in use for the dynamic control and stabilization of the servo loops themselves. That is, "direct digital control," which is quite common in the process-control field has not yet been much used in the machine-tool field. The main reason for this is that machine servo loops require much faster response than process temperature, flow, and the like loops, thus sampling rates in a digital controller must be very high to get adequate dynamic response. This made the cost of the digital control approach uncompetitive, however developments in microprocessor cost and performance are changing this picture.

Although digital computer technology is just starting to be used "inside" machine-tool servo loops, significant nonservo applications exist. Earlier numerically controlled (NC) machines used "hardwired" digital techniques rather than programmable computers. As minicomputer price and performance improved, it became feasible to replace many of the hardwired functions with their software-implemented equivalents, using a minicomputer as a built-in component of the machine tool. Replacement of paper-tape data storage with magnetic floppy-disk media is then also possible. This general approach has been called

---

[7]D. H. Sheingold, Ed., *Transducer Interfacing Handbook, 1980;* Analog Devices, Norwood, Mass., 1980; D. P. Burton and A. C. Dexter, *Microprocessor Systems Handbook, 1977,* Analog Devices, Norwood, Massachusetts, 1977.
[8]R. S. Pressman and J. E. Williams, *Numerical Control and Computer-Aided Manufacturing,* Wiley, New York, 1977.

Computerized Numerical Control (CNC).[9] Another development, Direct Numerical Control (DNC) (see footnote 9) uses a single (larger) minicomputer in a time-shared manner to simultaneously control several machines, eliminating the dedicated computer at each machine. Note that the meaning of the word *direct* in DNC is *different* from its meaning in the process industries, where in DDC the word *direct* means the computer is being used to implement the feedback loop control law.

The recent appearance of powerful and inexpensive microcomputers has made digital control practical for a wide variety of applications. Does this mean that all control system designers must now become expert in sampled-data system theory, microcomputer machine/assembly language programming, and electronic design of interface hardware? Certainly some control engineers will need and/or want to extend their expertise to some or all of these areas. Many others, however, will find that they can function well in the increasingly digital environment with a more modest investment in updating their skills. There are a number of reasons for this, which I will now explain.

First, as mentioned earlier, digital control contributes relatively few basic control concepts of its own to the designer's tool kit. The vast majority of digital systems in operation today are implementing well-known design principles conceived before digital computers were invented. Most of these principles were actually used in analog systems; whereas some were thought of, analyzed, and evaluated, but found impractical to implement in analog form and thus were shelved for the time being. I am thinking here of control modes such as on–off, proportional, integral, derivative, cancellation compensation, feedforward, ratio and cascade, decoupling for multivariable control, Smith predictors, intentional nonlinearities, various forms of adaptive control, and so forth. That is, when one starts the design of a new control system and is trying to conceive of several different approaches that appear initially feasible and worthy of further study in the interactive conceive–analyze–evaluate–decide design procedure, the store of ideas to be drawn on is found mainly in this list. There are, of course, some concepts that are *inherently* digital, much as deadbeat controllers,[10] but most of these provide little advantage over conventional methods and thus have as yet found little practical application. Therefore an in-depth study of digital control theory adds little to ones expertise in the initial conceptual phase of design.

Second, when we turn to the details of performance, we find that in the majority of digital systems the sampling rate must be set high enough so that the quality of the digital system response is not significantly worse than that of an equivalent analog system. Then the performance of the digital system can usually be adequately approximated by that of the equivalent analog system and special analysis techniques for sampled-data systems are not required. In designing digital systems under such circumstances, we can use our familiar

---

[9]Pressman and Williams, *Numerical Control,* Chapter 10.
[10]R. Isermann, *Digital Control Systems,* Springer, Berlin, 1981, Chap. 7.

analog techniques to determine an appropriate analog controller. (However, we *must* then know how to implement this controller digitally.) When our control computer has limited speed and/or memory we may be forced to use sampling rates so slow that performance can *not* be adequately approximated by an equivalent analog system. Then the specialized techniques of sampled-data system analysis are necessary. As the price and performance of microprocessors improves, and higher sampling rates become economical, the percentage of design problems requiring such treatment can be expected to be further reduced.

A third consideration relates to the "designer friendliness" of the control computer and associated software. When control system designers conceive new "good ideas" or control algorithms, they wish to implement them as directly as possible. (As a pertinent example from a related field, if one needs the function of voltage amplification for a sensor signal one should not have to "interrupt" the control design process to learn amplifier design; one selects an appropriate ready-made amplifier and installs it. One must know how to intelligently select an amplifier but it is much easier to learn this than to learn to design amplifiers.) One would hope that the user of computers as control "components" could adopt a similar philosophy. We might note here a similar situation with respect to the use of computers in analysis and design. Some engineers resist the use of "canned" programs such as CSMP, preferring to write their own software "from scratch." Although one can certainly find instances where a specially tailored program would be more effective than CSMP, this is *not* true in the vast majority of cases.

Fortunately, "designer friendly" control computers are available today in several forms and are appropriate for many, but not all, digital control tasks. It appears that an even more comprehensive selection of such devices will be available in the future, reducing further the need for specialized computer expertise on the part of control engineers. At the lowest level of control, many manufacturers now offer microprocessor-based single-loop PID process controllers that require *no* computer or sampled-data system knowledge for their application. These units include A/D and D/A converters at their input and output, thus one cannot really distinguish them functionally from their analog predecessors. The sampling rate is usually fixed, and fast enough to make closed-loop performance similar to that of an equivalent analog-controlled system. Although the dynamics of the digital PID action will generally be somewhat inferior to analog performance, the digital device excels in other areas. Numerical values for the various control modes can be set more precisely and have absolutely no drift. Auxiliary control functions such as logic operations for start-up, shut-down, or emergency conditions are more readily implemented as are special nonlinear effects such as square rooting and variable gain.

At a somewhat higher level of control, the most successful digital control device for factory automation has been the programmable logic controller (PLC), introduced in Chapter 4. Originally designed only for logic control (as a replacement for electromechanical relay systems) these machines are today much more versatile, some using several microcomputers to simultaneously perform logic functions and provide continuous control such as in servo systems and PID process control. Since most continuous control applications also require logic operations, this new dual capability of PLC's will make them even more popular. Their industrial success can be attributed largely to their user friend-liness. From their original invention to the present time they have been designed to "talk the user's language" and have never required any extensive computer expertise for their applications. In the field of factory automation they presently outnumber "ordinary" computers by about 10 to 1. They are especially appro-priate controllers for the many special-purpose machines needed in small quan-tities for factory automation. These machines are often developed by relatively small engineering firms who can afford neither to keep full-time computer spe-cialists on their staff nor to provide the expensive microprocessor development systems these people require in their work. Thus, although a general-purpose microprocessor *could* be adapted for use in such applications, a PLC may be a much more cost-effective solution. The initial cost of the PLC will be much more, but the engineering cost will be much less and the time schedule can be much shorter and more predictable. If the machines were to be produced by the thousands, then the situation would be more favorable to the "bare" mi-croprocessor; however when only a few will be made, the PLC will generally be the winner. An excellent example of this type of application, using a PLC with both logic and servo-drive capability to control a hydraulically actuated flying cutoff saw appeared recently in the literature.[11]

In the process-control field, between the single-loop controllers described earlier and the mainframe (or large mini) process computers used to control entire power plants, refineries, and the like, lies an application area devoted to multiloop processes with a modest number of loops. User-friendly control com-puters especially designed for such application are available from several man-ufacturers.[12] The Foxboro controller can handle up to 24 individual temperature, flow, and the like loops, providing both continuous (PID) and logic control functions. It can serve as an isolated autonomous controller but it also includes the necessary interface ports to make it a satellite controller whose set points and other adjustments are commanded in a supervisory mode by a large central computer. Programming uses a high-level language based on the interconnection of function blocks selected from a menu of basic functions. One can thus con-figure each control loop to meet its specific needs. These configurations (and associated parameter values) can be easily and quickly modified if process op-erating experience dictates such changes. *This ability to "redesign" the controller on line by changing software (rather than hardware) is one of the major advantages of digital control over analog control.* The function blocks include: standard PID modes; lead–lag; dead time; algebraic equations with up to 7 inputs using the

---

[11]R. B. Noha, Servo-Oriented PC Controls Flying Cutoff Saw, *Hyd. and Pneu.,* Nov. 1983, pp. 103–106.
[12]Product Spec. PSS 1-7G4A, Tech. Info. TI 016-605, Pub 511 15M 11/80, Foxboro Co., Foxboro, Mass., 1980.

operators $+$, $-$, $\times$, $\div$, square root, absolute value, and exponential; a nonlinear function generator using 16 line segments to approximate curves; input signal conditioning including gain, offset, and square rooting; contact inputs, high and low selectors, and Boolean equations using AND, NOT, BIT OR, OR, and EXCLUSIVE OR. Programming such a system is very similar to programming CSMP to *simulate* the system since we merely select the functions we want, adjust their numerical parameters, and connect the inputs and outputs to create the configuration we want. We might also note that such a controller offers a combination of logic and continuous control functions similar to those provided in some PLC's whose roots are in factory automation rather than process control. The PLC designers started with logic functions only, and when their machines begain using microprocessors, they found they could easily add continuous control functions to expand the market for their product. The process controller designers at first emphasized continuous control but found that once they began using microprocessors, they could easily include logic functions, which are almost always needed in their applications. Thus, starting from very different origins, the two sets of designers have today arrived at control computers with rather similar capabilities.

Finally, the most comprehensive use of digital control is found in the large-scale process computer systems designed to manage entire processing plants using a central mainframe or large minicomputer (see Fig. 1.12). Although a single computer could be time-shared among all the loops, a distributed-control concept using satellite computers to control portions of the overall operation gives the backup capability needed to maintain system operation when individual computer failures occur. The general configuration is thus one in which a large central computer is connected in a network with smaller peripheral computers. Within this general concept, many detailed variations are possible. Such large-scale systems have in the past and continue today to use high-level languages similar in concept to that just described for the 24-loop controller. Also, sampling rates will often be fast enough to allow approximation of behavior with the equivalent analog system. Thus this class of digital control applications does not usually require the control engineer to develop in-depth computer or sampled-data expertise.

The preceeding discussions are intended to show that many applications of digital control can be implemented with only a modest upgrading of computer control skills by the control engineer. The main application area where this may *not* be true involves the use of microcomputers of modest performance in mass-produced products. Here purchase cost of the computer must be kept low to maintain competitive product pricing, thus the computer will be one of modest performance, not able to support a high-level language. It is then necessary to use "every trick" of machine language/assembly programming to maximize computing efficiency and enable the somewhat primitive computer to perform the needed control functions. Also, since computer resources are being stretched to the limit, sampling rates may have to be set at barely acceptable low values; thus system performance can *not* be adequately approximated by that of an equivalent analog system. Under such conditions, the computer-engineering and control-engineering aspects of system design become more closely interwoven, requiring design personnel with a combination of these skills, either in individuals or in a design team. Fortunately the more-involved and time-consuming design and programming effort can be amortized over the large production run, giving a low per-unit cost for both hardware and engineering effort.

14.2   OPEN-LOOP DIGITAL CONTROL

We begin our discussion of some of the details of digital control with some simple open-loop systems that employ digital actuators. Commercially available digital actuators include electrical and electro-hydraulic stepping motors, digital valves, and digital piston/cylinders. Before describing the actuators in detail we first present the *thumbwheel switch,* a simple device, costing about $5, for entering numerical commands into a digital system. In an analog electrical system the reference input is usually a continuously variable dc voltage, say 0 to 10 V, which we could enter manually with a 10-turn potentiometer excited with a precise 10-V power supply. To enter commands in digital form requires that the digital "bits" used to define the numerical value be turned "on" or "off" in the proper pattern. In a typical digital electronic system a voltage anywhere between 0 and 4.5 V corresponds to "off" and one anywhere between 7.5 to 12 V corresponds to "on." Of the various coding schemes in use, I will describe a binary-coded-decimal (BCD) method in which each decimal digit of a number is specified by the on–off pattern of four digital signals. The least significant bit (LSB) has a weight of 1, the next 2, the next 4, and the last (most significant bit, MSB) 8. To represent decimal zero, none of the four bits are HI (on). To represent decimal 3, both the 1 bit and the 2 bit are HI; to represent decimal 9, the LSB and the MSB are HI, and so forth. Thus all decimal numbers between 0 and 9 can be represented by some combination of the four bits. For numerical values greater than 9 we need another set of four BCD lines to represent the 10's digit, another set of four for the 100's, and so forth. Thus, for example, to represent decimal numbers from 0 to 999 requires 12 lines of BCD digital data.

Figure 14.1 shows schematically a thumbwheel switch using our BCD code to enter manually decimal numbers between 0 and 9 into a digital control system. The switch is shown in the position needed to represent decimal 1; that is, line Q_1 is HI while Q_2, Q_4, and Q_8 are LO. By sliding the moving contacts successively through the 10 detented positions we can select all the combinations of Q_1, Q_2, Q_4, and Q_8 needed to represent decimals 0 through 9. For larger numbers we require one additional such switch for each decimal digit. Since codes other than our BCD are in use, switches are available in a variety of patterns.

Returning now to digital actuators, Fig. 14.2 shows a fluid piston/cylinder device using three binary-weighted stages to provide load positioning over the range 0 to 7 units with a resolution of 1/7 of full scale. For example if the LSB is a 0.01-in. motion, we can provide positions between 0.00 and 0.07 in. in 0.01-in. increments. Figure 14.2 shows the actuator with the load positioned at 6/7 of full scale. Note that each cylinder requires its own valve, however these valves

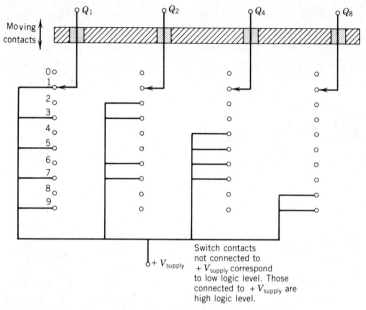

FIGURE 14.1 *Thumbwheel switch (BCD) for digital data entry.*

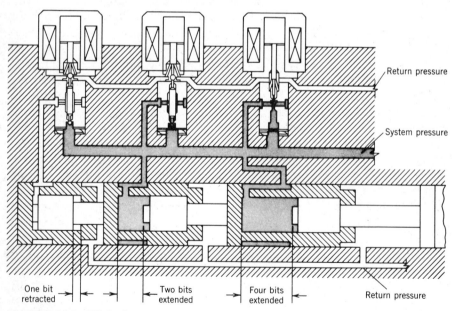

FIGURE 14.2 *Digital actuator.*

are inexpensive and reliable on–off solenoid valves, not precision servovalves. Also the accuracy of load positioning depends only on the cylinder full stroke, which is essentially fixed at the time of manufacture and can be held to close tolerances. No position-measuring transducer or feedback loop is necessary; this is an open-loop system. So long as fluid supply pressure is sufficiently high, external disturbing forces have *no* effect on positioning accuracy. We could use lines Q_1, Q_2, and Q_4 of the thumbwheel switch of Fig. 14.1 to turn the solenoid valves on and off, thereby creating a simple open-loop digital positioning system with manual command input.

Another commercially available digital "actuator" is the digital flow-control valve. Although it can be used for open-loop control it is more often used as an alternative to the pneumatically positioned proportional valve in process control feedback systems using DDC. Figure 14.3*a* compares these two methods of manipulating process fluid flow in response to digital commands from a control computer; whereas 14.3*b* shows some details of valve construction and operation. Simple on–off solenoid valves are manifolded to provide the desired combination of binary-weighted flow areas, the pictured example using six bits to provide a resolution of 1/63 of full scale. Digital valves have their own problems, including possible water-hammer transients due to sudden valve closures and flow coefficient uncertainties when flow is near the laminar/turbulent transition. Proper design and application can usually avoid such difficulties. Figure 14.4 shows two large digital valves (12 bits) that are used to control flow (5-500 gpm) of sodium hypochlorite in a heavy-water plant. The corrosive fluid required use of a special titanium as valve construction material. Some internal construction details of these valves are shown in Fig. 14.5.

By far the most common digital actuator is the electromechanical *stepping motor*,[13] widely used in open-loop and closed-loop position and speed control systems. One might argue that it is not really a digital actuator since it does not respond directly to a set of binary-weighted bits, however it is not usually difficult to program the computer to produce a pulse train with a certain pulse rate and total number of pulses, which is the type of input the stepping motor requires. Basically, a stepping motor plus its associated driver electronics accepts a pulse train input and produces an increment of rotary displacement, say 1.8° (200 pulses per revolution), for each pulse. Thus we can control motor average speed by manipulating pulse rate, and motor position by controlling total pulse count. We should immediately make clear that:

a. If the input pulse rate is too fast, the motor will "miss" steps, making speed and position inaccurate.

---

[13]J. Proctor, Stepping Motors Move In, *Pro. Eng.,* Vol. 34, 1963, pp. 74–78; Ninth Annual Symposium on Incremental Motor Control Systems and Devices, B.C. Kuo, Ed., 1981. (These annual symposia provide a wealth of information on stepping motors and associated devices and systems.); P.P. Acarnely, Stepping Motors: a Guide to Modern Theory and Practice, P. Peregrinus Ltd. N.Y., 1979.

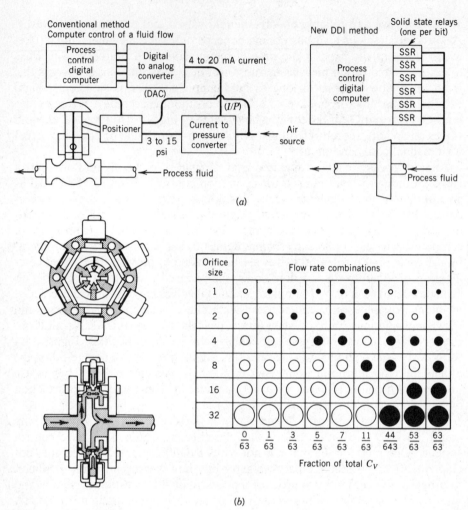

FIGURE 14.3 *Digital valve.*
Source: Digital Valve Engineering Handbook, Digital Dynamics Inc., Sunnyvale, California, 1977.

b. If the motor is at rest, an external disturbing torque greater than the motor's holding torque can twist the motor shaft away from its commanded position by any number of steps. When the disturbance is removed, the motor will *not* recover its correct position.

Two types of stepping motors are in common use: variable reluctance and permanent magnet. The permanent magnet[14] type is somewhat easier to understand so I will describe it briefly. In Fig. 14.6 the motor has a permanent

---

[14]D. J. Robinson, Dynamic Analysis of Permanent Magnet Stepping Motors, NASA TN D-5094, Mar. 1969.

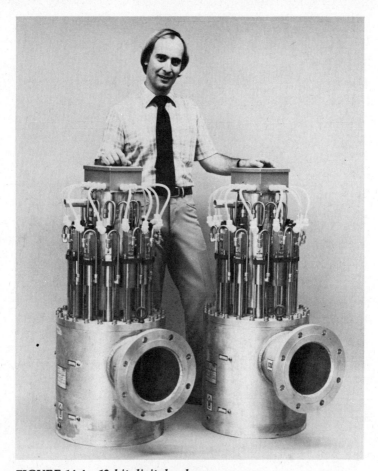

FIGURE 14.4   *12-bit digital valve.*
Source: Process System, Inc. Costa Mesa, California.

magnet rotor and the stator has four salient electromagnetic poles that are energized from a dc supply. If S3 is connect to positive polarity, S1 to negative polarity, and S2, S4 are open, an electromagnetic field is created between S3 and S1 with a north pole at S3 and a south pole at S1, attracting the rotor into the position shown. With this field maintained, if we try to twist the shaft away from its preferred position we feel a "magnetic spring" restoring torque; however a sufficiently large external torque can overcome the magnetic spring and we can turn shaft angle θ as far as we wish. When we let go, the rotor will "snap back" after some oscillation to the position of Fig. 14.6, however we have "lost" an integer number of revolutions. To command the rotor to turn 90° (one step) counterclockwise we connect positive to S2, negative to S4 (S1, S3 open). A further CCW 90° requires +S1, −S3, (S2, S4 open) and the final position requires +S4, −S2, (S1, S3 open). Obviously, clockwise rotation merely re-

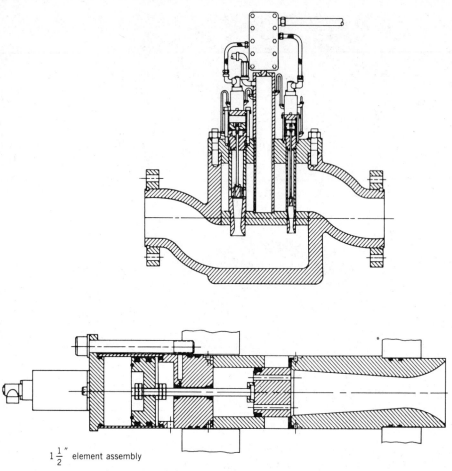

$1\frac{1}{2}''$ element assembly

FIGURE 14.5 *Construction details of digital valve.*
Source: Process System, Inc. Costa Mesa, California.

quires the proper sequence of stator excitations, and rotor average speed control is obtained by changing the rate at which stator excitation is switched. Although a microprocessor can be programmed to perform the required stator excitation switching, another approach uses available integrated circuit chips (stepper motor drivers) instead. These chips require as input only a pulse train at the stepping frequency and a logic signal to specify CW or CCW rotation. An adjustable-frequency pulse train is readily obtained from another IC chip, a voltage-controlled oscillator (VCO). The output frequency of such chips depends typically on the value of two resistors, one capacitor, and an input voltage (dc), allowing easy frequency control from manual inputs or dc control voltages.

A simple open-loop speed control can be assembled from a VCO, stepper motor driver, power supply, and stepper motor. For manual speed command

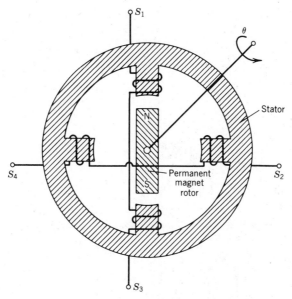

FIGURE 14.6 *Permanent-magnet stepping motor, 90°/step.*

we could use a 10-turn pot as a variable resistor in the VCO. As long as stepping rates and disturbing torques are low enough that no steps are missed, speed control will be accurate, a typical application being a chart-paper drive for a strip-chart recorder. By driving several motors from the same frequency source, synchronized motions at different points in a machine are easily obtained. Using standard frequency-divider chips we can drive a second motor at a precise fraction of another motor's speed, giving an "electronic gear train." For open-loop position control the arrangement of Fig. 14.7 can be used. Additional chips shown there are the counter, comparator, and flip flop. We set the desired position as a certain number of motor steps on the thumbwheel switches, and the variable resistor on the VCO selects the pulse rate (speed) with which we wish to make the move. As pulses are fed from the VCO to the motor, the counter counts them and sends this number to the comparator, which also receives the thumbwheel switch data and compares the two. When B (pulses counted) equals A (pulses wanted) the comparator $A = B$ logic signal goes HI and disables the VCO, stopping the pulses. If we now *change* the thumbwheel switch setting in either direction, the $A < B$, $A > B$ comparator outputs (through the flip-flop) switch the motor driver to the proper (CW/CCW) direction and since $A \neq B$, the $A = B$ line goes LO, enabling the VCO and producing pulses until the comparator is again satisfied. Note that motor *motion* is *not* sensed, thus if steps are missed due to excessive pulse rate and/or disturbing torques, the system does not know this and error results.

Since the cost and simplicity advantages of stepper-motor motion control

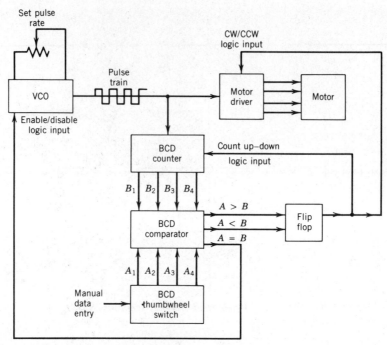

FIGURE 14.7 *Open-loop position control using step motor.*

systems erode when motion sensors and feedback loops are added, much effort has gone into improving the performance of open-loop systems. A particular motor/load inertia will exhibit an *error-free-start–stop rate* (EFSSR) that depends on load torque as in Fig. 14.8. As long as our commands do not require the system to exceed its EFSSR, there will be one motion pulse for each electrical command pulse and accuracy will be preserved. By accelerating and decelerating the load *gradually,* we can ensure such operation. This is called *ramping* and is easily achieved when we use a VCO as the pulse train source, since the frequency can be scheduled with a time-varying dc input voltage. Both exponential and straight-line "ramps" are used, exponential being preferred for acceleration and straight-line being adequate for deceleration. The circuitry for converting a command pulse rate to the proper sequence of motor phase switchings is called a *translator;* whereas the means for generating a proper pulse rate and count is called an *indexer.* For those users who must trade design time and skill for cost, ready-made systems of great versatility including motor, indexer, and translator are available.

The single-step response of stepper motors is often quite oscillatory and various schemes for improving damping are available. The Lanchester type of mechanical damper, which wastes no energy in friction for constant velocities, is a simple solution in some cases. A more complex approach uses the last few pulses of a move to apply magnetic counter-torque ("braking pulses") to reduce

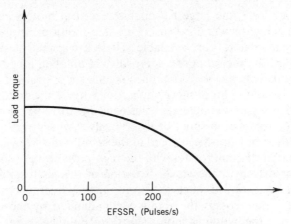

FIGURE 14.8   *Error-free-start-stop-rate for step motor.*

oscillation. Even more complex, but quite successful, is the concept of *microstepping*.[15] The motor of Fig. 14.6 has four steps per revolution, but it is more common to use additional stator and rotor magnetic poles to get, say, 200 steps per revolution. Microstepping provides 20,000 to 50,000 steps per revolution and is *not* accomplished by motor redesign but by driving a standard 200 steps per revolution motor with "smart" electronics. The basic concept can be understood in terms of Fig. 14.6. By energizing *both* S1/S3 and S2/S4 simultaneously, the rotor magnetic field can be oriented at 45°, causing the rotor to move 45°, rather than 90°. Using the proper combinations of individual and simultaneous stator pole excitations, we now have doubled the number of steps per revolution. If we now can also adjust the relative *strengths* of the stator pole excitation (current) for two adjacent poles, the angular orientation of the stator field (and thus rotor position) can be indexed in even finer steps, current practice providing the 20,000 to 50,000 steps per revolution mentioned earlier. This technique not only improves damping but gives finer angular motion resolution than expensive shaft encoders in competitive closed-loop systems.

Although the advantages of stepper drives in open-loop systems are most obvious, closed-loop applications also exist.[16] Ferris, Palombo, and Fortescue, compares several alternative solutions to a line printer paper-feed drive problem and show how a closed-loop stepper drive can be analyzed using classical techniques (root locus, etc.) employed for continuous-motion systems. Most stepper–motor drives (either open loop or closed loop) are relatively low power (a few

[15]E. Slingland, Microstepping for Fine Position Control, *Lasers and Applications,* Aug. 1983, pp. 55–58; Small Steps Turn into Big Improvements, *Powerconversion International,* Oct. 1983, pp. 20–25.

[16]T. A. Ferris, G. A. Palombo, and S. M. Fortescue, A Low-Cost High Performance Incremental Motion System Using a Closed Loop Stepping Motor, *Drive and Controls International,* Jan. 1982, pp. 18–26.

hundred watts or less), the largest available electrical motors currently being about 3 hp. When high-power drives using the step-motor principle are required, electrohydraulic step motors are available. These use a small electromechanical step motor to position the spool of a servovalve in an open-loop fashion. Using mechanical feedback in a closed-loop fashion, the spool position is reproduced at a high power level (10 hp or more) by a rotary hydraulic motor. Figure 14.9 shows two schemes used to implement this concept. In 14.9*a* a low-power electrical step motor rotates the spool of a servovalve, which also then translates (opening the valve) because the left end of the spool is the screw of a ball screw/ nut unit. The nut of this unit rotates with the high-power output shaft of a rotary hydraulic motor driven by the valve. Rotation of this shaft tends to translate the valve spool back toward its closed position, thus hydraulic motor rotation continues until it just equals the rotation commanded by the electrical step motor, whereupon the hydraulic motor stops, since the valve has now returned to null. This hydromechanical arrangement is a complete closed-loop servo

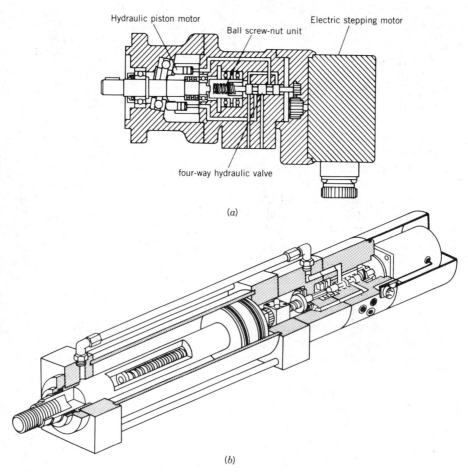

(a)

(b)

FIGURE 14.9 *Electro-hydraulic stepping actuators.*

system, thus hydraulic motor pulses duplicate those of the electrical motor. However, if the *overall* motion control system does not measure load motion, we can still have errors due to the electrical motor shaft not duplicating each electrical input pulse. Figure 14.9b shows a similar arrangement for translational load motion. Rotary motors up to about 20 hp are available, a typical 20 hp motor having a step size of 0.72°, maximum pulse rate of 13,000/s, EFSSR of 2000 pulses/s, with torque of 1150 in.-lb$_f$ at this pulse rate. A typical translational actuator might have a step size of 0.001 or 0.0005 in. with a maximum speed of 300 in/min.

14.3 SAMPLING, A/D AND D/A CONVERSION, QUANTIZATION, AND NOISE FILTERING IN COMPUTER CONTROL

Having briefly shown some (mainly open-loop) digital control systems using digital or pulsetype actuators, we now turn to the more general field of closed-loop control using a digital computer as the controller. Figure 14.10 shows a configuration useful for this discussion. In most cases the measuring transducer and final control elements are analog devices, requiring, respectively, A/D and D/A conversion at the computer input and output. There are, of course, exceptions; we saw in Fig. 14.3 how a digital valve combines the function of D/A converter and final control element. For motion control systems (servomechanisms) digital shaft angle encoders[17] are available which combine the

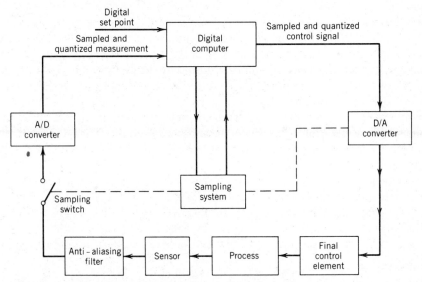

FIGURE 14.10 *General computer-control configuration.*

---

[17]E. O. Doebelin, *Measurement Systems*, 3rd Ed., McGraw-Hill, New York, 1983, pp. 294–299.

function of displacement sensing and A/D conversion, providing BCD output directly useable by the computer. In most cases, however, our sensors will provide an analog voltage output and our final control elements will accept an analog voltage input.

Two main types of electronic A/D converters,[18] successive approximation and dual slope, are in use. The dual slope is superior in most respects except speed; however, its speed limitation rules out the faster applications so successive-approximation types are quite common. They are available up to about 16 bits (conversion time is approximately 30 μs), an 8-bit unit can be as fast as 1 μs, and they often require a sample/hold amplifier (see footnote 18) at their inputs for accuracy. The dual-slope type has up to 17 bits, exhibits good noise rejection, can be easily auto zeroed to null drift, but is limited to about 30 conversions per second. Its averaging principle makes an external sample/hold unnecessary. Most D/A converters use the principle shown in three-bit form in Fig. 14.11 to convert the set of HI/LO digital signals provided at the computer output to a single analog voltage. Note that the output voltage E_o is analog but *is not* a smoothly varying signal; it changes in discrete steps. For the three-bit unit shown, if E_{ref} = 10 V, the step size is 1.25 V. This quantization could lead to limit cycling in a closed-loop control system since all final control element positions are not available and the system may hunt between two discrete positions in an attempt to meet the needs of a particular set-point and/or disturbance. This hunting problem is not of much practical importance since we usually use 10 or more bits in both the A/D and D/A converters, giving a resolution of 0.1% of full scale, or better.

Since the digital computer operates on discrete numbers rather than on continuously varying voltages, we must sample the analog sensor's voltage (usually periodically) using an appropriate sampling system synchronized with the computer's clock. When a single computer is time shared among several or many control loops, the sampling rate usually needs to be separately adjustable for each loop, to suit its needs. Simulation studies of each loop can be used to study the effect of various sampling times and select an optimum value. For process control systems, generalized studies[19] show that a sampling period of one tenth the sum of loop dead time plus first-order time constant often is a good compromise. Typically, flow loops (usually the fastest) require 0.5 to 1.0-s sampling intervals, level and pressure require about 5-s intervals, and temperature requires about 20-s intervals.

The analog feedback signal coming from the sensor in Fig. 14.10 contains useful information related to controllable disturbances (relatively low frequency) but also may often include higher frequency "noise" due to uncontrollable disturbances (too fast for control system correction), measurement noise, and stray electrical pickup. Such noise signals cause difficulties in analog systems,

[18] Ibid., pp. 767–775.
[19] P. C. Badavas, Microprocessor Simulation Reveals Control Sampling Differences, *Cont. Eng.* Apr., 1982, pp. 106–108.

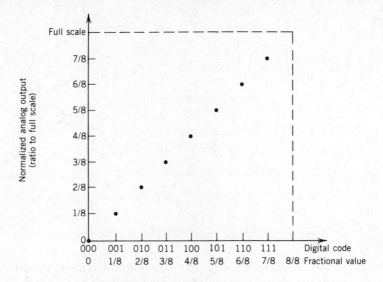

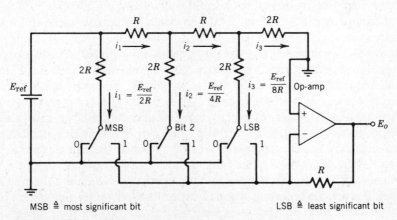

FIGURE 14.11 *Digital-to-analog converter, three-bit.*

MSB ≙ most significant bit   LSB ≙ least significant bit

and low-pass filtering is often needed to allow good control performance. In digital systems a phenomenon called *aliasing* introduces some new aspects to the area of noise problems.[20] If a signal containing high frequencies is sampled too infrequently, the output signal of the sampler contains low-frequency ("aliased") components not present in the signal before sampling. If we base our control actions on these false low-frequency components they will of course result in poor control. Figure 14.12 shows how a high-frequency signal, inadequately sampled, can produce a reconstructed function of a much lower fre-

---

[20]Smith, *Digital Computer Process Control*, pp. 128–135.

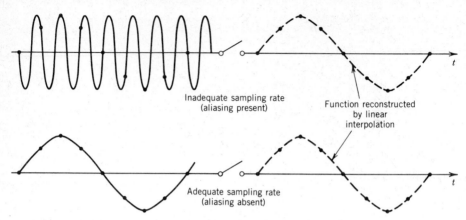

Inadequate sampling rate
(aliasing present)

Function reconstructed
by linear
interpolation

Adequate sampling rate
(aliasing absent)

FIGURE 14.12   *Concept of aliasing due to sampling.*

quency, which cannot be distinguished from that produced by adequate sampling of a low-frequency function. A theoretical study of this phenomenon shows (Shannon sampling theorem) that the absolute minimum sampling rate to prevent aliasing is two samples per cycle; however in practice, rates of 2.6 to 10 are more commonly used.

If a closed-loop system is to follow commands and/or reject disturbances over a certain range of frequencies (have a certain "bandwidth"), then we know the highest frequency that it must be able to handle. If, however, noise signals of frequency content *beyond* the desired system bandwidth are present, we must base our antialiasing measures on these noise frequencies. In some cases a simple analog low-pass filter, $1/(\tau_f s + 1)$, with τ_f chosen to attenuate sufficiently the lowest-frequency noise component, is used ahead of the sampler as an antialiasing device. Since such simple filters have a rather gradual cutoff, τ_f may sometimes have to be chosen somewhat smaller than desired for noise reduction in order that its dynamics not degrade control action within the system bandwidth. In such cases, an additional stage of *digital* low-pass filtering (*after* the sampler) may be programmed into the computer. Such digital filters (the simplest are the arithmetic average and the first-order lag[21]) can exhibit sharper cutoffs than the alternative analog approach, thus improving performance in those cases where the high end of the system bandwidth comes close to the low end of the noise frequency spectrum. (A *completely digital* filtering approach is technically possible if we sample fast enough to not alias the highest noise frequency. This is usually not practical since the very high sampling rate increases the load on the computer to an uneconomical degree).

Sometimes the analog filtered sensor signal is sampled at a *higher* rate[22] than is used to update the control algorithm. This meets the needs of aliasing pre-

[21]Ibid., pp. 128–135; K. W. Goff, Dynamics in Direct Digital Control, *ISA Jour.,* Vol. 13, No. 11, Nov. 1966, pp. 44–49 and Vol. 13, No. 12, Dec. 1966, pp. 44–54.
[22]A. B. Corripio, C. L. Smith, and P. W. Murrill, Filter Design for Digital Control Loops, *Inst. Tech.,* Jan. 1973, pp. 33–38.

vention and conserves computer resources, but may increase cost since the dual sampling scheme requires additional hardware. An analog filter ahead of the sampler can use a time constant of about $T_s/2$, where T_s is the "fast" sampling time interval at the sensor. The digital filter that follows should have a time constant of about $T/2$, where T is the time interval at which the control algorithm is updated. These rules of thumb are helpful in preliminary design but simulation studies of each specific system give more detailed information and should be run, if technically and economically feasible.

14.4 DIFFERENCE EQUATIONS AND Z TRANSFORMS

The sampling, A/D, and D/A functions can occur in various places and ways in a feedback control loop; we will use the configuration of Fig. 14.13 to develop some basic ideas. The operations between the analog actuating signal e and the input to the analog system $G(s)$ are conventionally treated as instantaneous for analysis purposes. The sampling, conversion, and computing delays present in a real system can all be lumped into a single overall dead time that we include in the process model $G(s)$. We describe the operation of the computer by a linear, time-invariant *difference equation*.

$$m(kT) = b_0 e(kT) + b_1 e[(k-1)T] + \cdots + b_n e[(k-n)T]$$
$$\quad - a_1 m[(k-1)T] - a_2 m[(k-2)T] - \cdots - a_n m[(k-n)T] \tag{14.1}$$

Here, m and e are, respectively, the output and input number sequences for the computer and $m(kT)$ and $e(kT)$ denote the values at the kth sampling interval, that is, at time kT, where T is the sampling interval. The symbols $e[(k-1)T]$, $m[(k-1)T]$, and the like represent previous values of e and m that might be stored in the computer. The coefficients $a_n, \cdots, a_1$ and $b_n, \cdots, b_0$ are fixed numbers whose presence and numerical value determine the input–output relation for the computer.

As an example of Eq. 14.1 suppose we wish to implement a digital version of PID control on our computer. An analog controller might have

$$\frac{M}{E}(s) = K_D s + K_P + \frac{K_I}{s} = \frac{1}{s}(K_D s^2 + K_P s + K_I) \tag{14.2}$$

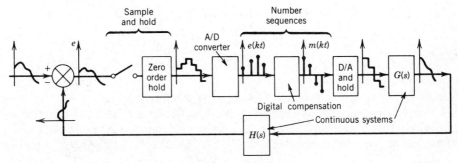

FIGURE 14.13 *Typical computer control configuration.*

$$\frac{dm}{dt} = K_D \frac{d^2e}{dt^2} + K_P \frac{de}{dt} + K_I e \tag{14.3}$$

To convert this to an equivalent difference equation we approximate the derivatives by finite differences. Of the various schemes possible, the simplest uses

$$\frac{de}{dt}(kT) \approx \frac{e(kT) - e[(k - 1)T]}{T} \tag{14.4}$$

which can be applied successively to approximate the higher derivatives.

$$\frac{d^2e}{dt^2}(kT) \approx \frac{\frac{de}{dt}(kT) - \frac{de}{dt}[(k - 1)T]}{T} \tag{14.5}$$

$$\approx \frac{e(kT) - 2e[(k - 1)T] + e[(k - 2)T]}{T^2} \tag{14.6}$$

Applying this to Eq. 14.3 we get

$$m(kT = m[(k - 1)T] + \left(\frac{K_D}{T} + K_P + K_I T\right)e(kT)$$

$$- \left(\frac{2K_D}{T} + K_P\right)e[(k - 1)T] + \frac{K_D}{T}e[(K - 2)T] \tag{14.7}$$

To see how equations of form of Eq. 14.1 are used to construct the sequence $m(kT)$ when the input $e(kT)$ is given, let us take in Eq. 14.7 the numerical values $K_I = K_P = T = 1$, $K_D = 10$ and let $e(kT)$ be a "unit step input." That is, let $e(kT)$ be zero for $k < 0$ and 1 for $k \geq 0$; also let $m(kT)$ be zero for $k < 0$, as an initial condition.

$$m(kT) = m[(k - 1)T] + 12e(kT) - 21e[(k - 1)T] + 10e[(k - 2)T] \tag{14.8}$$

Substitution of $k = 0, 1, 2, \ldots$ into Eq. 14.8 generates the $m(kT)$ sequence shown in Fig. 14.14. Note that although the digital response does not (and cannot) duplicate the analog response, the digital response does clearly exhibit the essential *characteristics* of a PID response to a step input:

a. A large initial "derivative" response that fades away.
b. A sustained "proportional" response.
c. A growing "integral" response.

Figure 14.15 shows the analog signal (produced by the D/A converter) that would be the input to the analog system $G(s)$. As mentioned earlier the digital delay shown can be included in $G(s)$ for analysis purposes.

I next introduce the concept of the *Z Transform* as applied to number sequences. This transform operates on a number sequence $f(kT)$ and produces a function of a complex variable $z = a + ib$. The one-sided (number sequence

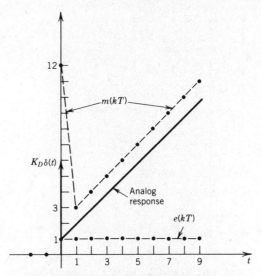

FIGURE 14.14 *Comparison of analog and digital versions of PID controller.*

defined for positive k only) Z transform is defined by[23]

$$\mathcal{Z}[f(kT)] \triangleq F(z) \triangleq \sum_{k=0}^{\infty} f(kT)z^{-k} \tag{14.9}$$

As defined by Eq. 14.9, $F(z)$ is a power series in z^{-1} and may or may not converge, depending on the sequence $f(kT)$ and the region of the complex plane where z falls. For most applications the convergence question is not a problem and we will pursue it no further. When $F(z)$ is convergent it can be put into a closed form

$$F(z) = \frac{b_0 z_m + b_1 z^{m-1} + \cdots + b_m}{z^n + a_1 z^{n-1} + \cdots + a_n} \tag{14.10}$$

By factoring the polynomials in z we can put Eq. 14.10 in the form

$$F(z) = b_0 \frac{(z - z_1)(z - z_2) \cdots (z - z_m)}{(z - p_1)(z - p_2) \cdots (z - p_n)} \tag{14.11}$$

where the numbers z_i are called the zeros of $f(z)$ and the p_i the poles. When z takes on the value of a zero, $F(z) = 0$ and when z takes on the value of a pole, $F(z) = \infty$. Note the similarity to Laplace transforms, which also appear as ratios of polynomials (in s) and have poles and zeros.

[23]Cadzow, J. A. and H. R. Martens, *Discrete-Time and Computer Control Systems*, Prentice–Hall, Engelwood Cliffs, New Jersey, 1970, Chap. 4.

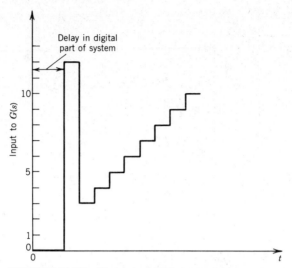

FIGURE 14.15 *Output of D/A converter.*

Let us apply definition 14.9 to determine the *digital transfer function* of the digital computer characterized by Eq. 14.1. In z transforming both sides of Eq. 14.1 a linearity theorem allows us to transform each term separately and to take multiplying constants $(b_0, b_1, \ldots)$ outside the transform. The terms $m(kT)$ and $b_0e(kT)$ transform directly to give $M(z)$ and $b_0E(z)$. To deal with the term $b_1e[(k - 1)T]$, note that if the sequence $e(kT)$ is, say, 3, 4, 2, -1, 6, $\ldots$ for $k = 0, 1, 2, 3, 4, \ldots$ (and zero for $k < 0$), then $e[(k - 1)T]$ is 0, 3, 4, 2, -1, 6, $\ldots$ This gives

$$\mathscr{Z}[e(kT)] = 3 + 4z^{-1} + 2z^{-2} - z^{-3} + 6z^{-4} + \cdots \qquad (14.12)$$

$$\mathscr{Z}[e[(k - 1)T]] = 3z^{-1} + 4z^{-2} + 2z^{-3} - z^{-4} + 6z^{-5} + \cdots \qquad (14.13)$$

and shows that $\mathscr{Z}[e[(k - 1)T]] = z^{-1}\mathscr{Z}[e(kT)]$. We can now write

$$M(z) = b_0E(z) + b_1z^{-1}E(z) + \cdots + b_nz^{-n}E(z) \qquad (14.14)$$
$$- a_1z^{-1}M(z) - a_2z^{-2}M(z) - \cdots - a_nz^{-n}M(z)$$

which gives

$$M(z) = \frac{b_0 + b_1z^{-1} + \cdots + b_nz^{-n}}{1 + a_1z^{-1} + a_2z^{-2} + \cdots a_nz^{-n}}E(z) \qquad (14.15)$$

This is in the form of a transfer function relationship $M(z) = D(z) E(z)$ where

$$D(z) \triangleq \frac{b_0 + b_1z^{-1} + \cdots + b_nz^{-n}}{1 + a_1z^{-1} + a_2z^{-2} + \cdots a_nz^{-n}} \qquad (14.16)$$

is called the *digital transfer function*. For a computer programmed for the PID algorithm as in Eq. 14.8 we would have

$$D(z) = \frac{12 - 21z^{-1} + 10z^{-2}}{1 - z^{-1}} \qquad (14.17)$$

We now need to show how the Z transform is applied to the analysis of the remaining portions of systems such as that of Fig. 14.13. Consider the sample/hold portion of the system, shown in Fig. 14.16. We assume the sampling switch picks off an instantaneous value of $g(t)$ every T seconds and the hold device then holds this value constant until the next sampling instant, at which time it is updated. (Other types of hold schemes are possible but this *zero-order hold* is the simplest and most common.) A real sample/hold and A/D[24] (successive approximation, 14 bits) might take about 13 μs between the command to sample and the availability of the digitized number to the computer.

The action of the zero-order hold is described mathematically (using delayed step functions) by

$$h(t) = \sum_{k=0}^{\infty} g(kT)[u(t - kT) - u(t - kT - T)] \qquad (14.18)$$

$$\mathscr{L}[h(t)] = H(s) = \sum_{k=0}^{\infty} g(kT)\left[\frac{e^{-skT} - e^{-s(k+1)T}}{s}\right] \qquad (14.19)$$

$$H(s) = \frac{(1 - e^{-sT})}{s} \sum_{k=0}^{\infty} g(kT)e^{-skT} \qquad (14.20)$$

This last result takes the form of a transfer function relation, output function = (system function) (input function).

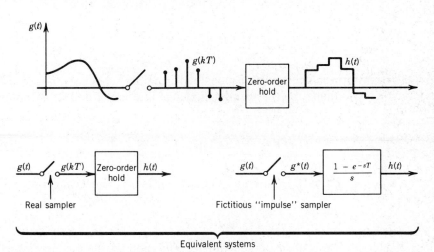

FIGURE 14.16 *Definition of impulse sampler.*

---

[24]Doebelin, *Measurement Systems*, pp. 772–775.

$$H(s) = G_m(s)G^*(s)$$

if we define

$$G_m(s) \triangleq \frac{1 - e^{-sT}}{s} \tag{14.21}$$

$$G^*(s) \triangleq \sum_{k=0}^{\infty} g(kT)e^{-skT} \tag{14.22}$$

If we have $G^*(s)$ we should be able to find its $g^*(t)$; actually

$$g^*(t) = \sum_{k=0}^{\infty} g(t)\,\delta(t - kT) \tag{14.23}$$

which can be verified by Laplace transforming to get Eq. 14.22. The function $g^*(t)$ is a train of impulses of strength (area) equal to $g(kT)$, occuring at the sampling instants $t = kT$. Even though these impulses do not physically occur anywhere in the system, we may think of, and analyze, the system in this way since the overall effect is identical with that of the real system. This allows us to treat the continuous ("analog") parts of the system by studying their response to impulse trains. Let a continuous system with transfer function $W(s)$ be subjected to an impulse-train input

$$r^*(t) = \sum_{k=0}^{\infty} r(t)\,\delta(t - kT) \tag{14.24}$$

If $w(t)$ is the system's impulse response, superposition gives for the system output $c(t)$

$$c(t) = \sum_{k=0}^{\infty} r(kT)w(t - kT) \tag{14.25}$$

To use Z-transform methods we must deal with number sequences, not continuous functions; so think of $c(t)$ as being sampled at intervals T and let us inquire as to the value of $c(t)$ at an arbitrary instant nT.

$$c(nT) = \sum_{k=0}^{\infty} r(kT)w(nT - kT) \tag{14.26}$$

We can now take Z transforms of both sides of Eq. 14.26, which can be shown[25] to give

$$C(z) = R(z)W(z) \tag{14.27}$$

where $R(z)$ is the Z transform of the sequence $r(kT)$ and $W(z)$ is the Z transform of $w(kT)$.

---

[25]Cadzow and Martens, *Discrete-Time* p. 210, p. 140.

To obtain $W(z)$ when $W(s)$ is given (we do *not* substitute z for s!) we evaluate $w(t)$ at the sampling instants and then use definition 14.9 to get $W(z)$. Let us take $W(s) = a/[s(s + a)]$ as an example; $w(t) = (1 - e^{-at})u(t)$. This gives $w(kT) = 1 - e^{-akT}$ for $k \geq 0$ and 0 for $k < 0$. This number sequence has the Z transform

$$W(z) = \sum_{k=0}^{\infty} (1 - e^{-akT})z^{-k} \tag{14.28}$$

$$= \sum_{k=0}^{\infty} z^{-k} - \sum_{k=0}^{\infty} z^{-k}e^{-akT} \tag{14.29}$$

By long division

$$\frac{z}{z - 1} = 1 + z^{-1} + z^{-2} + z^{-3} + \cdots \tag{14.30}$$

$$\frac{z}{z - e^{-aT}} = 1 + e^{-aT}z^{-1} + e^{-2aT}z^{-2} + e^{-3aT}z^{-3} + \cdots \tag{14.31}$$

so we get

$$W(z) = \frac{z}{z - 1} - \frac{z}{z - e^{-at}} = \frac{z^{-1}(1 - e^{-aT})}{(1 - z^{-1})(1 - e^{-aT}z^{-1})} \tag{14.32}$$

The use of z^{-1} in the right-hand side of Eq. 14.32 is preferred usage since it eases the transition to iterative equations when this is necessary. Since tables such as Table 14.1 are available, we can often go from $W(s)$ to $W(z)$ with little effort.

TABLE 14.1

$F(s)$	$\mathcal{Z}[F(s)]$
$\left(\dfrac{1 - e^{-sT}}{s}\right)F(s)$	$(1 - z^{-1})\mathcal{Z}\left(\dfrac{F(s)}{s}\right)$
$\dfrac{1}{s}$	$\dfrac{1}{1 - z^{-1}}$
$\dfrac{1}{s^2}$	$\dfrac{Tz^{-1}}{(1 - z^{-1})^2}$
$\dfrac{1}{s^3}$	$\dfrac{T^2(1 + z^{-1})z^{-1}}{2(1 - z^{-1})^3}$
$\dfrac{a}{s(s + a)}$	$\dfrac{(1 - e^{-aT})z^{-1}}{(1 - z^{-1})(1 - e^{-aT}z^{-1})}$
$\dfrac{a}{s^2(s + a)}$	$\dfrac{Tz^{-1}}{(1 - z^{-1})^2} - \dfrac{(1 - e^{-aT})z^{-1}}{a(1 - z^{-1})(1 - e^{-aT}z^{-1})}$

TABLE 14.1 (Cont.)

$F(s)$	$\mathscr{L}[F(s)]$
$\dfrac{ab}{s(s+a)(s+b)}$	$\dfrac{1}{1-z^{-1}} + \dfrac{(be^{-bT} - ae^{-aT})z^{-1} - (b-a)}{(b-a)(1 - e^{-aT}z^{-1})(1 - e^{-bT}z^{-1})}$
$\dfrac{a^2 b^2}{s^2(s+a)(s+b)}$	$\dfrac{abTz}{(z-1)^2} - \dfrac{(a+b)z}{z-1} - \dfrac{b^2 z}{(a-b)(z-e^{-aT})}$ $+ \dfrac{a^2 z}{(a-b)(z-e^{-bT})}$
$\dfrac{a^2+b^2}{s[(s+a)^2+b^2]}$	$\dfrac{z}{z-1} - \dfrac{z^2 - ze^{-aT}\sec\phi\cos(bT+\phi)}{z^2 - 2ze^{-aT}\cos bT + e^{-2aT}}$

$$\phi \triangleq \tan^{-1}(-a/b)$$

We are now in a position to discuss the analysis of complete systems such as that of Fig. 14.13, which we redraw as in Fig. 14.17, where any delays in the digital part of the system have been included as dead times in $G(s)$. It can be shown[26] that the closed-loop transfer function is given by

$$\frac{C(z)}{R(z)} = \frac{D(z)\,\mathscr{L}\left[\dfrac{1-e^{-sT}}{s}G(s)\right]}{1 + D(z)\,\mathscr{L}\left[\dfrac{1-e^{-sT}}{s}G(s)H(s)\right]} \tag{14.33}$$

Note that the evaluation of the Z transforms of the two bracketed quantities is facilitated by the first entry of Table 14.1. A numerical example will be of help at this point. Suppose we wish to control a system $G(s) = 1/[s(s+1)]$ using a measuring device $H(s) = 1$ and a PID controller with $K_D = 1.8$, $K_P = 2.6$, and $K_I = 2$; however the PID controller is to be realized with a digital computer,

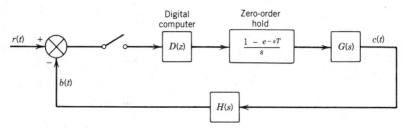

FIGURE 14.17   *Computer control system analysis diagram.*

[26]Ibid., p. 223.

using $T = 0.1$ s. From Eq. 14.7 and 14.16 we find

$$D(z) = \frac{20.8 - 38.6z^{-1} + 18.0z^{-2}}{1 - z^{-1}} \tag{14.34}$$

which gives, with the aid of Table 14.1

$$\frac{C(z)}{R(z)} = \frac{\dfrac{20.8z^2 - 38.6z + 18}{z^2 - z} \left[\dfrac{0.005(z + 0.9)}{(z - 1)(z - 0.905)} \right]}{1 + \dfrac{20.8z^2 - 38.6z + 18}{z^2 - z} \left[\dfrac{0.005(z + 0.9)}{(z - 1)(z - 0.905)} \right]} \tag{14.35}$$

$$\frac{C(z)}{R(z)} = \frac{0.104z^3 - 0.0995z^2 - 0.084z + 0.081}{z^4 - 2.801z^3 + 2.71z^2 - 0.989z + 0.081} \tag{14.36}$$

Equation 14.36 has a number of practical uses, two of which we will explore. It can be shown[27] that the closed-loop system characteristic equation

$$z^4 - 2.801z^3 + 2.71z^2 - 0.989z + 0.081 = 0 \tag{14.37}$$

formed from the denominator of Eq. 14.36 must have all its roots inside the unit circle $|z| = 1.0$ for closed-loop system absolute stability. This is analogous to requiring, for continuous systems, that all roots of the characteristic equation in s be in the left half plane (see Fig. 14.18). Although methods analogous to Routh criterion for detecting the presence of unstable roots without actually finding their numerical values are available for sampled-data systems, they are tedious and the ready availability of computerized root-finding routines (such as the SPEAKEASY POLYROOT we have used earlier) makes this direct

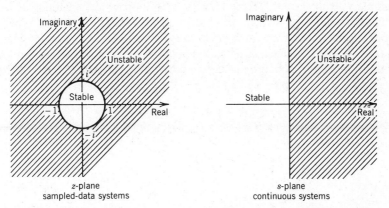

FIGURE 14.18 *Comparison of continuous and sampled-data stability criteria.*

---

[27]Ibid.

approach preferable. Only three statements are needed to find the roots of Eq. 14.37

```
DOMAIN COMPLEX
COFS=.081,-.989,2.71,-2.80,1
POLYROOT (COFS)
```

whereupon the computer returns the list of roots 0.11293, 0.82354, 0.93227 $\pm$ i 0.04243, all of which are inside the unit circle, so we have absolute stability.

With regard to relative stability, if the roots are all close to the origin (center of unit circle) then the system will generally not exhibit excessive oscillation; however if some roots are close to the circle itself, the system *may or may not* be oscillatory, thus the root positions themselves are not always a reliable guide to relative stability. This is particularly true if the sampling rate is high (T small) since then the roots tend to be close to the unit circle, whereas relative stability may actually be excellent. Intuitively, when T becomes small compared to system time constants, the sampled-data behavior should approach that of the continuous system obtained by ignoring the sample-and-hold operation and replacing the computer controller by its continuous equivalent. This *does* happen but the Z-transform stability analysis is misleading under these circumstances since the roots will be close to the instability boundary.

To get a more correct indication of system behavior, the actual time response to a specific input r (usually a step input) should be calculated. Fortunately, Z-transform methods allow this, although the basic method will find $c(t)$ *only* at the sampling instants, behavior *between* these instants remaining unknown. (To find the response between sampling instants the *modified Z transform* method is available.[28] It also must be used if the open-loop model includes any dead-time elements and they are not approximated by rational algebraic functions.) Although an inverse Z-transform method using partial-fraction expansions similar to those of the Laplace transform is available, we here show only the simpler "long division" approach. For a step input $r(t) = u(t)$, $r(kT)$ would be the sequence 1, 1, 1, . . . and

$$R(z) = \sum_{k=0}^{\infty} 1 + z^{-1} + z^{-2} + z^{-3} + \cdots$$

$$= \frac{z}{z-1}$$

(14.38)

From 14.36

$$C(z) = \frac{0.104z^3 - 0.0995z^2 - 0.084z + 0.081}{z^4 - 2.801z^3 + 2.71z^2 - 0.989z + 0.081} \left(\frac{z}{z-1} \right)$$

(14.39)

---

[28]C. L. Smith, "Digital Computer Process Control," pp. 101–106.

$$C(z) = \frac{0.104z^4 - 0.0995z^2 - 0.084z^2 + 0.081z}{z^5 - 3.80z^4 + 5.51z^3 - 3.70z^2 + 1.07z - 0.081} \quad (14.40)$$

An "ordinary" division of denominator into numerator gives

$$C(z) = 0.104z^{-1} + 0.295z^{-2} + 0.463z^{-3} + \cdots \quad (14.41)$$

which produces the values of $c(kT)$ for $k = 0, 1, \ldots$ from Eq. 14.9 as

$$c(0, 0.1, 0.2, 0.3, \ldots) = 0, 0.104, 0.295, 0.463, \ldots \quad (14.42)$$

14.5 DIGITAL SIMULATION OF SAMPLED-DATA SYSTEMS

Although the division in Eq. 14.40 can in principle be carried on indefinitely to produce $c(kT)$ values as far into time as we wish, this becomes both tedious and inaccurate. Partial-fraction expansion methods avoid these problems by giving an expression for the *general* term $c(kT)$ and the modified Z transform will even get values of c *between* sampling instants. However these analytical methods will often be passed over today in favor of a digital simulation of the system, using a language such as CSMP. We now study a program that compares the digital control example we have been carrying along with an equivalent continuous system.

```
METHOD RKSFX        use fixed-step-size integrator for precise timing
INIT                want to do some preliminary calculations
            C1 = KD/T + KP + KI*T ⎫  calculates constants
            C2 = 2.*KD/T + KP     ⎪  in Eq. 14.7; explores
            C3 = KD/T             ⎬  sampling times of
PARAM       T=(.1,.5,.7), KD=1.8  ⎪  0.1, 0.5, and 0.7 s
PARAM       KI = 2., KP = 2.6     ⎭
DYNAM
            R=STEP(0.0)       unit step command    ⎫
            RI = INTGRL(0.0,R)                     ⎪
            RII = INTGRL(0.0,RI)                   ⎪  model of
            CDOT=1.8*R+2.6*RI+2.*RII               ⎪  closed-loop
                  -2.8*C-2.6*CI-2.*CII             ⎬  continuous
            C = INTGRL(0.0, CDOT)                  ⎪  system
            CI = INTGRL (0.0,C)                    ⎪
            CII = INTGRL (0.0,CI)                  ⎭
            E1=R-CS        actuating signal of digital system
            SW = IMPULS (0.0,T)  ⎫  sampled and held
                                 ⎬  value of El, sampling
            ESH = ZHOLD(SW,E1)   ⎭  interval T
```

```
NOSORT              enters a FORTRAN subprogram
                    IF (T-.2)1,1,2     selects program path for T = 0.1
1                   ED1 = DELAY (20, .1, ESH)  ⎫ computes e[(k − 1)T]
                                               ⎬ and e[(k − 2)T] in
                    ED2 =DELAY(40,.2, ESH)     ⎭ Eq. 14.7
                    GO TO 20
2                   IF (T-.6)3,3,4     selects program path for T = 0.5
3                   ED1 =DELAY (100,.5,ESH)
                    ED2 =DELAY(200,1.0, ESH)
                    GO TO 20
4                   IF (T-.8)5,5,10    selects program path for T = 0.7
5                   ED1 =DELAY(140., .7, ESH
                    ED2 =DELAY(280,1.4,ESH)
                    GO TO 20
10                  CONTINUE
20                  CONTINUE
SORT                   leaves FORTRAN subprogram
NOSORT                 enters another FORTRAN subprogram
                    IF (TIME.GT.0.0) GO TO 40
                    CMPO =0.0     initializes computer output at 0.0 at t = 0
40                  IF(SW.EQ.1) GO TO 50     enables computer at sampling
                                      instants          ⎫ Compute
                                                        ⎪ controller
                    GO TO 60                             ⎪ output at
50                  CMPO = CMPO + C1*ESH − C2*ED1 + C3*ED2 ⎬ sampling
60                  CONTINUE                             ⎪ instants,
SORT                   leaves FORTRAN subprogram         ⎭ Eq. 14.7
                    M= INTGRL (0.0, CMPO)     applies computer output to
                                          process input
                    CA =REALPL (0.0, TAUPL,M) ⎫
                    CS =KPL*CA                 ⎬ process dynamics
PARAM               KPL =1.,TAUPL =1.         ⎭
TIMER               FINTIM =7.5, DELT = .005, OUTDEL = .02
```

In this program, METHOD RKSFX specifies use of a Runge–Kutta numerical integration algorithm with *fixed* step size, rather than the (default) variable-step-size Runge–Kutta used in most CSMP problems. This is necessary since the timing of the sampling switch must be precisely synchronized with the time steps of the numerical procedure in a sampled-data system simulation. The statements between INIT and DYNAM are used to compute some constants needed later and to set up multiple runs for $T = 0.1, 0.5,$ and 0.7 s. Next we simulate a continuous system with numerical values to match our sampled-data system. This will allow us to compare easily the behavior and thus see the effects of sampling and digital compensation. This continuous system has an open loop

$$\frac{C}{E}(s) = \frac{1.8s^2 + 2.6s + 2}{s^2(s + 1)} \tag{14.43}$$

giving a closed loop

$$\frac{C}{R}(s) = \frac{1.8s^2 + 2.6s + 2}{s^3 + 2.8s^2 + 2.6s + 2} \tag{14.44}$$

which we simulate by

$$\dot{C} = 1.8R + 2.6 \int R dt + 2 \int\int R dt - 2.8C - 2.6 \int C dt - 2 \int\int C dt \tag{14.45}$$

The sampled-data system has a continuous actuating signal E1 which is then sampled every T seconds (starting at 0.0) with the IMPULS statement and zero-order held with the ZHOLD statement. We next enter a FORTRAN subprogram to allow multiple runs for $T = 0.1, 0.5,$ and 0.7 s. The "past" values of ESH required in Eq. 14.7 are obtained using the DELAY (dead-time) statement, however CSMP's designer's did not provide for the delay to be a *letter*, it *must* be a number, thus making $T = (0.1, 0.5, 0.7)$ on a PARAM card is *not* sufficient. Our little FORTRAN subprogram overcomes this problem, the variables ED1 and ED2, respectively, being $e[(k-1)T]$ and $e[k-2)T]$ in Eq. 14.7. Another FORTRAN subprogram is then used to implement the digital control law of Eq. 14.7, using the values ED1 and ED2 computed earlier. We then return to "ordinary" CSMP programming for the analog part of the system, the controlled variable of the process being called CS.

We first run the program with $T = 0.1$ s only and request a graph of C (continuous system controlled variable) and CS (sampled-data system controlled variable) versus time for R a unit step input. Figure 14.19 confirms our often-repeated statement that digital control systems act much like their analog counterparts when sampling time T is sufficiently short. Evidently $T = 0.1$ is short enough in this example to make the two responses nearly identical. To show the effect of too-infrequent sampling we then make a run with $T = (0.1, 0.5, 0.7)$ and plot CS only, in Fig. 14.20. For $T = 0.5$ and 0.7 the sampled-data response is now much different from that of the corresponding analog system. Furthermore, $T = 0.5$ gives considerable overshoot and oscillation, whereas $T = 0.7$ is absolutely unstable. From just this brief example it should be clear to the reader that digital simulation languages (like CSMP) provide the same kinds of benefits in the analysis/design of digital control systems that we have oberved so often in our study of continuous systems. In fact, the benefits may be even more significant since the analytical treatment of sampled-data systems is considerably more complex whereas their simulation is really not much harder than that of continuous systems.

As another application of our simulation tool we consider briefly the class of systems that use sampled data but do *not* use digital control/compensation. That is, if we use a digital computer as controller, then sampling is inherent in our system; however sampling can be present for *other* reasons and we then may or may not decide to use digital control. One such example is found in control of chemical composition. Here the device that measures the controlled variable often does not provide a continuous readout, but rather takes intermittent samples of the process stream, performs its chemical analysis in a given time, and then reports this value to the controller. Another example is a rotating

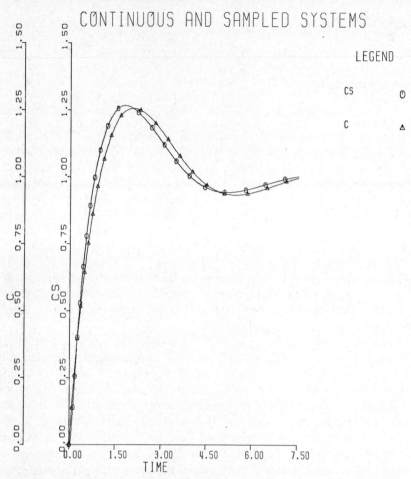

FIGURE 14.19 *Comparison of continuous and sampled-data systems for fast sampling rate.*

radar antenna for measuring position and range of aircraft. If it rotates at 1rps, data on the location of a given aircraft is available only once per second on a sampled basis.

Figure 14.21*a* shows a simple example of this class of systems. (Earlier we studied a related type of system, Fig. 13.24). Figure 14.21*b* shows that a useful analytical approximation for the sample/hold operation in such systems is a dead time equal to one half the sampling interval T. The main assumption here is that the low-pass-filter nature of most controlled processes will "smooth out" the steps in ESH, and the effect on C of a delayed ($T/2$) version of E will be nearly the same as the effect of ESH. If this is true, it is a useful approximation since analytical design of a continuous system with dead time is considerably

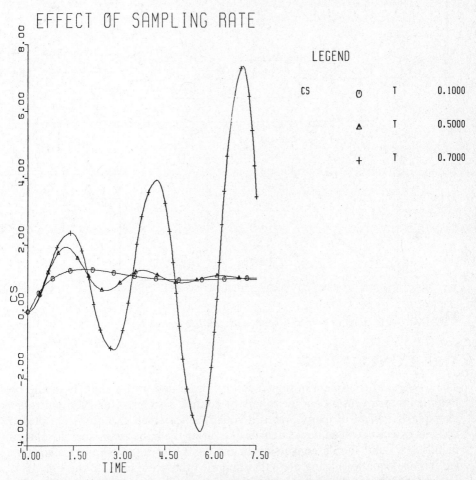

FIGURE 14.20   *Effect of sampling rate on system step response.*

easier than that of a sampled-data system. To check the validity of this approximation we write a CSMP simulation of the exact sampled-data system (controlled variable called *C*), our continuous approximation with an added dead time of *T*/2 (controlled variable *CD*), and a continuous system *without* added dead time (controlled variable *CMD*). Figure 14.22 shows that this approximation gives quite acceptable results, but simply *ignoring* the sampling (*CMD*) is very misleading. The reader is cautioned that this simple dead-time approximation can *not* be used with confidence when a digital control algorithm is employed, unless the sampling rate is so high that the system behaves like a continuous one. In that case, the dead time would be so small as to be negligible and the approximation would really not be necessary, we would just analyze the system as continuous.

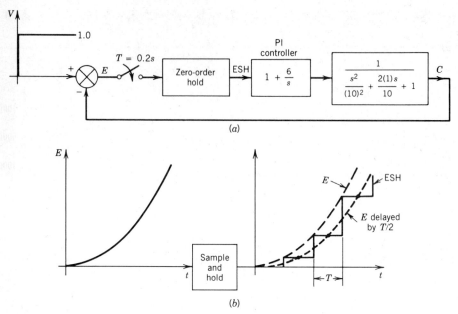

FIGURE 14.21   *Sampled-data system with analog controller.*

14.6   CONCLUSION

The condensed treatment of digital control presented in this chapter should be helpful for many users and designers of control systems. Our PID controller example is typical of many practical situations in which digital system design consists essentially of analog system design (using the basic techniques developed in this text) and then a conversion of the controller to a computer algorithm (difference equation) using methods similar to those of Eq. 14.2 through 14.7. Preliminary designs worked out in this way can then be "fine tuned" using CSMP (or other) digital simulation. This design approach does not exploit certain control algorithms that take advantage of characteristics inherent in digital control and not possible in analog control. One, treated in many books, is the *deadbeat controller.* Briefly, such controllers are designed so that in response to a definite command input (say a step input of desired value V), the controlled variable is brought *precisely* to its final value in a *finite* (usually small) number of sampling intervals. This is theoretically possible for any digitally controlled linear system and theoretically impossible for analog controllers. When one sets such performance as the design criterion, digital control analysis techniques lead one directly to a specific control algorithm that achieves the design goal. Controllers designed in this way are called deadbeat controllers.

Although deadbeat controllers are at first appealing because of their theoretical performance, they exhibit certain flaws that greatly restrict their practical use. First, perfect deadbeat performance will not be achieved in practice since the system models used are never precise, thus the controlled variable will *not*

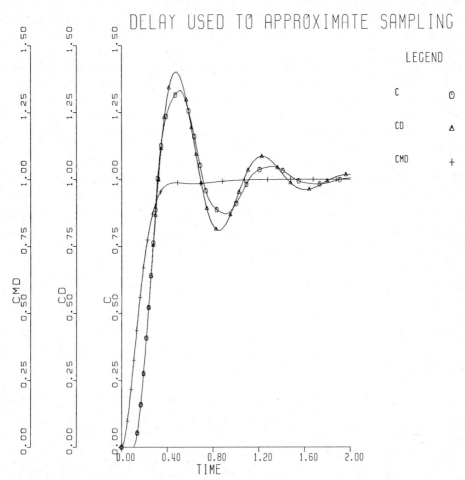

FIGURE 14.22 *Validation of dead-time approximation for sampling.*

come to rest in the predicted number of sampling intervals. Second, although analog controllers theoretically take an "infinite" length of time to settle after a step input, in practice they are so close to the desired value in a finite time that the difference is of no consequence, thus perfect deadbeat performance has no real practical advantage. Third, deadbeat controllers designed for a certain type of input are deadbeat *only* for that specific input; response to *other* forms of input may be terrible. Finally, although the *controlled* variable theoretically goes to its final value smoothly, *intermediate* variables (such as valve position in a process control system) may be *very* oscillatory in a deadbeat-controlled system, causing excessive actuator wear. Modifications to the basic deadbeat design philosophy can sometimes overcome these difficulties, but we then often find that system performance is really not better than that achieved by "conventional" (PID, phase-lag, phase-lead, etc.) methods.

Readers interested in exploring digital control techniques in greater breadth and depth are referred to the texts referenced in Section 14.1. They may also find useful some simplified design methods available in the literature.[29] The reference by Knowles is particularly interesting since it gives a comprehensive yet simple design approach based on frequency-response methods. Controllers designed by this method are claimed to be superior to those designed with time-domain approaches such as deadbeat or minimum variance techniques.

This chapter has concentrated on explaining the technical details of the effect of sampled-data operation on the performance of single-loop digital computer control. We should not loose sight of the computer's capability for integrating the management and operation of much larger-scale systems, as was discussed with reference to Fig. 1.12 (repeated as Fig. 14.23). Even more comprehensive

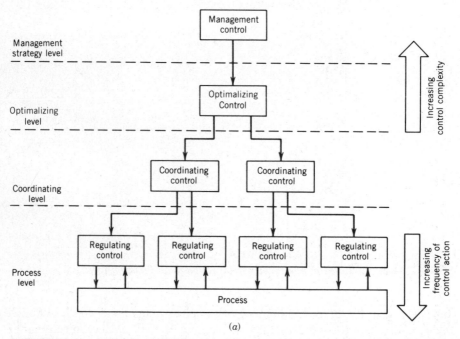

(a)

FIGURE 14.23   *Multi-level ("hierarchical") control strategy and example: (a) Multilevel hierarchical control structure. (b) Paper machine as an example of multilevel control structure.*

[29]J. B. Knowles, Some DDC System Design Procedures, in *Design of Modern Control Systems,* D. J. Bell, P.A. Cook, N. Munro, (Eds.), Peter Peregrinus Ltd., New York, 1982, pp. 223–245, C. M. Cundall and V. Latham, Designing Digital Computer Control Systems, *Cont. Eng.,* Oct. 1962, pp. 82–86, Jan. 1963, pp. 109–113; J. D. Fehr and D. C. Fosth, *Approximate Design Methods for High Response Digital Control Systems,* 1968 JACC Conf., R. R. DeBolt and B. E. Powell, A "Natural" 3-Mode Controller Algorithm for DDC, *ISA Jour.,* Sept. 1966, pp. 43–47; J. B. Cox et al., A Practical Spectrum of DDC Chemical-Process Control Algorithms, *ISA Jour.,* Oct. 1966, pp. 65–72.

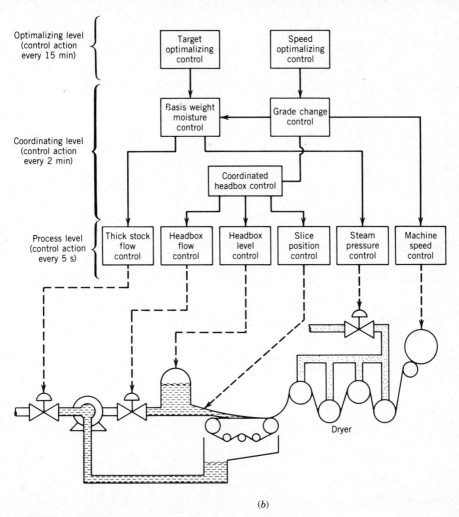

(b)

FIGURE 14.23 (Cont.)

computer control schemes are being planned and implemented. Figure 14.24 represents one viewpoint on the integrated control of an entire factory, using computers for a wide variety of tasks and linking them in networks to efficiently manage the whole enterprise. I quote from Taylor,

"Just about everyone is aware that productivity in our country has languished for years. Many manufacturers are making some productivity improvement by installing a robot here, a programmable controller there, or a computer aided design (CAD) unit across the street. But, with few exceptions, there has been little coordinated effort to plan the automated factory, a step at a time, from A to Z over a period of time, taking into consideration such

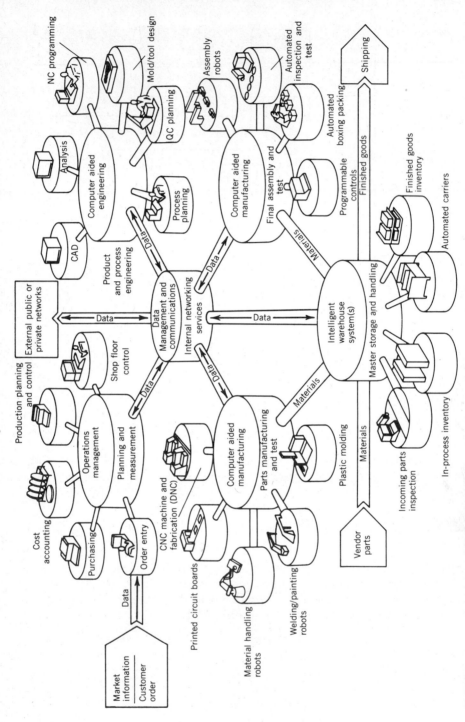

FIGURE 14.24   *Computer-aided engineering/manufacturing system.*
Source: A. P. Taylor, Improving Factory Productivity through Automation Planning, in "Automation in the Factory with a Future", Industrial Power Systems, Sept. 1982, General Electric Company, Schenectady, New York, 12345.

important "islands of automation" as operations management systems, computer aided engineering and manufacturing, incorporating not only robots, but programmable controls, computerized numerical controls, intelligent vision systems, intelligent warehouses, data management and communications. It is through this more comprehensive approach that the big productivity hit can be made, a hit that will pay back huge dividends in quality as well."

Operations management systems are the "up front" part of a manufacturing business that reduces costs and achieves optimum investment utilization. It includes such functions as order entry, purchasing, cost accounting, production planning and control, and shop floor control. *Computer-aided manufacturing* affords the greatest opportunity for immediate cost and quality improvement through the automation of such universal plant operations as component machining and fabrication, welding, painting, assembling, boxing and packing. Robotics systems are part of this area, as are systems for automatic test and inspection. *Intelligent warehouses* address the problem that in an average factory 50 percent of the floor space is used for material handling and storage. *Data management and communications* are the "brain and nerve system" that will link people, machines and processes together in a paperless environment."

PROBLEMS

14.1 How would you change the system of Fig. 14.7 to make it a *closed-loop* positioning system? Does the VCO pulse rate need to be fixed; can you think of useful ways to vary it? If your initial design turned out to be unstable when built, what might be done to fix this?

14.2 What would Eq. 14.1 be for the lead–lag device with $(M/E)(s) = (\tau_1 s + 1)/(\tau_2 s + 1)$?

14.3 Explain how the D/A converter of Fig. 14.11 works.

14.4 Write and run the CSMP (or other available simulation) program needed to produce the results of Fig. 14.22. Start the sampling at $t = 0.1$ (rather than the usual $t = 0.0$) to get the curves shown.

14.5 For the system simulated in Section 14.5 (Fig. 14.19, 14.20), add a sinusoidal noise signal to the sensor output to study aliasing effects. Then add an anti-aliasing filter to try to correct the problem.

14.6 Derive Eq. 14.37 for the case $T = 0.7$ and check stability with a computer root-finder. Do these analytical results agree with the simulation results of Fig. 14.20?

14.7 Change Eq. 14.34 so that $K_D = 0$. Check stability analytically and compute the first four values of $c(kT)$ for a unit step input of $r(t)$. Then check your results by running a simulation.

CHAPTER 15
Multivariable Control Systems

15.1 Introduction

Complex processes and machines often have several variables (outputs) that we wish to control, and several manipulated inputs available to provide this control. Sometimes it is obvious which input should be changed to cause a desired change in a certain controlled variable, but in other cases a single input causes *several* controlled variables to change. Then it may not be clear which output should be paired with which input in a feedback control scheme. Furthermore, if we do choose certain input–output pairs, and implement control loops, there may be significant interactions between these loops, leading to poor control. In practice, single input–single output (SISO) control strategies are often used where theory might indicate some advantage for the more comprehensive multiple-input–multiple output (MIMO) approach. This might be due to:

a. Lack of awareness of MIMO techniques.

b. Technical complexity of MIMO analysis and implementation.

c. Economic feasibility of SISO relative to MIMO.

Multivariable control methods have been under study for many years and are part of both "classical" and "modern" control theory. Modern control theory has especially concentrated on this area since classical methods have proven, for the most part, quite satisfactory for SISO problems. An application area that has received a great deal of attention, and after many years of effort has started to provide some practical payoff, is that of control of aircraft turbine engines.[1] (Automotive piston engine control has more recently been the subject of similar effort[2].) Early (1950) engines (J47) had a single controlled variable (engine speed) and a single manipulated variable (fuel flow rate). More recent engines control high compressor speed, fan speed, fan turbine inlet temperature,

---

[1] J. Zeller, B. Lehtinen, and W. Merrill, The Role of Modern Control Theory in the Design of Controls for Aircraft Turbine Engines, NASA TM 82815, 1982.

[2] W. F. Powers, *Internal Combustion Engine Control System Research at Ford,* 20th IEEE Conf. on Decision and Control, 1981, pp. 1447–1452 (Idle-speed control, using "modern control" approach); F. E. Coats, Jr. and R. D. Fruechte, Dynamic Engine Models for Control Development Part II: Application to idle-speed control, *Genl. Mtrs. Res. Labs Dept. GMR-3789,* 1982 (The same problem treated by Powers, but using instead largely "classical control" methods.)

main burner pressure, and afterburner pressure using as manipulated variables main-burner fuel flow, jet exhaust area, fan inlet guide-vane position, and rear compressor vane position.

Zeller, Lehtinen, and Merrill describe the historical development of this area, including a list of 118 papers and reports devoted to the technical details of various aspects of the problem. Although further work is needed and is continuing, significant progress has been made. My main purpose in briefly discussing this topic is to suggest that successful application of modern control theory to real-world MIMO problems seems to require a level of effort far beyond what I might provide in a single chapter. I will thus limit the presentation to some classical techniques which are fairly easy to understand and implement, but yet have had considerable success in practice.

15.2 SPECIFYING THE AMOUNT OF INTERACTION IN MIMO PROCESSES

Before suggesting control schemes specifically intended to deal with multivariable control problems, it is first desirable to characterize the amount of interaction present in the process to be controlled. If all interactions are slight, SISO control strategies can be applied to each individual loop with good success. If some interactions are significant but others are not, a proper pairing of inputs and outputs in control loops may still yield satisfactory control using only SISO concepts. If this is not possible, then the MIMO design technique called *decoupling control* may be helpful.

A means of specifying process interaction called the *relative gain*[3] method is in wide use. Although the method is applicable to processes with any number of inputs and outputs we will use a two-input, two-output process as an example. We also initially consider only steady-state behavior, leaving the generalization to dynamics for later. (The decoupling control scheme to be presented shortly has often provided significant control improvement in actual practice when only steady-state relations were enforced.) To allow application to nonlinear (real-world) processes, we assume the system of Fig. 15.1a is at an equilibrium operating point when small changes Δm_1, Δm_2 in the inputs cause small changes Δc_1, Δc_2 in the outputs.

$$\Delta c_1 \approx \left. \frac{\partial c_1}{\partial m_1} \right|_{m_2 \text{ constant}} \Delta m_1 + \left. \frac{\partial c_1}{\partial m_2} \right|_{m_1 \text{ constant}} \Delta m_2 \triangleq K_{11}\,\Delta m_1 + K_{12}\Delta m_2 \quad (15.1)$$

---

[3]E. H. Bristol, On a New Measure of Interaction for Multivariable Process Control, *IEEE Trans. on Auto. Control*, AC-11, 1966, pp. 133–134.

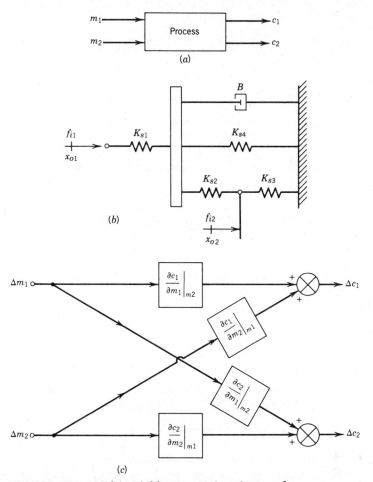

FIGURE 15.1 *Multivariable process, 2 × 2 example.*

$$\Delta c_2 \approx \frac{\partial c_2}{\partial m_1}\bigg|_{m_2 \text{ constant}} \Delta m_1 + \frac{\partial c_2}{\partial m_2}\bigg|_{m_1 \text{ constant}} \Delta m_2 \triangleq K_{21}\Delta m_1 + K_{22}\Delta m_2 \quad (15.2)$$

The K's are called the *open-loop steady-state gains* and could be either estimated from theory, or found by experimental test if the process has been constructed.

Figure 15.1b shows a physical example chosen to facilitate easy visualization of basic concepts. Note that if K_{s4} is very stiff (relative to the other springs), then input f_{i1} has little effect on output x_{o2} and input f_{i2} has little effect on output x_{o1}. If, however, K_{s4}'s stiffness is comparable to (or less than) that of the other springs, then significant interactions among f_{i1}, f_{i2}, x_{o1}, x_{o2} will occur. Analysis gives

$$x_{o1} = \frac{[B(K_{s2} + K_{s3})s + K_{s4}(K_{s2} + K_{s3}) + K_{s1}(K_{s2} + K_{s3}) + K_{s2}K_{s3}]}{[BK_{s1}(K_{s2} + K_{s3})s + K_{s1}K_{s4}(K_{s2} + K_{s3}) + K_{s1}K_{s2}K_{s3}]} f_{i1}$$

$$+ \frac{K_{s1}K_{s2}}{[BK_{s1}(K_{s2} + K_{s3})s + K_{s1}K_{s4}(K_{s2} + K_{s3}) + K_{s1}K_{s2}K_{s3}]} f_{i2} \qquad (15.3)$$

$$x_{o2} = \frac{K_{s2}}{[BK_{s1}(K_{s2} + K_{s3})s + K_{s1}K_{s4}(K_{s2} + K_{s3}) + K_{s1}K_{s2}K_{s3}]} f_{i1}$$

$$+ \frac{BK_{s1} s + K_{s1}(K_{s2} + K_{s4})}{[BK_{s1}(K_{s2} + K_{s3})s + K_{s1}K_{s4}(K_{s2} + K_{s3}) + K_{s1}K_{s2}K_{s3}]} f_{i2} \qquad (15.4)$$

In this example, considering steady state only ($s \equiv 0$) Eq. 15.1 and 15.2 become

$$\Delta x_{o1} = \frac{K_{s4}(K_{s2} + K_{s3}) + K_{s1}(K_{s2} + K_{s3}) + K_{s2}K_{s3}}{K_{s1}K_{s4}(K_{s2} + K_{s3}) + K_{s1}K_{s2}K_{s3}} \Delta f_{i1}$$

$$+ \frac{K_{s1}K_{s2}}{K_{s1}K_{s4}(K_{s2} + K_{s3}) + K_{s1}K_{s2}K_{s3})} \Delta f_{i2}$$

$$\Delta x_{o2} = \frac{K_{s2}}{K_{s1}K_{s4}(K_{s2} + K_{s3}) + K_{s1}K_{s2}K_{s3}} \Delta f_{i1} \qquad (15.5)$$

$$+ \frac{K_{s1}(K_{s2} + K_{s4})}{K_{s1s4}(K_{s2} + K_{s3}) + K_{s1}K_{s2}K_{s3}} \Delta f_{i2} \qquad (15.6)$$

For $K_{s1} = K_{s2} = K_{s3} = 1.0$ and $K_{s4} = 10$ these become

$$\Delta x_{o1} = 1.09\ \Delta f_{i1} + 0.0476\ \Delta f_{i2} \qquad (15.7)$$
$$\Delta x_{o2} = 0.0476\ \Delta f_{i1} + 0.524\ \Delta f_{i2}$$

whereas $K_{s1} = K_{s2} = K_{s3} = 1.0$, $K_{s4} = 0.1$ gives

$$\Delta x_{o1} = 2.667\ \Delta f_{i1} + 0.833\ \Delta f_{i2} \qquad (15.8)$$
$$\Delta x_{o2} = 0.833\ \Delta f_{i1} + 0.917\ \Delta f_{i2}$$

When feedback controllers are applied to some of the outputs of a process the open loop gains for the uncontrolled variables will in general change. For example, consider the situation in Fig. 15.2. We assume the controller to be perfect in steady state, thus $\Delta c_2 = 0.0$ and we define a new steady-state gain relating to c_1 to m_1 by

$$a_{11} \triangleq \frac{\partial c_1}{\partial m_1}\bigg|_{c_2 \text{ constant}} \qquad (15.9)$$

The gain a_{11} shows how much c_1 is affected by m_1 if the other variable is under closed-loop control. (For the general case of n inputs and n outputs a_{ij} would be $\partial c_i / \partial m_j$, evaluated with all other c's held constant.) We now define the *relative gain* λ_{11} as

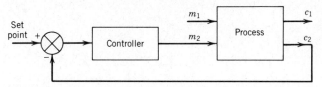

FIGURE 15.2 *Definition of open-loop gains when other loops are closed.*

$$\lambda_{11} \triangleq \frac{K_{11}}{a_{11}} = \frac{\left.\dfrac{\partial c_1}{\partial m_1}\right|_{m_2 \text{ constant}}}{\left.\dfrac{\partial c_1}{\partial m_1}\right|_{c_2 \text{ constant}}} \tag{15.10}$$

or in the general case

$$\lambda_{ij} \triangleq \frac{K_{ij}}{a_{ij}} = \frac{\left.\dfrac{\partial c_i}{\partial m_j}\right|_{\text{all other } m\text{'s constant}}}{\left.\dfrac{\partial c_i}{\partial m_j}\right|_{\text{all other } c\text{'s constant}}} \tag{15.11}$$

The gains a_{ij} could be calculated or measured from the definition given, however a more convenient approach is available that relates them to the K_{ij} as follows. In Eq. 15.2, since $a_{11} = \Delta c_1/\Delta m_1$ with $\Delta c_2 = 0$, we have

$$\Delta c_2 = 0 = K_{21}\Delta m_1 + K_{22}\Delta m_2 \tag{15.12}$$

$$\Delta m_2 = -\frac{K_{21}}{K_{22}}\Delta m_1 \tag{15.13}$$

Substituting into Eq. 15.1 gives

$$\Delta c_1 = K_{11}\,\Delta m_1 - \frac{K_{12}K_{21}}{K_{22}}\Delta m_1 = \left(K_{11} - \frac{K_{12}K_{21}}{K_{22}}\right)\Delta m_1 \tag{15.14}$$

and thus

$$a_{11} = \left.\frac{\Delta c_1}{\Delta m_1}\right|_{\Delta c_2 = 0} = K_{11} - \frac{K_{12}K_{21}}{K_{22}} = \frac{K_{11}K_{22} - K_{12}K_{21}}{K_{22}} \tag{15.15}$$

So finally

$$\lambda_{11} = \frac{K_{11}}{a_{11}} = \frac{K_{11}K_{22}}{K_{11}K_{22} - K_{12}K_{21}} \tag{15.16}$$

Similar analysis gives

$$\lambda_{12} = \frac{K_{12}K_{21}}{K_{12}K_{21} - K_{11}K_{22}} \tag{15.17}$$

$$\lambda_{21} = \frac{K_{12}K_{21}}{K_{12}K_{21} - K_{11}K_{22}} \tag{15.18}$$

$$\lambda_{22} = \frac{K_{11}K_{22}}{K_{11}K_{22} - K_{12}K_{21}} \tag{15.19}$$

For the general case it can be shown that if the K's are given in matrix form as

$$K = \begin{bmatrix} K_{11} & K_{12} & \cdots & K_{1n} \\ K_{21} & K_{22} & \cdots & K_{2n} \\ \vdots & & & \\ K_{n1} & K_{n2} & \cdots & K_{nn} \end{bmatrix} \tag{15.20}$$

and we compute a matrix C by first inverting and then transposing K

$$C = (K^{-1})^T = \begin{bmatrix} C_{11} & C_{12} & \cdots & C_{1n} \\ C_{21} & C_{22} & \cdots & C_{2n} \\ \vdots & & & \\ C_{n1} & C_{n2} & \cdots & C_{nn} \end{bmatrix} \tag{15.21}$$

then the λ_{ij} are given by

$$\lambda_{ij} = K_{ij}C_{ij} \tag{15.22}$$

It is now possible and useful to display the input and output variables in the form

	m_1	m_2	$\cdots$	m_n	
c_1	λ_{11}	λ_{12}	$\cdots$	λ_{1n}	
c_2	λ_{21}	λ_{22}	$\cdots$	λ_{2n}	
$\vdots$	$\vdots$				(15.23)
c_n	λ_{n1}	λ_{n2}	$\cdots$	λ_{nn}	

from which judgements can be reached as to the best control pairings and the degree of interaction.[4] It can be shown that each row of the λ array adds to 1.0 as does each column, thus if any λ's are greater than 1.0, then some others must be negative. For a 2 × 2 array if $\lambda_{11} = 1.0$ then $\lambda_{22} = 1.0$ and $\lambda_{12} = \lambda_{21} = 0$. This is the case of zero interaction (either K_{12} or K_{21} is zero) and then m_1 should

---

[4]F. G. Shinskey, *Controlling Multivariable Processes*, ISA, Research Triangle Park, North Carolina 27709, 1981, pp. 97–133; P. B. Deshpande and R. H. Ash, *Computer Process Control*, ISA, Research Triangle Park, North Carolina 27709, 1981, p. 293–313; C. L. Smith, *Digital Computer Process Control*, Intext, Scranton, Pennsylvania, 1982, pp. 212–221.

control c_1 and m_2, c_2. For the opposite situation ($\lambda_{12} = 1.0$), m_2 should control c_1 and m_1, c_2. Negative values of relative gain indicate difficult control problems and such pairs should not be selected for control. A negative value for a λ indicates that the sense of the output–input relation for that pair of variables is *reversed* when all other loops are closed, which may cause stability problems. Note that a λ of 1.0 means that, for that particular pairing of c and m, the response of c to m is the same when all loops are open as when all other loops are closed, that is, the presence of the other closed loops does not disturb the open-loop behavior of the c,m pair under consideration, a desirable condition. We thus prefer λ's near 1.0 but since each row and column sums to 1.0, we often must accept pairs with λ's less than 1.0. In selecting which m is to be used to control which c, we thus scan across each row in array 15.23 and choose the m that has the λ closest to 1, avoiding λ's that are negative or much larger than 1.0, large negative values being particularly undesirable. Shinskey[5] indicates that λ's larger than 5 exhibit severe interaction, sometimes preventing simultaneous control of multiple loops.

If the λ's are all of similar magnitude, this indicates strong interaction and the pairing of m's and c's cannot be decided by the relative gain array. Examination of *dynamic* behavior may then be of some help. If one loop is much faster than the other, good control may be achievable even when the λ's are comparable. When the λ array clearly suggests a definite pairing scheme, this should be tried out with single-loop controllers, using simulation first (if possible) to see if satisfactory performance can be achieved. If not, or if the λ's did not indicate a clear choice of pairs, it may be time to consider a decoupling controller scheme.

Before introducing the decoupling control strategy, let us get some experience with the mechanical system of Fig. 15.1b. When K_{s4} is stiff (Eq. 15.7) we get

$$
\begin{array}{c|cc}
 & f_{i1} & f_{i2} \\
\hline
x_{o1} & 1.004 & -0.004 \\
x_{o2} & -0.004 & 1.004
\end{array}
\qquad (15.24)
$$

whereas soft K_{s4} (eq. 15.8) gives

$$
\begin{array}{c|cc}
 & f_{i1} & f_{i2} \\
\hline
x_{o1} & 1.392 & -0.392 \\
x_{o2} & -0.392 & 1.392
\end{array}
\qquad (15.24)
$$

As expected, stiff K_{s4}, which we can see physically reduces the interaction between the two loops, gives a λ array close to the noninteracting ideal (1's and 0's), whereas soft K_{s4} deviates considerably from this ideal. In both cases, however, the same pairings (x_{o1}/f_{i1}, x_{o2}/f_{i2}) are suggested by the λ array. Let us now implement feedback loops on each controlled variable, as in Fig. 15.3 where

---

[5]F. G. Shinskey, *Controlling Multivariable Processes,* p. 101.

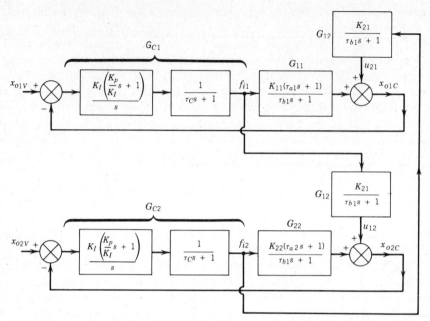

FIGURE 15.3 *Conventional single-loop control approach applied to 2 × 2 process.*

we have used PI controllers and a first-order lag (τ_c) for the force-producing hardware. Analysis gives

$$[(1 + G_{C2}G_{22})(1 + G_{C1}G_{11}) - G_{C1}G_{C2}G_{12}^2]\, x_{o1C} \qquad (15.26)$$
$$= [(1 + G_{C2}G_{22})G_{C1}G_{11} - G_{C1}G_{C2}G_{12}^2]\, x_{o1V} + (G_{C2}G_{12})\, x_{o2V}$$

The interaction between the two loops is manifested in two ways. First the characteristic equation (and thus stability and response) for x_{o1C} is influenced by G_{C2}, G_{22}, and G_{12}. Second, the x_{o1C} equation has *both* x_{o1V} and x_{o2V} as input terms. (Similar results are obtained for x_{o2C}.) Note, however, that as $G_{12} \to 0$, the x_{o1C} equation collapses into familiar single-loop form.

The complexity of Eq. 15.26 plus the need to consider *both* loops "simultaneously" should make clear the potential difficulties involved in designing multivariable systems. Practical systems have, however, been successfully implemented for many years. Often, interaction is not severe and individual loops, designed without regard for the presence of other loops, work well when the complete system is activated. One reason for this is the basic power of the feedback principle in following commands and rejecting disturbances. In Fig. 15.3, for example, one could consider the effect of loop 2 on loop 1 as simply a disturbance injected as the signal u_{21}. We know from past experience that a properly designed controller is very effective in isolating the controlled variable from just such disturbances. If this is the case, then loop 2 does not really have much effect on loop 1. The signal u_{21} is of course *not* an "outside" independent input but is actually an "internal" variable (an unknown in the system equations, not a given input) thus we should not push our "disturbance analogy" too far. However, for a given set of commands and/or outside disturbances in the two loops, $u_{21}(t)$ would be some definite time function, and if loop 1 is capable of fighting out such a disturbance, then loop 1 may operate in the presence of loop 2 with nearly the same behavior as it does in its absence.

We can use CSMP simulation to study such questions, and in fact, a viable design procedure can in many cases be an analytical preliminary design (Nichols chart, root locus, etc.) for each single loop considered isolated, followed by a simulation study to check for interaction problems and to fine tune parameter values. Let us look first at the case (Eq. 15.7) where K_{s4} is stiff and there is little interaction. Since the simulation has no trouble with dynamics we give B a numerical value, say 1.0, whereupon Eq. 15.3 and 15.4 become

$$x_{o1} = \frac{1.09\,(0.0870s + 1)}{0.0952s + 1}f_{i1} + \frac{0.0476}{0.0952s + 1}f_{i2}$$
$$x_{o2} = \frac{0.0476}{0.0952s + 1}f_{i1} + \frac{0.524(0.0909s + 1)}{0.0952s + 1}f_{i2}$$
(15.27)

Choosing a value for τ_c, say 0.2, then allows an analytical preliminary design of each loop separately and we find that $K_p = 1.0$ and $K_I = 20.0$ give a desirable response (about 25% overshoot) for a step command, in each case. We now use CSMP to simulate the coupled systems and (simultaneously) the x_{o1} loop as a *separate* system (signal u_{21} disabled in Fig. 15.3). Applying a unit step at x_{o1V} and letting $x_{o2V} \equiv 0$ gives the results of Fig. 15.4, where $X1 \triangleq x_{o1C}$, $X2 \triangleq x_{o2C}$, and $W \triangleq x_{o1C}$ for the isolated system. Clearly $X1$ and W are nearly identical and $X2$ shows almost no effect from the activity in loop 1. Note that setting $x_{o2V} \equiv 0$ does *not* make $f_{i2} = 0$. The loop 2 controller is definitely active in adjusting f_{i2} so as to keep x_{o2C} near its desired zero value. A rerun with $x_{o1V} \equiv 0$ and x_{o2V} a unit step gives similar results in Fig. 15.4*b*, where $Z \triangleq x_{o2C}$ for the isolated loop 2. This example shows that when interaction is slight, we can use our "ordinary" SISO design technique with good success for MIMO systems.

Changing K_{s4} to 0.1 (eq. 15.8) gives

$$x_{o1} = \frac{2.67(0.625s + 1)}{1.67s + 1}f_{i1} + \frac{0.833}{1.67s + 1}f_{i2}$$
$$x_{o2} = \frac{0.833}{1.67s + 1}f_{i1} + \frac{0.917(0.909s + 1)}{1.67s + 1}f_{i2}$$
(15.28)

which had the λ array of (15.25) and significant open-loop interaction. A rerun of the simulation, however, shows almost no change from Fig. 15.4, the multivariable system still responds almost like two isolated loops, demonstrating again the power of the basic feedback to give uniform performance in the face of varying conditions. If the two loops are "poorly" designed, then interaction may be more noticeable. In our present system, let us make loop 1 "slow" by reducing K_I from 20 to 2 and K_p from 1.0 to 0.1 (in both the multivariable and single-loop versions) and keeping loop 2 "fast" with its present values. Also, let x_{o1V} be a unit step at $t = 0$ and x_{o2V} a unit step at $t = 1.0$. Figure 15.5*a* now

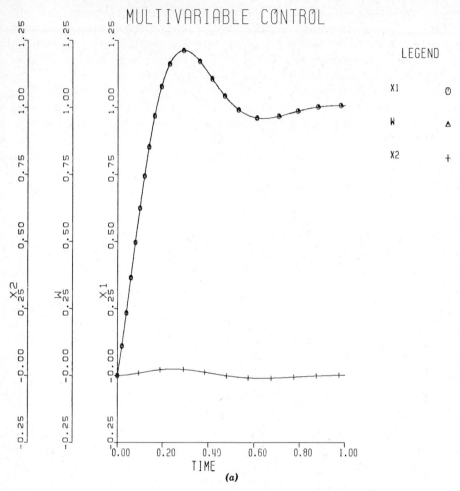

FIGURE 15.4 *Single-loop controllers show negligible interaction effects.*

shows a significant difference between $X1$ (multivariable system) and W (single loop system), due to the interaction with loop 2. If we now return loop 1's K_I to 20, K_p to 1, and repeat the run (the step command of x_{o2V} is made at $t =$ 0.1 because the systems are now faster), Fig. 15.5b shows that there is now little effect of loop 2 on loop 1, $X1$ and W are very similar.

15.3 DESIGN OF NONINTERACTING (DECOUPLED) CONTROLLERS

When conventional single-loop controllers are unable to provide the required performance in a multivariable control system, the classical technique of non-interacting control may be of help. Here the controllers for each loop send

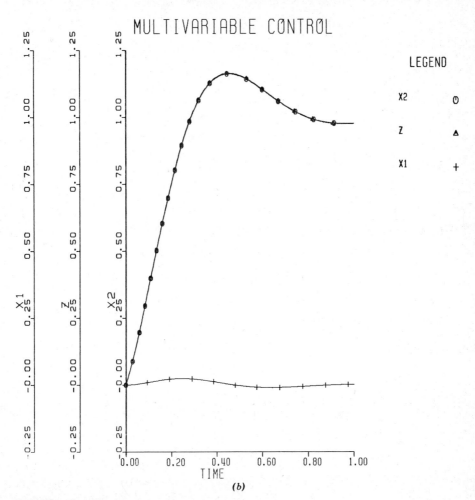

FIGURE 15.4 (Cont.)

control signals not only to "their own" manipulated variable but also to the manipulated variables in all other loops. These signals are designed so that they exactly cancel the interactions inherent in the process, making each loop behave as an isolated system, whose performance can then be designed on a single-loop basis. If all the math models were perfect and we were able to implement perfectly whatever relations the decoupling procedure requires, then there would be absolutely no interaction. In real systems, there will always be some mismatch between assumed models and actuality and we are also unable to implement the dynamics sometimes required for decoupling (such as perfect differentiators), so some interaction will always remain. However, significant improvements have been obtained in practical systems, sometimes using only steady-state decoupling or simple dynamics like lead-lags.

We will develop the procedure for a two-variable system and then generalize

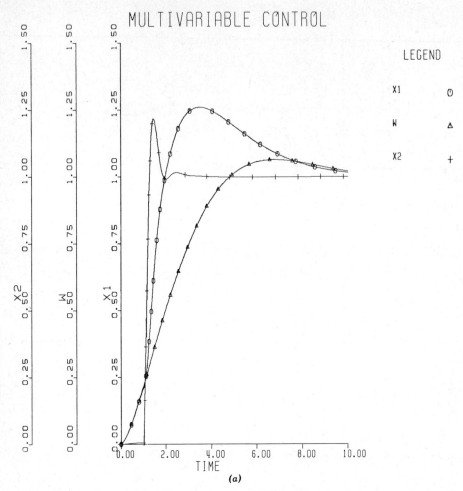

FIGURE 15.5   *Interaction effects become noticeable when one loop is poorly tuned.*

to n variables. Writing Eq. 15.1 and 15.2 in more general form (including dynamics) we have

$$\Delta C_1(s) = G_{11}(s)\Delta M_1(s) + G_{12}(s)\Delta M_2(s)$$
$$\Delta C_2(s) = G_{21}(s)\Delta M_1(s) + G_{22}(s)\Delta M_2(s) \tag{15.29}$$

The structure of the decoupling controller is shown in Fig. 15.6. Our example takes $D_{11} = D_{22} = 1.0$ for simplicity, however these two blocks are included in Fig. 15.6 in preparation for the general case to be treated later. To decouple the two loops, the signal going from M_1^* to C_2 by the path $D_{21}G_{22}$ should just cancel that going by the path $D_{11} G_{21}$.

$$M_1^* D_{21} G_{22} = -M_1^* G_{21} \tag{15.30}$$

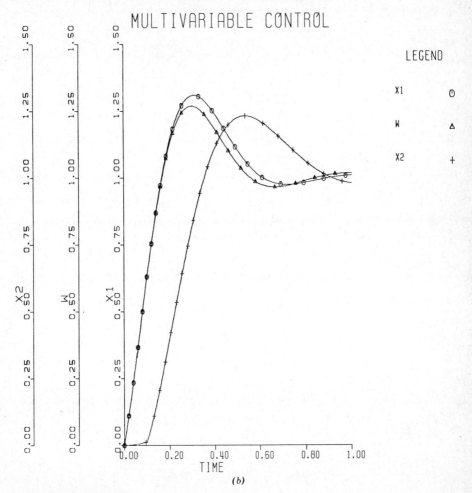

FIGURE 15.5 (Cont.)

and the signal going from M_2^* to C_1 by the path $D_{12}\,G_{11}$ should just cancel that going by the path $D_{22}\,G_{12}$.

$$M_2^*\,D_{12}\,G_{11} = -\,M_2^*\,G_{12} \qquad (15.31)$$

(Note the resemblance of this principle to *feedforward* control, discussed earlier.) These two equations define the decoupler as

$$D_{21} = -\frac{G_{21}}{G_{22}} \qquad D_{12} = -\frac{G_{12}}{G_{11}} \qquad (15.32)$$

Now M_1^* affects only C_1 and M_2^* only C_2, however the transfer functions are *not* G_{11} and G_{22} but are given by

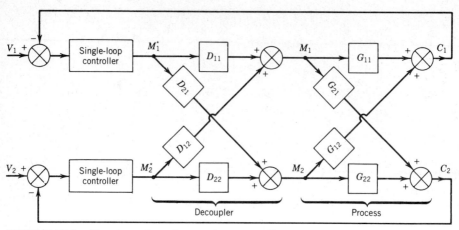

FIGURE 15.6 *Structure of noninteracting controllers.*

$$M_1^* G_{11} + M_1^* D_{21} G_{12} = C_1 \tag{15.33}$$

$$\frac{C_1}{M_1^*} = G_{11} - \frac{G_{21} G_{12}}{G_{22}} \tag{15.34}$$

and

$$M_2^* G_{22} + M_2^* D_{12} G_{21} = C_2 \tag{15.35}$$

$$\frac{C_2}{M_2^*} = G_{22} - \frac{G_{21} G_{12}}{G_{11}} \tag{15.36}$$

Using our earlier example (Eq. 15.28)

$$D_{21} = - \frac{0.908}{0.909s + 1} \qquad D_{12} = - \frac{0.312}{0.625s + 1} \tag{15.37}$$

$$\frac{C_1}{M_1^*} = \frac{1.91(0.478s + 1)}{0.909s + 1} \qquad \frac{C_2}{M_2^*} = \frac{0.657(0.478s + 1)}{0.909s + 1} \tag{15.38}$$

We could now design controllers for each loop separately, using conventional SISO methods, since the loops no longer interact. Rather than designing new controllers, to demonstrate the success of the noninteracting design let us return our simulation to the conditions of Fig. 15.5a, where loop 1 had $K_p = 0.1$, $K_I = 2$, and loop 2 had $K_p = 1$, $K_I = 20$, except we add the decouplers D_{21} and D_{12} as in Fig. 15.6. We now apply a unit step command of x_{o2V} with $x_{o1V} \equiv 0$. Without the decoupler, $x_{o1C}(X1)$ is significantly disturbed from its desired value by the activity in loop 2 (see Fig. 15.7a), however the decoupled system (Fig. 15.7b) shows no effect at all.

The decoupling technique just presented can be extended[6] to the general case of a process with n manipulated variables $M_1, M_2, \ldots, M_n$ and n controlled

---

[6]P. B. Deshpande and R. H. Ash, *Computer Process Control*, pp. 307–313.

variables $C_1, C_2, \ldots, C_n$. Such a process can be described by a matrix of transfer functions.

$$G = \begin{bmatrix} G_{11}(s) & G_{12}(s) & \cdots & G_{1n}(s) \\ G_{21}(s) & & & \\ \cdot & & & \\ \cdot & & & \\ \cdot & & & \\ G_{n1}(s) & G_{n2}(s) & \cdots & G_{nn}(s) \end{bmatrix} \qquad (15.39)$$

The decoupler now takes the form

$$D = \begin{bmatrix} D_{11}(s) & D_{12}(s) & \cdots & D_{1n}(s) \\ D_{21}(s) & & & \\ \cdot & & & \\ \cdot & & & \\ \cdot & & & \\ D_{n1}(s) & D_{n2}(s) & \cdots & D_{nn}(s) \end{bmatrix} \qquad (15.40)$$

where

$$\begin{bmatrix} M_1 \\ M_2 \\ \cdot \\ \cdot \\ \cdot \\ M_n \end{bmatrix} = \begin{bmatrix} D_{11}(s) & D_{12}(s) & \cdots & D_{1n}(s) \\ D_{21}(s) & & & \\ \cdot & & & \\ \cdot & & & \\ \cdot & & & \\ D_{n1}(s) & D_{n2}(s) & \cdots & D_{nn}(s) \end{bmatrix} \begin{bmatrix} M_1^* \\ M_2^* \\ \cdot \\ \cdot \\ \cdot \\ M_n^* \end{bmatrix} \qquad (15.41)$$

and

$$\begin{bmatrix} C_1 \\ C_2 \\ \cdot \\ \cdot \\ \cdot \\ C_n \end{bmatrix} = \begin{bmatrix} G_{11}(s) & G_{12}(s) & \cdots & G_{1n}(s) \\ G_{21}(s) & & & \\ \cdot & & & \\ \cdot & & & \\ \cdot & & & \\ G_{n1}(s) & G_{n2}(s) & \cdots & G_{nn}(s) \end{bmatrix} \begin{bmatrix} M_1 \\ M_2 \\ \cdot \\ \cdot \\ \cdot \\ M_n \end{bmatrix} \qquad (15.42)$$

Thus, for example,

$$M_1 = D_{11}(s)M_1^* + D_{12}(s)M_2^* + \cdots D_{1n}(s)M_n^* \qquad (15.43)$$

and

$$C_1 = G_{11}(s)M_1 + G_{12}(s)M_2 + \cdots + G_{1n}(s)M_n \qquad (15.44)$$

To decouple the system we require

$$\begin{bmatrix} C_1 \\ C_2 \\ \cdot \\ \cdot \\ \cdot \\ C_n \end{bmatrix} = \begin{bmatrix} H_{11}(s) & & 0 & \cdots & 0 \\ 0 & H_{22}(s) & & & \\ \cdot & & & & \\ \cdot & & & & \\ \cdot & & & & \\ 0 & 0 & & \cdots & H_{nn}(s) \end{bmatrix} \begin{bmatrix} M_1 \\ M_2 \\ \cdot \\ \cdot \\ \cdot \\ M_n^* \end{bmatrix} \qquad (15.45)$$

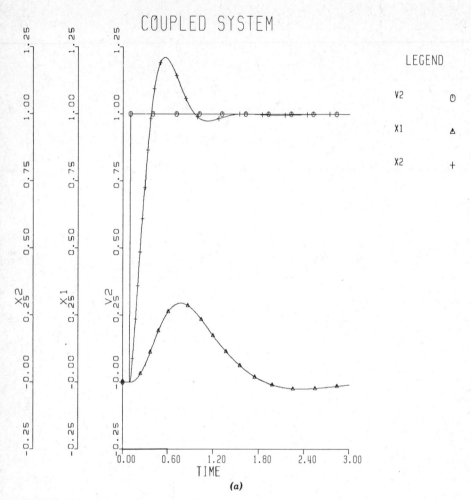

FIGURE 15.7 *Decoupling controller completely removes interaction.*

From Eq. 15.43 and 15.44 we see that

$$
\begin{aligned}
C_1 = \; & G_{11} \, (D_{11}M_1^* + D_{12}M_2^* + \cdots + D_{1n}M_n^*) \\
& + G_{12}(D_{21}M_1^* + D_{22}M_2^* + \cdots + D_{2n}M_n^*) \\
& + \cdots + G_{1n}(D_{n1}M_1^* + D_{n2}M_2^* + \cdots + D_{nn}M_n^*)
\end{aligned}
\tag{15.46}
$$

and from Eq. 15.45

$$
C_1 = H_{11}M_1^*
$$

so

$$
H_{11}(s) = G_{11}(s)D_{11}(s) + G_{12}(s)D_{21}(s) + \cdots + G_{1n}(s)D_{n1}(s) \tag{15.47}
$$

with a similar analysis for $H_{22}(s)$, $H_{33}(s)$, . . . , $H_{nn}(s)$. If we are successful in achieving Eq. 15.45, then we have

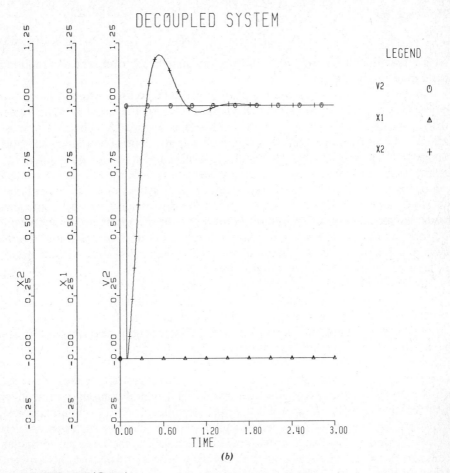

FIGURE 15.7 (Cont.)

$$C_1 = H_{11}(s)M_1^* \qquad C_2 = H_{22}(s)M_2^* \qquad \cdots \qquad C_n = H_{nn}(s)M_n^* \quad (15.48)$$

To find the D matrix that will cause the desired decoupling Eq. 15.48 we write

$$[C] = [G][M] \qquad [M] = [D][M^*] \qquad (15.49)$$

and thus

$$[C] = [G][D][M^*] \qquad (15.50)$$

but we require

$$[C] = [H][M^*] \qquad (15.51)$$

and thus

$$[G][D] = [H] \qquad (15.52)$$

This requires

$$[D] = [G]^{-1}[H] \qquad (15.53)$$

so we find that we must *invert* the process matrix G to obtain the required decoupler matrix D.

This theory has been usefully applied in practical control systems, though of course we cannot expect perfect decoupling, due to the usual nonlinearities, drifts, and the like associated with real systems. Figure 15.8 shows a few examples of multivariable processes where decoupling control has been profitably applied. In the paper machine headbox, level and outflow of pulp slurry are controlled by manipulating air pressure and slurry inflow. The direct contact water heater controls discharge temperature and outflow rate by manipulating the flow rates of cold water and steam. In the distillation column,[7] controlled variables include column pressure, liquid levels in the reflux accumulator and bottom, distillate composition or temperature, and bottom composition or temperature. Manipulated variables include cooling water flow rate, distillate rate, reflux rate, bottom rate, and boilup rate.

We should also note that the decoupling *structure* assumed by this theory is not the only one possible, and others may display certain advantages in specific instances. Shinskey[8] gives a particularly good discussion of such structural variations of the decoupling scheme. He also shows how nonlinear decouplers can be used for strongly nonlinear processes, letting the process equations suggest the structure needed in the decoupler. In closing we should also note that the complexity of decoupling controllers was one factor restricting their practical application when they were first conceived. With digital computer controllers now readily available, this factor is today less significant. Also, computing power may be helpful in alleviating other difficulties of these schemes, such as mismatch between assumed and actual process characteristics. In some cases the computer can periodically "run tests" on the operating process to discover its current behavior and use this information to update the process model resident in the computer, thus improving control performance.

BIBLIOGRAPHY

Books
1. H. S. Tsien, *Engineering Cybernetics*, McGraw-Hill, New York, 1954, Chap. 5.
2. H. H. Rosenbrock, *Computer Aided Control System Design*, Academic Press, London, 1974.
3. T. J. McAvoy, *Interaction Analysis*, ISA, Research Triangle Park, N.C., 1984.

---

[7]J. C. Wang, Compute Relative Gain Matrices for Better Distillation Control, *In. Tech.*, March 1980, pp. 40–44.
[8]F. G. Shinskey, "Controlling Multivariable Processes," pp. 123–131.

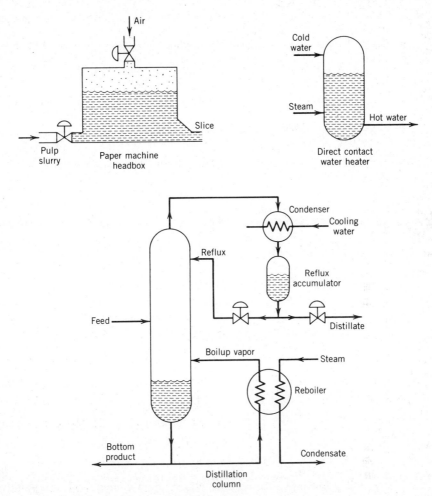

FIGURE 15.8 *Practical examples of multivariable processes.*

Journals

3. K. L. Chien, et al., The Noninteracting Controller for a Steam-Generating System, *Cont. Eng.,* Oct. 1958, pp. 95–101.

4. C. S. Zalkind, Practical Approach to Noninteracting Control, *Inst. and Cont. Syst.,* March 1967, pp. 89–93.

5. D. M. Wills, Simple Multiloop Control Systems, *ISA Jour.,* March 1963, pp. 67–70.

6. P.V. Webb, Reducing Process Disturbances with Cascade Control, *Cont. Eng.,* Aug. 1961, pp. 73–76.

7. S. G. Lloyd, Basic Concepts of Multivariable Control, *Instr. Tech.,* Dec. 1973, pp. 31–37.

8. P. S. Buckley, Distillation Column Design Using Multivariable Control, *Instr. Tech.,* Sep. 1978, pp. 115–122, Oct. 1978, pp. 49–53.

9. P. F. Woolverton, How to Use Relative Gain Analysis in Systems with Integrating Variables, InTech, Sep., 1980, pp. 63–65.

10. J. R. Borer, A Multi-Variable Master Controller, *Meas. and Cont.*, Vol. 13, Apr. 1980, pp. 126–132.

11. A. S. Bokensbom and R. Hood, General Algebraic Method Applied to Control Analysis of Complex Engine Types, NACA Rep. 980, 1950.

12. R. K. Pearson *Modern Control: Why Don't We Use It?*, InTech, Nov. 1984, pp. 47–49.

13. R. K. Pearson, *Obstacles to the Practical Application of Nontraditional Control*, ISA/84 Conf., Oct. 22–25, 1984, Houston, Texas.

14. J. L. Tyler and J. E. Purviance, *Optimal Control for Industrial Plants: Has Its Time Come?*, InTech, Sept. 1984, pp. 103–107.

PROBLEMS

15.1 Derive Eq. 15.3 and 15.4.

15.2 Derive Eq. 15.17 through 15.19.

15.3 Show that, for a two-input, two-output process, negative K's are needed to give positive λ's less than 1.0. What relation must there be among the K's to give the "worst" interaction (λ's = 0.5)?

15.4 Derive a result comparable to Eq. 15.26 for x_{o2C}.

15.5 Write a CSMP (or other available simulation) program to produce the results of Fig. 15.4.

15.6 Write a CSMP (or other available simulation) program to produce the results of Fig. 15.7.

15.7 Derive an expression for $H_{22}(s)$ analogous to Eq. 15.47.

15.8 For the paper machine headbox of Fig. 15.8, use linearized analysis to obtain a process model analogous to Eq. 15.3 and 15.4 relating level and volume outflow rate of slurry to air pressure and volume inflow rate of slurry. Assuming numerical values were available for this model, explain how you would decide to pair the manipulated and controlled variables. For each of the possible pairings, find the form of the decoupler needed to achieve noninteracting control.

Index

571